Computational Intelligence and Optimization Methods for Sustainable Water Management

Yassine Ezaier
Hassan II University, Casablanca, Morocco

Rajae Gaamouche
Moroccan School of Engineering Sciences, Rabat, Morocco

Mohamed Lahby
Hassan II University, Casablanca, Morocco

IGI Global Scientific Publishing
Publishing Tomorrow's Research Today

Vice President of Editorial	Melissa Wagner
Director of Acquisitions	Mikaela Felty
Director of Book Development	Jocelynn Hessler
Production Manager	Mike Brehm
Cover Design	Jose Rosado

Published in the United States of America by
IGI Global Scientific Publishing
701 East Chocolate Avenue
Hershey, PA, 17033, USA
Tel: 717-533-8845 | Fax: 717-533-7115
Website: https://www.igi-global.com E-mail: cust@igi-global.com

Library of Congress Cataloging-in-Publication Data

Names: Ezaier, Yassine, 1996- editor | Gaamouche, Rajae, 1985- editor | Lahby, Mohamed editor

Title: Computational intelligence and optimization methods for sustainable water management / edited by Yassine Ezaier, Rajae Gaamouche, and Mohamed Lahby.

Description: Hershey, PA : IGI Global Scientific Publishing, [2026] | Includes bibliographical references and index. | Summary: "This book explores how modern solutions can be integrated with traditional methods to achieve sustainable and effective water management"-- Provided by publisher.

Identifiers: LCCN 2025026936 (print) | LCCN 2025026937 (ebook) | ISBN 9798337327006 hardcover | ISBN 9798337327013 paperback | ISBN 9798337327020 ebook

Subjects: LCSH: Water-supply--Management--Data processing | Water resources development--Data processing | Sustainable development--Data processing | Sustainable engineering

Classification: LCC TD353 .C627 2026 (print) | LCC TD353 (ebook)

LC record available at https://lccn.loc.gov/2025026936

LC ebook record available at https://lccn.loc.gov/2025026937

British Cataloguing in Publication Data
A Cataloguing in Publication record for this book is available from the British Library.

Table of Contents

Detailed Table of Contents

data. This chapter explores neutrosophic sets—a generalization of fuzzy logic that independently models truth (T), indeterminacy (I), and falsity (F)—as a robust framework for disaster response and recovery. We demonstrate its applications in flood risk assessment, wildfire management, and sustainable technology selection, highlighting how neutrosophic logic outperforms classical methods by quantifying ambiguity and conflicts. Case studies illustrate its integration with multi-criteria decision-making (MCDM) to optimize resource allocation and policy planning. Challenges like computational complexity and future directions for real-time systems and interdisciplinary adoption are discussed. Neutrosophic sets offer a transformative approach to resilience in an era of climate volatility, bridging advanced mathematics with practical crisis management.

Artificial Intelligence (AI) is revolutionizing water management by addressing critical challenges such as water scarcity, pollution, and inefficient resource allocation. This chapter explores the transformative role of AI technologies, including Machine Learning (ML), Deep Learning (DL), Internet of Things (IoT), Predictive Analytics, and Computer Vision, in optimising water systems. Key applications include demand forecasting, leak detection, water quality monitoring, smart irrigation, and flood prediction, all of which enhance efficiency, sustainability, and resilience. Case studies, such as Barcelona's AI-powered water management system, demonstrate real-world successes in reducing water loss and improving operational performance. However, challenges must be addressed to ensure equitable and scalable solutions. The chapter also highlights future trends to further integrate AI into water governance. By fostering collaboration across sectors, AI can play a pivotal role in achieving water sustainability goals, ensuring safe and access to this vital resource for future generations.

Faced with the intensification of water shortages caused by climate change and human pressures, this chapter presents a hybrid architecture combining predictive modelling and extreme scenario generation to strengthen territorial resilience. The approach is based on the integration of time series transformers and deep generative models to anticipate the evolution of hydrological variables and simulate critical events. In the absence of sufficiently rich real data, a set of realistic synthetic data was constructed from hydrological distributions, seasonal cycles and stochastic processes, then qualitatively validated. This game allows a rigorous evaluation of predictive and generative performance. Results demonstrate the complementarity of the two dimensions to improve risk visualization and activation of early responses. The chapter also discusses technical limitations, ethical issues, and proposes concrete perspectives for integration into digital twins, warning systems or decision support tools for climate adaptation.

This chapter emphasizes the potential of integrating green infrastructure (GI) with computational techniques to revolutionize urban stormwater management. Conventional gray infrastructure is inadequate in mitigating urban flooding and pollution, whereas green infrastructure such as rain gardens and bioswales provides nature-based remedies. Nonetheless, their efficacy is contingent upon local context, rendering models such as SWMM and GIS indispensable for planning. Machine learning, IoT sensors, and digital twins significantly improve forecasting, control, and optimization. Despite ongoing issues related to data quality and fairness, the amalgamation of intelligent technology with ecological design offers a robust and flexible strategy for constructing sustainable, flood-resilient, and inclusive urban water systems in a dynamic environment.

Desalination is an innovative solution to the global water shortage; however, scheduling and resource allocation in these systems present significant challenges. To address these challenges, this paper investigates the application of metaheuristic optimization techniques to enhance the performance of distributed desalination systems. The problem is formulated as a distributed flow shop scheduling model, with the primary objective of minimizing total flow time (TFT) to ensure timely water production and operational efficiency. To solve this, three metaheuristics are proposed: Iterated Greedy, Iterative Local Search, and Variable Neighborhood Search. The effectiveness of these metaheuristics is enhanced using two initialization techniques: NEH and GRASP, yielding a total of six approaches. A comprehensive simulation study was conducted on a set of benchmark instances. The results show that all proposed metaheuristics provide valuable solutions for minimizing TFT and optimizing resource allocation, with varying levels of performance across different instances.

Climate change poses existential risks to water management using a sustainable approach, particularly in agricultural-dependent areas such as Andhra Pradesh, India. In this context, using Machine Intelligence, which is Artificial Neural Networks (ANNs), this research investigates the application of Machine Intelligence in aiding climate adaptation for water management. The method involves the integration of multi-source data sets, including climate parameters (2020–2024), agricultural-based data on paddy farming, and geographic information for irrigation and topography. Preprocessing procedures maintained uniformity and compatibility among such datasets, after which Particle Swarm Optimization (PSO) was applied for feature selection to identify the most impactful factors influencing water supply and crop yield. The ANN model trained well and precisely forecasted irrigation demand and pinpointed possible periods of water stress, allowing localized real-time decision-making support.

This chapter introduces a technique that combines simulations using Ground Penetrating Radar (GPR), advanced processing of signals and deep learning to determine water content in plant roots as well as nearby soil. Using dielectric mixing models, computer programs were used to create synthetic radargrams of root-soil interactions under various moisture levels. Radargrams were improved by using methods such as Singular Value Decomposition and Inverse Amplitude Decay. Convolutional Neural Networks (CNNs), including InceptionV3, MobileNetV2 and ResNet50V2, were adapted for using in classifying both root and soil water content. InceptionV3 got the top results among tested models for estimating plant roots. For easier soil moisture classification, MobileNetV2 was able to run quickly and efficiently. ResNet50V2 managed to perform with similar accuracy and level of stability for both types of tasks.

Life depends on water, yet its management is under more stress than ever. Besides, contamination and the scarcity of water made it impossible to fulfill humans' basic needs. Artificial intelligence (AI) provides revolutionary possibilities through real-time monitoring, statistical analysis, and machine learning, while more than 80% of polluted water worldwide is released untreated. In this chapter, the strategies behind the application of AI in wastewater treatment and recycling are explained. It addresses both opportunities and constraints while striking a balance between creativity and reality. The chapter bridges the gap between technical depth and readable storytelling by referencing industry data, scholarly research, and expert perspectives. It draws attention to how AI is changing water sustainability by transforming wastewater from a liability into a resource.

The intensifying global water crisis demands innovative approaches to wastewater treatment and recycling as core strategies for sustainable water management. The chapter discusses how artificial intelligence transforms the process of modernizing wastewater systems offering improvement in the operational effectiveness of the system, optimal utilization of resources, and adaptive process control. It starts with a survey of the traditional treatment steps and its limitations and proceeds to show a detailed report of the Artificial Intelligence (AI) application in the field

Preface

INTRODUCTION

Water is life—yet the systems that sustain and deliver this essential resource are under unprecedented strain. Climate change, population growth, urban expansion, and mounting environmental pressures are converging to create a global water management crisis. In the face of these challenges, the urgent question we must ask is: how can we rethink water governance and management for a sustainable future?

This book emerges from the recognition that traditional approaches, while foundational, are no longer sufficient on their own. New solutions must be integrated—particularly those enabled by modern computational technologies. With this in mind, *Computational Intelligence and Optimization Methods for Sustainable Water Management* brings together a diverse range of insights at the intersection of water science and advanced machine intelligence.

The chapters in this volume explore the transformative potential of artificial intelligence, deep learning, and generative AI in addressing critical water-related challenges. From optimizing resource allocation and improving wastewater treatment, to enhancing desalination processes and building adaptive infrastructure for climate resilience, the contributions in this book showcase how computational intelligence is redefining what is possible in water management.

Our aim is to provide a comprehensive and interdisciplinary guide that bridges theory and practice. In doing so, we emphasize real-world applications—supported by case studies and research evidence—that demonstrate the integration of intelligent systems into various facets of water management. We highlight how generative AI, in particular, offers novel pathways for modeling, decision support, and system optimization, and we explore its emerging role in advancing sustainability goals across urban and rural contexts alike.

This book also addresses the environmental and governance implications of deploying AI-driven solutions. As we consider the evolving demands of climate adaptation and sustainable development, we underscore the need for intelligent frameworks that are not only technologically sound but also socially and environmentally responsible.

We envision this volume as a resource for a wide audience: from researchers and academics working at the forefront of AI and environmental science, to policymakers, water resource managers, and urban planners seeking practical innovations for real-world application. Our hope is that this book not only informs but also inspires a new wave of interdisciplinary collaboration in pursuit of resilient, equitable, and sustainable water futures.

We extend our deepest gratitude to the contributors whose expertise and dedication have shaped this work, and to the readers who share our commitment to solving one of the most pressing challenges of our time.

ORGANIZATION OF THE BOOK

This volume brings together pioneering research and practical applications at the intersection of water management, sustainability, and computational intelligence. The chapters collectively address historical legacies, current global challenges, and innovative solutions powered by artificial intelligence, machine learning, optimization, and modeling techniques. Below, we provide an editorial overview of each chapter to guide the reader through the breadth and depth of contributions presented in this reference work.

Chapter 1: History of Water Management: Historical Perspectives and Technological Dynamics

The opening chapter sets the stage by tracing the evolution of water management from ancient hydraulic societies to the present digital age. Through a cross-disciplinary lens, the chapter examines a range of traditional techniques and governance structures—such as khettaras, seguias, and community-based management systems—and contrasts them with the rise of centralized, technocratic models. The case study of Morocco serves as a compelling illustration of the tension between preserving historical water heritage and embracing modernity. The chapter concludes by arguing for the integration of historical wisdom into contemporary digital and AI-driven strategies, urging a balanced approach to future water governance.

Chapter 2: Global Water Resource Challenges

This chapter explores the multifaceted global water crisis, highlighting the compounding effects of climate change, population growth, and poor governance. Despite water's physical abundance, access to clean freshwater remains limited, inequitable, and increasingly at risk due to contamination, overuse, and salinity. Using real-world case studies, the authors detail the human and environmental costs of mismanagement and emphasize the need for transboundary cooperation. The chapter advocates for inclusive, adaptive, and interdisciplinary frameworks that combine scientific research, policy innovation, and local knowledge.

Chapter 3: Fuzzy Multi-Criteria Decision-Making Approach for Optimizing Sustainable Water Management

Addressing the complexity and uncertainty inherent in water-related decisions, this chapter introduces a Fuzzy Multi-Criteria Decision-Making (FMCDM) framework to support more rational and robust planning. The method integrates fuzzy set and soft set theories to evaluate strategies based on diverse criteria such as environmental sustainability, cost efficiency, and social acceptability. A case study illustrates its practical utility, offering structured decision support for multi-stakeholder scenarios. The chapter also points to areas where FMCDM can be further enhanced for real-time applications in sustainable water planning.

Chapter 4: Neutrosophic Sets for Decision-Making in Natural Disasters and Environmental Recovery

This chapter delves into neutrosophic logic as a novel mathematical approach to manage uncertainty in disaster response and environmental recovery. Building on fuzzy logic, neutrosophic sets uniquely represent truth, indeterminacy, and falsity—making them ideal for modeling complex, ambiguous situations like flood risk and wildfire management. The chapter features compelling case studies and discusses integration with MCDM methods. It offers a forward-looking discussion on computational challenges, interdisciplinary adoption, and the value of neutrosophy in enhancing resilience amidst environmental volatility.

Chapter 5: Role of Artificial Intelligence in Water Management

Artificial Intelligence (AI) is rapidly reshaping water systems worldwide. This chapter provides a comprehensive overview of AI's applications, from leak detection and quality monitoring to smart irrigation and demand forecasting. It examines real-world implementations, such as Barcelona's AI-enhanced infrastructure, while also acknowledging the challenges of scalability and fairness. The chapter underscores the transformative potential of AI in ensuring resource efficiency and resilience and calls for cross-sector collaboration to drive equitable adoption.

Chapter 6: Predictive Modeling and Generative AI for Anticipating Water Shortages

This chapter introduces a hybrid framework that leverages time-series transformers and generative AI models to forecast water shortages under climate stress. By synthesizing synthetic data from hydrological patterns and stochastic processes, the method overcomes gaps in real-world data availability. The approach enhances both visualization and response capabilities for critical scenarios, offering a new toolset for early warning systems and digital twin environments. The chapter concludes with a discussion on ethical and technical limitations and proposes concrete strategies for broader implementation.

Chapter 7: Smart Stormwater Systems: AI-Driven Forecasting, Optimization, and Real-Time Control

Focusing on urban resilience, this chapter explores how AI-powered systems can revolutionize stormwater management when combined with green infrastructure. The integration of models such as SWMM and GIS with machine learning, IoT sensors, and digital twins enables real-time optimization and control of flood-prone areas. While challenges in data fairness and model accuracy persist, the chapter demonstrates that intelligent technologies, when paired with ecological design, offer promising pathways to sustainable, inclusive, and adaptive urban water systems.

Chapter 8: Mathematical Modeling and Metaheuristic Optimization for Distributed No-Wait Flow Shop Scheduling in Desalination Systems

This chapter addresses the operational complexity of desalination systems through advanced scheduling algorithms. Using a distributed no-wait flow shop model, the authors apply six metaheuristic techniques—based on Iterated Greedy, Iterative Local Search, and Variable Neighborhood Search—enhanced by NEH and GRASP initialization. Simulation results across benchmark instances validate the effectiveness of these methods in minimizing total flow time and optimizing resource allocation. The chapter provides valuable insights for improving desalination system efficiency in water-scarce regions.

Chapter 9: Machine Learning Approaches to Climate-Smart Irrigation Planning

With a focus on agricultural regions like Andhra Pradesh, India, this chapter highlights how machine intelligence—specifically Artificial Neural Networks (ANNs) and Particle Swarm Optimization (PSO)—can inform climate-smart irrigation practices. By integrating multisource datasets, the model forecasts irrigation demand and identifies potential water stress periods. This data-driven approach enables localized, real-time decision support for farmers and planners, illustrating the value of AI in adapting agriculture to climate variability.

Chapter 10: A Non-Invasive Revolution in Soil and Root Water Content Monitoring by Integrating Deep Learning and GPR

This chapter presents a novel, non-invasive method for assessing soil and root water content using Ground Penetrating Radar (GPR) enhanced with deep learning models. Simulated radargrams processed through CNNs such as InceptionV3, MobileNetV2, and ResNet50V2 offer precise classification of moisture content. The study demonstrates how different models perform across accuracy, speed, and stability metrics, showcasing the potential for smart agricultural monitoring and sustainable water use.

Chapter 11: AI-Driven Wastewater Treatment and Recycling: Smart Water

Recognizing the critical need for efficient wastewater management, this chapter explores how AI is transforming water treatment from a reactive to a predictive discipline. From real-time monitoring to anomaly detection and system optimization, AI enables more sustainable reuse of treated water. Blending technical depth with accessible narrative, the chapter makes a compelling case for AI's role in making wastewater a viable resource, supported by data, industry trends, and expert insights.

Chapter 12: Wastewater Treatment and Recycling with AI

This chapter extends the discussion on AI in wastewater management, offering a systematic overview of how machine learning can improve monitoring, microbial analysis, and anomaly detection. The authors chart a roadmap from traditional methods to AI-enhanced systems capable of decentralized, smart water reuse. By reducing energy and chemical usage while ensuring reliability, the integration of AI contributes meaningfully to sustainable urban and industrial water cycles.

Chapter 13: Wastewater Treatment and Recycling Using Adsorption Enhanced by Artificial Neural Networks

The final chapter examines the synergy between adsorption techniques and Artificial Neural Networks (ANNs) for wastewater treatment. As an economically feasible method, adsorption benefits from the predictive and optimization power of ANNs to increase efficiency and water quality. Through practical applications in municipal and industrial settings, the chapter shows how this integration offers scalable, environmentally friendly, and cost-effective solutions for addressing global water scarcity.

Together, these chapters reflect a multidisciplinary convergence of knowledge, innovation, and purpose. We believe the insights offered here will support researchers, practitioners, and policymakers in advancing sustainable, intelligent water management for a changing world.

CONCLUSION

As we bring this volume to a close, we reflect on the pressing reality that water—an essential element of life—is under increasing strain across the globe. The challenges are complex and multifaceted: from historical legacies and geopolitical tensions to climate change, aging infrastructure, and uneven access. Yet, the chapters in this book offer a hopeful and forward-looking narrative: that through the intelligent application of computational tools and innovative thinking, we can reimagine and reshape how we manage, govern, and sustain our most precious resource.

Throughout this reference work, our contributors have demonstrated how artificial intelligence, deep learning, optimization techniques, and hybrid modeling systems can move water management from reactive practices toward predictive, adaptive, and resilient frameworks. Whether it is through the use of generative AI for early warning systems, machine learning for climate-smart agriculture, or neural networks optimizing wastewater treatment, the message is clear—computational intelligence is not only relevant, but indispensable in meeting the water challenges of the 21st century.

Yet, technology alone is not a panacea. Several chapters underscore the importance of integrating digital innovation with historical knowledge, inclusive governance, ethical design, and socio-environmental awareness. A future-oriented approach to water sustainability must remain interdisciplinary, human-centered, and grounded in both scientific rigor and community needs.

We hope that this book serves not only as a repository of advanced research and case studies, but also as a catalyst for continued dialogue and collaboration across disciplines, institutions, and regions. The journey toward sustainable water management is ongoing, but with the tools and insights shared in this volume, we are better equipped to face it—together.

Chapter 1
History of Water Management:
Historical Perspectives and Technological Dynamics

Taoufik Ait Bouchgour
https://orcid.org/0000-0001-5633-020X
Faculty of Arts and Humanities Ain Chock, Hassan II University of Casablanca, Morocco

Khalid Lahyani
https://orcid.org/0000-0002-0035-1615
ENSAM Casablanca, Hassan II University of Casablanca, Morocco

Ghizlane Moukhliss
https://orcid.org/0000-0003-3699-2008
ENS Casablanca, Hassan II University of Casablanca, Morocco

ABSTRACT

Since ancient times, water management has shaped human societies, their economies, political structures, and imaginations. This chapter traces the major stages of this history, from ancient hydraulic civilizations to contemporary policies, including medieval innovations and colonial transformations. Through a cross-disciplinary analysis, it highlights the diversity of techniques (khettaras, seguias, dams, desalination) and governance systems (community-based, piecemeal, technocratic). The case of Morocco, studied over a long period of time, illustrates the tensions between local heritage and modernization. The recent rise of digital technologies and artificial intelligence opens up new prospects for more sustainable, equitable, and resilient management. This work argues for a revaluation of historical knowledge in future water policies.

DOI: 10.4018/979-8-3373-2700-6.ch001

INTRODUCTION

Water is at the heart of human history. As a vital resource, it determines the survival of societies, shapes their territorial settlements, and determines their economic and political models. Since antiquity, the ways communities obtained, used, and shared water have reflected not only their technical capacities but also their systems of power, belief, and governance. Studying the history of water management therefore means examining the forms taken by the relationship between humans, nature, and technology through both local and global dynamics (Delli Priscolli and Wolf 2009). More recently, several researchers have emphasized the need to re-examine this history in light of contemporary issues of sustainability and environmental justice (Loftus, 2018; Mehta et al., 2012).

From the earliest hydraulic civilizations of the Fertile Crescent—Mesopotamia, Egypt, and the Indus Valley—access to water has been a strategic issue. These societies developed sophisticated infrastructure: irrigation canals, dams, cisterns, and aqueducts. These devices were not mere technical instruments; they embodied a hierarchical social organization in which control over water was equivalent to the exercise of power. As early as 1957, sinologist Karl Wittfogel formulated his famous theory of hydraulic despotism, arguing that states such as those of imperial China had built their power on centralized control of water networks (Wittfogel 1957). However, this interpretation has been qualified by more recent approaches that emphasize the diversity of local trajectories and the importance of community institutions in water management (Barraqué, 2025; Boelens & Seemann, 2014).

In fact, hydraulic history cannot be reduced to a dichotomy between centralization and autonomy. In the Andes, *acequias* were co-managed by peasant assemblies; in Southeast Asia, Balinese *subak* networks are based on cooperative logic linked to temples. This research highlights the diversity of water management systems, calling for a move away from a linear model of modernization (Molle & Mollinga, 2003; Shah & Ballabh, 1997). Today, these studies resonate particularly strongly in contemporary debates on water governance and the resilience of local systems to climate change (Pahl-Wostl, 2020). This research highlights the diversity of historical trajectories, calling for a move away from a linear model of modernization.

The medieval Arab-Muslim world, particularly between the 8th and 14th centuries, is a particularly rich example of this complexity. Cities such as Fez, Cordoba, and Baghdad developed impressive water distribution infrastructures, combining scientific knowledge, hydraulic techniques, and religious norms. Techniques such as the *noria,* the Persian *qanat*, and the Maghreb *saguia* ensured precise irrigation in often arid environments. Al-Karaji, al-Razi, and Ibn al-Haytham, in their treatises, discussed the physical properties of water, collection systems, and the medical and urban uses of this resource (Nasr 2006; Hill 1994).

In Europe, medieval water management followed other logics: feudal, monastic, municipal. Cistercian abbeys, for example, played a major role in the hydraulic development of the countryside (Guillerme 1983). Italian cities and Flemish-, meanwhile, developed canal systems for industrial and commercial purposes. Water was also used to demarcate areas of power, structure sanitation, and fuel popular beliefs (Duby 2011). As such, it was perceived as both a divine blessing and a potential threat—flood, disease, or divine punishment.

The transition to the modern era, and even more so to the industrial era, introduced a new rationality into water management systems. From the 18th century onwards, water became an economic resource integrated into the logic of growth and national power. Colonies, particularly in the Maghreb, served as laboratories for these technical policies (Bonneuil & Fressoz, 2013). This history highlights lasting North-South asymmetries, which continue to shape the distribution of resources and inequalities of access.

But hydraulic modernity has revealed its limitations: pollution, depletion of aquifers, conflicts over use, and marginalization of rural populations. Since the 1990s, a new paradigm—integrated water resources management (IWRM)—has emerged under the impetus of international institutions. However, several recent studies show that this approach, which is supposed to reconcile sustainability, participation, and economic efficiency, remains marked by tensions between discourse and practice (Molle & Sanchis-Ibor, 2019; Zeitoun et al., 2013).

Colonies have often served as laboratories for these technical policies. In the Maghreb, the French colonial authorities promoted vast dam, irrigation, and water capture projects designed to serve the colonists and restructure local agriculture in accordance with the interests of the empire. This history reveals structural asymmetries between North and South, between colonial science and vernacular knowledge (Bonneuil and Fressoz 2013). The colonial history of water is thus also a history of exclusion, dispossession, and resistance.

But this modern approach to water management is now reaching its limits. It has generated numerous negative externalities: pollution, depletion of aquifers, conflicts over use, and marginalization of rural populations. Since the 1990s, under the impetus of international institutions (World Bank, UN), a new paradigm has emerged: integrated water resource management (IWRM), which aims to reconcile sustainability, citizen participation, and economic efficiency. Sustainable Development Goal 6, adopted in 2015, enshrines water as a fundamental human right and promotes inclusive governance (United Nations 2015).

However, this shift is not without its ambiguities. Researchers such as Erik Swyngedouw (2009) and Jamie Linton (2010) point out that the concepts of "sustainability," "scarcity," and even "resource" are socially constructed. They reflect political choices, power relations, and ideological priorities. Thus, certain so-called "participatory" policies actually mask a technocratic recentralization of decision-making. The critique of *political ecology* is to reveal the interests at stake behind the official narratives of the water crisis.

Water is now at the heart of conflicts over use (agriculture vs. industry, cities vs. rural areas), jurisdictional conflicts, and cross-border tensions. The concept of "water wars," popularized by Peter Gleick (1993), refers to these growing tensions around basins such as the Nile, Euphrates, and Jordan rivers. Added to this are crucial contemporary dynamics such as the privatization of urban water services, the financialization of the resource, and social inequalities in access (Bakker, 2014; Loftus, 2018). These debates confirm that water management is never neutral: it is always intertwined with political, economic, and cultural power relations.

In addition to these technical and geopolitical dimensions, there is also a symbolic and cultural interpretation of water. In many traditions—Islamic, Christian, Hindu—water is a sacred substance, purifying, a source of life and regeneration. These beliefs influence social practices, customary rights, and access rituals. Anthropologist Veronica Strang (2004) shows that water is never neutral: it is imbued with multiple meanings, often gendered, localized, and linked to the historical trajectories of communities.

Africa, long marginalized in global water historiography, deserves special attention. Researchers such as Sara Berry (1993) and Allen and Smits (2016) have shown that forms of water governance on the continent often rely on hybrid arrangements involving chiefdoms, the state, NGOs, and private actors. Morocco, in particular, offers a fertile ground for analysis, combining pre-colonial legacies, post-independence reforms, and privatization dynamics since the 1990s.

The history of water is therefore fundamentally interdisciplinary. It draws on environmental history, the sociology of technology, political geography, anthropology, law, and economics. It also involves reflection on sources: hydraulic archives, ancient maps, travel accounts, colonial archives, and oral traditions. It is a rapidly expanding field, driven by contemporary issues such as climate change, inequality, and ecological transition.

Viewed in this way, the history of water management is not just a narrow academic field. It is a critical tool for analyzing development trajectories, questioning past and present models of governance, and rethinking the relationship between human societies and the natural environment. In a context marked by climate change, population growth, and rapid urbanization, a historical perspective allows us to connect long-term trends with contemporary emergencies, while revaluing often marginalized knowledge.

With this in mind, this chapter offers a cross-disciplinary and critical reading of water management through time, drawing on environmental history, political anthropology, and development studies. It is organized around four main themes that trace the continuities, ruptures, and recompositions in the forms of appropriation, regulation, and symbolization of water:

– Hydraulic systems in ancient civilizations, between engineering and social organization;
– Medieval innovations, at the crossroads of knowledge, beliefs, and regional practices;
– The Moroccan case, analyzed over the long term, from pre-colonial to contemporary times;
– Current issues, marked by resource scarcity, inequalities of access and global reconfigurations.

The aim here is not to provide a simple technical chronology of hydraulic infrastructure, but to understand how water is, in each context, a historical, political, and symbolic construct at the heart of tensions between nature, society, and power.

WATER AND EARLY CIVILIZATIONS: CONTROLLING RESOURCES TO STRUCTURE SOCIETY

The history of water management in early civilizations is one of the most striking examples of the interdependence between the natural environment, technical innovation, and social organization. Far beyond functional developments, ancient hydraulic systems shaped the political, economic, and symbolic dynamics of societies, embedding the resource in power structures and culture. From the floodplains of Mesopotamia to the fertile valleys of the Nile, Indus, and Yellow Rivers, early communities developed centralized institutions, complex social hierarchies, and cosmological representations in which water became both a material resource and a symbolic foundation of power.

This interpretation has long been dominated by Wittfogel's (1957) theory of "Oriental despotism," which linked hydraulic infrastructure with authoritarian centralization. While this perspective has left its mark on historiography, it has since been greatly qualified. Research in historical anthropology (Lansing, 1991; Scarborough, 2003) and environmental history (Tvedt & Oestigaard, 2016) shows the diversity of trajectories, where water management could be either community-based or technocratic. More recently, interdisciplinary approaches (Boelens & Seemann, 2014) have emphasized the importance of placing these ancient experiences in the context of a reflection on the commons, water justice, and resilience to climate hazards. This work invites us to consider ancient hydraulic systems not as mere technical curiosities, but as laboratories of social and political organization, whose lessons still resonate in contemporary debates on resource sustainability.

Hydraulic Systems and their Impact on Social Organization

The emergence of the first great civilizations—in Mesopotamia, Egypt, the Indus Valley, and China—cannot be understood without analyzing the decisive role played by water. Each geographical context posed specific water challenges, to which societies responded with diverse technical solutions that converged in their ambition to control the environment. These technical choices were not neutral: they conditioned distinct forms of social hierarchies, governance, and collective imaginaries.

In Mesopotamia, the Sumerians are considered the pioneers of large-scale irrigation. The Tigris and Euphrates rivers, although generous in water, were unpredictable in their flooding and could be as destructive as they were fertile. As early as the 4th millennium BC, Sumerian communities built a dense network of canals, dykes, reservoirs, and basins to divert water to agricultural fields. This mastery fostered intensive agriculture, the basis of sedentarization and urban growth. Butzer (1976) shows that maintaining these infrastructures required strict collective organization, paving the way for early bureaucratic systems. Hydraulic works were coordinated by local authorities, often linked to temples, which monitored water distribution and settled disputes between farmers. Water, as a common good, thus became a vehicle for social organization, encouraging the establishment of norms and hierarchies designed to ensure equitable access to this vital resource. Studies compiled by Mays (2010) confirm this analysis by showing how Mesopotamian systems, such as the Nimrud dams and gravity-fed canal networks, required sustained inter-community coordination, heralding the first forms of public engineering.

This model of social organization based on water management also emerged in Egypt, where the Nile dictated the rhythm of agricultural life. Unlike the Mesopotamian rivers, the Nile had regular and predictable floods. The Egyptians were able to exploit its benefits by developing irrigation systems based on the collection and redistribution of floodwaters. Philipe Leveau (2008) points out that the role of the pharaoh was closely linked to this water management: as the guarantor of cosmic order (ma'at), the sovereign embodied the state's ability to control the elements. The flooding of the Nile was seen as a divine gift, but its channeling depended on organized and effective human power. Thus, water management reinforced royal centralization, offering the pharaoh both an economic lever and a symbolic instrument of power. As Hassan (2011) suggests, control over water made it possible to unify a vast territory around a common project of agricultural development and redistribution. This organization was also accompanied by measuring devices, such as nilometers, which made it possible to predict harvests and anticipate crises, illustrating these societies' ability to integrate environmental forecasting into their governance.

In the Indus Valley, the planned cities of Mohenjo-Daro and Harappa demonstrate a different approach to water management. Here, the emphasis was on urban sanitation and collective water supply. Archaeologists have uncovered wells, public reservoirs, covered sewers, and elaborate drainage systems, revealing a keen concern for hygiene and public health (Kenoyer, 1998). This technical culture coincided with a seemingly egalitarian urban organization: infrastructure was evenly distributed, reflecting a collective approach to water use. According to Possehl (2002), this architectural uniformity reflects centralized planning geared toward the common good. It is also remarkable that these systems demonstrate an anticipation of modern principles of sustainable urban planning and public health, long before these concepts were formalized in Western thought. Mays (2010) emphasizes that the hydraulic urban planning of Mohenjo-Daro, with its 700 individual wells, bathing platforms, and masonry sewers, constitutes an early model of integrated water management. Recent paleoenvironmental research (Madella & Fuller, 2006) suggests that the collapse of this civilization may have been linked to increased climate variability, already linking the history of hydraulic systems to resilience to climate hazards.

In China, water management took on an even more strategic dimension. The Yellow River, nicknamed "the sorrow of China," caused devastating floods, but its basin was also the cradle of Chinese civilization. Under the Xia, Shang and Zhou dynasties, the first hydraulic management institutions were established, mobilizing thousands of workers in the construction of dykes, canals and reservoirs. These efforts reached their peak with the construction of the Grand Canal, which began in the 5th century BC and was developed under the Sui and Tang dynasties. James Lightfoot (1996) emphasizes the role of hydraulic engineering in consolidating imperial power, which used water to connect economic centers, transport grain, and integrate distant provinces. More recently, Pietz (2015) has highlighted the importance of the Grand Canal as an instrument of environmental governance, linking hydraulic engineering, fiscal control, and political integration over the long term.

These various examples show that water management was much more than just a factor in agricultural productivity: it structured social relations, legitimized power, and shaped the political imagination. They also show that hydraulic systems were not only technical responses to the challenges of water scarcity, excess, or uncertainty, but also social and political ones. This dialectic between ecological constraints and institutional innovation remains highly relevant in the era of climate change.

Finally, echoes of these dynamics persist to this day. Research conducted by Haddache (2022) on the Toudgha oasis shows how traditional irrigation systems (khettaras, nouba, oral conventions) embody shared and resilient governance rooted in a long social memory. These local forms of management, often ignored by modern technocratic approaches, nevertheless offer relevant responses to current challenges in terms of water equity and sustainability.

Governance Models and Political Issues

Water systems are not limited to technical infrastructure. They also involve political and legal models of collective or centralized management. In ancient societies, the need to maintain, organize, and distribute water flows led to the invention of specific governance practices, often intertwined with monarchical, religious, or community power systems. Water thus became a field of political legitimization, an instrument of economic regulation, and a testing ground for institutional experimentation.

Mesopotamia offers an early example of water legislation. The famous Code of Hammurabi (circa 1750 BC) contains several articles on water distribution, canal maintenance, and responsibilities in the event of water damage. As Mays (2010) explains, these laws demonstrate a primitive environmental awareness and a desire to ensure the sustainability of irrigation systems through legal regulation. They anticipate the principles of integrated water management, where law, technology, and collective responsibility are interlinked. For Hillel (1994), these provisions also reveal an ancient awareness of water scarcity and the need for its rational and equitable management.

In Egypt, the pharaonic authorities mobilized centralized institutions to supervise the flooding of the Nile and the maintenance of hydraulic networks. Scribes and administrators monitored water levels and planned public work. Taxation was directly linked to the fertility of irrigated land, which shows how water management was central to the pharaonic political economy (Butzer, 1980). Control over water was not only an agricultural necessity, but also a means of sanctifying power by reinforcing the role of the pharaoh as the guarantor of cosmic balance.

In China, the administrative tradition of large-scale hydraulic works is part of a long history of state expertise. The Ministry of Public Works centralized technical knowledge and organized construction sites. This imperial bureaucracy strongly linked political legitimacy to the ability to control the symbolically dangerous waters of the Yellow River. For Elvin (2004), this hydraulic technocracy illustrates the structuring role of engineers in the Chinese imperial state. Pietz (2015) also points out that the construction and maintenance of the Grand Canal were a tool for fiscal control and economic redistribution, strengthening territorial cohesion around the waterway.

But alongside these centralized examples, other forms of governance emerged in ancient societies. In Bali (Lansing, 1991), *subak* irrigation systems were managed by farmer-priests according to ritual calendars and collective rules. These systems illustrated decentralized but robust coordination, where water was considered a common good. Similarly, some ancient Mediterranean societies used forms of shared customary management (Molle & Sanchis-Ibor, 2019). These experiences demonstrate that water management does not necessarily lead to authoritarian centralization, but can also strengthen community cohesion and horizontal solidarity.

Ultimately, these different models of governance—authoritarian, technocratic, or community-based—shed light on the diversity of water regimes throughout history. They also show how water management has been used to legitimize power, stabilize societies, and strengthen collective solidarity. The historical plurality of water governance still resonates today in debates on the commons and the necessary balance between centralized management and local participation.

Technical, Cultural, and Symbolic Legacies

The legacy of hydraulic civilizations is not limited to their infrastructure. These systems have had a lasting influence on material cultures, collective representations, development models, and imaginaries of power. Mays (2010) emphasizes that these ancient technologies were not mere artifacts of the past, but have often survived the centuries, influenced contemporary knowledge, and continue to be used or restored in several regions of the world. They thus constitute a living technological heritage, whose resilience still inspires current approaches to sustainability.

On a technical level, many ancient hydraulic infrastructures have survived for centuries. In Egypt, some ancient dykes still serve as the basis for modern installations. The Grand Canal in China, which has been restored several times, continues to link the north and south of the country. In the Euphrates Valley, some canals dating back to Babylonian times were still in use in the 20th century (Tvedt, 2016). These material continuities illustrate not only the effectiveness of ancient knowledge, buly its integration into a long period of coevolution between society and the environment.

On a symbolic level, water plays a central role in cosmologies and founding narratives. In Egypt, the flooding of the Nile was associated with the god Osiris, symbolizing the cycle of life, death, and renewal. In Mesopotamia, the god Enki represented fertility and wisdom, linked to underground waters. In India, river water, particularly that of the Ganges, is sacred. In China, Feng Shui philosophy gives water a structuring role in cosmic harmony. These representations have served as ideological support for political power, which presented itself as a mediator between the cosmic order and the social order. As Strang (2004) observes, water is a fundamental vehicle for social values and symbolic meanings, and its management reflects prevailing worldviews.

Finally, the legacy of ancient hydraulic practices is echoed in contemporary community models. The example of the Toudgha oasis (Haddache, 2022) shows how local knowledge, customary institutions, and solidarity mechanisms still structure access to water in semi-arid contexts today. These practices, such as the management of khettaras, water towers (nouba) and oral agreements (orf), reflect a horizontal, distributive and resilient form of governance in the face of climate uncertainty. They constitute a long social memory that offers avenues for rethinking sustainability in current water policies.

Contemporary approaches are enriched in particular by research on commons and forms of territorialized governance. Boelens and Seemann (2014) argue for greater recognition of indigenous hydraulic knowledge, which is often more sustainable and equitable than large, centralized modern projects. The collective work edited by Eymard and Barraqué (2021), *Les ressources en eau: enjeux et gouvernance (Water Resources: Challenges and Governance),* illustrates this evolution by drawing on a variety of case studies in which local knowledge, multi-level cooperation, and collaborative management practices are valued in the face of rising climate and social risks. Thus, the symbolic and technical legacies of the past directly feed into contemporary debates on water justice and sustainable governance.

MEDIEVAL INNOVATIONS AND THE IMPACT OF TECHNOLOGICAL REVOLUTIONS

The medieval era was a decisive stage in water management, marked by technical and organizational innovations that reshaped agriculture, cities, and trade. Unlike the large hydraulic systems of antiquity, which were often associated with centralized structures and state power, the medieval period was distinguished by the spread of more flexible and adaptable technologies that promoted both local autonomy and regional economic integration.

The irrigation canals of the Islamic world, European water mills, urban water supply networks, and community management systems in East Asia illustrate the diversity and richness of the solutions experimented with between the eighth and fifteenth centuries. These innovations had a considerable impact on agricultural productivity, urban growth, trade, and the structuring of collective institutions.

Several recent studies (Squatriti, 2002; El Faïz, 2005; Tvedt & Oestigaard, 2016) have shown that this period was marked by a dual movement: on the one hand, the reinvention and improvement of tools inherited from Antiquity; and on the other hand, the emergence of new forms of water governance adapted to rapidly changing societies. In this sense, the Middle Ages were not simply an interlude, but a pivotal stage in which the technical, social, and symbolic foundations of hydraulic modernity were laid.

The Legacy and Transformation of Ancient Techniques

The Middle Ages did not mark a complete break with Antiquity, but rather a period of transmission, adaptation, and reinvention. Medieval hydraulic innovations were based on a foundation inherited from the great ancient civilizations, which they transformed in response to new social, political, and cultural contexts. This dynamic illustrates what historian Donald Hill (1994) calls "creative continuity": knowledge is not simply reproduced, but integrated into new configurations, responding to other needs and other forms of governance.

The Continuity of Large Hydraulic Systems

Persian qanats, invented in the first millennium BC, perfectly illustrate this logic of transmission. Known for their ability to capture groundwater and transport it over long distances through sloping galleries, they were adopted throughout the Islamic world under local names: khettaras in the Maghreb, foggaras in the Sahara, and falajs in the Arabian Peninsula. Their efficiency was based not only on engineering but also on collective organization of maintenance and distribution, which ensured equitable distribution among community members (Goblot, 1979; El Faïz, 2005).

Similarly, Roman aqueducts, symbols of ancient engineering, continued to be used in several European cities in the early Middle Ages. In Rome, Nîmes, and Toledo, ancient sections were maintained and integrated into medieval networks. This continuity demonstrates that, despite the weakening of the Roman state, hydraulic infrastructure retained its relevance over time and was adapted to new social contexts (Squatriti, 2002).

Transformations and Improvements

The Middle Ages also saw a series of technical improvements that profoundly transformed traditional practices.

– Water mills, known since Greco-Roman times, multiplied spectacularly in Europe between the 9th and 12th centuries. Jean Gimpel (1977) refers to this as a veritable "medieval industrial revolution." These mills were not only used to grind grain, but also to power forges, fulling mills, and sawmills. They thus contributed to economic growth and a new organization of labor. Their spread reflects the integration of water into an increasingly diversified and pre-industrial economy.

– In the Islamic world, waterwheels were significantly improved. These large hydraulic wheels, installed on rivers, were used to lift water and irrigate orchards. Examples from Al-Andalus, particularly in Cordoba, show how these devices combined hydraulic efficiency with urban aesthetics. Water became both a tool for agricultural production and an element of architectural prestige, integrated into gardens and palaces (El Faïz, 2005).

These improvements reflect a major change: water was no longer managed solely for agricultural survival, but was integrated into a diversified economy and into urban, aesthetic, and symbolic projects.

Intercultural Exchanges

The medieval period was also marked by an intense circulation of hydraulic knowledge. Mediterranean contacts, Arab conquests, the Crusades, and trade exchanges promoted the spread of techniques.

– Persian and Arab knowledge was transmitted to Europe via Sicily and Muslim Spain. The Moroccan khettaras and Algerian foggaras thus have their origins in techniques from the East.
– Conversely, some European solutions were adapted in Islamic territories, demonstrating the early globalization of hydraulic knowledge (Hill, 1994).
– In East Asia, China developed its own systems, but contacts existed via the Silk Road, which facilitated the transfer of knowledge.

This interregional circulation invites us to think of the Middle Ages not as a series of isolated worlds, but as a connected space where techniques traveled and adapted.

Social and Symbolic Legacies

The ancient legacies were not only technical. They also took on a renewed social and symbolic dimension.

– In the Islamic world, Andalusian gardens (such as those of the Alhambra and Madinat al-Zahra) made water a symbol of cosmic and spiritual harmony (Brookes, 1987).
– In Christian Europe, monastic fountains became places of purification and order, integrating water into a theology of purity.
– In the Maghreb and the Sahara, khettaras and foggaras were regulated by customary rules and local institutions, embodying forms of community and solidarity-based governance (El Faïz, 2005).

Thus, medieval transformations were not limited to the technical sphere: they also produced new social, religious, and political representations of water.

Table 1. Ancient legacies and medieval innovations in water management

Ancient legacies	Medieval transformations	Social and cultural impact
Persian qanats	Khettaras (Maghreb), Foggaras (Sahara)	Equitable collective distribution, customary governance
Roman aqueducts	Adaptation in European cities	Urban supply, post-Roman continuity
Roman mills	Widespread diffusion in Europe (9th–12th centuries)	Increased productivity, economic diversification
Ancient waterwheels	Perfected in Al-Andalus	Irrigation + urban aesthetics (gardens, palaces)
Cosmic symbolism (Egypt, Mesopotamia)	Medieval religious and spiritual symbolism (Islam, Christianity)	Legitimization of power, identity dimension

This table shows how ancient foundations were reinterpreted in medieval contexts, producing new economic and social dynamics.

Far from being a mere passive legacy, ancient techniques were transformed, hybridized, and reinvested in various social and political contexts during the Middle Ages. They helped to boost agricultural productivity, stimulate urban growth, and redefine the relationship between water, power, and culture. This dynamic illustrates the ability of societies to incorporate ancient knowledge into new temporalities and give it renewed meaning.

Social and Political Organization Around Water Management

In the Middle Ages, water was not only a technical or agricultural resource: it was an essential factor in social and political organization. Hydraulic infrastructure required ongoing maintenance, precise rules of use, and conflict resolution mechanisms, which necessitated the establishment of institutions, norms, and collective practices. Water management thus became an instrument of power, a lever for cohesion, but also a source of tension between social groups.

In the Islamic world, water governance was based on a combination of local structures and central authority. The khettaras and foggaras of the Maghreb, descendants of the Persian qanats, were collectively administered by village assemblies, the jma'a, which distributed water according to rights proportional to each family's investment in the construction and maintenance of the galleries. These rights were often recorded in oral or written agreements, known as orf, which guaranteed the intergenerational transmission of uses. Water thus became a shared resource, but one governed by a customary hierarchy. At the same time, political power used water as an instrument of legitimization. In Al-Andalus, the Umayyads of Cordoba and the Nasrids of Granada invested in sophisticated hydraulic networks, including waterwheels and gardens, whose splendor testified to the prosperity of their reign. In the palaces of the Alhambra, water was staged in pools and canals to symbolize the cosmic order and divine justice embodied by the sovereign (Brookes, 1987). Added to this was the framework of Islamic law (fiqh), which established rules for sharing and prioritizing agricultural, domestic, and religious uses. These principles, based on the idea that water is a divine gift, aimed to ensure equity and community solidarity. They have been widely studied in contemporary literature on water management in Islam (Faruqui et al., 2001; Caponera & Nanni, 2019)."

In medieval Europe, the organization of water management reflected the diversity of local powers and the gradual rise of municipalities. In rural areas, water mills were generally controlled by lords, who forced peasants to use them in exchange for a fee. As Gimpel (1976) points out, these infrastructures became symbols of feudal domination, reinforcing economic and social dependence. But water could also be at the heart of religious experiences. Monastic orders, particularly the Cistercians, developed vast networks of canals, ponds, and mills, where water was used to irrigate fields, produce energy, and feed fish farms. These rationalized landscapes reflected spiritual discipline as much as technical mastery and were part of a regional hydraulic economy (Aston, 2000). With urban growth, municipal institutions gradually took over the supply of drinking water and sanitation. In Italian cities such as Florence and Siena, aqueducts and public fountains were financed by local taxes and designed as civic assets, embodying a new form of collective governance (Squatriti, 2002).

In China, water management was part of an imperial strategy of territorial control. The Grand Canal, whose construction was consolidated during the Sui and Tang dynasties, served as both a transport route and an instrument of political cohesion, linking the north and south of the country. Controlling the floods of the Yellow River was a real test of legitimacy: an emperor unable to control this capricious river symbolically lost the "mandate of heaven" (Pietz, 2015). To manage these colossal projects, a specialized bureaucracy attached to the Ministry of Public Works was established. Engineers and technicians were mobilized to design and maintain the infrastructure, illustrating the centrality of hydraulic knowledge to the legitimacy of imperial power (Elvin, 2004).

In all these societies, water was simultaneously a vehicle for cooperation and a source of conflict. Water rights could give rise to tensions between families, villages, or lordships, requiring the intervention of customs or courts. But the collective management of systems, whether by village assemblies in the Maghreb, European monastic communities, or urban municipalities, also strengthened solidarity and gave rise to forms of shared governance. The monopolization of water by elites, whether seigneurial, imperial, or tribal, illustrates its role as a resource for social domination; conversely, community cooperation highlights its potential as a common good.

These medieval models still offer valuable lessons today. They shed light on the ongoing tension between centralized management and local governance, which is always at the heart of debates on water. They remind us that this resource is a common good, subject to social rules of sharing, as Elinor Ostrom (1990) had already highlighted in her work on the commons. Finally, they show how water, beyond its materiality, was and remains an object of political legitimation, conflict, but also cooperation, whose legacy continues to influence contemporary sustainable management policies.

Table 2. Social and political organization around water management in the Middle Ages: a comparison between the Islamic world, medieval Europe, and imperial China

Region	**Institutions and modes of governance**	**Role of central power**	**Place of the community**	**Symbolic and cultural dimension**
Islamic world (Maghreb, Al-, Al-Andalus, the East)	Local assemblies (*jma'a*), customary conventions (*urf*), Islamic law (*fiqh*)	Sovereigns (Umayyads, Nasrids) legitimize their power through Hydraulic infrastructure (norias, gardens, canals)	Collective management of khettaras and foggaras, water rights proportional to investment	Palatine gardens (Alhambra) = cosmic order, water as a divine gift and symbol of justice
Medieval Europe (Christian West)	Seigneuries (water mills), monasteries (Cistercians), urban municipalities (Italy, Spain)	Lords exercised economic control; urban authorities financed aqueducts and fountains	Peasants subject to taxes, but community solidarity in villages and religious orders	Water = spiritual purity (monasteries), civic good (public fountains)
Imperial China	Imperial bureaucracy (Ministry of Public Works), corps of engineers	Emperor judged on his ability to control floods (Yellow River)	Mass mobilization of peasants for hydraulic works	Water linked to the "mandate of heaven," a Confucian symbol of order and harmony

This comparative table highlights the diversity of forms of water governance in the Middle Ages. Whether it be the combination of community management and political legitimation in the Islamic world, the intertwining of seigneurial, monastic, and urban powers in Europe, or bureaucratic centralization in China, water appears to be a structuring factor in societies.

These differences also highlight a convergence: in all contexts, hydraulic innovation had major social and economic effects. It is precisely this link between technical progress and economic transformation that the following section will address.

Legacy and Transition to the Modern Era

Medieval hydraulic innovations should not be seen as a mere intermediate step between antiquity and modernity: they formed the foundation on which the technical, economic, and political transformations that marked the entry into the modern era were built. The history of water reveals a process of continuity and reinvention, in which inherited knowledge was gradually integrated into changing social and economic structures.

In the Islamic world, irrigation networks established in the Middle Ages continued to supply rural societies long after the decline of the great empires. The khettaras of the Maghreb, the falajs of Oman, and the foggaras of the Sahara, descendants of Persian qanats, illustrate this persistence. These systems, often managed by local communities according to customary rules, embodied a form of resilience in the face of political and climatic change. Their legacy was decisive when, from the 16th century onwards, the colonial powers set about transforming agriculture in arid regions: by modernizing or marginalizing these structures, they drew on a medieval technical and social memory that was still very much alive (El Faïz, 2005). Similarly, in Al-Andalus, the hydraulic innovations of the Muslim period—waterwheels, basins, orchard irrigation networks—were largely taken up by the Christian kingdoms after the Reconquista, they served as the basis for the development of intensive and diversified agriculture, which contributed to the economic development of the Iberian Peninsula in modern times (Harvey, 2005).

In Western Europe, medieval mills marked a decisive turning point. Their widespread use between the 11th and 13th centuries paved the way for a diversification of hydraulic energy uses, heralding certain features of the industrial revolution. Mills were no longer limited to grinding grain: they powered sawmills, forges, tanneries, and fulling mills, multiplying their productive applications. This versatility contributed to the concentration of economic activities in cities, reinforcing their role as dynamic hubs. As Gimpel (1976) has shown, the "industrial revolution of the Middle Ages" was no exaggeration: water became a fundamental driving force of the pre-industrial economy. This model, reinforced during the Renaissance by the rise of cartography and applied sciences, created the conditions for the transition to a mechanized and systematic exploitation of natural forces.

In China, the medieval hydraulic heritage was even more visible in the transition to modernity. The Grand Canal, consolidated under the Sui and Tang dynasties, remained a vital artery until modern times, feeding the capital and maintaining territorial unity. Controlling the Yellow River and maintaining the dykes remained ongoing challenges, mobilizing the imperial bureaucracy and affirming the political legitimacy of the rulers. As Elvin (2004) has pointed out, Chinese water management illustrates a "long environmental history," in which each technical innovation is part of a cumulative dynamic of centralization and rationalization. This technocratic model would influence Chinese state practices into the contemporary era, where hydraulic engineering remains at the heart of the political project.

A common element in these trajectories was the persistence of a symbolic imagery of water, transmitted from the Middle Ages to the modern era. In Europe, medieval urban fountains, initially designed as water supply devices, were transformed during the Renaissance into artistic monuments, associating water with civic order and the glory of patrons. In the Islamic world, Andalusian and Persian gardens, where water evoked the Koranic paradise, inspired the hydraulic architecture of the Ottomans, Mughals, and other modern dynasties (Brookes, 1987). In China, the Confucian and Taoist symbolism of water as a vehicle for harmony and balance continued to permeate the imperial imagination and guide governance practices.

However, this transition to the modern era should not be understood as a simple extension. It was also marked by profound ruptures. From the 16th century onwards, the rise of global trade, agricultural intensification, and urban growth imposed new hydraulic demands. The inherited systems, while remaining functional, were often reorganized by more centralized states or by private actors seeking to maximize profits. In the Ottoman Empire, the inherited systems were often reconfigured within state mechanisms combining taxation, corvée labor, and public works, particularly in Egypt and the major capitals. Network maintenance and water distribution contributed to the legitimization of power and territorial integration (Mikhail, 2011). In Europe, hydraulic engineering experienced an unprecedented boom with the scientific Renaissance, marked by the emergence of figures such as Leonardo da Vinci, who combined naturalistic observation and mechanical experimentation.

Thus, the medieval legacy proved ambivalent: it offered solid technical and institutional models, but it also constituted a constraint in the face of the new needs of modernity. Khettaras, for example, remained effective for irrigating oases, but their maintenance became difficult in the face of rural exodus and the introduction of modern pumping techniques. Similarly, medieval mills, once drivers of innovation, were gradually replaced by steam engines.

In short, the medieval legacy was not limited to a transfer of technology: it had a lasting impact on water cultures, governance practices, and political imaginaries. But as the modern era dawned, these legacies were reinterpreted, adapted, and sometimes marginalized, revealing that water management is a historical process that is always in flux, oscillating between continuity and rupture.

THE CASE OF MOROCCO: BETWEEN HERITAGE AND MODERNIZATION

Morocco offers a particularly rich field of observation for analyzing the historical evolution of water management, due to its contrasting geography, ancient hydraulic traditions, and the institutional transformations that have marked its contemporary history. Between the southern oases shaped by ancestral community know-how, the major hydraulic works imposed by the French colonial administration, and the modern public policies implemented since independence, the country illustrates a complex trajectory marked by tension between local practices and centralized logic. This section examines three defining moments in this trajectory: traditional forms of water management and their social roots, colonial centralization and hydro-agricultural developments, and finally, post-independence policies in a context of resource scarcity and growing environmental challenges.

Traditional Water Management: Khettaras, Séguias, and Ancestral Dams

The history of water management in Morocco cannot be understood without an in-depth analysis of the traditional techniques that shaped the organization of rural and urban societies for centuries. Khettaras, séguias, and early dams constitute an exceptional hydraulic heritage, both technical and social, whose operating principles continue to inform contemporary debates on sustainability and equity. These systems, which are the legacy of both ancient knowledge and medieval reinventions, demonstrate an ingenuity adapted to the ecological constraints of the arid and semi-arid Maghreb, but also a sophisticated social organization that placed water at the heart of local governance structures.

Khettaras - known, as foggaras in Algeria or qanats in Iran - are underground channels designed to capture groundwater and convey it by gravity to cultivated land. Their widespread use in Morocco, particularly in the Tafilalet, Drâa and Haouz regions, dates back to the Middle Ages, probably under the influence of trans-Saharan trade and the Muslim dynasties that encouraged their construction (El Faïz, 2005). The system was based on a combination of engineering and solidarity: the several-kilometer-long tunnels required a large workforce, mobilized through collective obligations. Maintenance was carried out by the village community, which came together in the *jma'a* and determined water rights in proportion to each family's contribution to the construction. As Haddache (2022) points out, khettaras were not only hydraulic infrastructures, but also social institutions, as they embodied a mode of collective governance based on equity, social memory, and intergenerational sustainability. Recent research confirms their historical relevance and current fragility in the face of motorized pumping and declining water tables (Beraaouz, 2022).

Seguias, or *saguias*, are another emblematic technique. These are open irrigation channels that divert water from wadis, springs, or khettaras to irrigate agricultural land. The distinctive feature of these systems is their close link to customary law (*orf*), which organized water sharing according to turns called *nouba*. Each family or group of users had a time slot proportional to their rights, measured by rudimentary but accurate devices such as clepsydras or shadows. Ftaïta (2011) showed that this organization reflected genuine local legal knowledge, where the regulation of sharing prevented conflicts and ensured the continuity of agricultural exploitation. These systems also had a remarkable ecological dimension: by promoting infiltration, the seguias recharged the water table and contributed to the water regulation of oasis ecosystems.

Alongside these horizontal systems, Moroccan societies also built traditional dams to store floodwater and regulate seasonal flows. These structures, often modest but effective, were built with local materials (stone, clay, wood) and were used to divert water to the seguias. Several medieval chronicles mention the existence of these dams in the Souss and Haouz regions, areas of intensive agriculture (El Bouzidi, 1999). Their role went beyond irrigation alone: they protected villages from flooding, ensured a more stable distribution of water, and embodied a collective project mobilizing local labor. These dams were both technical structures and political symbols, as their construction and maintenance required the authority of local elites or the central government.

The effectiveness of these traditional systems should not obscure their limitations. The khettaras, for example, required constant maintenance and became problematic in the event of rural exodus or prolonged drought. Their flow could decrease drastically with the lowering of the water table, a phenomenon that is exacerbated today by the widespread use of motorized pumping. Seguias, meanwhile, were vulnerable to evaporation and losses and remained susceptible to climatic hazards. Nevertheless, their resilience over several centuries testifies to a remarkable capacity for adaptation, based on flexible social rules and complementary techniques.

One of the major contributions of these systems is the way they shaped social institutions and local culture. Water sharing in oases was not just a technical issue: it structured hierarchies, alliances, and sometimes conflicts. As Zoubir (2019) has shown, water management was central to social and political organization, determining land rights, consolidating the authority of local leaders, and organizing community life around a water calendar. Water thus became a common good, but also an instrument of power, reflecting tensions between customary equality and socio-economic inequalities.

Even today, despite the expansion of large dams and motorized drilling, khettaras and séguias continue to operate in several regions, often thanks to heritage preservation programs. Their value goes beyond the strictly agricultural: they embody a model of community governance that several researchers, such as Boelens (2014), consider a source of inspiration for modern commons policies. In a context of increasing water stress—Morocco had 2,500 m^3 of water per capita in the 1960s, compared to less than 600 m^3 today (Plan Bleu, 2020)—the reevaluation of these traditional systems is a relevant avenue for reconciling ecological sustainability and social equity.

In short, khettaras, séguias, and ancestral dams are much more than vestiges of a bygone era. They bear witness to a unique combination of technical knowledge, social organization, and customary regulation that has enabled Moroccan societies to adapt to arid environments. Studying them highlights the strength of their heritage, but also the need to rethink their place in contemporary water policies at a time when scarcity and unequal access threaten social cohesion and food security.

Colonial Centralization and Major Water Projects

The establishment of the French Protectorate in Morocco in 1912 marked a turning point in the history of water resource management in the country. This change was not limited to a modification of infrastructure, but led to a profound reconfiguration of the relationship between society, territory, and water. Where traditional systems were based on community-based, endogenous, and customary forms of management, the colonial regime introduced a technocratic logic of centralization focused on intensive agricultural development, territorial control, and export. This process, far from neutral, reshaped social hierarchies, altered ecological balances, and created new vulnerabilities.

A State-Led Approach to Water Management

From the early decades of the Protectorate, colonial engineers and administrators adopted a centralized approach inspired by French land-use planning doctrines. The objective was twofold: to rationalize the exploitation of natural resources in order to promote export-oriented capitalist agriculture, and to strengthen political control over the territory by securing access to water. This state-run water model—close to what James Scott (1998) terms 'high modernism'—relied on dams, canalized wadis, irrigated perimeters, and a new legal-cadastral framework for water.

According to Ariane Salem (2017), "Morocco is one of the African countries where colonial water policy has been most developed," with projects aimed at developing more than 500,000 hectares of agricultural land since the 1930s. These developments are concentrated in the plains of Gharb, Tadla, and Haouz, considered to be the areas with the highest agricultural potential. Large dams, such as the Bin el Ouidane dam (begun under the Protectorate but completed in 1953), illustrate this desire to create a sustainable infrastructure for water storage and distribution.

Marginalization of Traditional Systems

This centralized strategy did not replace existing systems: it marginalized them. As Mohamed Houbaida showed in his work on the Moulouya Valley, the khettara-, seguias, and small community dams were gradually abandoned in favor of technical solutions considered more modern and efficient (Houbaida, 1999). The colonial state thus replaced rural communities in water management, depriving them of their knowledge and customary rights.

In many regions, this transition has led to conflicts over water use. Ancestral water rights, often unwritten, were invalidated or not recognized by the new legal system established by the colonial administration. Surface water is now subject to administrative authorizations issued by the water services of the Rural Engineering Department. This has led to the gradual exclusion of small farmers and the concentration of water resources in the hands of an agricultural elite, often linked to the administration or European capital.

The summary report "The Future of Water in Morocco" (CESE, 2014) recalls that this period saw the rise of a dual system: a modern, irrigated sector regulated by the state, and a traditional sector that was marginalized and unrecognized, reproducing social and spatial inequalities.

A Productivist and Export-Oriented Vision

Colonial water policy in Morocco was based on an explicitly productivist logic. Water was not seen as a resource to be managed equitably or sustainably, but as an economic lever serving the agricultural and commercial interests of the metropolis. Irrigated areas were therefore designed to produce cash crops—sugar cane, citrus fruits, sugar beet—for export, rather than to meet the food needs of local populations (Salem, 2017).

This approach had lasting consequences. On the one hand, it encouraged water-intensive agriculture dependent on heavy infrastructure. On the other hand, it widened the gap between "developed" and "neglected" areas, particularly the high valleys, pre-Saharan areas, and mountainous regions, which continued to operate according to traditional models without administrative support.

Towards Technocratic Governance

Colonial centralization also changed water governance. Far from community-based decision-making (*jmaa, nadhir,* etc.), management became the preserve of experts, engineers, and senior civil servants. While bringing new tools (mapping, hydrometry, civil engineering), this technocratization contributed to distancing users from the resource and establishing a top-down approach. Local knowledge has been discredited in favor of technical knowledge. This epistemological break is part of a "civilizing mission" ideology that views hydraulic traditions as archaic and inefficient.

Zoubir's analysis (2019) echoes this interpretation, emphasizing that the colonial system not only transformed infrastructure but also social relations to water by imposing a new hierarchy of knowledge and power.

Legacy and Discontinuities

The Protectorate's influence on water management in Morocco is marked by a dual dynamic: on the one hand, the introduction of modern techniques (dams, large irrigated areas, centralized codification), and on the other hand, the marginalization of traditional and community-based systems. This hydraulic modernization, based on a logic of yield and state contro, certainly made it possible to control floods and increase irrigated areas, but it also resulted in a loss of autonomy for rural populations, a worsening of social imbalances, and a break with local water regulations.

After independence, Morocco continued along this path with the Great Dams Plan of the 1960s, consolidating a state-run, technocratic management model. However, this model is now showing its limitations: resource scarcity, conflicts over use, excessive centralization, and low user participation. As highlighted by Zaki (2002) and the CESE (2014), these flaws call for structural reform in favor of more equitable, decentralized, and sustainable governance.

In short, the hydraulic heritage of contemporary Morocco remains marked by historical contradictions. Recognizing the value of local knowledge, integrating the principles of hydraulic justice, and reinventing governance have become imperative in the face of the challenges of the 21st century.

Post-Independence Policies and Modern Water Management

Morocco's independence in 1956 marked a decisive step in the evolution of national water policies. From the very beginning, the Moroccan state embarked on a proactive strategy of land use planning based on water, inheriting both a centralized management system established under the Protectorate and traditional water management expertise that had been largely marginalized. This political choice responded to multiple imperatives: food security, rural stabilization, agricultural modernization, and industrial development. Water thus became a fundamental lever of the postcolonial national project.

The Great Dams Plan: A Pillar of State Voluntarism

One of the most emblematic initiatives is undoubtedly the Great Dams Plan, launched in the 1960s under the impetus of King Hassan II. Inspired by technocratic models inherited from the colonial period, this plan aimed to provide the country with a vast network of dams capable of ensuring irrigation, drinking water supply, and hydroelectric power generation.

According to El Faïz (2015), the construction of large dams such as Al Massira, Idriss I, Hassan II, and Oued El Makhazine reflects a desire to control Morocco's unpredictable hydrological cycles and strengthen food sovereignty. In 2023, there were more than 140 large-capacity dams, representing a storage capacity of more than 17 billion m^3. These structures irrigate more than 1.5 million hectares, mainly in the plains of Gharb, Haouz, Souss, and Tadla.

However, this strategy also had its limitations. As Houbaida (1999) points out, it was built according to a top-down approach, focused on large, irrigated areas and often to the detriment of subsistence farming or peripheral regions. The CESE (2014) notes that these choices have reinforced territorial imbalances and accentuated social inequalities in access to water.

Modernization of Irrigation Techniques and Demand Management

Faced with the limitations of traditional gravity irrigation, which is often inefficient, Morocco adopted a policy of modernizing irrigation in the 2000s, notably through the National Irrigation Water Conservation Program (PNEEI). The PNEEI promotes localized irrigation, particularly drip irrigation, which can save up to 40% of the water used.

However, several studies (Molle & Sanchis-Ibor, 2019) show that this modernization benefits large farms more than small family farms, accentuating disparities in access to subsidies and technologies. This technological shift is part of a demand-side management approach, breaking with the purely infrastructure-based vision of previous decades.

Desalination and Resource Diversification

One of the striking innovations of contemporary water policy is the development of seawater desalination, particularly for coastal areas subject to acute water stress. Desalination plants in Agadir, Laâyoune, Safi, and soon Casablanca ensure a regular supply of drinking water while reducing pressure on overexploited aquifers.

According to the Ministry of Equipment and Water (2022), Morocco currently operates 17 desalination plants, is building four more, and plans to add nine additional plants to reach an estimated total capacity of 1.7 billion m^3/year by 2030.

Legislative Reforms and Integrated Governance

The modernization of water management has not been limited to technical aspects alone. It has also involved a profound institutional and legal reform, initiated with the enactment of Law 10-95 on water. This law enshrines the principle of integrated water resource management at the watershed level, entrusts planning to Water Basin Agencies (ABH), and introduces a logic of local participation and cost recovery.

Nevertheless, this governance remains largely centralized and technocratic, as noted by the Economic, Social and Environmental Council (CESE) in its 2014 report. The CESE calls for an overhaul of the institutional architecture of water management, recommending in particular the creation of a National Agency for Strategic Water Planning and better coordination between the ministries concerned, local authorities, and users. In the same vein, Houdret (2012) showed that irrigation policies in the Souss region mainly benefit agricultural elites at the expense of smallholders, highlighting the need for more inclusive governance.

The National Water Program: Towards More Inclusive Management?

Adopted in 2020, the National Program for Drinking Water Supply and Irrigation (PNAEPI) represents an attempt to reconcile the imperatives of sustainability, citizen participation, and water sovereignty. The program aims to strengthen trans transparency in water management, improve intersectoral coordination, and promote innovation, particularly through the use of digital technologies (smart sensors, remote sensing, AI) to optimize water networks.

The program also introduces a desire to rehabilitate certain local knowledge—khettaras, séguias, customary co-management—by linking it to modern policies. This convergence between hydraulic heritage and technological innovations is seen by several researchers (Boelens, 2014) as a lever for strengthening water justice and resilience to climate change.

Morocco's post-independence water policy reflects an ambition to achieve technical and territorial control over resources, based on proactive and centralized choices. While progress is undeniable in terms of infrastructure, water security, and irrigation, challenges related to sustainability, equity, and participation remain pressing. In the era of climate change, it is now imperative to transition to a more inclusive, adaptive, and environmentally sustainable form of governance that can reconcile the engineering genius of the past with the demands of the future.

CONTEMPORARY CHALLENGES AND FUTURE PROSPECTS FOR WATER MANAGEMENT

Water management is currently undergoing a profound transformation, driven by the climate emergency, growing demographic pressure and rising inequalities in access to this vital resource. Morocco, like many semi-arid countries, is facing a multifaceted water crisis, marked by resource scarcity, the obsolescence of certain systems, and the need to reinvent its water policies. This last section offers a critical analysis of the main current challenges and the responses being considered, combining technical innovations, institutional reforms, and sustainability imperatives.

The Impact of Climate Change and Demographic Pressure on Water Management

In Mediterranean regions, of which Morocco is a particularly vulnerable example, water management is currently facing the converging dynamics of climate change and demographic pressure. This dual constraint has a decisive influence on the availability, quality, and distribution of water resources, requiring a complete overhaul of governance models.

Climate change is radically altering hydrological patterns: according to IPCC projections (2023) for the Maghreb, there could be a 10 to 30% decrease in precipitation by 2050, accompanied by an increase in average temperatures of between 2 and 4°C, with more extreme weather events. At the same time, the World Bank (2023) warns that Morocco could reach absolute scarcity, i.e., less than 500 m^3 of renewable water per capita per year, before the end of the decade. Today, the available water resource per person is estimated at 565 m^3/year, compared to nearly 2,560 m^3 in the 1960s—and is expected to decrease to 480 m^3/year by 2030 according to the WRI (2023).

This scarcity is confirmed by the World Bank Country Profile 2021, which indicates a decline of nearly 60% in renewable resources per person since 1960, due to increased demand, intensive irrigation, and urbanization. These trends make

Morocco one of the countries most exposed to global water stress, ranked 27th according to the WRI (2023).

At the same time, the 2024 census indicates a population of over 37 million, more than 60% of whom live in urban areas, compared to only 12 million in 1960. This growth, combined with uncontrolled urbanization, is leading to a drastic reduction in groundwater recharge and an increase in losses through evaporation and runoff.

The cumulative effect of these structural changes is destabilizing watersheds—such as those of the Oum Er-Rbia, Tensift, and Souss-Massa—now classified as "closed" basins, where all usable water is exploited or even exceeded.

This crisis is not limited to a technical situation: it is deeply social and environmental. Rural communities, particularly small farmers, are bearing the brunt of the depletion of water sources. Traditional systems—khettaras, séguias—based on collective management are increasingly being abandoned in favor of individual solutions that are often more costly and less sustainable. These crises are not only the result of climate imbalances but also of limitations in governance and multi-level coordination (IRES, 2022; World Bank, 2023).

Technological Innovations and Alternative Solutions

In a global context of increasing water stress, technological innovations appear to be essential levers for revisiting water management models in Morocco. For several years, the kingdom has been exploring disruptive solutions—ranging from advanced engineering to agroecology—to build a water system that is efficient, sustainable, and resilient.

In the context of Morocco's structural water stress, technological innovation is a strategic lever for responding to climate and demographic constraints. Recent reforms are moving towards a combination of advanced engineering, energy transition, and integrated management in order to move beyond the extensive mobilization logic inherited from the large dam paradigm.

Seawater desalination, long marginalized, is now emerging as an essential solution for securing water supplies in heavily urbanized coastal areas. The Chtouka-Aït Baha plant, designed to reach a capacity of 275,000 m^3/day, illustrates this turning point. According to Kettani and Bandelier (2020), the integration of solar photovoltaic energy into this type of project reduces operating costs by around 15%, thereby strengthening the economic viability of desalination units. On a broader scale, Temsamani, Al Mers, and Filali Baba (2025) highlight that Morocco is among the pioneering countries in the MENA region in experimenting with renewable energy-powered desalination systems, which is a decisive step towards decarbonizing water policy.

These efforts are part of a broader regional trend. A study by Sayed (2023) published in *Environmental Science and Pollution Research* shows that MENA countries are exploring growing synergies between desalination and renewable energy, while seeking to limit negative externalities such as brine discharge and pressure on marine ecosystems. Al-Addous (2024) emphasizes the need to compare different solar desalination processes (photovoltaic and thermal) in order to select the configurations best suited to local contexts in terms of cost, energy efficiency, and environmental impact.

Beyond desalination, Morocco is also banking on the rise of digital and smart technologies to improve water management efficiency. Remote sensing, connected sensors, and artificial intelligence are helping to optimize resource distribution, detect losses in urban networks, and improve land-use planning. These innovations are in line with the recommendations of the Economic, Social, and Environmental Council (2014), which already emphasized the need to reduce water losses in the supply system, which can reach up to 40% in some cities.

Finally, agricultural irrigation, the main source of water consumption in Morocco, is undergoing rapid modernization. While localized irrigation (drip irrigation) is now widely subsidized, recent research emphasizes the need to combine this transformation with regulation of usage. Without this, efficiency gains at the plot level risk being offset by an overall increase in demand (Molle & Sanchis-Ibor, 2019). Pilot programs integrating moisture sensors, precision irrigation, and crop diversification illustrate an attempt to link agricultural productivity and water efficiency.

These developments show that technological innovations cannot be considered as isolated solutions, but as components of a systemic strategy. Their success depends as much on their technical relevance as on their integration into inclusive governance mechanisms capable of limiting inequalities of access and strengthening water justice.

Governance and Sustainable Water Management Policies

Water governance is a key issue for ensuring sustainable and equitable management of water resources, particularly in countries facing structural water stress such as Morocco. In this context, institutional reform, adaptation of the legal framework, capacity building for local actors, and citizen participation appear to be essential levers for responding to the challenges of water scarcity, worsening territorial inequalities, and the growing complexity of water uses.

In Morocco, the water management model has long been based on a centralized and technocratic approach, inherited from the post-independence period and consolidated by the large dam policy. This approach has led to significant progress in water infrastructure, with a drinking water coverage rate of over 97% in rural areas and irrigation covering approximately two million hectares. However, this model, based primarily on supply, is now showing its limitations in the face of increasingly constrained resources and growing demand. Recent studies (Ben-Daoud et al., 2021;

Silva-Novoa Sánchez, 2022) confirm that centralization and top-down approaches reduce policy effectiveness by neglecting local dynamics.

This finding is part of a broader trend observed in many Mediterranean countries, where irrigation policies have historically been based on centralized planning, often disconnected from socio-environmental dynamics. Molle and Sanchis-Ibor (2019) emphasize that these developmentalist approaches have resulted in systems that are poorly adapted and difficult to coordinate. The Moroccan case illustrates these limitations and confirms the need for a shift towards more polycentric, integrated, and contextualized governance.

Several critical issues are highlighted. First, the multiplicity of institutional actors—ministries, basin agencies, local authorities, planning offices—creates fragmentation of responsibilities and undermines the effectiveness of public policies. The lack of intersectoral coordination hinders the implementation of coherent strategies and prevents the optimization of resources. Studies on the Drâa Valley (Silva-Novoa Sánchez et al., 2025) show that this fragmentation generates conflicts of use and weakens participatory mechanisms.

On the legal front, the enactment of Law 36-15 on water in 2016 marks an important step towards integrated water resource management (IWRM). This law introduces principles of resource protection, citizen participation, polluter pays, and decentralization. It also provides for planning mechanisms at the river basin level. However, Ben-Daoud et al. (2021) reveal that its implementation remains limited: participatory mechanisms such as basin councils suffer from a lack of representativeness and effectiveness.

The involvement of local authorities is another blind spot in current governance. Despite their key role in managing local drinking water, sanitation, and irrigation services, municipalities often lack the technical and financial resources to develop coherent action plans. Laamari, Laamrani, and Khoali (2022) emphasize the need to strengthen local capacities and create consultation platforms involving institutions, research, and users.

Citizen participation, although enshrined in Law 36-15, remains marginal. Water users—farmers, industrialists, associations—are rarely involved in decision-making processes, despite their central role in management. However, several local experiences show that user involvement improves the efficiency of resource allocation and promotes collective ownership of issues. Silva-Novoa Sánchez (2022) demonstrates that community initiatives in the Drâa offer hybrid models of governance that could be replicated elsewhere.

Finally, sustainable water governance requires better coordination with global issues such as the Sustainable Development Goals (SDGs), climate change, and agroecological transition. El Haddadi (2025) emphasizes that Morocco can only meet its water challenges by integrating water governance into a comprehensive socioeconomic development strategy.

In short, Morocco has a relatively comprehensive legal and institutional framework. However, the current challenges—resource scarcity, unequal access, weak coordination, and lack of information—require an ambitious and inclusive overhaul. This overhaul requires shared governance that is adapted to local realities, open to innovation, and attentive to social and environmental balances.

Future Prospects: Towards More Sustainable and Resilient Management

Faced with the limitations of the post-colonial water management paradigm—based on centralization, massive resource mobilization, and heavy infrastructure—contemporary thinking on water management in Morocco is increasingly turning to more sober, equitable, and resilient models. This transition calls for the simultaneous integration of environmental sustainability, social justice, and institutional adaptation (Silva-Novoa Sánchez et al., 2025).

First, a strategic shift towards demand management is needed. Kaiss et al. (2025) emphasize that approaches based exclusively on supply—building more dams and intensively mobilizing aquifers—no longer respond to the realities of climate change and resource scarcity. The future lies in policies promoting water efficiency, progressive water pricing, and integrated basin-wide planning to reduce pressures and optimize allocations.

Secondly, participatory governance is an essential lever. El Mansoum et al. (2025) show, in a study on participatory groundwater management, that involving farmers in collective decision-making improves institutional resilience and promotes local ownership of public policies. Strengthening user associations, developing "basin contracts" and supporting community co-management mechanisms therefore appear to be priority areas.

Thirdly, technological resilience must be accompanied by a rigorous assessment of sustainability. Saghiry (2025), through his analysis of the water transfer project between the Beni Mansour and Mohammed V dams, proposes a systemic assessment framework that takes into account hydrological, ecological, and socioeconomic impacts. This type of integrated approach makes it possible to limit the negative externalities of megaprojects and ensure their viability in a context of increased climate variability.

Finally, international support and policy coherence strengthen Morocco's capacity to secure its water future. The World Bank (2023) recommends coordinating wastewater reuse, groundwater protection, the modernization of basin agencies, and the digitization of hydrological information systems. These measures could enable the reuse of more than 50 million m^3 of treated water per year and diversify supply sources in a resilient manner.

In short, the sustainability of water management in Morocco will depend less on the construction of new infrastructure than on the ability to rethink uses, strengthen participatory governance, and combine technological innovations with local knowledge. The future of water must be conceived as a social and territorial pact, integrating equity, resilience, and shared responsibility.

Figure 1. Conceptual diagram of strategic priorities for sustainable and resilient water management in Morocco

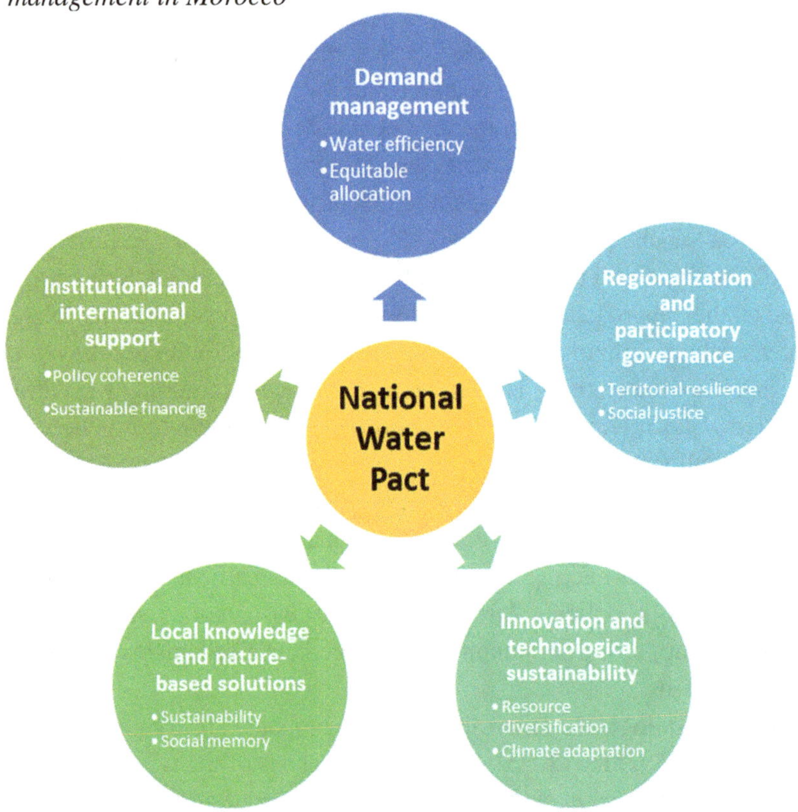

This diagram illustrates the main levers identified to strengthen national water governance: demand management (progressive pricing, precision irrigation), regionalization and local participation (basin contracts, user associations), technological innovation (desalination, wastewater reuse, digitization), the promotion of endogenous knowledge (khettaras, séguias, nature-based solutions), and institutional and international support (reactivation of the Higher Water Council, global partnerships).

All converge towards a "national water pact" that guarantees resilience, equity, and sustainability.

CONCLUSION

Towards Fair, Resilient and Pluralistic Hydro-Social Governance

What does water say about the civilizations that shape it? Much more than a natural resource, it is a mirror that reflects technical choices, power relations, collective imaginaries, and social tensions. By tracing the long history of water management—from the first hydraulic societies to the contemporary challenges of climate change—this chapter has explored the multiple dimensions of water, both material and symbolic, local and global, ancestral and eminently contemporary. The aim was to understand how societies have sought to capture, channel, share, or monopolize this elusive resource, and how these practices have structured, sometimes permanently, their economic, political, and cultural dynamics.

From the first hydraulic civilizations to contemporary challenges, via medieval systems and the specific dynamics of the Moroccan case, the analysis highlighted how water is at once a vital resource, a vector of power, a cultural object, and a strategic issue. This historical journey, informed by multidisciplinary contributions, reveals a constant: water management reflects and shapes the social structures, governance systems, and power relations of each era.

In ancient societies, control over water was never reduced to a purely functional dimension. On the contrary, it was part of a logic of territorial domination, political organization, and symbolic structuring. The large irrigation systems, canals, and dams of Mesopotamia, Pharaonic Egypt, imperial China, and the pre-Columbian Andes were as much technical tools as they were instruments of power. Water was seen as a sacred resource, entrusted to the authority of a priest-king or emperor who guaranteed the cosmic order. In this sense, it became the foundation of political legitimacy, the lever of administrative centralization, and the support of a collective vision of prosperity.

This approach, long associated with Karl Wittfogel's theory of "hydraulic despotism," has, however, been largely qualified by contemporary research. Numerous studies have highlighted the diversity of management regimes, the participation of local communities, and the complexity of institutional arrangements within ancient societies. Far from a monolithic and authoritarian vision, the history of water in antiquity reveals a mosaic of practices, norms, and knowledge rooted in specific ecological and cultural contexts.

The medieval period, meanwhile, was a turning point in this history. It saw the emergence of more diffuse and sometimes more egalitarian forms of water management. In the Islamic world in particular, systems such as khettaras, seguias, hydraulic habous, and customary practices testify to a deep connection between religion, technology, and community organization. Water was considered a common good, subject to rules of sharing, mediation procedures, and an ethic of fairness. It also played a role in the moral economy of villages, where its distribution reflected principles of balance between families, seasons, crops, and acquired rights.

In Europe, the rise of water mills, the management of rivers for agricultural and artisanal needs, and the first municipal regulations also reflect a reconfiguration of the relationship with water, under the influence of urban growth, agricultural changes, and transformations in seigneurial authority. These dynamics should not be interpreted solely as transitions to modernity, but as expressions of technical and social intelligence adapted to the diversity of local environments and issues.

The case of Morocco, viewed from this long-term perspective, provides particularly fertile ground for observing the tensions between tradition and modernity, between centralization and vernacular practices. The khettaras of southern Morocco, the seguias systems in the Atlas valleys, the collective basins of the Souss and the customary conventions for managing springs and watercourses embody a unique heritage. These systems are based on a detailed knowledge of ecosystems, forms of concerted governance and deep roots in the local agrarian culture.

However, the arrival of the protectorate, followed by the water policies implemented by the post-independence state, profoundly disrupted this balance. By introducing a technocratic, centralized, and planned approach—embodied in particular by the "dam policy"—the public authorities imposed a modernist vision of water management based on large-scale hydraulics, intensive irrigation, and territorial control. While this approach has led to significant gains in terms of storage, resource security and infrastructure, it has also had adverse effects: exclusion of local communities, disruption of hydrological balances, concentration of investment in certain regions to the detriment of others, and marginalization of traditional knowledge.

Today, in a global context marked by climate change, resource scarcity, population growth and exacerbated inequalities, the issue of water is taking on a critical dimension. Morocco, like many countries in the global South, faces a range of vulnerabilities: overexploited aquifers, recurrent droughts, conflicts over land use between agriculture, industry and housing, and fragmented governance. Recent public policies, such as the National Drinking Water Supply and Irrigation Program (PNA-EPI 2020–2027), reflect a desire for reform, diversification of sources (particularly through desalination and wastewater reuse), and adaptation to the crisis. However, their implementation remains dependent on a more profound paradigm shift.

The work of Arianne Salem and other researchers highlights the need to break with a purely utilitarian and technocratic view of water. They invite us to think of water as a hydrosocial object, i.e., as a resource that is at once material and political, technical and symbolic, contested and shared. With this in mind, water governance must open up to a plurality of actors, recognize differentiated rights, value indigenous knowledge, and embed policies in the territories where people live. This implies strengthening environmental democracy, establishing mechanisms for transparency and accountability, and promoting a culture of cooperation between users, institutions, and experts.

It is also becoming urgent to reconsider the place of women in water systems. Although they are often responsible for domestic water supply, daily management, and the transmission of water-related knowledge, they remain largely absent from decision-making bodies. Integrating a gender perspective into water management is not only a matter of social justice, it is a prerequisite for sustainability, efficiency, and resilience. Similarly, young people, rural populations, and marginalized groups must be recognized as sources of solutions and not simply as passive beneficiaries.

The history of water management, as traced in this chapter, teaches us that water is a revealing indicator of societal choices. It forces us to question our development models, our relationship with nature, and our forms of power. It confronts us with ethical and political dilemmas: how can we reconcile equity and efficiency? How can we arbitrate between present needs and the rights of future generations? How can we combine water sovereignty and cross-border cooperation?

Faced with these challenges, we need to adopt a reflective and responsible stance. The water transition cannot be imposed from above or reduced to technological engineering. It requires a new collective imagination based on sobriety, solidarity, and ecological humbl . It calls for a reinvention of the links between humans and their environments, between institutions and citizens, between memory and innovation.

Ultimately, thinking about water means thinking about civilization. And it is undoubtedly in the way we choose to manage this resource collectively that one of the great ethical, social, and ecological battles of the 21st century will be fought. Water represents a double challenge: that of material survival and that of meaning. Between the drop that irrigates, the drop that heals, the drop that is missing and the drop that breaks the camel's back, it is up to us to chart a fairer, more sustainable and more humane path.

REFERENCES

Al-Addous, M. (2024). Innovations in solar-powered desalination in the MENA region. *Water (Basel)*, *16*(13), 1877. DOI: 10.3390/w16131877

. Bakker, K. (2014). The business of water: Market environmentalism in the water sector. Annual Review of Environment and Resources, 39, 1.1–1.26. https://doi.org/ DOI: 10.1146/annurev-environ-012220-122552

. Barraqué, B. (2025). Wateringues yesterday and today, in France and the Netherlands. Flux, Advance publications (4), I16–XVI. https://doi.org/DOI: 10.3917/ flux1.pr1.0016

Ben-Daoud, M., El Mahrad, B., Elhassnaoui, I., Moumen, A., Sayad, A., & Elbouhadioui, M. & Eljaafari, S. (2021). Integrated Water Resources Management: An indicator framework for water management system assessment in the R'Dom Sub-basin, Morocco. *Environmental Challenges*, *3*, 100062. DOI: 10.1016/j. envc.2021.100062

Beraaouz, M. (2022). Khettaras in the Tafilalet oasis (Morocco). *Built Heritage*, *6*(24). Advance online publication. DOI: 10.1186/s43238-022-00052-2

Berry, S. (1993). *No condition is permanent: The social dynamics of agrarian change in Sub-Saharan Africa*. University of Wisconsin Press.

Boelens, R. (2014). Cultural politics and the hydrosocial cycle: Water, power and identity in the Andean highlands. *Geoforum*, *57*, 234–247. DOI: 10.1016/j.geoforum.2013.02.008

Boelens, R., & Seemann, M. (2014). Forced engagements: Water security and local rights formalization in Yanque, Colca Valley, Peru. *Human Organization*, *73*(1), 1–12. DOI: 10.17730/humo.73.1.v544h37k38445065

Bonneuil, C., & Fressoz, J.-B. (2013). *The Anthropocene Event: The Earth, History, and Us. Le Seuil*. King's College London.

Brookes, J. (1987). *Gardens of paradise: The history and design of the great Islamic gardens*. New Amsterdam.

. Caponera, D. A., & Nanni, M. (2019). *Principles of Water Law and Administration: National and International* (3rd ed.). Routledge. https://doi.org/DOI: 10.1201/9780429465703

. Delli Priscolli, J., & Wolf, A. T. (2009). Managing and transforming water conflicts. Cambridge University Press. https://doi.org/DOI: 10.1017/CBO9780511551536

Duby, G. (2011). *Europe in the Middle Ages*. Flammarion.

. Economic, Social and Environmental Council (CESE). (2014). Governance through integrated water resource management in Morocco: A fundamental lever for sustainable development. Self-referral No. 15/2014.

El Bouzidi, A. (1999). Qa āyā tawzīʿ al-māʾ bi-wā at Draʿa min khilāl al-wathāʾiq al-ma alliyy. In Proceedings of the symposium on Water in the history of Morocco (Hassan II University, Ain Chock), No. 11. FLSH publication.

El Faïz, M. (2005). *The Masters of Water: A History of Arab Hydraulics*. Actes Sud.

El Haddadi, M. (2025). The nexus between economic growth and water stress in Morocco. *Sustainability*, *17*(15), 6990. DOI: 10.3390/su17156990

Elvin, M. (2004). *The retreat of the elephants: An environmental history of China*. Yale University Press.

Faruqui, N. I., Biswas, A. K., & Bino, M. J. (Eds.). (2001). Water Management in Islam. United Nations University Press & IDRC.

Ftaïta, T. (2011). Community water management. *Anuário Antropológico*, *36*(1), 195–212. DOI: 10.4000/aa.1161

Gimpel, J. (1977). *The Medieval Machine: The Industrial Revolution of the Middle Ages*. Penguin.

Gleick, P. H. (1993). Water and conflict: Fresh water resources and international security. *International Security*, *18*(1), 79–112. DOI: 10.2307/2539033

Goblot, H. (1979). *The qanats: A technique for acquiring water*. Mouton.

Guillerme, A. (1983). *Les temps de l'eau: La cité, l'eau et les techniques*. Champ Vallon.

Haddache, M. (2022). Hydraulic knowledge and socioeconomic changes in the Toudgha oasis (southeastern Morocco). *Asinag*, *7*, 111–122.

Harvey, L. P. (2005). *Muslims in Spain, 1500 to 1614*. University of Chicago Press.

Hill, D. R. (1994). *Islamic science and engineering*. Edinburgh University Press.

Hillel, D. (1994). *Rivers of Eden: The struggle for water and the quest for peace in the Middle East*. Oxford University Press.

Houdret, A. (2012). The water connection: Irrigation, water grabbing and politics in southern Morocco. *Water Alternatives*, *5*(2), 284–303. https://www.water-alternatives.org/index.php/alldoc/articles/vol5/v5issue2/170-a5-2-6

Kaiss, R., Benjouid, Z., Faiz, M., Ech Chahed, H., Rakhimi, A., Zarouali, S. S., Maimoun, A., Hmid, A., & Cherkaoui, M. (2025). Water Stress and Regional Governance in Morocco. *Research on World Agricultural Economy*, *6*(3), 957–972. DOI: 10.36956/rwae.v6i3.2173

Kenoyer, J. M. (1998). *Ancient cities of the Indus Valley Civilization*. Oxford University Press.

Kettani, M., & Bandelier, P. (2020). Techno-economic assessment of solar energy coupling with large-scale desalination plant: The case of Morocco. *Desalination*, *501*, 114627. DOI: 10.1016/j.desal.2020.114627 PMID: 32863401

Laamari, A., Laamrani, H., & Khoali, S. (2022). Smart water governance in Moroccan agriculture. *Open Journal of Social Sciences*, *10*, 11–27. DOI: 10.4236/jss.2022.109002

Lansing, J. S. (1991). *Priests and programmers: Technologies of power in the engineered landscape of Bali*. Princeton University Press.

Leveau, P. (2008). Integrated water management in environmental history: Traditional knowledge and modern practices. *Nature Sciences Sociétés*, *16*(2), 173–175. DOI: 10.1051/nss:2008045

Linton, J. (2010). *What is water? The history of a modern abstraction*. UBC Press.

Loftus, A. (2018). Political ecology II: Whither the state? *Progress in Human Geography*, *44*(1), 139–149. DOI: 10.1177/0309132518803421

Madella, M., & Fuller, D. Q. (2006). Paleoecology and the Harappan civilisation of South Asia: A reconsideration. *Quaternary Science Reviews*, *25*(11–12), 1283–1301.

Mansoum, R., & Chfadi, T. (2025). Innovation without participation? Moroccan farmers' perspectives on Haouz Mejjat aquifer contract. *Groundwater for Sustainable Development*. Advance online publication. DOI: 10.1016/j.gsd.2025.101477

Mays, L. W. (Ed.). (2010). *Ancient water technologies*. Springer., DOI: 10.1007/978-90-481-8632-7

Mehta, L., & Veldwisch, G. J., & Franco, J. (2012). Introduction to the Special Issue: Water grabbing? Focus on the (re)appropriation of finite water resources. *Water Alternatives*, *5*(2), 193–207.

Molle, F., & Mollinga, P. (2003). Water poverty indicators: Conceptual problems and policy issues. *Water Policy*, *5*(5–6), 529–544.

Molle, F., & Sanchis-Ibor, C. (2019). Irrigation policies in the Mediterranean: trends and challenges. In Molle, F., Sanchis-Ibor, C., & Avellà-Reus, L. (Eds.), *Irrigation in the Mediterranean. Global challenges for water policy* (Vol. 22). Springer., DOI: 10.1007/978-3-030-03698-0_10

Nasr, S. H. (2006). *Science and civilization in Islam*. Harvard University Press.

Pietz, D. A. (2015). *The Yellow River: The problem of water in modern China*. Harvard University Press.

Royal Institute for Strategic Studies (IRES). (2022). What future for water in Morocco? Summary report. Rabat: IRES. https://www.ires.ma/sites/default/files/docs _publications/Avenir_de_leau_au_Maroc-Rapport_de_synthese.pdf

. Saghiry, S. (2025). An integrated conceptual framework for interconnected dam-reservoir systems (IDRS). Journal of Sustainability Research. HAPres.

Saghiry, S., Loudyi, D., & Laassilia, O. (2025). An Integrated Conceptual Framework for Interconnected Dam Reservoirs Systems. *Journal of Sustainability Research, 7*. Advance online publication. DOI: 10.20900/jsr20250017

Salem, A. (2017). Sharing water in Morocco: management of a common resource or public property? Communication, CAGE Summer School, University of Warwick. https://warwick.ac.uk/fac/soc/economics/research/centres/cage/events/11-07-17 -cage_summer_school_2017/details/arianne_salem_article.pdf

Shah, T., & Ballabh, V. (1997). Water markets in North Bihar: Six village studies. *Economic and Political Weekly, 32*(52), A183–A190.

Silva-Novoa Sánchez, L. M. (2022). Governance and sustainability challenges in the water policy of Morocco (1995–2020). *Water (Basel), 14*(18), 2932. DOI: 10.3390/w14182932

Silva-Novoa Sánchez, L. M., Mahjoubi, I., Frör, O., & Bossenbroek, L., & Schilling, J. (2025). What groundwater governance challenges and opportunities arise in arid regions? *Sustainable Water Resources Management, 11*(42). Advance online publication. DOI: 10.1007/s40899-025-01210-x

Squatriti, P. (2002). *Water and society in early medieval Italy, AD 400–1000*. Cambridge University Press.

Strang, V. (2004). *The Meaning of Water*. Routledge., DOI: 10.4324/9781003087090

Swyngedouw, E. (2009). The political economy and political ecology of the hydro-social cycle. *Journal of Contemporary Water Research & Education, 142*(1), 56–60. DOI: 10.1111/j.1936-704X.2009.00054.x

Tvedt, T. (2016). *The River Nile in the age of the British: Political ecology and the quest for economic power.* I.B. Tauris.

. Tvedt, T., & Oestigaard, T. (2016). A history of water: Series III, Volume 3: Water and food: From hunter-gatherers to global production in Africa. I.B. Tauris.

Wittfogel, K. A. (1957). *Oriental despotism: A comparative study of total power.* Yale University Press., DOI: 10.30541/v56i4pp.393-395

World Bank. (2023). Morocco – Country Climate and Development Report. Washington, DC. https://documents.worldbank.org/curated/en/099052223171017467

World Bank. (2023). New World Bank Program in Morocco Supports Efforts to Boost Water Security and Resilience for All. Washington, DC. https://www.worldbank.org/en/news/press-release/2023/07/24/new-world-bank-program-in-morocco-supports-efforts-to-boost-water-security-and-resilience-for-all

World Resources Institute (WRI). (2023, August 16). 25 Countries Face Extremely High Water Stress. https://www.wri.org/insights/highest-water-stressed-countries

Zeitoun, M., & Warner, J. (2006). Hydro-hegemony: A framework for analysis of trans-boundary water conflicts. *Water Policy*, *8*(5), 435–460. DOI: 10.2166/wp.2006.054

. Zeitoun, M., Lankford, B., Bakker, K., & Conway, D. (2013). Introduction: A battle of ideas for water security. In B. Lankford, K. Bakker, M. Zeitoun, & D. Conway (Eds.), Water security: principles, perspectives, practices (pp. 3–10).

Zoubir, L. (2019). Al-Māʾ Dākhil al- u ūn al-Makhzanīya bi-l-Maghrib al-Wasī . *Hesperis Tamuda*, *54*(2), 101–128. https://revues.imist.ma/index.php/H2/article/view/17640

Chapter 2
Global Water Resource Challenges

Samanta Alam
Chicago State University, USA

Likhon Chandra Roy
https://orcid.org/0009-0003-6943-6777
Rajshahi University of Engineering and Technology, Bangladesh

Tekleab Gala
Chicago State University, USA

Dhananjay Ray
Centre for the Development of Human Initiatives, India

ABSTRACT

Water is a fundamental resource for human survival, health, and development, yet it faces mounting pressures from rapid population growth, urbanization, industrialization, and climate change. Despite its abundance, covering 71% of the Earth's surface, freshwater constitutes less than 1% of the world's accessible water, and its availability is increasingly compromised by pollution, overextraction, and uneven distribution. This chapter explores the multifaceted nature of global water challenges, focusing on the interrelated issues of scarcity, contamination, climate variability, and governance. It highlights the impacts of salinity, groundwater depletion, waterborne diseases, and inequitable access. Case studies illustrate the socio-environmental consequences of poor water management and a lack of transboundary cooperation. The chapter further examines technological and policy solutions. By integrating scientific evidence, global frameworks, and local practices, the chapter advocates for inclusive, interdisciplinary, and adaptive approaches to water governance.

DOI: 10.4018/979-8-3373-2700-6.ch002

INTRODUCTION

Water is one of the most essential natural resources for sustaining life, ecosystems, and human civilization. Of all the water on Earth, about 97.5% is saline and located in oceans, while less than 1% is accessible for direct human use (UNEP, 2016). With population growth, industrial expansion, and rapid urbanization, the demand for freshwater is increasing dramatically. By 2050, global water demand is expected to rise by 20–30% (WWAP, 2019). This increase is exacerbated by pollution, climate change, and inadequate governance, making water scarcity a growing concern.

Water management has shifted from supply-side solutions (dams, wells, diversions) to IWRM, balancing economic, social, and environmental objectives (GWP, 2010). Access to safe water and sanitation is also a critical human rights issue. According to the WHO/UNICEF Joint Monitoring Program (2023), over 2 billion people lack safely managed drinking water services, and nearly 3.6 billion lack safely managed sanitation. These disparities are especially pronounced in Sub-Saharan Africa, parts of South Asia, and rural and conflict-affected areas. Women and girls often bear the burden of water collection, limiting education and economic opportunities. Regional disparities in water availability further complicate the global picture. While some countries, such as Canada, enjoy abundant renewable water resources, others, including Jordan and many African nations, face chronic water scarcity. This uneven distribution creates geopolitical tensions, impacts food and energy security, and exacerbates poverty and inequality.

Water is also central to the achievement of multiple Sustainable Development Goals (SDGs), particularly SDG 6, which aims to "ensure availability and sustainable management of water and sanitation for all." Yet, progress toward this goal remains off track in many regions. Without accelerated efforts and increased investment, billions of people will continue to face water-related hardships in the decades to come. Climate change is introducing a new dimension to water insecurity by altering precipitation patterns, intensifying extreme weather events, and affecting water quality. Droughts, floods, and glacier melt are already impacting water availability in both developed and developing countries.

Moreover, water issues are inherently interdisciplinary, intersecting with public health, agriculture, energy production, urban planning, and biodiversity conservation. For instance, contaminated water sources contribute to the spread of waterborne diseases, while unsustainable irrigation practices can degrade soil and reduce agricultural productivity. Urban water management must address stormwater runoff, pollution, and infrastructure aging to safeguard both human health and environmental quality.

This chapter explores the global water resource challenge in depth, examining the drivers, consequences, and regional disparities of water scarcity. It also highlights the interconnections between water, climate, health, and socio-economic development, and proposes pathways for sustainable, equitable, and resilient water management.

Figure 1. Global distribution of water resources

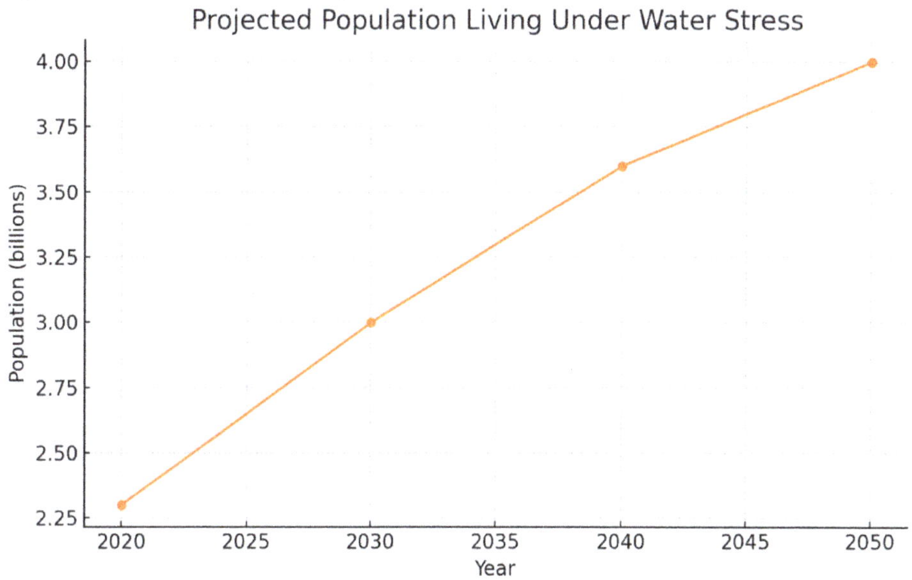

(Shiklomanov, 1993; UN Water, 2021)

Figure 2. Projected global water demand by sector

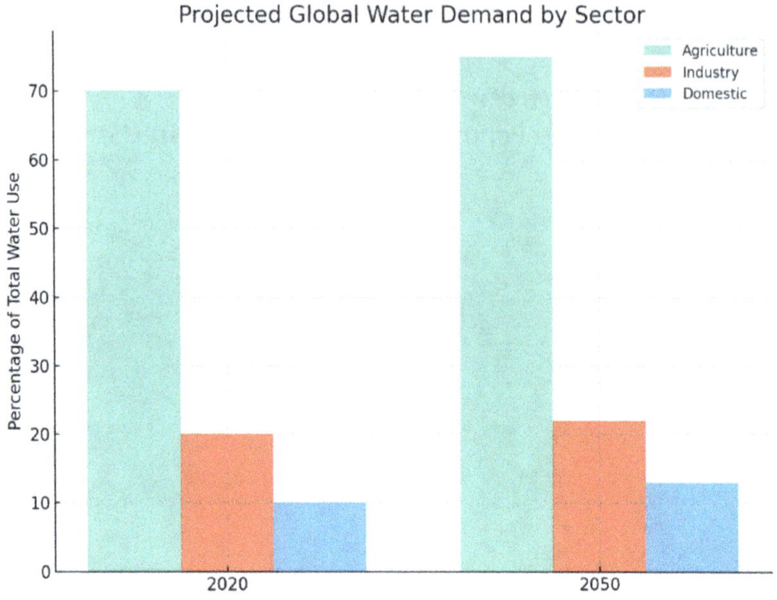

(WWAP, 2019; FAO, 2017)

Figure 3. Projected population living under water stress

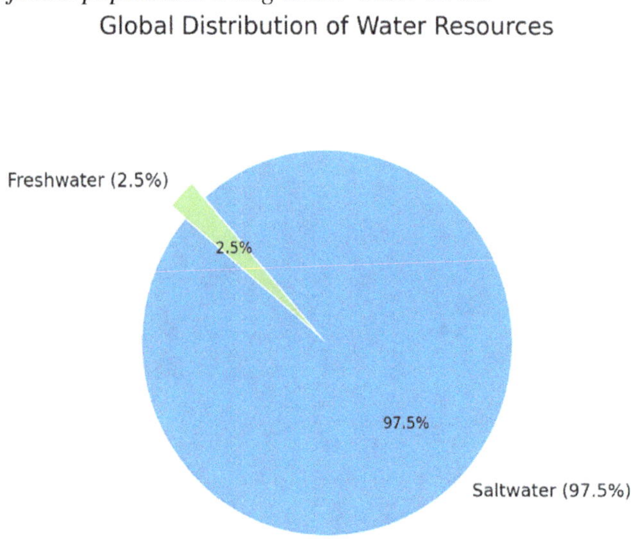

(UN DESA, 2018; WWAP, 2019; OECD, 2012)

Contribution of this Chapter

This chapter moves beyond descriptive assessments of scarcity to highlight the social, institutional, and governance dimensions of global water challenges. While much of the prior literature has focused primarily on hydrological stress, population growth, or the impacts of climate change, this chapter integrates underexplored themes, such as institutional corruption, governance inefficiencies, and Indigenous water rights, into the global discourse.

By drawing on diverse case studies from different regions—including Africa, Asia, North America, and Oceania—the analysis emphasizes how weak institutions, inequitable water allocations, and contested rights exacerbate physical scarcity and undermine resilience. For instance, the Mseilha dam in Lebanon, plagued by mismanagement, cost overruns, and environmental risks, illustrates how corruption and poor planning undermine water resilience. Environmental activists argue that the location was unsuitable due to the permeable nature of the Lebanese karstic mountains and seismic risks posed by the Batroun fault. The failed dam project and others like it reflect ongoing issues with Lebanon's water management policies, which are criticized for mismanagement and political corruption. Despite halting some projects, the Lebanese government continues its ambitions for dam construction, aiming to implement a new water management strategy by 2035.

In addition, this chapter contributes by offering a comparative evaluation of governance models—contrasting top-down, community-led, technology-based, and nature-based approaches—to assess their relative effectiveness in ensuring water security. The integration of recent scholarship (2022–2025) and global policy documents ensures that the discussion reflects cutting-edge debates, such as the role of AI-driven monitoring, climate adaptation, and transboundary conflict resolution.

Taken together, these contributions provide new insights for scholars, practitioners, and policymakers seeking innovative, inclusive, and resilient pathways for addressing global water challenges

LITERATURE REVIEW

Foundational assessments of global water availability (e.g., Shiklomanov, 1993) established volumes and spatial patterns of freshwater, while more recent work emphasizes the coupled dynamics of hydrology, governance, and socio-economic drivers (World Bank, 2023; IPCC, 2022).

Process-based and data-driven studies document significant shifts in precipitation regimes, evapotranspiration, cryosphere loss, and groundwater storage under climate change (IPCC, 2022). Modeling shows climate-driven TWS reductions could more than double the people exposed to extreme drought by the late century (Pokhrel et al., 2021). Machine-learning reconstructions and satellite-anchored datasets extend trend detection beyond the GRACE era, confirming persistent global TWS decline and increasing detectability of forced hydrological signals since 2016 (Yin et al., 2023; Huo et al., 2024; Huang et al., 2025).

A parallel literature demonstrates that weak accountability, opaque allocation, and procurement fraud can negate technical gains and exacerbate inequities (Transparency International, 2023; World Bank, 2023). Comparative work on community-based and polycentric arrangements finds improved maintenance, legitimacy, and equity, but notes scaling and coordination challenges in multi-jurisdictional basins (Boelens et al., 2022).

Applications include leak detection, anomaly flagging, groundwater monitoring, and early warning (Zhang et al., 2024; Singh et al., 2023). Nonetheless, adoption gaps persist where data governance, financing, and local capacity are limited (UNESCO, 2021/2022). Case studies from North America and Australia show how infrastructure siting, basin allocations, and licensing regimes often marginalize Indigenous water rights, despite UNDRIP and FPIC norms (Smith, 2023; Jackson & Head, 2022).

Evidence on Integrated Water Resources Management (IWRM) indicates uneven performance: strong enabling law and basin authorities correlate with better outcomes, while capacity constraints, policy incoherence, and political interference limit impacts elsewhere (World Bank, 2023; IPCC, 2022). Nature-based solutions (e.g., wetland restoration) deliver multi-benefit adaptation but can be slower to implement; technology-centric options (e.g., reuse, desalination, smart irrigation) scale quickly where institutions and finance are robust, yet raise equity and environmental externality concerns (WWF, 2023; Zhang et al., 2024).

Some gaps are: (1) Limited linkage of *quantified* hydrological signals (e.g., TWS anomalies) to *specific* institutional responses and outcomes; (2) few comparative evaluations of tech-based versus nature-based approaches across distinct political-economic contexts; (3) under-integration of Indigenous law and knowledge into basin-scale decision rules; (4) data transparency and anti-corruption safeguards are uneven, constraining evidence-based allocation.

Building on these strands, this chapter advances the field by connecting TWS trends with governance performance and policy design; top-down vs. community-led and tech- vs. nature-based solutions with cross-regional cases (2022–2025 evidence); explicit treatment of corruption risks and Indigenous/legal dimensions; actionable policy and managerial applications (pricing, participation, data governance) tied to SDG 6 and IWRM.

GLOBAL WATER RESOURCES: BACKGROUND AND OVERVIEW

Projections indicate that water stress will intensify in Africa and Asia by 2050 due to population growth, climate variability, and unsustainable withdrawal rates (UNESCO, 2024; World Bank, 2024).

Table 1. Key hydrological challenges and implications for water security

Hydrological challenge	Description (simplified)	Implications
Groundwater depletion	Over-pumping of aquifers beyond recharge rates	Declining water tables, reduced irrigation capacity
Eutrophication	Nutrient overload in lakes and rivers	Fish kills, reduced water quality, loss of biodiversity
Agricultural runoff	Pesticides and fertilizers entering water systems	Contamination of drinking water, ecosystem stress
Declining glacier melt	Reduced snowpack and glacier-fed flows	Seasonal water shortages in Asia, South America
Urban water pollution	Untreated wastewater discharge	Public health risks, high treatment costs

Source: Adapted from UNESCO (2024); IPCC (2022); World Bank (2024).

Water stress is not only a physical phenomenon but also a socio-political challenge. Disparities in access exacerbate inequalities and contribute to conflicts over water allocation, particularly in regions with rapid urbanization or transboundary river systems. Integrating physical assessments with governance perspectives allows for a holistic understanding of water security, situating global resources within frameworks of equity, resilience, and sustainability.

Figure 4. Total internal renewable freshwater resources by region

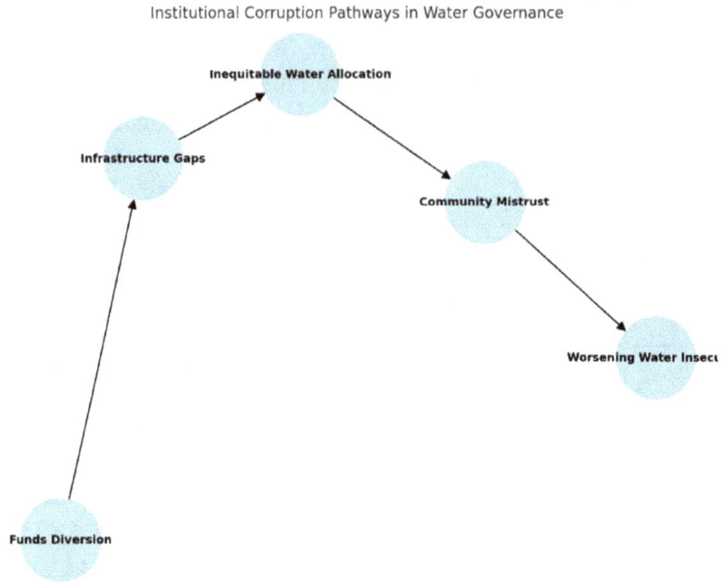

(World Bank, 2025)

Figure 5. Water stress index by continent in 2024 (baseline) and projected 2050 values

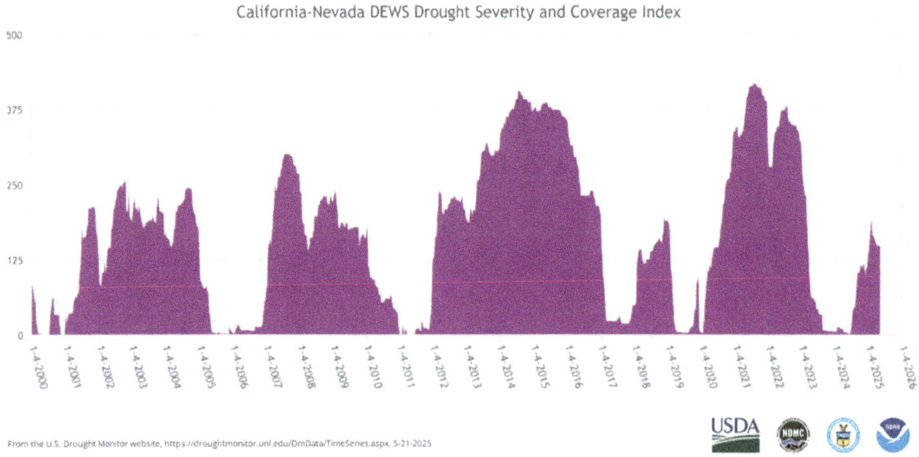

(UNESCO World Water Development Report, 2024; World Bank World Development Indicators, 2024)

This background provides a foundation for subsequent sections that examine institutional challenges, governance models, regional case studies, and future policy directions, emphasizing both technical and socio-institutional dimensions of water management.

MAJOR CHALLENGES

Population Growth & Urbanization

As the world population expands, freshwater requirements for agriculture, industry, and domestic use escalate. FAO (2018) projects that agricultural water demand alone could rise by 20–30% by 2050, intensifying stress on limited freshwater resources. In developing regions, especially sub-Saharan Africa, infrastructure constraints further hinder reliable water access (UN Water, 2021).

Currently, over 55% of the global population resides in urban areas—a proportion expected to reach 68% by 2050 (World Bank, 2020). Urban expansion often leads to water pollution, as wastewater management systems struggle to accommodate higher population densities and industrial activity (OECD, 2019). The urban heat island (UHI) effect further increases water demand by accelerating evaporation and reducing water retention in metropolitan areas (Gaffin et al., 2008).

Climate Change

Climate change disrupts the hydrological cycle, altering precipitation patterns, intensifying droughts, and increasing flood frequency and severity (IPCC, 2021). Tropical and high-latitude regions may experience increased rainfall, whereas mid-latitude and subtropical arid zones—such as the Mediterranean and southwestern United States—face declines (Trenberth, 2011; IPCC, 2021).

Glacier retreat compounds seasonal water challenges. Between 2000 and 2019, glaciers lost approximately 267 billion metric tons of ice per year (Hugonnet et al., 2021), affecting Himalayan rivers (Indus, Ganges, Brahmaputra) and Andean River systems critical to cities like La Paz and Quito.

Drought intensity has increased globally, driven by higher temperatures and accelerated evapotranspiration. Notable examples include California's 2012–2016 drought, which caused over $3 billion in agricultural losses (Diffenbaugh et al., 2015), and recurrent East African droughts impacting millions (UNICEF, 2020). Increased reliance on groundwater during droughts often leads to overextraction and long-term aquifer depletion (Wada et al., 2010).

Pollution

Water pollution is a critical threat to sustainability, primarily from agricultural runoff and industrial discharge (UNEP, 2016). Excess nutrients from fertilizers and pesticides cause eutrophication, hypoxia, and aquatic biodiversity loss, exemplified by the Gulf of Mexico Dead Zone and Lake Erie (Smith & Schindler, 2009; Scavia et al., 2014).

Industrial effluents containing heavy metals, solvents, pharmaceuticals, and thermal pollutants degrade water quality and pose serious ecological and public health risks, especially where regulatory enforcement is weak (UNESCO, 2020). Examples include the Citarum River (Indonesia) and India's Ganges River Basin (Central Pollution Control Board, 2020).

Emerging contaminants such as pharmaceuticals, personal care products, endocrine-disrupting chemicals, and microplastics are increasingly documented in freshwater systems worldwide, with potential long-term impacts on ecosystems and human health (Kümmerer et al., 2018; Schwarzenbach et al., 2006; Eerkes-Medrano et al., 2015; Smith et al., 2018).

Water and Public Health

Over 2 billion people lack safely managed drinking water, and nearly 3.6 billion lack basic sanitation (WHO/UNICEF, 2023). Waterborne illnesses, including diarrhea, cholera, typhoid, and hepatitis A, remain major causes of mortality, particularly among children (WHO, 2022).

The COVID-19 pandemic highlighted the importance of hygiene facilities, yet one in four health care facilities globally lack basic water services (WHO/UNICEF, 2021). Chronic exposure to contaminants such as arsenic, fluoride, and lead further contributes to long-term illnesses, necessitating integrated water-health policies and targeted WASH investments.

Groundwater Depletion

Groundwater, supplying drinking water for over two billion people and 40% of irrigated agriculture, is being depleted at unsustainable rates (>1,000 km^3/year) (WWAP, 2022; UNESCO, 2022). Major aquifers affected include the North China Plain, Indo-Gangetic Basin, High Plains (Ogallala), Mexico's Central Valley, and parts of the Middle East. Overextraction causes declining water tables, higher pumping costs, land subsidence, and saltwater intrusion in coastal aquifers (Rodell et al., 2009; Scanlon et al., 2017; Barlow & Reichard, 2010).

Mitigation strategies include efficient irrigation (e.g., drip systems), managed aquifer recharge, groundwater governance, and community-led water management programs (Dillon et al., 2019; Narain, 2010).

Figure 6. Groundwater storage trends across Earth's 37 largest aquifers

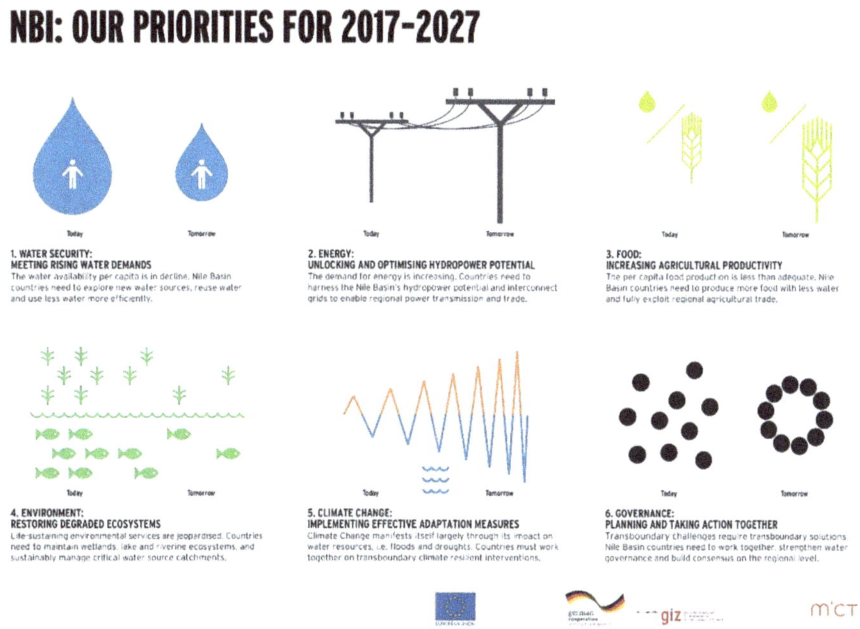

(GRACE data) (http://jpl.nasa.gov/)

Twenty-one aquifers have exceeded sustainability tipping points; 13 are significantly distressed.

Inequitable Access

Safe and affordable water access remains highly unequal. Rural populations constitute 80% of those without basic water services, often relying on unimproved sources vulnerable to contamination (JMP, 2023). Urban inequalities persist in informal settlements, where residents pay disproportionately high rates for water or depend on unsafe sources (UNDP, 2006).

Institutional Corruption & Inefficiency in Water Management

Corruption in water utilities—manifested as procurement fraud, nepotism, and misallocation of permits—undermines service delivery, particularly in sub-Saharan Africa and South Asia (Transparency International, 2023; Kumar et al., 2024). Strengthening transparency, enforcing anti-corruption laws, and promoting citizen oversight are essential for improving sectoral efficiency (World Bank, 2024).

Indigenous Water Rights & Legal Conflicts

Indigenous communities continue to face legal and governance challenges over water rights. Notable cases include the Dakota Access Pipeline (USA) and Murray–Darling Basin disputes (Smith, 2023; Jackson & Head, 2022). Key barriers include exclusion from decision-making, inadequate legal recognition, and insufficient integration of Indigenous ecological knowledge. Embedding free, prior, and informed consent (FPIC) into water governance is critical (United Nations, 2024).

Transboundary Water Conflicts

Transboundary water basins, covering nearly half of Earth's land and supporting 40% of the global population, are increasingly stressed by climate change, population growth, and rising water demand (UNESCO, 2019). Weak institutional frameworks exacerbate disputes.

Examples include the Grand Ethiopian Renaissance Dam (GERD) on the Blue Nile, which has heightened tensions between Ethiopia, Sudan, and Egypt (Almesafri et al., 2024), and the Indus Waters Treaty (India-Pakistan), currently strained due to climate variability and infrastructure changes (The Economic Times, 2025; Time, 2025). By 2050, approximately 920 million people could reside in high-risk basins under current management practices (Water Journalists Africa, 2023).

COMPARATIVE ANALYSIS OF WATER GOVERNANCE AND MANAGEMENT APPROACHES

This section synthesizes recent scholarship (2022–2025) to evaluate differences in governance structures, solution types, and regional Integrated Water Resources Management (IWRM) outcomes.

Governance Models: Top-Down vs. Community-Led

Examples include the Netherlands' Delta Programme and Singapore's Public Utilities Board. These approaches excel in large-scale infrastructure delivery and disaster preparedness but may lack flexibility in addressing local needs.

By contrast, community-led governance—as seen in Bolivia's water committees and Nepal's rural irrigation user groups—often fosters higher participation, stronger maintenance outcomes, and culturally adapted practices (Boelens et al., 2022). The most resilient systems often integrate centralized regulatory oversight with decentralized, participatory mechanisms, as illustrated by Australia's Murray–Darling Basin reforms post-2012 (Jackson & Head, 2022).

Technology-Based vs. Nature-Based Solutions

Technology-based solutions—including AI-powered leak detection in Israel, smart irrigation in California, and satellite-based groundwater monitoring—offer precision, efficiency, and scalability (Zhang et al., 2024; Huang et al., 2025).

Nature-based solutions, such as wetland restoration in the Mekong Delta or watershed reforestation in Kenya, tend to be lower cost, enhance biodiversity, and provide co-benefits such as flood mitigation and carbon sequestration (WWF, 2023).

Regional Variations in IWRM Outcomes

- Africa: South Africa has adopted IWRM principles into national law, but implementation is often constrained by limited resources, governance gaps, and political interference (World Bank, 2023).
- Europe: The EU Water Framework Directive provides a strong institutional and legal framework, resulting in measurable improvements in water quality. However, agricultural runoff and cross-border coordination remain ongoing challenges (European Environment Agency, 2022).
- Asia: In India and China, IWRM has facilitated large-scale basin planning, yet social equity issues persist. Marginalized communities are often affected by uneven enforcement and conflicting sectoral priorities (Asian Development Bank, 2023).

Table 2. Comparative governance approaches in water management

Governance Approach	Key Characteristics	Advantages	Limitations	Example Cases
Top-down (centralized)	Decisions made by national/ state authorities; strong regulatory framework	Large-scale coordination; resource mobilization; infrastructure investment	Can ignore local needs; risk of bureaucratic inefficiency & corruption	India's river basin management; China's South–North Water Transfer Project
Community-led (bottom-up)	Local communities manage water resources; participatory governance	Enhances accountability; context-specific; empowers marginalized groups	Limited funding; scalability challenges	Nepal's farmer-managed irrigation systems; Kenya's community water associations
Technology-based	Use of AI, IoT, smart irrigation, desalination, wastewater reuse	Data-driven efficiency; predictive management; resilience to variability	High cost; reliance on technical expertise; equity concerns	Israel's desalination plants; Singapore's NEWater recycling program
Nature-based	Ecosystem-based solutions (wetlands restoration, green infrastructure)	Cost-effective; multifunctional (climate adaptation + biodiversity)	Slower implementation; effectiveness depends on ecological context	China's "Sponge Cities"; U.S. wetland restoration projects

Source: Adapted from UNESCO (2024); World Bank (2024); IPCC (2022); Nature Water (2023).

CASE STUDIES: REAL-WORLD IMPLICATIONS

This section explores eight critical case studies from different continents, highlighting key challenges and lessons for sustainable water governance.

Nile Basin Conflict – Africa

In September 2023, Ethiopia completed the fourth and final filling of the GERD reservoir, sparking strong opposition from Egypt and Sudan. Egypt, which relies on the Nile for over 90% of its freshwater, condemned the unilateral filling as illegal and a violation of international agreements. Sudan expressed concerns about potential risks to its water infrastructure and regional stability (RFI, 2023; Tamazuj, 2023).

While Ethiopia views the GERD as crucial for economic development and energy generation, downstream countries fear reduced water flow and lack a comprehensive operational agreement. Despite numerous negotiation rounds and African Union mediation, a legally binding agreement on water flow management and drought mitigation remains elusive. This case underscores how unilateral infrastructure development on shared watercourses, compounded by outdated colonial-era treaties, can exacerbate regional tensions. It highlights the need for legally binding, equitable, and sustainable transboundary water governance frameworks (AP News, 2023).

Figure 7. Nile Basin initiative (2017-2027)

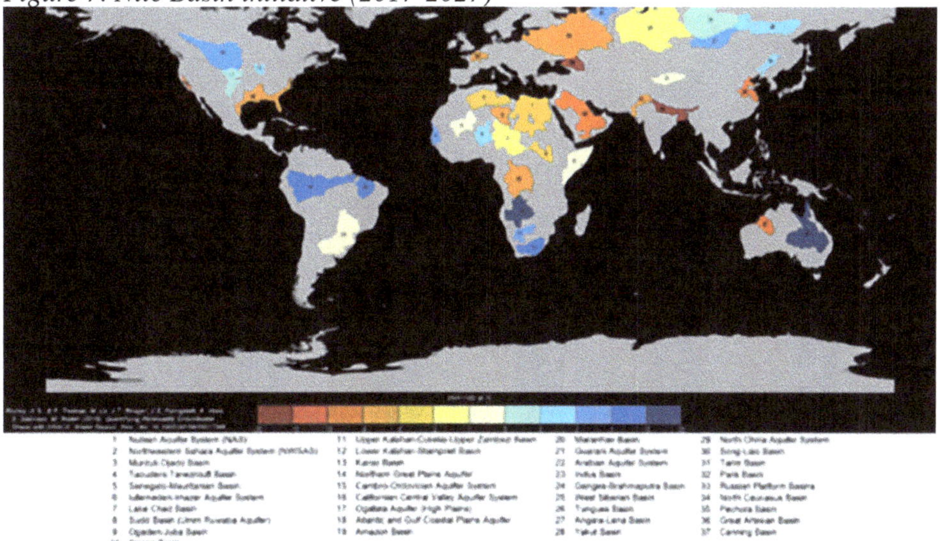

(https://www.nilebasin.org/)

California Drought – North America

California's recent droughts, particularly 2012–2016 and 2020–2022, have been driven by declining Sierra Nevada snowpacks, rising temperatures, and excessive groundwater extraction (OEHHA, 2022; CWW, 2024). The 2012–2016 drought caused billions in agricultural losses and fueled catastrophic wildfires, while the 2020–2022 drought illustrated "weather whiplash"—extreme fluctuations between dry and wet conditions (CDFA, 2024; Sierra Nevada Conservancy, 2024).

Climate change is central, altering precipitation patterns, reducing snow accumulation, and increasing evaporation. Urban centers have implemented mandatory water restrictions to cope with these challenges (LAO, 2022).

Figure 8. California–Nevada DEWS drought severity and coverage index (2000–2025)

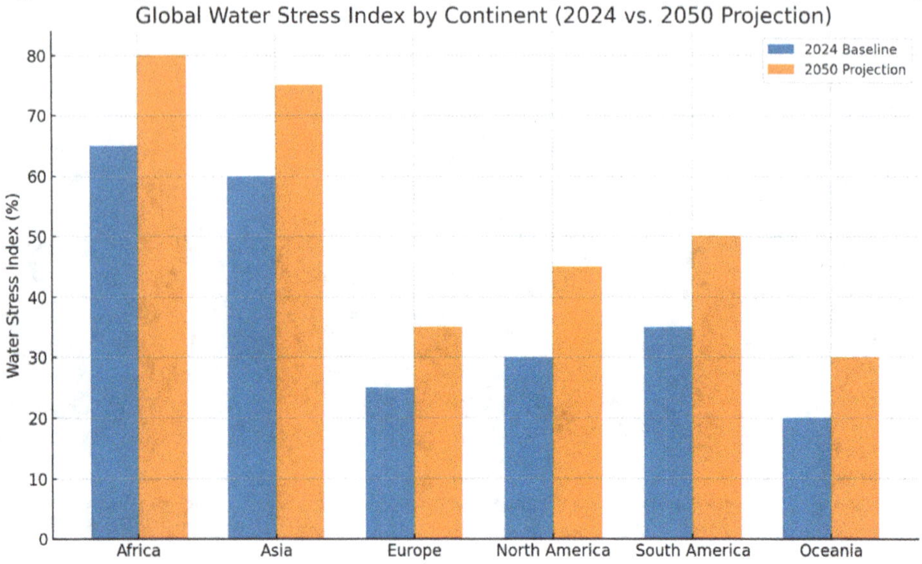

(U.S. Drought Monitor)

The DSCI highlights increasingly frequent and intense drought cycles, with major peaks in 2002–2004, 2007–2008, 2015–2016, and 2021–2022. Shorter recovery periods since 2010 indicate persistent dryness, emphasizing the importance of quantitative tools for informed water management, policy-making, and irrigation planning.

Aral Sea Disaster – Central Asia

The Aral Sea, once the fourth-largest inland lake globally, has shrunk by over 80% since the 1960s due to Soviet-era diversion of the Amu Darya and Syr Darya rivers for cotton irrigation (Micklin, 2007). Restoration efforts have succeeded in the North Aral Sea, where Kazakhstan's Kok-Aral Dam has revived fisheries and improved local livelihoods. However, the Southern Aral Sea remains largely desiccated. This case illustrates the dangers of unsustainable water management and the importance of integrated, transboundary river basin planning.

Ganges-Brahmaputra Basin – South Asia

Spanning Bangladesh, India, Nepal, and China, the Ganges-Brahmaputra basin supports over 600 million people. Despite heavy monsoonal rainfall, the region experiences extreme seasonal variability, leading to floods in the wet season and water scarcity in dry months (Parajuli et al., 2021). Over-extraction of groundwater, particularly in northern India, threatens long-term water security (Maina et al., 2024).

While the 1996 Ganges Water Sharing Treaty between India and Bangladesh represents a milestone, its effectiveness is limited by data opacity, unilateral actions, and weak enforcement. This case highlights the need for improved hydro diplomacy, transparent data sharing, and adaptive governance.

Cape Town Water Crisis – Africa

Cape Town's 2017–2018 water crisis, known as "Day Zero," was driven by severe drought, population growth, and inadequate planning. Stringent water rationing limited usage to 50 liters per person per day. Emergency conservation efforts, public awareness campaigns, and investment in alternative water sources averted a total collapse (Bailey, 2018).

The case demonstrates the importance of proactive planning, demand management, diversified water sources (desalination, wastewater reuse, stormwater capture), and community engagement in urban water resilience.

Israel's Water Management Model – Middle East

Israel has achieved water security in an arid region through technological innovation, policy reform, and public-private collaboration. Nearly 90% of wastewater is treated and reused, while desalination plants, like Sorek, supply a large portion of drinking water (Rinot, 2022).

However, the Oslo II Accord (1995) established an imbalanced water-sharing arrangement with the Palestinian Authority. Israelis have unrestricted access to the Mountain Aquifer, while Palestinian areas often receive below the WHO minimum standards. Jordan cooperates with Israel on projects like the Red Sea–Dead Sea pipeline, but regional instability delays implementation. Gaza faces critical water stress, with over 95% of water unfit for consumption (UNICEF, 2022).

Colorado River Basin Disputes – North America

The Colorado River serves over 40 million people in the U.S. and Mexico. Disputes among seven U.S. states and Mexico intensified under climate stress, requiring federal intervention in 2023 to negotiate reduced water use (U.S. Bureau of Reclamation, 2023). This case underscores the necessity of adaptive, basin-wide governance and equitable allocation under changing hydrological realities.

Jakarta Flooding – Southeast Asia

Jakarta faces chronic flooding due to poor drainage, land subsidence, heavy rainfall, and sea-level rise. Over 40% of the city lies below sea level, with unregulated groundwater extraction accelerating subsidence by up to 25 cm/year (Abidin et al., 2011). Flood events in 2007, 2013, and 2020 displaced thousands, caused fatalities, and damaged infrastructure (World Bank, 2021).

The National Capital Integrated Coastal Development (NCICD) plan aims to build sea walls, improve drainage, and regulate groundwater. Long-term success requires addressing informal settlements, coordinated urban planning, and integrating water resource management with climate adaptation and community resilience.

SOLUTIONS AND RECOMMENDATIONS

This section explores some key technologies, such as drip irrigation. Desalination, wastewater reuse, real-time monitoring and data analytics, green infrastructure and nature-based solutions, and community-level innovations.

Drip Irrigation

Drip irrigation is one of the most efficient agricultural water-use technologies, delivering water directly to plant roots through a network of valves, pipes, and emitters. Unlike traditional irrigation methods, drip systems minimize evaporation and surface runoff, achieving water-use efficiency of up to 90% (Camp, 1998).

Studies indicate that drip irrigation can reduce water use by 30–70% compared to conventional sprinkler systems while simultaneously improving crop yields (Narayanamoorthy, 2006). Countries like Israel and India have successfully scaled this technique through government subsidies and technology transfer programs.

- Israel: Leads the world in water-use efficiency, recycling approximately 87% of its wastewater for agriculture. Drip irrigation is applied on over 75% of irrigated land, contributing to efficient water use (Rinot, 2022; OECD, 2021).
- India: States like Maharashtra and Gujarat have achieved 30–50% water savings and up to 25% yield increases. However, national coverage of drip irrigation remains below 15% (FAO, 2018).

Desalination

Desalination removes dissolved salts and impurities from seawater or brackish water to produce potable water. Over 21,000 desalination plants operate worldwide, supplying freshwater to more than 300 million people (Jones et al., 2019).

The two main types of desalination are:

- Thermal desalination
- Membrane desalination

Despite its benefits, desalination has challenges, including high energy consumption and environmental concerns related to brine discharge. Innovations such as renewable energy integration and zero-liquid discharge systems are being explored to mitigate these impacts (Ghaffour et al., 2013).

Wastewater Reuse

Singapore's NEWater program is a global benchmark, recycling treated wastewater to meet up to 40% of the country's water needs (PUB Singapore, 2020). Recycled water is increasingly recognized as a reliable "drought-proof" source, particularly in urban areas facing rapid population growth and declining freshwater availability. Advanced treatment technologies—such as membrane bioreactors (MBR), ultrafiltration, and advanced oxidation—have made potable reuse both safe and socially acceptable in several regions (Asano & Cotruvo, 2004).

Real-Time Monitoring and Data Analytics

Digital technologies are transforming water management. Real-time monitoring using remote sensors, drones, and satellite imagery provides continuous data on water quality, groundwater levels, and infrastructure performance.

Integration with Geographic Information Systems (GIS) and artificial intelligence (AI) enables predictive modeling, early warning systems for floods and droughts, and informed policy-making (UNESCO, 2021; Jain et al., 2023).

Green Infrastructure and Nature-Based Solutions

Green infrastructure solutions—including rain gardens, bioswales, permeable pavements, and restored wetlands—complement engineered systems. These nature-based solutions:

- Enhance stormwater management
- Reduce urban runoff
- Improve groundwater recharge
- Contribute to urban cooling and biodiversity conservation

Such approaches align water security with broader sustainability objectives (Benedict & McMahon, 2012; IUCN, 2020).

Community-Level Innovations

Technological solutions must be accessible, adaptable, and context-specific, especially in developing regions. Examples include:

- Low-cost water filtration systems
- Solar-powered pumps
- Mobile water testing kits

Encouraging local innovation and participatory design enhances the long-term success, ownership, and sustainability of water technologies (Peter-Varbanets et al., 2009; WHO, 2022).

In sum, neither technology-only nor nature-based approaches are sufficient on their own. The most reliable gains come from context-specific hybrids, embedded in transparent institutions and sustained O&M funding.

ROLE OF INDIGENOUS AND TRADITIONAL KNOWLEDGE

In the face of global water crises, Indigenous and traditional knowledge systems offer time-tested, community-driven approaches to sustainable water management.

- India: The Zabo systems in Nagaland and Johads in Rajasthan exemplify effective community-managed rainwater harvesting in arid landscapes (Agarwal & Narain, 1997).
- Middle East: Qanats, underground canals developed over 3,000 years ago in Iran and Afghanistan, enable sustainable groundwater tapping by minimizing evaporation in desert climates (Lightfoot, 1996).
- Oman: The Aflaj systems, recognized by UNESCO, provide enduring models of equitable water sharing and recharge-based irrigation.

Indigenous peoples, who manage approximately 25% of the world's land surface and steward many vital watersheds, often maintain a spiritual and reciprocal relationship with water.

- New Zealand: The Māori regard water (*wai*) as a living ancestor, guiding a conservation ethic rooted in respect and guardianship (*kaitiakitanga*) (Harmsworth et al., 2016).
- Canada & U.S.: Many Indigenous groups uphold water ceremonies emphasizing intergenerational responsibility and ecosystem integrity (Chief et al., 2016).

Global policy frameworks, including Sustainable Development Goal 6 (SDG 6) and the United Nations Declaration on the Rights of Indigenous Peoples (UNDRIP), increasingly advocate for the inclusion of Indigenous perspectives in water governance. Integrating these perspectives is not only a matter of justice but also a strategic approach to designing climate-resilient water systems that align with local values, enhance biodiversity conservation, and ensure long-term sustainability.

SUSTAINABLE WATER MANAGEMENT PRACTICES: THE ROLE OF INTEGRATED WATER RESOURCES MANAGEMENT

The Global Water Partnership (2000) defines Integrated Water Resources Management (IWRM) as "a process that promotes the coordinated development and management of water, land, and related resources to maximize economic and social welfare equitably without compromising the sustainability of vital ecosystems."

IWRM is grounded in four key principles articulated in the 1992 Dublin Statement on Water and Sustainable Development:

a. Freshwater is a finite and vulnerable resource, essential for sustaining life, development, and the environment.
b. Participatory approaches—involving users, planners, and policymakers at all levels—are essential for effective water management.
c. Women play a central role in the provision, management, and safeguarding of water resources.
d. Water has an economic value in all its competing uses and should be recognized as an economic good (UN, 1992).

Effective IWRM requires institutional reform, stakeholder participation, regulatory mechanisms, and robust data management systems. Several countries provide illustrative examples of progress:

- South Africa: The National Water Act of 1998 institutionalized IWRM by creating catchment management agencies to oversee equitable water allocation and environmental conservation (Schreiner & van Koppen, 2001).
- Brazil: Basin committees and participatory councils have facilitated multi-stakeholder decision-making at the watershed level, balancing agricultural, urban, and ecological demands (UN-Water, 2021).
- Denmark: Strong regulatory enforcement, water pricing mechanisms, and investment in monitoring have supported sustainable water use across sectors (UN-Water, 2021).
- European Union: The Water Framework Directive (2000/60/EC) mandates river basin planning and public participation, requiring all member states to achieve "good status"—a classification encompassing both ecological integrity and chemical safety (European Commission, 2021).

Taken together, these cases highlight how IWRM operationalizes sustainability by linking governance, participation, and ecological stewardship. However, challenges remain in scaling these models, especially in regions with limited institutional capacity or political instability.

INTERNATIONAL COOPERATION AND AGREEMENTS

With more than 60% of the world's freshwater resources shared by two or more countries, international cooperation is essential for managing transboundary water systems peacefully and sustainably (UN-Water, 2013).

Transboundary Water Governance

Strengthening global frameworks, such as the 1997 UN Convention on the Law of the Non-Navigational Uses of International Watercourses, and supporting regional initiatives like the Nile Basin Initiative and the Mekong River Commission, remain critical for equitable water sharing and conflict prevention (UN-Water, 2020).

Notable examples include:

- Indus Waters Treaty (1960): Despite enduring political tensions, this agreement between India and Pakistan has remained functional for more than six decades.
- Senegal River Basin Development Organization (OMVS): Demonstrates how joint infrastructure, benefit-sharing, and equitable management can underpin regional stability (World Bank, 2021).

Emerging platforms such as the UNESCO International Hydrological Programme (IHP) and the UNECE Water Convention provide technical support, capacity building, and dialogue facilitation among riparian states (UNECE, 2015).

United Nations Sustainable Development Goal 6 (SDG 6)

SDG 6 commits to *"ensuring availability and sustainable management of water and sanitation for all"* by 2030. Target 6.5 explicitly emphasizes the implementation of IWRM at all levels, including transboundary cooperation.

Table 3. SDG 6 indicators and custodian agencies

SDG 6 Indicators	Custodians
6.4.1 Change in water use efficiency over time	FAO
6.4.2 Level of water stress: freshwater withdrawal as a proportion of available freshwater resources	FAO
6.5.2 Proportion of transboundary basin area with an operational arrangement for water cooperation	UNECE, UNESCO
6.3.2 Proportion of bodies of water with good ambient water quality	UNEP
6.5.1 Degree of integrated water resources management	UNEP
6.6.1 Change in the extent of water-related ecosystems over time	UNEP, Ramsar
6.a.1 Amount of water- and sanitation-related official development assistance that is part of a government-coordinated spending plan	WHO, OECD
6.b.1 Proportion of local administrative units with established and operational policies and procedures for participation of local communities in water and sanitation management	WHO, OECD
6.3.1 Proportion of domestic and industrial wastewater flows safely treated	WHO, UN-Habitat, UNSD
6.1.1 Proportion of population using safely managed drinking water services	WHO, UNICEF
6.2.1 Proportion of population using (a) safely managed sanitation services and (b) a hand-washing facility with soap and water	WHO, UNICEF

Source: *UNESCO (2024)*

Helsinki Rules (1966)

The Helsinki Rules on the Uses of the Waters of International Rivers, developed by the International Law Association in 1966, represented one of the first codifications of principles for managing shared watercourses. Although non-binding, they influenced subsequent treaties by emphasizing:

- Equitable and reasonable utilization of shared watercourses.
- Obligation not to cause significant harm to other riparian states.
- Prior notification and consultation in the case of planned measures affecting shared waters (ILA, 1966).

UN Watercourses Convention (1997)

The UN Convention on the Law of the Non-Navigational Uses of International Watercourses, adopted in 1997 and entered into force in 2014, is the most comprehensive global treaty governing shared freshwater resources. It builds upon the Helsinki Rules and aims to:

- Strengthen cooperation among riparian states.
- Prevent water-related conflicts.
- Ensure protection and conservation of ecosystems.

Though only ratified by around 40 countries, the Convention is recognized as customary international law and a guiding framework for transboundary water governance (Salman, 2007).

Successful Regional Agreements

Indus Waters Treaty (1960)

A landmark agreement between India and Pakistan brokered by the World Bank, the treaty has largely withstood wars and political instability. It allocates the eastern rivers to India and the western rivers to Pakistan, while allowing for limited use by each side under agreed rules.

Mekong River Commission (1995)

Involving Laos, Cambodia, Thailand, and Vietnam, the commission promotes data sharing, environmental monitoring, and coordinated development in the lower Mekong basin.

Senegal River Basin Development Organization (OMVS)

An exemplary African institution where member states jointly own and manage shared water infrastructure, ensuring equal benefits and compensation mechanisms (UNDESA, 2013).

Successful cooperation hinges on enforceable rules, open data, and shared benefits. Where these conditions exist, treaties endure even under geopolitical strain; where they do not, basin risk rises.

GLOBAL WATER OUTLOOK: CHALLENGES AND OPPORTUNITIES

By 2050, global water demand is projected to increase by 20–25%, largely driven by agricultural expansion, industrialization, and rapid urban growth (WWAP, 2019).

Emerging Opportunities in Water Management

- Smart water systems: Deployment of Internet of Things (IoT) sensors, automated leak detection, and remote monitoring are enhancing operational efficiency and minimizing distribution losses (Koo et al., 2020).
- Climate-smart agriculture (CSA): Adoption of drought-resistant crop varieties, precision irrigation, and soil moisture monitoring is optimizing water use while sustaining yields in climate-stressed regions (FAO, 2013).
- Water-energy-food (WEF) nexus tools: Integrated modeling platforms enable policymakers to assess cross-sectoral trade-offs, promoting more holistic resource allocation (Biggs et al., 2015).

Governance Priorities for Resilient Futures

Future progress hinges on robust and adaptive institutions. Key priorities include:

- Decentralized decision-making: Empowering basin-level authorities and local user associations ensures context-specific solutions (Ostrom, 1990).

- Water pricing and subsidies: Reforms that balance efficiency with social equity can discourage overuse while safeguarding the poor (Rogers et al., 2002).
- Inclusive governance: Integrating women, Indigenous peoples, and marginalized communities into planning reduces vulnerabilities and strengthens justice (UN-Water, 2021).

Institutional Corruption and Weak Governance

A persistent challenge in global water management is the role of corruption, inefficiency, and lack of accountability. Studies indicate that misallocation of funds, favoritism in water rights, and manipulation of hydrological data undermine both technical progress and community trust (Transparency International, 2023; World Bank, 2023). In countries facing acute water stress, political interference in procurement and licensing often skews resource allocation toward powerful elites, exacerbating inequities for small farmers and marginalized communities.

Examples include irrigation subsidies diverted to large landholders in South Asia, unauthorized groundwater pumping in parts of Sub-Saharan Africa, and inflated infrastructure contracts in Latin America (Boelens et al., 2022). These practices not only drain public budgets but also erode legitimacy, slowing implementation of otherwise robust policies such as IWRM.

Global watchdogs, including Transparency International and the Water Integrity Network, argue that anti-corruption safeguards—such as transparent procurement, participatory budgeting, and open-access water data—are essential for achieving SDG 6 targets. Without such safeguards, even advanced technologies like smart irrigation or desalination risk entrenching inequities rather than solving scarcity.

Figure 9. Institutional corruption pathways in water governance

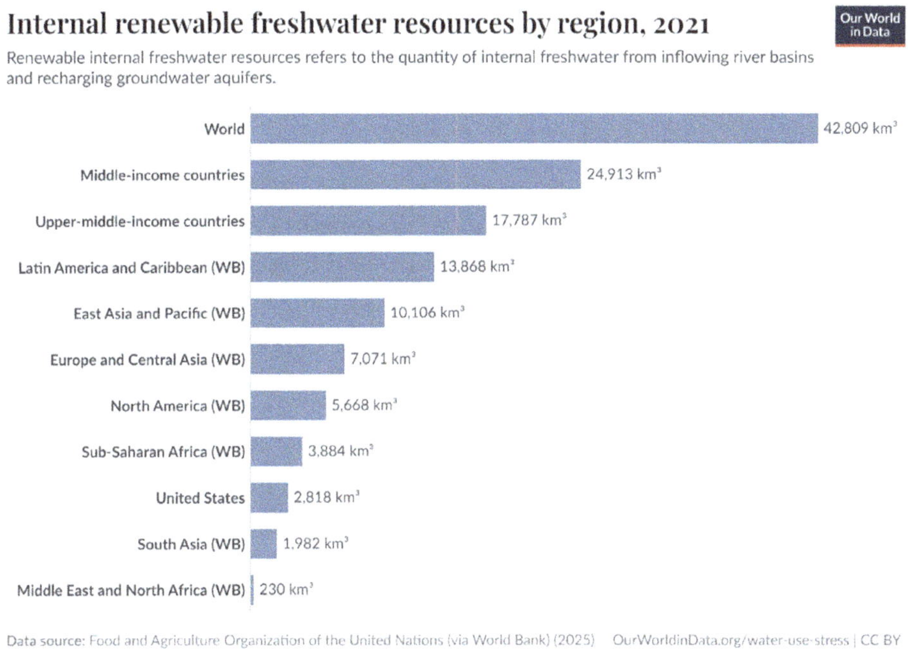

Cross-Sectoral Interdependencies

Water intersects critically with other global systems:

- Food security: Agriculture consumes ~70% of freshwater withdrawals; without improved irrigation and crop planning, demand risks exceeding supply (FAO, 2018).
- Energy production: Hydropower, thermal power cooling, and biofuels all rely heavily on water; climate-induced droughts threaten generation and reliability (IEA, 2021).
- Public health: Investments in clean water and sanitation lower disease burdens and improve economic productivity (WHO, 2022).

Building Climate Resilience

Water systems are among the most climate-sensitive. Nature-based solutions (NBS) are increasingly recognized as complements to "gray" infrastructure:

- Wetland restoration and afforestation for flood control and recharge.
- Green infrastructure (e.g., rain gardens, permeable pavements) to manage urban runoff.
- Managed aquifer recharge (MAR) to store excess rainfall and prevent over-extraction.

Evidence shows NBS not only bolsters water security but also provides biodiversity, carbon sequestration, and livelihood co-benefits (WWAP, 2018).

The outlook is challenging but tractable: institutions that are accountable, data-literate, and inclusive can convert uncertainty into manageable risk. Aligning investments with these governance capabilities is the decisive factor for achieving durable water security.

FUTURE RESEARCH DIRECTIONS

Future research must embrace interdisciplinary integration, combining hydrology, climatology, ecology, sociology, and economics to address complex water challenges (Vörösmarty et al., 2010).

Key Research Frontiers

- Data and monitoring: Many regions, particularly in the Global South, face fragmented monitoring. Satellite technologies, IoT-enabled sensors, and open-access platforms can strengthen transparency and early warning (UNESCO, 2022).
- Groundwater management: Research on aquifer recharge dynamics, pollution pathways, and sustainable withdrawal strategies is essential in climate-vulnerable regions (Taylor et al., 2013).
- Equity and participation: Gender roles, social norms, and power dynamics shape water access. Participatory research centering women and marginalized groups can yield culturally relevant solutions (UN Women, 2020).
- Low-cost, scalable technologies: Innovations such as constructed wetlands, fog harvesting, and decentralized treatment systems hold promise for improving availability and quality (WWAP, 2018).
- Capacity building: Training future leaders through interdisciplinary curricula, exchange programs, and South–South cooperation will ensure continuity in water governance (GWP, 2020).

Policy and Managerial Applications

Research insights translate into several policy implications:

1. Hybrid governance strategies: Combining centralized regulation with community-led initiatives enhances both efficiency and equity.
2. Technological adoption: AI-driven monitoring, recycling, and reuse systems can improve planning and resilience, provided there is transparent data governance and capacity building.
3. Institutional integrity: Addressing corruption, strengthening legal protections for Indigenous and marginalized groups, and embedding anti-corruption safeguards are critical for sustainable outcomes.

By embedding inclusivity, transparency, and adaptive governance, water managers can leverage both science and community knowledge for more resilient futures.

CONCLUSION

The global water crisis remains one of the most pressing challenges of the 21st century. Population growth, urbanization, climate change, and pollution continue to threaten availability, quality, and equitable distribution of freshwater. The case studies explored—from the Nile Basin to the Aral Sea and Jakarta—illustrate the diverse nature of water stress, as well as the interdependence of ecological, social, and political systems.

Sustainable solutions require systems-based approaches that integrate:

- Scientific innovation (e.g., smart irrigation, desalination, wastewater reuse).
- Traditional and Indigenous knowledge (e.g., community-led water harvesting and stewardship).
- Inclusive governance frameworks that prioritize equity, accountability, and participation.

International cooperation remains central. Experiences such as the Indus Waters Treaty and the OMVS in West Africa demonstrate that collaboration is achievable even amid political tensions. Strengthening transboundary governance and data-sharing mechanisms will be essential to prevent conflict and foster cooperation.

Finally, research and education are the foundation of a water-secure future. Empowering the next generation of water leaders with interdisciplinary expertise, ethical decision-making skills, and participatory approaches will be key.

Water security is not just a technical or environmental issue—it is fundamentally a development and human rights concern. Achieving resilience requires collective action at local, national, and global levels. With coordinated strategies grounded in science, justice, and cooperation, the global community can build a more sustainable and equitable water future.

REFERENCES

Abidin, H. Z., Andreas, H., Gumilar, I., Sidiq, T. P., & Fukuda, Y. (2011). Land subsidence of Jakarta (Indonesia) and its relation with urban development. *Natural Hazards*, *59*(3), 1753–1771. DOI: 10.1007/s11069-011-9866-9

. Africa, W. J. (2023, July 3). 920 million people could face conflict over the world's rivers by 2050: What our study found in Africa. *Water, climate change, land, research.*

Agarwal, A., & Narain, S. (1997). *Dying wisdom: Rise, fall and potential of India's traditional water harvesting systems.* Centre for Science and Environment.

Almesafri, A., Abdulsattar, S., Alblooshi, A., Al-Juboori, R. A., Jephson, N., & Hilal, N. (2024). Waters of contention: The GERD and its impact on Nile Basin cooperation and conflict. *Water (Basel)*, *16*(15). Advance online publication. DOI: 10.3390/w16152174

Asano, T., & Cotruvo, J. A. (2004). Groundwater recharge with reclaimed municipal wastewater: Health and regulatory considerations. *Water Research*, *38*(8), 1941–1951. PMID: 15087175

Asian Development Bank. (2023). *Asian water development outlook 2023: Advancing water security for sustainable development.* ADB. https://www.adb.org/publications/asian-water-development-outlook-2023

Bailey, K. (2018). Day Zero: Lessons from Cape Town's water crisis. *Journal of Urban Sustainability*, *4*(1), 12–20.

Barlow, P. M., & Reichard, E. G. (2010). Saltwater intrusion in coastal regions of North America. *Hydrogeology Journal*, *18*(1), 247–260. DOI: 10.1007/s10040-009-0514-3

Benedict, M. A., & McMahon, E. T. (2012). *Green infrastructure: Linking landscapes and communities.* Island Press.

Biggs, E. M., Bruce, E., Boruff, B., Duncan, J., Horsley, J., Pauli, N., McNeill, K., Neef, A., Van Ogtrop, F., Curnow, J., Haworth, B., & Duce, S. (2015). Sustainable development and the water–energy–food nexus: A perspective on livelihoods. *Environmental Science & Policy*, *54*, 389–397.

Boelens, R., Perreault, T., & Vos, J. (2022). Water justice and the politics of scale. *Water International*, *47*(4), 561–579. DOI: 10.1080/02508060.2022.2071764

California Department of Food and Agriculture. (2024). *Report on the 2012–2016 California drought years.* https://plantingseedsblog.cdfa.ca.gov/wordpress/?p=22505

California Water Watch. (2024). *Drought indicator.* https://cww.water.ca.gov/droughtindicator

Camp, C. R. (1998). Subsurface drip irrigation: A review. *Transactions of the ASAE. American Society of Agricultural Engineers, 41*(5), 1353–1367.

Central Pollution Control Board (CPCB). (2020). *National water quality monitoring programme.* Ministry of Environment, Forest and Climate Change, Government of India.

Chief, K., Meadow, A. M., & Whyte, K. P. (2016). Engaging southwestern Indigenous communities in sustainable water research. *Water (Basel), 8*(8), 350. DOI: 10.3390/w8080350

Diffenbaugh, N. S., Swain, D. L., & Touma, D. (2015). Anthropogenic warming has increased drought risk in California. *Proceedings of the National Academy of Sciences of the United States of America, 112*(13), 3931–3936. DOI: 10.1073/pnas.1422385112 PMID: 25733875

Dillon, P., Arshad, M., Fernandez, E., & Pavelic, P. (2019). Managed aquifer recharge in integrated water resource management. In Villholth, S. (Eds.), *Advances in groundwater governance* (pp. 435–459). CRC Press.

Eerkes-Medrano, D., Thompson, R. C., & Aldridge, D. C. (2015). Microplastics in freshwater systems: A review of the emerging threats, identification of knowledge gaps and prioritisation of research needs. *Water Research, 75*, 63–82. DOI: 10.1016/j.watres.2015.02.012 PMID: 25746963

European Commission. (2021). *Water Framework Directive & floods directive: Overview.* https://environment.ec.europa.eu/topics/water/water-framework-directive_en

European Environment Agency. (2022). *Water quality in Europe — 2022 status report.* EEA Report No. 14/2022. https://www.eea.europa.eu/publications/water-quality-in-europe-2022

. Food and Agriculture Organization of the United Nations. (2013). *Climate-smart agriculture sourcebook.*

. Food and Agriculture Organization of the United Nations. (2018). *The state of the world's water resources for food and agriculture.*

Food and Agriculture Organization of the United Nations. (2025). Renewable internal freshwater resources, total (billion cubic meters). [Data set] In World Bank (Ed.), *World Development Indicators.* Processed by Our World in Data.

Food and Agriculture Organization of the United Nations (FAO). (2016). *AQUASTAT main database*. https://www.fao.org/aquastat/

Food and Agriculture Organization of the United Nations (FAO). (2017). *The future of food and agriculture – Trends and challenges*. FAO.

Food and Agriculture Organization of the United Nations (FAO). (2018). *The state of the world's water resources for food and agriculture*. FAO.

Gaffin, S. R., Rosenzweig, C., & Parshall, L. (2008). Impact of urban heat islands on water availability and usage. *Environmental Research Letters*, *3*(4), 1–7. DOI: 10.1088/1748-9326/3/4/044003

Ghaffour, N., Missimer, T. M., & Amy, G. L. (2013). Technical review and evaluation of the economics of water desalination: Current and future challenges for better water supply sustainability. *Desalination*, *309*, 197–207.

. Global Water Partnership. (2000). *Integrated water resources management* (TAC Background Papers No. 4).

Global Water Partnership. (2000). *Integrated water resources management* (TAC Background Papers No. 4). Stockholm, Sweden.

Global Water Partnership (GWP). (2010). *Integrated water resources management*. GWP.

Harmsworth, G., Awatere, S., & Robb, M. (2016). Indigenous Māori values and perspectives to inform freshwater management in Aotearoa-New Zealand. *Ecological Applications*, *26*(7), 2417–2430. DOI: 10.1002/eap.1371

Huang, H., Cai, X., Li, L., Wu, X., Zhao, Z., & Tan, X. (2025). Interpreting global terrestrial water storage dynamics and drivers with explainable deep learning. *Remote Sensing*, *17*(13), 2118. DOI: 10.3390/rs17132118

Hugonnet, R., McNabb, R., Berthier, E., Menounos, B., Nuth, C., Girod, L., & Kääb, A. (2021). Accelerated global glacier mass loss in the early 21st century. *Nature*, *592*, 726–731. DOI: 10.1038/s41586-021-03436-z PMID: 33911269

. Huo, F., Xu, L., Li, Z. *et al.* Can climate change signals be detected from the terrestrial water storage at daily timescale?. *npj Clim Atmos Sci* 7, 158 (2024). https://doi.org/.DOI: 10.1038/s41612-024-00646-w

Intergovernmental Panel on Climate Change (IPCC). (2021). *Climate change 2021: The physical science basis. Contribution of Working Group I to the Sixth Assessment Report of the Intergovernmental Panel on Climate Change*. Cambridge University Press.

Intergovernmental Panel on Climate Change (IPCC). (2022). Climate change 2022: Impacts, adaptation and vulnerability. Sixth Assessment Report. Cambridge University Press. https://www.ipcc.ch/report/ar6/wg2/

International Energy Agency. (2021). *Water-energy nexus*. https://www.iea.org

International Law Association. (1966). *The Helsinki Rules on the Uses of the Waters of International Rivers*.

International Union for Conservation of Nature. (2020). *Nature-based solutions for water*. https://www.iucn.org

IPCC. (2022). *Climate Change 2022: Impacts, adaptation, and vulnerability*. Contribution of Working Group II to the Sixth Assessment Report of the Intergovernmental Panel on Climate Change. Cambridge University Press. https://www.ipcc.ch/report/ar6/wg2/

Jackson, S., & Head, L. (2022). Indigenous water rights and climate adaptation in Australia. *Journal of Hydrology (Amsterdam)*, *613*, 128–145. DOI: 10.1016/j.jhydrol.2022.128145

Jain, S., Kumar, P., & Rathore, V. (2023). Smart water monitoring systems using IoT: Trends and technologies. *Environmental Monitoring and Assessment*, *195*(4), 1–17.

Joint Monitoring Programme (JMP). (2023). *Progress on household drinking water, sanitation, and hygiene 2000–2022: Special focus on inequalities*. WHO & UNICEF. https://washdata.org

Jones, E., Qadir, M., van Vliet, M. T. H., Smakhtin, V., & Kang, S. (2019). The state of desalination and brine production: A global outlook. *The Science of the Total Environment*, *657*, 1343–1356. PMID: 30677901

Koo, C., Kim, J., Kim, S., & Shin, D. (2020). Smart water grid: The next generation water management platform. *Water (Basel)*, *12*(11), 3222.

Kumar, P., Singh, R., & Chatterjee, A. (2024). Governance challenges in South Asia's urban water supply systems. *Water Policy*, *26*(1), 45–62. DOI: 10.2166/wp.2023.100

Kümmerer, K., Dionysiou, D. D., Olsson, O., & Fatta-Kassinos, D. (2018). A path to clean water. *Science*, *361*(6399), 222–224. DOI: 10.1126/science.aau2405 PMID: 30026210

Legislative Analyst's Office. (2022). *Climate change impacts across sectors*. https://lao.ca.gov/reports/2022/4575/Climate-Change-Impacts-Crosscutting-Issues-040522.pdf

Lightfoot, D. R. (1996). Moroccan khettara: Traditional irrigation and progressive desiccation. *Geoforum, 27*(2), 261–273. DOI: 10.1016/0016-7185(96)00011-7

Maina, F. Z., Getirana, A., Kumar, S. V., Saharia, M., Biswas, N. K., McLarty, S., & Appana, R. (2024). Irrigation-driven groundwater depletion in the Ganges-Brahmaputra basin decreases the streamflow in the Bay of Bengal. *Communications Earth & Environment, 5*, 1348. Advance online publication. DOI: 10.1038/s43247-024-01348-0

Micklin, P. (2007). The Aral Sea disaster. *Annual Review of Earth and Planetary Sciences, 35*, 47–72.

Narain, V. (2010). Gone with the aquifer: Coping with groundwater depletion in western India. *Local Environment, 15*(2), 109–120. DOI: 10.1080/13549830903575508

Narayanamoorthy, A. (2006). *Potential for drip and sprinkler irrigation in India* (Working Paper No. 7). Gokhale Institute of Politics and Economics.

Nature Water. (2023). Selected articles on governance, technology, and water security. *Nature Water*. https://www.nature.com/natwater

Nile Basin Initiative. (2017). *Nile Basin Initiative 10-year strategy: 2017–2027.* NBI Secretariat. https://nilebasin.org

OECD. (2021). *OECD studies on water: Water governance in Israel.* OECD Publishing., DOI: 10.1787/0f315e3f-

Office of Environmental Health Hazard Assessment. (2022). *Changes in drought.* https://oehha.ca.gov/climate-change/epic-2022/changes-climate/drought

Organisation for Economic Co-operation and Development (OECD). (2012). *Environmental outlook to 2050: The consequences of inaction.* OECD Publishing., DOI: 10.1787/9789264122246-

Organisation for Economic Co-operation and Development (OECD). (2019). *Managing water for all: An OECD perspective on pricing and governance.* OECD Publishing.

Ostrom, E. (1990). *Governing the commons: The evolution of institutions for collective action.* Cambridge University Press.

Parajuli, B., Zhang, X., Deuja, S., & Liu, Y. (2021). Regional and seasonal precipitation and drought trends in Ganga–Brahmaputra Basin. *Water (Basel), 13*(16), 2218. DOI: 10.3390/w13162218

Peter-Varbanets, M., Zurbrügg, C., Swartz, C., & Pronk, W. (2009). Decentralized systems for potable water and the potential of membrane technology. *Water Research, 43*(2), 245–265. PMID: 19010511

Pokhrel, Y., Felfelani, F., Satoh, Y., Boulange, J., Burek, P., Gädeke, A., Gerten, D., Gosling, S. N., Grillakis, M., Gudmundsson, L., Hanasaki, N., Kim, H., Koutroulis, A. G., Liu, J., Papadimitriou, L., Schewe, J., & Veldkamp, T. I. E. (2021). Global terrestrial water storage and drought severity under climate change. *Nature Climate Change, 11*, 226–233. DOI: 10.1038/s41558-020-00972-w

Radio France Internationale (RFI). (2023, September 11). Egypt brands Ethiopia's move to fill Nile mega dam as 'illegal'. https://www.rfi.fr/en/africa/20230911-egypt-brands-ethiopia-s-move-to-fill-nile-mega-dam-as-illegal

Radio Tamazuj. (2023, September 12). Egypt writes to UN Security Council over Ethiopia filling dam, threatens action. https://www.radiotamazuj.org/en/news/article/egypt-writes-to-un-security-council-over-ethiopia-filling-dam-threatens-action

Rinot, Y. (2022). Israel's water strategy: Turning scarcity into abundance. *Water International, 47*(3), 250–265.

Rodell, M., Velicogna, I., & Famiglietti, J. S. (2009). Satellite-based estimates of groundwater depletion in India. *Nature, 460*(7258), 999–1002. DOI: 10.1038/nature08238 PMID: 19675570

Rogers, P., de Silva, R., & Bhatia, R. (2002). Water is an economic good: How to use prices to promote equity, efficiency, and sustainability. *Water Policy, 4*(1), 1–17.

Salman, S. M. A. (2007). The United Nations Watercourses Convention ten years later: Why has its entry into force proven difficult? *Water International, 32*(1), 1–15.

Scanlon, B. R., Ruddell, B. L., Reed, P. M., Hook, R. I., Zheng, C., Tidwell, V. C., & Siebert, S. (2017). The food-energy-water nexus: Transforming science for society. *Water Resources Research, 53*(5), 3550–3564. DOI: 10.1002/2017WR020889

Scavia, D., Allan, J. D., Arend, K. K., Bartell, S., Beletsky, D., Bosch, N. S., & Zhou, Y. (2014). Assessing and addressing the re-eutrophication of Lake Erie: Central basin hypoxia. *Journal of Great Lakes Research, 40*(2), 226–246. DOI: 10.1016/j.jglr.2014.02.004

Schreiner, B., & van Koppen, B. (2001). Catchment management agencies for poverty eradication in South Africa. *Physics and Chemistry of the Earth. Part B: Hydrology, Oceans and Atmosphere, 26*(11–12), 969–976.

Schwarzenbach, R. P., Escher, B. I., Fenner, K., Hofstetter, T. B., Johnson, C. A., von Gunten, U., & Wehrli, B. (2006). The challenge of micropollutants in aquatic systems. *Science, 313*(5790), 1072–1077. DOI: 10.1126/science.1127291 PMID: 16931750

Shiklomanov, I. A. (1993). World fresh water resources. In Gleick, P. H. (Ed.), *Water in crisis: A guide to the world's fresh water resources* (pp. 13–24). Oxford University Press.

Sierra Nevada Conservancy. (2024). *Weather whiplash in the Sierra Cascade.* https://sierranevada.ca.gov/weather-whiplash-in-the-sierra-cascade-and-the-need-to-accelerate-resilience

Singapore, P. U. B. (2020). *NEWater.* https://www.pub.gov.sg/watersupply/fournationaltaps/newater

Singh, R.. (2023). AI and big data for water resource management. *Water Resources Management, 37*(2), 587–604.

Smith, D. (2023). Indigenous resistance and environmental justice in the Dakota Access Pipeline. *Environmental Sociology, 9*(1), 77–92. DOI: 10.1080/23251042.2022.2066173

Smith, M., Love, D. C., Rochman, C. M., & Neff, R. A. (2018). Microplastics in seafood and the implications for human health. *Current Environmental Health Reports, 5*(3), 375–386. DOI: 10.1007/s40572-018-0206-z PMID: 30116998

Smith, V. H., & Schindler, D. W. (2009). Eutrophication science: Where do we go from here? *Trends in Ecology & Evolution, 24*(4), 201–207. DOI: 10.1016/j.tree.2008.11.009 PMID: 19246117

Taylor, R. G., Scanlon, B., Döll, P., Rodell, M., van Beek, R., Wada, Y., & Treidel, H. (2013). Groundwater and climate change. *Nature Climate Change, 3*(4), 322–329. DOI: 10.1038/nclimate1744

The Economic Times. (2025, May 8). India opens three gates of Salal dam in Reasi after Indus Waters Treaty freeze and military strikes on Pakistan.

Transparency International. (2023). *Water sector integrity global outlook.* Transparency International Secretariat. https://www.transparency.org/en/publications/water-sector-integrity-global-outlook

Transparency International. (2023). Global corruption report. *Water (Basel), 2023,* •••. https://www.transparency.org/en/publications/global-corruption-report-water-2023

Trenberth, K. E. (2011). Changes in precipitation with climate change. *Climate Research, 47*(1–2), 123–138. DOI: 10.3354/cr00953

UN-Water. (2013). *Water security and the global water agenda.* United Nations University.

UN-Water. (2020). *The United Nations World Water Development Report 2020: Water and Climate Change.* United Nations Educational, Scientific and Cultural Organization (UNESCO).

UN-Water. (2021). *Summary progress update 2021 – SDG 6: Water and sanitation for all.* UN-Water.

UNESCO. (2020). *United Nations world water development report 2020: Water and climate change.* https://unesdoc.unesco.org/ark:/48223/pf0000372985

UNESCO. (2021). *The United Nations World Water Development Report 2021: Valuing Water.* United Nations Educational, Scientific and Cultural Organization.

UNESCO. (2021). *United Nations World Water Development Report 2021: Valuing Water.* UNESCO Publishing.

. UNESCO. (2022). *The United Nations world water development report 2022: Groundwater Making the invisible visible.*

UNESCO. (2022). *The United Nations world water development report 2022: Groundwater – Making the invisible visible.* United Nations Educational, Scientific, and Cultural Organization.

UNESCO. (2022). *International Hydrological Programme (IHP): Enhancing transboundary cooperation.* https://en.unesco.org/themes/water-security/hydrology

UNESCO. (2024). *The United Nations world water development report 2024: Water for prosperity and peace.* United Nations Educational, Scientific and Cultural Organization. https://www.unesco.org/reports/wwdr/en/2024

UNESCO WWAP. (2019). *The United Nations world water development report 2019: Leaving no one behind.* UNESCO.

UNICEF. (2020). *Water scarcity in East Africa.* United Nations Children's Fund.

United Nations. (1992). *The Dublin Statement on Water and Sustainable Development. International Conference on Water and the Environment*, Dublin, Ireland.

United Nations. (2024). *Report on the implementation of the United Nations Declaration on the Rights of Indigenous Peoples (UNDRIP) in environmental governance*. United Nations Department of Economic and Social Affairs. https://www.un.org/development/desa/indigenous-peoples

United Nations Department of Economic and Social Affairs. (2013). *International Decade for Action 'Water for Life' 2005–2015*. https://www.un.org/waterforlifedecade

United Nations, Department of Economic and Social Affairs. (UN DESA). (2018). *World urbanization prospects: The 2018 revision*. United Nations.

United Nations Development Programme (UNDP). (2006). *Human development report 2006: Beyond scarcity – Power, poverty and the global water crisis*. UNDP.

United Nations Economic Commission for Europe. (2015). *Water Convention: Promoting transboundary water cooperation*. https://unece.org/env/water

. United Nations Educational, Scientific and Cultural Organization. (2022). *World Water Assessment Programme: Facts and figures*.

United Nations Environment Programme (UNEP). (2016). *A snapshot of the world's water quality: Towards a global assessment*. UNEP.

. United Nations Environment Programme (UNEP). (2022). *Bracing for super pollutants: Emerging contaminants in water and wastewater*.

United Nations World Water Assessment Programme (WWAP). (2019). *The United Nations world water development report 2019: Leaving no one behind*. UNESCO.

. United Nations World Water Assessment Programme (WWAP). (2022). *United Nations world water development report*.

U.S. Bureau of Reclamation. (2023). *Colorado River Basin drought response actions*. U.S. Department of the Interior. https://www.usbr.gov/dcp/

Vörösmarty, C. J., McIntyre, P. B., Gessner, M. O., Dudgeon, D., Prusevich, A., Green, P., & Davies, P. M. (2010). Global threats to human water security and river biodiversity. *Nature*, *467*(7315), 555–561. DOI: 10.1038/nature09440 PMID: 20882010

Wada, Y., Van Beek, L. P. H., & Bierkens, M. F. P. (2010). Modelling global water stress of the recent past: On the relative importance of trends in water demand and climate variability. *Hydrology and Earth System Sciences*, *15*(1), 3785–3808. DOI: 10.5194/hess-15-3785-2011

Women, U. N. (2020). *Gender equality and water security*. https://www.unwomen.org/en/digital-library/publications/2020/02/gender-equality-and-water-security

World Bank. (2019). *Quality unknown: The invisible water crisis*. https://www.worldbank.org/en/news/feature/2019/08/20/quality-unknown-the-invisible-water-crisis

World Bank. (2020). *Urbanization and water management: The case for change*. World Bank.

World Bank. (2021). *Jakarta flood management: Protecting the people and the economy*. https://www.worldbank.org/en/news/feature/2021/02/25/jakarta-flood-management

World Bank. (2021). *Senegal River Basin: Collaborative water management and regional development*. https://www.worldbank.org/en/topic/waterresources management

World Bank. (2023). Water governance and policy in a changing climate. The World Bank. https://openknowledge.worldbank.org/entities/publication/

World Bank. (2023). *Water governance and policy in a changing climate*. The World Bank. https://openknowledge.worldbank.org/handle/10986/39754

World Bank. (2024). *Strengthening water governance for climate resilience: Policy and institutional frameworks*. World Bank Group. https://www.worldbank.org/water

World Bank. (2024). *World development indicators: Renewable internal freshwater resources and water stress data*. The World Bank. https://databank.worldbank.org/source/world-development-indicators

World Health Organization. (2022). *Drinking water: Key facts*. https://www.who.int/news-room/fact-sheets/detail/drinking-water

. World Health Organization (WHO), & United Nations Children's Fund (UNICEF). (2021). *Global progress report on WASH in health care facilities: Fundamentals first*. Geneva: WHO and UNICEF.

. World Health Organization (WHO). (2022). *Water, sanitation, and hygiene: The essential elements for achieving health*.

. World Health Organization (WHO) & United Nations Children's Fund (UNICEF). (2023). *Progress on household drinking water, sanitation and hygiene: 2000–2022 data update*. Geneva: WHO and UNICEF

World Water Assessment Programme. (2018). *The United Nations World Water Development Report 2018: Nature-based solutions for water*. UNESCO.

WWF. (2023). *Nature-based solutions for water security*. World Wide Fund for Nature. https://wwf.panda.org/discover/our_focus/freshwater_practice/nature_based_solutions/

Yin, J., Slater, L. J., Khouakhi, A., Yu, L., Liu, P., Li, F., Pokhrel, Y., & Gentine, P. (2023). GTWS-MLrec: Global terrestrial water storage reconstruction by machine learning from 1940 to present. *Earth System Science Data*, *15*(12), 5597–5615. DOI: 10.5194/essd-15-5597-2023

Zhang, W., Li, Y., & Chen, H. (2024). Remote sensing and AI integration for global water monitoring. *Remote Sensing of Environment*, *298*, 113828. DOI: 10.1016/j.rse.2023.113828

KEY TERMS AND DEFINITIONS

Renewable Freshwater Resources: Naturally replenished supplies from precipitation, runoff, and groundwater recharge that are essential for supporting ecosystems, people, and agriculture.

Integrated Water Resources Management (IWRM): A holistic water management approach integrates all parts of the water cycle to optimize social, economic, and environmental outcomes.

Water Stress: A condition marked by a significant lack of freshwater, causing resource depletion and declining water quality, making it difficult to meet essential water needs in a given area.

Transboundary Water Cooperation: The collaborative efforts made by two or more nations to sustainably manage and use shared water bodies like rivers, lakes, and aquifers for mutual benefit.

Desalination: A process of extracting salts and minerals from water to make it usable for drinking, agriculture, and industrial processes.

Wastewater Reuse: The process of treating used water for non-drinking purposes like irrigation, industry, or with advanced treatment, even for drinking.

Waterborne Diseases: Infections caused by bacteria, viruses, or parasites are present in contaminated water through ingestion or contact.

Groundwater Depletion: Groundwater is depleted when extraction from an aquifer exceeds its natural recharge, causing a long-term drop in water volume and levels.

Climate-Smart Agriculture: A strategy aimed at transforming agri-food systems toward environmentally sustainable and climate-resilient practices.

Nature-Based Solutions (NBS): Actions designed to protect, restore, and sustainably manage ecosystems to tackle issues such as climate change, disaster risks, and water security, while supporting biodiversity and human well-being.

Chapter 3
Fuzzy Multi-Criteria Decision-Making Approach for Optimizing Sustainable Water Management

Ajoy Kanti Das
https://orcid.org/0000-0002-9326-1677
Tripura University, India

Nandini Gupta
https://orcid.org/0009-0000-5450-9783
Bir Bikram Memorial College, India

Hani Amir Aouissi
https://orcid.org/0009-0004-2258-5764
Environmental Research Center (CRE), Algeria

Bishnupada Debnath
https://orcid.org/0009-0008-3162-4082
Rajiv Gandhi University, India

ABSTRACT

This chapter explores the application of Fuzzy Multi-Criteria Decision-Making (FMCDM) in Optimizing Sustainable Water Management (SWM). FMCDM integrates fuzzy set theory and soft set theory to address the complexities and uncertainties involved in decision-making, especially in scenarios with multiple criteria and conflicting objectives. The chapter presents a methodological framework for applying

DOI: 10.4018/979-8-3373-2700-6.ch003

FMCDM, which includes steps such as problem definition, fuzzification of data, fuzzy soft set (FSS) construction, weight assignment, aggregation, and ranking of alternatives. A case study demonstrates how this framework can be applied to assess water management strategies based on criteria like Water Quality Improvement, Cost Efficiency, Technological Feasibility, Environmental Sustainability, and Social Acceptability. The chapter highlights the benefits of FMCDM and provides structured decision support, while also addressing its limitations and suggesting future directions for enhancing its applicability in real-time, multi-stakeholder, and sustainability-focused decision-making contexts.

INTRODUCTION

The increasing demand for Sustainable Water Management (SWM) has led to the development of innovative methods for optimizing water resources, which is vital for ensuring the long-term availability of water. Water management is inherently complex, involving assessing various environmental, social, and economic factors. Thus, an integrated decision-making approach is required. Multi-Criteria Decision-Making (MCDM) methods have become essential in addressing these challenges by allowing stakeholders to evaluate and compare different water management strategies based on multiple criteria. These methods have been successfully applied to various water management contexts, including agricultural water management (Radmehr et al., 2022), irrigation water allocation (Elleuch et al., 2019), and wastewater treatment (Sharma et al., 2023).

MCDM approaches are beneficial when making uncertain decisions, as they allow for incorporating fuzzy logic to handle imprecise data (Zarghami & Szidarovszky, 2009). For instance, fuzzy techniques, such as the Fuzzy Analytic Hierarchy Process (AHP) and the Fuzzy Technique for Order of Preference by Similarity to Ideal Solution (TOPSIS), have been widely adopted in water resources management for evaluating water quality (Zyoud et al., 2016), optimizing water supply choices (Noori et al., 2020), and addressing the challenges of water loss management (Zyoud et al., 2016). Integrating expert judgment and stakeholder preferences makes decision-making more robust.

Furthermore, the emergence of hybrid MCDM approaches, which combine fuzzy logic with other techniques like AHP and TOPSIS, has proven effective in improving decision outcomes in complex scenarios, such as water resources planning (Afshar et al., 2011) and water quality assessment (Nourani & Najafi, 2022). These hybrid models enhance the accuracy of decision-making and provide a comprehensive framework for assessing the sustainability of water management practices, considering short-term and long-term impacts.

This chapter explores the role of FMCDM approaches in optimizing SWM practices. By examining relevant applications in water resources management, wastewater treatment, and water loss management, this chapter highlights the potential of FMCDM in providing valuable insights for decision-makers. It also discusses the challenges and opportunities presented by the integration of fuzzy logic in MCDM models. It presents a novel approach for improving water quality assessment using FSSs (Maji et al., 2001), aiming to contribute to the ongoing efforts in SWM.

Literature Review

Sustainable development relies on effective water management, especially in water scarcity, pollution, and environmental changes. The application of MCDM techniques, particularly fuzzy decision-making models, has emerged as an effective tool for optimizing water resources. These techniques help address the uncertainty and vagueness inherent in water quality assessments and management decisions. The following review explores recent advances in using FMCDM approaches for optimizing SWM, focusing on water quality indices, pollution assessment, and water resource allocation. The use of fuzzy logic in MCDM allows for incorporating linguistic variables and imprecise data, making it a robust tool in situations where decision criteria are uncertain or vague. FMCDM methods have been applied to various challenges in water management, from water supply choice optimization to wastewater treatment and irrigation water allocation. Noori et al. (2020) propose a group MCDM method for optimizing water supply choices, considering multiple conflicting criteria such as cost, availability, and environmental impact. They combine fuzzy logic with AHP and TOPSIS (Technique for Order of Preference by Similarity to Ideal Solution), demonstrating the applicability of FMCDM in water supply systems (Noori et al., 2020). This approach helps select optimal water sources under uncertain conditions by considering qualitative and quantitative criteria. Similarly, Radmehr et al. (2022) present an integrated strategic planning framework for agricultural water management using FMCDM. They integrate a range of criteria, including water availability, irrigation efficiency, and economic factors, to optimize agricultural water use in regions facing water stress. The FMCDM approach comprehensively assesses these factors, aiding in more informed decision-making for sustainable farming practices (Radmehr et al., 2022). Zarghami and Szidarovszky (2009) explore using stochastic-fuzzy decision-making for robust water resources management. They address the uncertainties in hydrological data, emphasizing the importance of stochastic models for random water resource variability. Their work shows how combining fuzzy logic with stochastic models enhances the robustness of decision-making in water resource management under uncertainty (Zarghami & Szidarovszky, 2009).

Water quality assessment is another area where FMCDM methods have proven useful (Hashim et al., 2015). Water quality indices are commonly used to assess the suitability of water for various purposes (Al-Falal et al., 2025). Still, they often rely on multiple criteria that must be evaluated simultaneously. In these cases, FMCDM methods help to integrate diverse and conflicting factors, such as chemical parameters, physical properties, and biological indicators, to provide a comprehensive water quality rating. Sharma et al. (2023) extensively review various FMCDM methods applied to wastewater treatment. They highlight the use of fuzzy AHP and fuzzy TOPSIS for evaluating treatment options, considering multiple factors such as cost, efficiency, and environmental impact. These methods are instrumental when the available data is uncertain or imprecise, enabling decision-makers to select the most suitable wastewater treatment technology (Sharma et al., 2023). Elleuch et al. (2019) address Tunisia's irrigation water allocation problem using a hybrid FMCDM approach. By integrating fuzzy logic with AHP and TOPSIS, they develop a model to optimize water allocation under uncertainty. This model benefits regions with limited water resources, and competing demands must be balanced. Their approach offers a more nuanced solution by considering the imprecision and vagueness inherent in water allocation decisions (Elleuch et al., 2019). Zyoud et al. (2016) propose a framework for water loss management in developing countries using fuzzy AHP combined with fuzzy TOPSIS. The framework helps to identify areas with high water loss and prioritize interventions based on fuzzy criteria. Integrating fuzzy decision-making models with water loss data provides a valuable tool for improving water distribution systems in resource-constrained regions (Zyoud et al., 2016).

The application of FMCDM to groundwater quality assessment and pollution control is gaining significant attention due to the increasing challenges posed by urbanization, industrialization, and climate change. Water quality indices, such as the Entropy Weighted Water Quality Index (EWQI) and the Pollution Index of Groundwater (PIG), have been used to assess water quality in various regions. Das et al. (2025b) developed an FMCDM model for human influence assessment on urban river water quality, a significant environmental MCDM work by their team. The 2025 research uses weighted hesitant FSS and FSS models (Maji et al., 2001) to rate water pollution by analyzing the Gomati and Haora Rivers (Das et al., 2024; 2025a). Das et al. (2024) propose an innovative approach using weighted FSSs for water pollution rating in the Gomati River, Tripura, India. This model integrates FSSs with weighted indices to evaluate the water quality, accounting for anthropogenic and natural influences. Their approach highlights the importance of using advanced fuzzy methods for assessing water pollution's complex and multidimensional nature (Das et al., 2024). Similarly, Adimalla (2021) uses the EWQI and PIG to assess groundwater quality in Telangana, India, for drinking purposes. By applying a fuzzy logic approach to the water quality data, the study accurately assesses groundwater suitability for various uses. Integrating FMCDM techniques allows for incorporating uncertainty and variability in water quality data, providing more reliable assessments (Adimalla, 2021).

Molodtsov (1999) developed the soft set theory, which presents an alternative framework to help decision-making in uncertain situations through flexible parameter definition. The soft set is well-suited for dealing with realistic decision tasks that encounter uncertain or incomplete information since it is an effective instrument in groundwater risk assessments. Extending the soft set theory introduced by Molodtsov in 1999 with operations for subsets, unions, intersections, and complements was discussed by Maji et al. (2003). This development made soft sets effective for MCDM applications in environmental analyses. Soft set theory (Molodtsov, 1999) gained significant relevance for practical decision-making, according to Maji et al. (2002), in their examination of real-world applications. Ali et al. (2009) and Alcantud and Santos-García (2017) applied soft set theory to different decision-making scenarios beyond insufficient or ambiguous situations. The research by Chen et al. (2020) developed soft set theory through their work on generalized vague N-soft sets. These sets provide complex tools for decision-making in doubtful environmental conditions.

Molodtsov (1999) developed soft set theory, which presents an alternative framework to help decision-making under uncertain situations through flexible parameter definition. The soft set is well suited for dealing with realistic decision tasks that encounter uncertain or incomplete information since it serves as an effective instrument in groundwater risk assessments. Extending the soft set theory introduced by Molodtsov in 1999 with operations for subsets, unions, intersections, and complements was discussed by Maji et al. (2003). This development made soft sets effective for MCDM applications in environmental analyses. Soft set theory (Molodtsov, 1999) gained significant relevance for practical decision-making, according to Maji et al. (2002), in their examination of real-world applications. Ali et al. (2009) and Alcantud and Santos-García (2017) applied soft set theory to different decision-making scenarios beyond insufficient or ambiguous situations. The research by Chen et al. (2020) developed soft set theory through their work on generalized vague N-soft sets. These sets provide complex tools for decision-making in doubtful environmental conditions.

Soft set (Molodtsov, 1999) and fuzzy set (Zadeh, 1965) integration has resulted in FSSs (Maji et al., 2001) being used effectively for uncertain and imprecise decision problems. FSSs emerged as a concept in 2001 from Maji et al. (2001) and have since shown effectiveness in managing complex decisions requiring unclear information across diverse application areas. FSSs (Maji et al., 2001) are a strong theoretical basis for uncertainty management. Roy and Maji (2007) established the initial approach for FSS utilization in decision-making through their framework development before subsequent progressions. The approach gained new dimensions when Feng et al. (2010) presented weighted FSSs because these structures enabled the management of uncertainties and decision-maker preference prioritization. The weighted assessment method proves beneficial for groundwater assessments because it allows different parameters to be weighted according to their significance levels from local pollution sources. Bhardwaj and Sharma (2021) developed an advanced uncertainty assessment method by integrating FSS theory, which Maji et al. (2001) introduced to enhance decision-making under uncertainty conditions. Researchers Das and Granados (2022a) established a fuzzy parameterized intuitionistic FSMS theory for MCDM applications. The authors of Das and Granados (2022b) developed FP-intuitionistic multi-fuzzy N-soft sets as model adaptability improvements for MCDM methods alongside their research (Das and Granados 2022b). Their presentation of IFP-intuitionistic multi-fuzzy N-soft sets expanded this research field, making them suitable for MCDM applications (Das and Granados, 2023a). Weighted FSMSs serve as a new MCDM method according to Das (2018), alongside the partial inclusion research done by Das (2016) and Das & Granados (2023b) on intuitionistic fuzzy rough relations and preference intuitionistic fuzzy rough relations. Das and Granados (2021) presented a weighted average rating fuzzy soft group MCDM model, enhancing theoretical knowledge and practical understanding of decision-making analysis. Das et al. (2022) investigated new operations on FSSs for MCDM applications, while Das et al. (2023) developed weighted hesitant bipolar-valued FSSs as an advanced method to enhance alternative evaluation procedures. The application of FSMS theories in decision-making received significant advancements from the research conducted by Mukherjee and Das (2016a). The authors combined interval-valued intuitionistic FSMSs with practical MCDM problems in 2015 (Mukherjee & Das, 2015b) before applying interval-valued intuitionistic FSSs to investment decision-making (Mukherjee & Das, 2015a). Mukherjee and Das developed Einstein operations for FSMSs in 2022 as an advancement of decision analysis tools (Mukherjee & Das, 2022). The authors have published a 2023 book examining theoretical foundations and practical implementations of FSMSs (Mukherjee & Das, 2023). The investigation of fuzzy and rough set models regarding relational and topological structures is presented in Mukherjee & Das (2015c; 2016b). The authors Granados et al. (2022) developed continuous neutrosophic distributions with neutrosophic parameters to model probabilistic uncertainty. The research work by Granados et al. (2023) created a weighted neutrosophic soft multiset solution that helps resolve decision problems involving various uncertainty levels, while Das et al. (2025c) focused on the algebraic importance of neutrosophic SuperHyper BCI-semigroups. Das et al. (2025e) proposed an innovative weighted hypersoft expert system framework for group decision-making, advancing neutrosophic and plithogenic models to better handle uncertainty and complexity in applied mathematics contexts. These investigations significantly contribute to the development of fuzzy set (Zadeh, 1965), soft set (Molodtsov, 1999), as well as FSS (Maji et al., 2001) theories that support applications in computational intelligence, environmental sustainability, and MCDM. This chapter aims to develop and demonstrate an advanced data-driven FMCDM framework for optimizing SWM, particularly in urban environments facing complex environmental challenges. By leveraging fuzzy and soft set theories, the research presents valuable methods to manage uncertainties and enhance decision accuracy in real-world applications. This approach constructs a robust, flexible methodology grounded in established theoretical foundations to address one of the foremost environmental concerns in urban settings, the sustainable and efficient management of water resources.

Recent studies highlight the growing role of fuzzy and multi-criteria decision-making approaches in addressing sustainability challenges across diverse domains. Dabral et al. (2025) employed MCDM techniques to optimize reservoir operations for sustainable water management, demonstrating how multi-criteria frameworks can balance ecological, social, and operational objectives. Keshavarz-Ghorabaee et al. (2025) proposed a fuzzy OPARA-based group decision-making method to assess sustainable solutions, emphasizing the importance of consensus-building in complex decision contexts. Extending these applications beyond water systems, Hašková et al. (2023) developed a fuzzy evaluation model for assessing manufacturing machinery in terms of sustainability, thereby showcasing the adaptability of fuzzy models in industrial decision-making. Together, these contributions reinforce the relevance of fuzzy MCDM methods in supporting sustainability-driven decisions in environmental and business settings.

Advancements of the Fuzzy Multi-Criteria Decision-Making Approach in Sustainable Water Management

The chapter "FMCDM Approach for Optimizing Sustainable Water Management" advances existing research and practice by integrating fuzzy set theory with soft set theory to form FSSs, offering a more comprehensive framework for handling vagueness and parameter uncertainty in decision-making. Unlike traditional MCDM approaches or standalone fuzzy models, this framework systematically addresses the complexities of sustainable water management by combining fuzzification, weight assignment, aggregation, and ranking into a structured methodology. Its novelty lies in extending the application of FMCDM beyond narrow technical contexts such as irrigation or wastewater treatment, toward sustainability-focused evaluation that incorporates environmental, economic, technological, and social dimensions, aligning with global SDG targets. Including a practical case study comparing rainwater harvesting, smart irrigation, and wastewater recycling bridges the gap between theory and practice, demonstrating the real-world utility of the model for policymakers and planners. Moreover, the chapter critically addresses the shortcomings of conventional models—such as rigid independence assumptions, lack of uncertainty handling, and limited stakeholder integration—by enabling expert-driven, linguistically grounded evaluations. Finally, it sets a future research agenda by highlighting opportunities for integrating FMCDM with AI, IoT-based real-time monitoring, and hybrid neutrosophic models, thereby positioning the framework as a forward-looking decision-support tool in sustainable water management.

Motivation and Challenges

Water is a finite and indispensable resource central to public health, food security, industrial development, and ecosystem services. Nevertheless, the world faces growing water stress due to increasing demand, pollution, mismanagement, and climate variability. Addressing these complex challenges requires SWM, which integrates economic, environmental, and social goals.

Decision-making in water management involves multiple stakeholders, conflicting interests, diverse criteria, and ambiguous information. For example, a wastewater treatment technology may be environmentally superior but economically unviable, or a water supply project may benefit urban populations while undermining rural water rights. Conventional decision-making models often fail to address such complexities effectively.

Role of Computational Intelligence

Computational intelligence methods, particularly fuzzy set theory (Zadeh, 1965) and its extensions, offer robust tools to model uncertainty, incorporate human judgments, and support multi-criteria evaluations. FMCDM methods provide flexibility and adaptability to capture the linguistic nuances and imprecise judgments that characterize real-world water management scenarios.

This chapter introduces an FMCDM framework that integrates:

- Fuzzy Set Theory (Zadeh, 1965): To model vagueness and partial truth.
- Soft Set Theory (Molodtsov, 1999): To handle parameterized uncertainty.
- FSSs (Maji et al., 2001): Combining the strengths of fuzzy (Zadeh, 1965) and soft sets (Molodtsov, 1999).
- Weighted Functions: To prioritize decision criteria based on importance.

The methodology is applied to evaluate sustainable alternatives in a water management context, highlighting its effectiveness in handling complexity and uncertainty.

Applications of FMCDM in Water Management

The versatility of Fuzzy Multi-Criteria Decision-Making (FMCDM) makes it highly suitable for a wide range of applications in sustainable water management. By integrating fuzzy and soft set theories, FMCDM enables decision-makers to evaluate alternatives under uncertainty, incorporate expert judgment, and balance conflicting criteria. Some of the key application areas are outlined below:

- Water Quality Assessment and Pollution Control: FMCDM can be used to develop advanced water quality indices by combining multiple physical, chemical, and biological parameters. It allows policymakers to prioritize pollution sources, design remediation strategies, and monitor urban and rural water bodies more accurately than traditional WQI models.
- Groundwater Management: Decision-making related to groundwater extraction, recharge, and contamination control often involves uncertain and incomplete data. FMCDM provides a structured framework to assess groundwater suitability for drinking, irrigation, and industrial uses while considering environmental and socio-economic impacts.
- Wastewater Treatment Technology Selection: Selecting appropriate treatment technologies involves evaluating multiple criteria such as cost, efficiency, energy consumption, and social acceptability. FMCDM helps stakeholders rank alternatives under uncertain conditions, ensuring sustainable and cost-effective wastewater management.
- Irrigation and Agricultural Water Allocation: In agriculture, FMCDM supports optimizing irrigation schedules and the equitable allocation of limited water resources among competing users. Integrating expert knowledge and environmental constraints helps balance productivity, water conservation, and sustainability.
- Urban Water Supply Planning: Urban water supply systems face challenges regarding demand growth, resource availability, and infrastructure limitations. FMCDM assists planners in selecting optimal supply options, incorporating technical feasibility and stakeholder preferences.
- Flood Risk and Water Loss Management: FMCDM can be applied to assess urban flood risks and prioritize interventions in drainage infrastructure. Similarly, it can be used for water loss management in distribution networks by evaluating leakage hotspots and identifying the most effective mitigation strategies.

- Sustainability-Oriented Decision Support: By integrating economic, environmental, technological, and social criteria, FMCDM provides holistic decision support for long-term water sustainability strategies. This aligns with the Sustainable Development Goals (SDGs), particularly Goal 6 (Clean Water and Sanitation).

THEORETICAL FOUNDATIONS

Fuzzy Set Theory

Fuzzy set theory, introduced by Zadeh (1965), is a classical (crisp) set theory generalization. Unlike classical sets, where an element belongs or does not belong to a set, fuzzy set theory (Zadeh, 1965) allows partial membership characterized by a membership function. This is particularly effective for modeling vague or imprecise concepts such as "high water quality" or "low maintenance cost". Let:

- X be a universe of discourse, i.e., the set of all possible elements under consideration.
- A be a fuzzy set on X.

Then, the fuzzy set A is mathematically represented as:

$$A = \{(x, \mu A(x)) \mid x \in X, \mu A(x) \in [0, 1]\} \quad (1)$$

where:

- x is an element of the universe X,
- $\mu A(x)$ is the membership function of the fuzzy set A,
- $\mu A(x): X \rightarrow [0,1]$,
- $\mu A(x)$ denotes the degree of membership of element x in fuzzy set A,
    - o If $\mu A(x)=1$, then x fully belongs to set A,
    - o If $A(x)=0$, then x does not belong to set A,
    - o If $0< \mu A(x)<1$, then x partially belongs to set A to the specified degree.

Fuzzy numbers, such as triangular and trapezoidal fuzzy numbers, represent linguistic variables (e.g., low, medium, high).

Fuzzy Numbers and Linguistic Variables

In practical applications, fuzzy numbers are used to model linguistic terms such as "low", "medium", and "high". Two commonly used fuzzy numbers are:

Triangular Fuzzy Number (TFN)

A TFN $A=(a,b,c)$ has the membership function:

$$\mu A\left(x\right)=\begin{cases}0, & x < a\\ \frac{x-a}{b-a}, & a \leq x \leq b\\ \frac{c-x}{c-b}, & b < x \leq c\\ 0, & x > c\end{cases} \tag{2}$$

where $a<b<c$, and b is the most plausible value.

Trapezoidal Fuzzy Number (TRFN)

A TRFN $B=(a,b,c,d)$ has the membership function:

$$\mu B\left(x\right)=\begin{cases}0, & x < a\\ \frac{x-a}{b-a}, & a \leq x < b\\ 1, & b \leq x \leq c\\ \frac{d-x}{d-c}, & c < x \leq d\\ 0, & x > d\end{cases} \tag{3}$$

where $a<b\leq c<d$, and the interval $[b,c]$ represents the core (fully possible values).

These fuzzy numbers are frequently used in FMCDM to express preferences or evaluations in a human-centric and linguistically meaningful manner.

Fuzzy Set for Water Quality In SWM

In SWM, water quality is a critical factor. However, terms like "high water quality" or "poor water quality" are inherently subjective and cannot be accurately modeled using binary logic (i.e., either acceptable or not). To address this, we use fuzzy sets (Zadeh, 1965) to represent water quality based on quantitative indicators such as Dissolved Oxygen (DO).

Let the universe of discourse be:

X=[0,14] (measured in mg/L of DO)

This range covers all possible DO levels in water samples.
Define a fuzzy set A for "High Water Quality" based on DO level:

$$A = \{(x, \mu A(x)) \mid x \in X, \mu A(x) \in [0, 1]\} \quad (4)$$

where:

- x is the DO concentration,
- $\mu A(x)$ is the degree to which the water is considered high quality.

We define a TFN A=(6,8,10) to represent "High Water Quality":

$$\mu A\left(x\right) = \begin{cases} 0, \ x \leq 6 \\ \frac{x-6}{2}, 6 < x \leq 8 \\ \frac{10-x}{2}, 8 < x < 10 \\ 0, x \geq 10 \end{cases} \quad (5)$$

This means:

- $DO \leq$ 6mg/L: not considered high quality (membership = 0),
- DO= 8mg/L: fully high quality (membership = 1),
- DO between 6 and 10 mg/L: partially high quality (membership between 0 and 1),
- $DO \geq$ 10mg/L: quality deteriorates due to supersaturation and other concerns.

Suppose we measure DO values at four locations in a river as in Table 1.

Table 1. Fuzzy evaluation of water quality based on DO levels

Location	DO (mg/L)	A(x) (High Quality)
Site A	5.5	0
Site B	7.0	(7 – 6) / 2 = 0.5
Site C	8.0	1
Site D	9.5	(10 – 9.5) / 2 = 0.25

Remark:

- Site C has optimal water quality concerning DO.
- Site B has moderate water quality.
- Site A is not suitable under the "high quality" standard.
- Despite high DO, Site D is less ideal, possibly due to over-aeration or algal activity.

Such fuzzy sets (Zadeh, 1965) help decision-makers quantify linguistic judgments, combine multiple water quality indicators, and apply FMCDM techniques, e.g., Fuzzy AHP, Fuzzy TOPSIS (Sharma et al., 2023; Zyoud et al., 2016) to select optimal water management policies, prioritize remediation sites, or assess sustainability levels.

Soft Set Theory

Soft set theory, proposed by Molodtsov (1999), is a mathematical tool for handling uncertainty and vagueness in decision-making problems, especially when the available data is imprecise or insufficient. Unlike fuzzy set theory or probability theory, soft set theory (Molodtsov, 1999) does not require predefined membership grades or probability distributions, making it a more flexible framework for multi-parameter environments.

Let:

- U be a universe of discourse (a set of objects under consideration),
- E be a set of parameters (attributes, features, or criteria relevant to the decision problem),
- $P(U)$ denotes the power set of U, i.e., the set of all subsets of U.

A soft set (F, E) over U is defined as a set-valued function:

$$F: E \rightarrow P(U)$$

That is, for each parameter $e \in E$, the function F assigns a subset $F(e) \subseteq U$, which contains the elements of U that are compatible with or satisfy the parameter E.

The soft set can thus be written as a collection of ordered pairs:

$$F = \{(e, F(e)) \mid e \in E, F(e) \subseteq U\} \tag{6}$$

Remark:

- Each pair $(e, F(e))$ represents the approximate description of the elements of U under parameter E.
- There is no restriction on the nature of the subsets $F(e)$; they may overlap, be disjoint, or even be empty.
- Uncertainty is captured by considering multiple parameters, each potentially identifying a different subset of relevant elements.

Soft Set for Evaluating Water Treatment Plants

In SWM, selecting the best water treatment plant involves evaluating multiple criteria (parameters) such as cost, energy usage, and purification efficiency. However, the available data may be incomplete or imprecise. Soft set theory provides a way to model such decision-making problems without relying on probabilistic or fuzzy membership values.

Let the universe of discourse $U = \{u1, u2, u3, u4\}$ be a set of water treatment plants. Each $ui \in U$ represents a distinct plant being evaluated.

Let $E = \{e1, e2, e3\}$ be the set of parameters (attributes) relevant to sustainable water treatment, where:

- $e1$: "low operational cost"
- $e2$: "low energy consumption"
- $e3$: "high purification efficiency"

We define a soft set (F, E) over U. For each parameter $e \in E$, we associate a subset of U containing the plants that satisfy that parameter (based on expert judgment or qualitative analysis).

Let:

- $F(e1) = \{u1, u2\}$
  (Plants $u1$ and $u2$ are known to have low operational costs)
- $F(e2) = \{u2, u3\}$
  (Plants $u2$ and $u3$ are energy efficient)
- $F(e3) = \{u1, u3, u4\}$
  (Plants $u1, u3$, and $u4$ have high purification efficiency)
- Thus, the soft set F can be written as:

$$F = \{(e1, \{u1, u2\}), (e2, \{u2, u3\}), (e3, \{u1, u3, u4\})\} \quad (7)$$

Table 2 summarizes the association between each water treatment plant and the parameters defined in the soft set, highlighting which plants satisfy which criteria.

Table 2. Soft set evaluation of water treatment plants

Plant	Low Cost ($e1$)	Low Energy ($e2$)	High Efficiency ($e3$)
$u1$	1	0	1
$u2$	1	1	0
$u3$	0	1	1
$u4$	0	0	1

- Plant $u1$ meets $e1$ and $e3$: economically and effectively.
- Plant $u2$ meets $e1$ and $e2$ economically and energy-efficiently.
- Plant $u3$ meets $e2$ and $e3$: energy-efficient and highly effective.
- Plant $u3$ meets only $e3$: high efficiency, but lacks cost and energy performance.

Soft sets allow decision-makers to:

- Analyze which plants meet which combination of criteria,
- Apply filters or priority weightings later (e.g., through soft set-based decision rules or hybrid methods),
- Make flexible, parameter-driven decisions without requiring numeric scores or membership functions.

Fuzzy Soft Sets

Fuzzy soft set (FSS) (Maji et al., 2001) theory integrates the advantages of fuzzy sets (Zadeh, 1965) and soft sets (Molodtsov, 1999). It allows us to handle uncertainty arising from the gradual nature of membership (fuzziness) and the parameterization of the data (softness). This is especially useful in complex decision-making scenarios like SWM, where the criteria and data precision vary.

Let:

- X be a universe of discourse (e.g., a set of water treatment plants),
- E be a set of parameters (e.g., cost, efficiency, environmental impact),
- $F(x)$ denotes the set of all fuzzy subsets of X.

A FSS (F,E) over X is defined as a function:

$$F: E \rightarrow F(X)$$

where each parameter $e \in E$ is associated with a fuzzy subset $F(e)$ of X.

In other words, for each $e \in E$, $F(e)$ is a fuzzy set in X, i.e.,

$$F(e) = \{(x, \mu F(e)(x)) \mid x \in X, \mu F(e)(x) \in [0,1]\} \tag{8}$$

So, the FSS (F,E) can be written as a collection of fuzzy sets indexed by parameters:

$$(F, E) = \{(e, \{(x, \mu F(e)(x)) \mid x \in X\}) \mid e \in E\} \tag{9}$$

Remark:

- $x \in X$ is an element of the universe (e.g., a water treatment plant),
- $e \in E$ is a parameter (e.g., low energy use),
- $\mu F(e)(x)$ is the membership degree of element x to fuzzy set $F(e)$, representing the degree to which x satisfies parameter E.

Fuzzy Soft Sets in SWM

Let:

- $X = \{u1, u2, u3\}$ be water treatment plants,
- $E = \{e1, e2\}$, where:
    - o $e1$: "low energy consumption"
    - o $e2$: "high purification efficiency"

We define the FSS (F,E) as follows:
For $e1$ (low energy consumption):

$$F(e1) = \{(u1, 0.8), (u2, 0.6), (u3, 0.3)\} \tag{10}$$

For $e2$ (high purification efficiency):

$$F(e2) = \{(u1, 0.5), (u2, 0.9), (u3, 0.7)\} \tag{11}$$

So, the full FSS is:

$$(F, E) = \{(e1, \{(u1, 0.8), (u2, 0.6), (u3, 0.3)\}), (e2, \{(u1, 0.5), (u2, 0.9), (u3, 0.7)\})\} \tag{12}$$

Remark:

- Under the low energy criterion ($e1$), $u1$ is highly suitable (0.8), while $u3$ is less suitable (0.3).
- Under the high purification efficiency criterion ($e2$), $u2$ is the most effective (0.9).

This structure enables graded, parameterized evaluations of alternatives and can be directly used in FSS-based decision-making models.

Weighted Functions in MCDM

In MCDM, multiple conflicting criteria are used to evaluate alternatives. Since not all criteria are equally important, weights are assigned to reflect their relative significance.

Weight Vector

Let:

- $C = \{C1, C2, \ldots, C_n\}$ be the set of decision criteria, where n is the number of criteria.
- $W = \{w1, w2, \ldots, wn\}$ be the weight vector, where wj is the weight of criterion Cj

Each weight satisfies:

$wj \in [0,1]$ for all j = 1, 2, ..., n and the total sum of all weights is:

$$\sum_{j=1}^{n} wj = 1 \tag{13}$$

Remark

- wj represents the relative importance of criterion Cj in the decision-making process.
- A higher value of wj implies greater importance of the corresponding criterion.
- The constraint $\sum_{j=1}^{n} wj = 1$ ensures a normalized importance distribution.

Weights can be derived using various methods:

- Expert Judgment: Subjective assignment based on expert opinions.
- Entropy Method: An objective method that uses data variability to compute weights.
- AHP: Structured pairwise comparison approach based on Saaty's scale.
- Fuzzy AHP / Fuzzy Entropy: Extensions for handling uncertainty in judgment.

Example 2.8.2. SWM

Suppose we are evaluating water treatment plants using 3 criteria:

- $C1$: Operational cost
- $C2$: Energy efficiency
- $C3$: Purification efficiency

Let the weights be:

$$w = \{w1 = 0.4, w2 = 0.3, w3 = 0.3\}$$

This means:

- Operational cost is the most important criterion (40%),
- Energy and purification efficiency are equally important (30% each).

Use in Decision Score Calculation

Let ai be an alternative (e.g., a water treatment plant), and sij be the performance score of ai concerning criterion Cj (e.g., membership value or utility).

Then, the overall weighted score of alternative ai is:

$$S\left(ai\right) = \sum_{j=1}^{n} wj \cdot sij \tag{14}$$

This score $S(ai)$ is used to rank the alternatives.

Methodological Framework

The proposed FMCDM model for SWM consists of the following mathematically defined steps:

Step 1: Problem Definition

Let:

- $A = \{a1, a2, \ldots, am\}$ be the set of alternatives (e.g., water treatment methods or projects),
- $C = \{c1, c2, \ldots, cn\}$ be the set of evaluation criteria.

For SWM, typical criteria may include:

$$C = \{WQI, CE, TF, ES, SA\}$$

Where:

- WQI: Water Quality Improvement
- CE: Cost Efficiency
- TF: Technological Feasibility
- ES: Environmental Sustainability
- SA: Social Acceptability

Step 2: Fuzzification of Data

Let expert linguistic evaluations (e.g., "low", "medium", "high") be mapped to fuzzy numbers, often using triangular or trapezoidal membership functions.

Define the fuzzy scale as:

- $Low \rightarrow L = (0, 0, 0.25)$
- $Medium \rightarrow M = (0.25, 0.5, 0.75)$
- $High \rightarrow H = (0.75, 1, 1)$

These are examples of TFN $x=(l,,m,u)$, with lower, modal, and upper bounds, respectively.

Step 3: Construct FSS

For each criterion $cj \in C$, construct a FSS:

$$F(cj) = \{(ai, \mu ij) \mid i = 1,2,\ldots,m\} \tag{15}$$

where $\mu ij \in [0,1]$ is the fuzzy membership value (or defuzzified score) indicating how well alternative *ai* satisfies criterion *Cj*.

Thus, the complete FSS is:

$$F = \{(cj, F(cj)) \mid j = 1,2,\ldots,n\} \tag{16}$$

Or more explicitly as a matrix:

$$[\mu ij]_{m\times n} = \begin{bmatrix} \mu 11 & \mu 12 & \cdots & \mu 1n \\ \mu 21 & \mu 22 & & \mu 23 \\ \vdots & \vdots & \ddots & \vdots \\ \mu m1 & \mu m2 & \cdots & \mu m3 \end{bmatrix} \tag{17}$$

Step 4: Weight Assignment

Assign a weight vector:

$$w = \left\{w1, w2, \ldots, wn\right\}, wj \in \left[0, 1\right], \sum_{j=1}^{n} wj = 1 \tag{18}$$

Weights can be derived via:

- Expert input
- Entropy method
- AHP (Analytic Hierarchy Process)
- Fuzzy AHP or other hybrid models

Step 5: Aggregation

Compute the overall fuzzy score (or defuzzified crisp score) $S(ai)$ for each alternative $ai \in A$ using weighted averaging:

$$S(ai) = \sum_{j=1}^{n} wj \cdot sij, \forall i = 1, 2, \ldots, m \tag{19}$$

Where:

- μij is the performance of ai under criterion *Cj*,
- *wj* is the weight of criterion *Cj*.

This yields a scalar aggregated score $S(ai) \in [0,1]$ for each alternative.

Step 6: Ranking

Rank all alternatives $ai \in A$ in descending order of $S(ai)$:

$$S(a(1)) \geq S(a(2)) \geq \cdots \geq S(a(m)) \quad (20)$$

Where $a(1)$ is the best-ranked alternative.

Optionally, perform a sensitivity analysis by varying weights wj or membership values μij to test the robustness and stability of the ranking.

Practical Illustrations of FMCDM in Water Management

- Addressing Water Crises through FMCDM: Real-world water crises such as prolonged droughts in Sub-Saharan Africa, groundwater depletion in India, and inequitable access in urban slums highlight the need for structured decision-making under uncertainty. FMCDM offers practical solutions by allowing decision-makers to weigh multiple factors—such as water availability, cost, and social equity—under fuzzy environments. For instance, during drought conditions, FMCDM can help prioritize between groundwater recharge projects, rainwater harvesting systems, or desalination plants by balancing economic feasibility with sustainability.
- Managing Multi-Stakeholder Conflicts: Water allocation often involves conflicting interests among farmers, urban consumers, industries, and environmental groups. For example, irrigation may maximize crop yield but reduce water availability for domestic use or harm aquatic ecosystems. FMCDM incorporates diverse perspectives using fuzzy group decision-making and consensus-building techniques such as participatory MCDM or Delphi methods. Each stakeholder's preferences can be represented as fuzzy weights, enabling a negotiated solution that reflects collective priorities rather than a single dominant viewpoint.
- Illustrative Case Study: Consider a hypothetical urban river system facing pollution from industries, declining groundwater tables, and frequent flooding. Three management strategies are proposed:
    - Rainwater harvesting for recharge
    - Upgrading wastewater treatment plants
    - Smart irrigation for peri-urban agriculture

Using FMCDM, criteria such as water quality improvement (WQI), cost efficiency (CE), technological feasibility (TF), environmental sustainability (ES), and social acceptability (SA) are fuzzified and evaluated. After aggregation and ranking, rainwater harvesting emerges as the top option due to its low cost, high social acceptability, and positive environmental impact. At the same time, wastewater treatment ranks lower due to high costs despite strong water quality benefits. This example demonstrates how FMCDM translates complex, competing challenges into transparent, evidence-based rankings, guiding policymakers toward feasible and sustainable solutions.

Example 2.11: Select the most suitable SWM strategy from a set of alternatives, using FSSs and weighted criteria.

We are evaluating three SWM alternatives $A= \{a1,a1,a3\}$ where:

$va1$: Rainwater Harvesting System
$va2$: Advanced Wastewater Recycling Plant
$va3$: Smart Irrigation System

Criteria $C= \{c1,c2,c3,c4,c5\}$ where:

$vc1$: Water Quality Imporvement (WQI)
$vc2$: Cost Efficiency (CE)
$vc3$: Technological Feasibility (TF)
$vc4$: Environmental Sustainability (ES)
$vc5$: Social Acceptability (SA)

Experts rate the alternatives using linguistic terms mapped to TFN for each alternative against the criteria as presented in Table 3

Table 3. Linguistic ratings of alternatives

Linguistic Term	TFN
Low	(0.0, 0.2, 0.4)
Medium	(0.3, 0.5, 0.7)
High	(0.6, 0.8, 1.0)

For simplicity, we will defuzzify the triangular numbers using the centroid method:

$$Defuzzified\ value = (l + m + u) / 3 \tag{21}$$

Based on expert evaluations, each alternative is assessed under the five criteria using linguistic terms presented in Table 4. The FSS is represented as a matrix [

$\mu ij]_{3\times5}$ where each entry corresponds to the defuzzified membership value of an alternative under a specific criterion.

Table 4. FSS – Defuzzified evaluations of SWM alternatives

Alternative	*WQI*(*c*1)	*CE*(*c*2)	*TF*(*c*3)	*ES*(*c*4)	*SA*(*c*5)
*a*1	High → 0.8	Medium → 0.5	Medium → 0.5	High → 0.8	High → 0.8
*a*2	High → 0.8	Low → 0.2	High → 0.8	Medium → 0.5	Medium → 0.5
*a*3	Medium → 0.5	High → 0.8	High → 0.8	Medium → 0.5	High → 0.8

Matrix $[\mu ij]_{3\times5}$:

$$[\mu ij]_{3\times5} = \begin{bmatrix} 0.8 & 0.5 & 0.5 & 0.8 & 0.8 \\ 0.8 & 0.2 & 0.8 & 0.5 & 0.5 \\ 0.5 & 0.8 & 0.8 & 0.5 & 0.8 \end{bmatrix} \tag{22}$$

Let stakeholder-derived weights be:

$w = \{w1 = 0.25, w2 = 0.2, w3 = 0.2, w4 = 0.2, w5 = 0.15\}$

These satisfy:

$$\sum_{j=1}^{5} wj = 1$$

.00 (23)

Calculate aggregated scores

$$S\left(ai\right) = \sum_{j=1}^{n} wj \cdot sij \tag{24}$$

For $a1$

$S(a1)$= 0.25·0.8 + 0.2·0.5 + 0.2·0.5 + 0.2·0.8 + 0.15·0.8 = 0.20 + 0.10 + 0.10 + 0.16 + 0.12

= 0.68

For *ai2*

$S(a2)= 0.25 \cdot 0.8 + 0.2 \cdot 0.2 + 0.2 \cdot 0.8 + 0.2 \cdot 0.5 + 0.15 \cdot 0.5$

$= 0.20 + 0.04 + 0.16 + 0.10 + 0.075$

$= 0.575$

For $a3$

$S(a3)= 0.25 \cdot 0.5 + 0.2 \cdot 0.8 + 0.2 \cdot 0.8 + 0.2 \cdot 0.5 + 0.15 \cdot 0.8 = 0.125 + 0.16 + 0.16 + 0.10 + 0.12$

$= 0.665$

Final ranking based on scores:

1. $a1$: 0.68→ Rainwater Harvesting System
2. $a3$: 0.665→ Smart Irrigation System
3. $a2$: 0.575→ Advanced Wastewater Recycling Plant

The FMCDM method suggests Rainwater Harvesting as the optimal SWM strategy based on expert-driven fuzzy assessments and stakeholder-weighted criteria.

RESULTS AND DISCUSSION

This study applied an FMCDM approach integrated with FSSs (Maji et al., 2001) to evaluate and prioritize SWM strategies. Three alternatives, $a1$, $a2$, and $a3$, were assessed against five decision criteria: WQI, CE, TF, ES, and SA.

Results Summary

The performance of each alternative was evaluated using fuzzy linguistic assessments provided by experts. These assessments were transformed into fuzzy numbers and defuzzified using the centroid method. The criteria weights, derived through expert judgment, reflected their relative importance in decision-making.

The final aggregated scores for each alternative were computed using a weighted averaging method:

- $S(a1)= 0.680$ → Rainwater Harvesting System
- $S(a3)= 0.665$ → Smart Irrigation System

- $S(a2)$= 0.575 → Advanced Wastewater Recycling Plant

These results indicate that the Rainwater Harvesting System is the most favorable alternative, offering a balanced performance across all five sustainability-oriented criteria.

Implications

Rainwater Harvesting ($a1$) emerged as the most optimal solution, primarily due to its strong performance in ES and SA. It is low-tech, cost-effective, and highly feasible in urban and rural settings, explaining its broad appeal.

- While nearly as effective, Smart Irrigation ($a3$) excelled in TF and CE. It represents a strong candidate, especially in agriculture-driven regions, where water conservation through precision irrigation is critical.
- Advanced Wastewater Recycling ($a2$) although scoring high in WQI and TF, received lower evaluations for Cost Efficiency and Social Acceptability. These results highlight practical barriers to large-scale implementation, such as high capital costs and public perception challenges.

Benefits of FMCDM

The FMCDM approach offers several advantages, especially when applied to complex decision-making problems involving uncertainty, vagueness, and subjective judgments. Some of the key benefits of FMCDM are as follows:

Handling Uncertainty and Vagueness

FMCDM allows for the inclusion of fuzzy logic, which enables the representation of uncertainty, vagueness, and imprecision inherent in human decision-making. Many real-world problems, such as those in SWM, involve qualitative data (e.g., "high," "low," "medium") that cannot be precisely quantified. Using fuzzy numbers and membership functions, FMCDM helps model such uncertainties and makes decision-making more realistic.

Flexibility in Criteria Evaluation

FMCDM provides flexibility in evaluating a range of decision criteria, which can be quantitative (e.g., costs) or qualitative (e.g., social acceptability). Linguistic variables can be used to describe criteria in a manner easily interpretable by stakeholders, without the need for exact numeric values. This is particularly useful when dealing with subjective or expert judgments, as it aligns with human reasoning more closely than traditional crisp sets.

Incorporation of Expert Judgment

FMCDM allows for the inclusion of expert opinions in the decision-making process. Decision criteria are typically weighted based on the perceived importance, and experts can provide fuzzy evaluations of alternatives based on their experience and knowledge. Incorporating expert knowledge makes the decision process more robust, especially when hard data is unavailable.

Support for Multiple Decision-Makers

FMCDM can easily accommodate multiple decision-makers or stakeholders, each with their own preferences and judgments. Using fuzzy sets and weighted functions, it is possible to aggregate and harmonize the preferences of different decision-makers, ensuring that the final decision reflects the collective opinion.

Enhanced Decision Quality

By incorporating fuzzy logic and multi-criteria analysis, FMCDM enhances the quality of the final decision. It accounts for the quantitative and qualitative dimensions of alternatives, providing a more comprehensive evaluation of the options. This leads to more informed, balanced decisions, especially in complex fields like SWM, where trade-offs between environmental, social, economic, and technological factors must be considered.

Ability to Model Complex Problems

FMCDM is particularly beneficial for complex, multi-dimensional decision problems, where multiple conflicting objectives need to be optimized simultaneously. In the context of SWM, factors such as water quality, cost efficiency, environmental sustainability, and social acceptability need to be considered together. FMCDM's ability to process these diverse criteria enables decision-makers to identify solutions that best satisfy all objectives.

Improved Robustness of Decisions

FMCDM provides a robust framework that helps mitigate the effects of small uncertainties in the decision-making process. Through fuzzy logic, minor changes in input data or criteria weights typically have little impact on the final decision, making the process more stable and reliable.

Transparency in Decision-Making

FMCDM makes the decision process more transparent by clearly showing how alternatives are evaluated against each criterion and how the final scores are aggregated. This transparency is crucial in fields such as water management, where public and stakeholder trust in the decision-making process is essential.

Adaptability to Different Contexts

FMCDM is highly adaptable to various decision-making scenarios and can be tailored to meet the specific needs of different problems. Whether the decision-making process involves environmental assessments, policy decisions, or strategic planning, FMCDM provides a flexible tool for integrating diverse types of information and preferences.

FMCDM is a powerful approach that facilitates more informed, reliable, and transparent decision-making. It is especially beneficial in complex, multi-faceted problems where uncertainty, subjectivity, and conflicting criteria must be addressed.

Incorporating Multi-Stakeholder Preferences in FMCDM

Decision-making in sustainable water management often involves multiple stakeholders with differing, and sometimes conflicting, objectives—such as policymakers, engineers, farmers, environmental groups, and community representatives. FMCDM can be extended to incorporate multi-stakeholder preferences through fuzzy group decision-making approaches, where linguistic assessments from diverse actors are aggregated into a collective evaluation. Consensus-building techniques such as the Delphi method and participatory MCDM frameworks can be integrated to iteratively refine judgments, reduce bias, and align decisions with collective priorities. By assigning weights not only to criteria but also to stakeholder groups, FMCDM models ensure that marginalized voices (e.g., rural communities or indigenous populations) are included in the decision process. This participatory integration makes FMCDM more democratic, transparent, and context-sensitive, ultimately leading to more socially acceptable and implementable water management strategies.

Enhancing FMCDM with GIS-Based Spatial Analysis

Another promising direction for FMCDM in water management is its integration with Geographic Information Systems (GIS). Many SWM problems—such as site selection for treatment plants, flood risk assessment, or spatial prioritization of conservation efforts—are inherently spatial. By embedding GIS data layers (e.g., land use, hydrology, pollution sources, socio-economic maps) into FMCDM models, decision-makers can visualize and analyze spatial trade-offs more effectively. Approaches like GIS-Fuzzy AHP and GIS-Fuzzy TOPSIS have already been applied in environmental and urban planning to support geospatial sustainability mapping, demonstrating their ability to handle location-specific criteria and fuzzy uncertainty. In the context of SWM, such GIS-FMCDM integration would allow stakeholders to identify optimal sites for rainwater harvesting structures, prioritize vulnerable groundwater recharge zones, or map socially acceptable regions for wastewater reuse projects. This spatial dimension strengthens FMCDM by making its recommendations multidimensional and geographically grounded, providing actionable insights for planners and policymakers.

Emerging Technologies and Future Trends in FMCDM

While FMCDM offers a robust framework for managing uncertainty in water-related decision-making, its effectiveness can be significantly enhanced by integrating emerging digital technologies. AI-augmented decision-making enables automated tuning of membership functions, adaptive learning from new data, and predictive modeling of water demand or pollution scenarios. Similarly, big data and IoT integration can allow FMCDM models to incorporate real-time sensor data on water quality, rainfall, and consumption, transforming static evaluations into dynamic, adaptive decision systems. Climate-resilient infrastructure planning is another domain where FMCDM can be extended by integrating hydrological forecasts, urban climate models, and resilience indicators into the criteria set. Combining FMCDM with machine learning, neural networks, or blockchain-enabled water governance could increase transparency, adaptability, and reliability. Such hybrid systems move beyond traditional decision support, evolving toward next-generation intelligent water management platforms.

Alignment with Global Frameworks and Policies

Another important direction for FMCDM is its alignment with global sustainability and policy frameworks. Sustainable Water Management is central to UN Sustainable Development Goal 6 (Clean Water and Sanitation), which emphasizes universal access to safe water, efficient use, and protection of ecosystems. FMCDM models can incorporate SDG indicators directly into their evaluation criteria, ensuring that selected strategies contribute to global sustainability benchmarks. Similarly, Integrated Water Resources Management (IWRM) principles—balancing social, economic, and environmental needs—are inherently multi-criteria in nature and can be operationalized through FMCDM. In climate change adaptation, FMCDM can support water managers in prioritizing resilient infrastructure projects, identifying vulnerable communities, and evaluating low-carbon water supply technologies. Explicitly linking FMCDM applications to SDGs, IWRM frameworks, and national climate adaptation strategies not only strengthens their policy relevance but also increases their acceptance among international organizations, funding agencies, and local governments.

Limitations

While FMCDM is a valuable tool for handling complex, multi-criteria decisions under uncertainty, it has several limitations:

- Model Complexity: Constructing fuzzy sets, membership functions, and aggregating multiple criteria can be computationally intensive and time-consuming.
- Subjectivity: Defining membership functions and assigning weights relies heavily on expert judgment, introducing potential bias and inconsistency.
- Fuzzification Challenges: Converting qualitative data into fuzzy values can be ambiguous, especially when expert opinions vary.
- Interdependencies Ignored: FMCDM assumes that the criteria are independent and may not hold in real-world scenarios.
- Lack of Standardization: There are no universal rules for fuzzy aggregation, leading to variability in implementation and results.
- Sensitivity Issues: Small changes in fuzzy parameters or weights can significantly affect outcomes.
- Interpretability: Final results may be difficult to interpret without further defuzzification or analysis, reducing clarity for decision-makers.

Despite these challenges, FMCDM remains a robust decision-support tool when carefully applied.

Future Directions in FMCDM

The FMCDM framework is evolving to meet the growing need for more adaptive and intelligent decision-support systems. Key future directions include:

- AI and Machine Learning Integration: Enhancing fuzzification, optimizing membership functions, and enabling data-driven decision-making.
- Hybrid Models: Combining FMCDM with advanced soft set theories (e.g., Neutrosophic, Intuitionistic, Bipolar fuzzy sets) for richer uncertainty handling.
- Real-Time Applications: Leveraging IoT and big data to develop dynamic, real-time decision support systems in domains like water resource management.
- Multi-Stakeholder & Multi-Objective Support: Addressing conflicting goals across diverse stakeholder groups with better consensus and negotiation mechanisms.
- Improved Sensitivity and Robustness Analysis: Strengthening decision reliability by comprehensively evaluating parameter impact and uncertainty.
- Sustainability Integration: Aligning FMCDM with global frameworks like the SDGs and Life Cycle Assessment for holistic sustainability evaluation.
- Advanced Defuzzification: Developing more refined methods to translate fuzzy outputs into actionable insights with minimal information loss.
- New Application Domains: Expanding FMCDM use in climate policy, smart cities, healthcare, and other emerging fields requiring complex trade-offs.
- User-Friendly Tools: Creating intuitive software platforms to make FMCDM accessible to non-experts, facilitating broader adoption.

These directions will help FMCDM become a more robust, adaptable, and widely used approach for sustainable MCDM.

CONCLUSION

This chapter aimed to explore the application of FMCDM in Optimizing SWM. We have provided a detailed understanding of FMCDM, beginning with the foundational concepts of fuzzy set theory, soft set theory, and their integration into FSSs. These theories form the backbone of FMCDM, enabling decision-makers to model uncertainty, subjectivity, and multiple criteria when making complex decisions.

By highlighting the step-by-step methodology for applying FMCDM, including problem definition, fuzzification, FSS construction, weight assignment, aggregation, and ranking, we outlined a practical approach to evaluate alternatives in water management. We discussed a case study application to demonstrate how FMCDM can assess various water management alternatives based on criteria such as WQI, CE, TF, Environmental Sustainability, and SA.

The results demonstrate that FMCDM accommodates qualitative judgments alongside quantitative metrics and yields more balanced recommendations for SWM strategies. The framework's flexibility in incorporating expert judgment, handling uncertainty, and aggregating various criteria makes it a powerful tool for optimizing SWM strategies.

However, as with any decision-making framework, FMCDM has its limitations, including the complexity of model construction, defining membership functions, subjectivity in weight assignment, and sensitivity to fuzzy parameters. Despite these challenges, the benefits of FMCDM, such as its ability to deal with uncertainty, provide insights into complex decision scenarios, and support multi-criteria evaluations, make it a valuable tool in real-world applications.

Future directions for FMCDM include its integration with machine learning and AI, hybridization with other soft set theories, the development of real-time decision support systems, and the application to new and emerging domains like climate change adaptation and smart cities. As FMCDM continues to evolve, its ability to handle complex, multi-dimensional decision-making problems will become even more vital, enabling more effective, sustainable, and informed decision-making across various sectors.

REFERENCES

Adimalla, N. (2021). Application of the Entropy Weighted Water Quality Index (EWQI) and the Pollution Index of Groundwater (PIG) to Assess Groundwater Quality for Drinking Purposes: A Case Study in a Rural Area of Telangana State, India. *Archives of Environmental Contamination and Toxicology*, *80*, 31–40. PMID: 33388839

Afshar, A., Mariño, M., Saadatpour, M., & Afshar, A. (2011). Fuzzy TOPSIS Multi-Criteria Decision Analysis Applied to Karun Reservoirs System. *Water Resources Management*, *25*, 545–563. DOI: 10.1007/S11269-010-9713-X

Al-Falal, A. N. A., Elsayed, S., El Fadaly, E. A., Gaagai, A., & Aouissi, H. A., Abd El-baki, M. S., Eid, M. H., Elwakeel, A. E. E., Yaseen, Z. M., Elsherbiny, O., Eltahir, E. I., & Gad, M. (2025). Aquatic system assessment of potentially toxic elements in El Manzala Lake, Egypt: A statistical and machine learning approach. *Results in Engineering*, *26*, 105027. DOI: 10.1016/j.rineng.2025.105027

Alcantud, J. C. R., & Santos-García, G. (2017). A new criterion for soft set based decision-making problems under incomplete information. *International Journal of Computational Intelligence Systems*, *10*, 394–404.

Ali, M. I., Feng, F., Liu, X. Y., Min, W. K., & Shabir, M. (2009). On some new operations in soft set theory. *Computers & Mathematics with Applications (Oxford, England)*, *57*(9), 1547–1553.

Bhardwaj, N., & Sharma, P. (2021). An advanced uncertainty measure using fuzzy soft sets: Application to decision-making problems. *Big Data Mining and Analytics*, *4*(2), 94–103.

Chen, Y., Liu, J., Chen, Z., & Zhang, Y. (2020). Group decision-making method based on generalized vague N-soft sets. In: *Chinese Control and Decision Conference*, 4010–4015.

Dabral, R., Sandhu, H., Trivedi, M., & Cherubini, C. (2025). Optimal operation of reservoir for sustainable water management using multi-criteria decision-making approach. *Journal of Hydroinformatics*. Advance online publication.

Das, A. K. (2016). On partially included intuitionistic fuzzy rough relations. *Afrika Matematika*, *27*, 993–1001. DOI: 10.1007/s13370-016-0395-2

Das, A. K. (2018). Weighted fuzzy soft multiset and decision-making. *International Journal of Machine Learning and Cybernetics*, *9*, 787–794. DOI: 10.1007/s13042-016-0607-y

Das, A. K., Das, R., Das, S., Debnath, B. K., Granados, C., Shil, B., & Das, R. (2025c). A comprehensive study of neutrosophic SuperHyper BCI-semigroups and their algebraic significance. *Transactions on Fuzzy Sets and Systems*, *4*(2), 80–101.

Das, A. K., Das, R., Datta, R., & Granados, C. (2025d). An Innovative Approach to Group Decision-Making Based on Weighted Hypersoft Expert System. In Smarandache, F., Jdid, M., & Leyva-Vazquez, M. (Eds.), *Neutrosophic and Plithogenic Inventory Models for Applied Mathematics* (pp. 483–518). IGI Global Scientific Publishing., DOI: 10.4018/979-8-3693-3204-7.ch019

Das, A. K., & Granados, C. (2021). An advanced approach to fuzzy soft group decision-making using weighted average ratings. *SN Computer Science*, *2*, 471. DOI: 10.1007/s42979-021-00873-5

Das, A. K., & Granados, C. (2022a). A new fuzzy parameterized intuitionistic fuzzy soft multiset theory and group decision-making. *Journal of Current Science and Technology*, *12*(3), 547–567. DOI: 10.14456/jcst.2022.42

Das, A. K., & Granados, C. (2022b). FP-intuitionistic multi fuzzy N-soft set and its induced FP-Hesitant N soft set in decision-making. *Decision Making: Applications in Management and Engineering*, *5*(1), 67–89. DOI: 10.31181/dmame181221045d

Das, A. K., & Granados, C. (2023a). IFP-intuitionistic multi fuzzy N-soft set and its induced IFP-hesitant N-soft set in decision-making. *Journal of Ambient Intelligence and Humanized Computing*, *14*, 10143–10152. DOI: 10.1007/s12652-021-03677-w

Das, A. K., & Granados, C. (2023b). Preference intuitionistic fuzzy rough relation and its theoretical approach. *Journal of the Indian Mathematical Society*, *90*(3-4), 199–212.

Das, A. K., Granados, C., & Bhattacharya, J. (2022). Some new operations on fuzzy soft sets and their applications in decision-making. *Songklanakarin Journal of Science and Technology*, *44*(2), 440–449. DOI: 10.14456/sjst-psu.2022.61

Das, A. K., Gupta, N., & Granados, C. (2023). Weighted hesitant bipolar-valued fuzzy soft set in decision-making. *Songklanakarin Journal of Science and Technology*, *45*(6), 681–690. https://sjst.psu.ac.th/journal/45-6/10.pdf

Das, A. K., Gupta, N., Mahmood, T., Tripathy, B. C., Das, R., & Das, S. (2024). Assessing anthropogenic influences on the water quality of Gomati River using an innovative weighted fuzzy soft set based water pollution rating system. *Discover Water*, *4*, 73. DOI: 10.1007/s43832-024-00136-3

Das, A. K., Gupta, N., Mahmood, T., Tripathy, B. C., Das, R., & Das, S. (2025a). An efficient water quality evaluation model using weighted hesitant fuzzy soft sets for water pollution rating. In Kumar, A., Kumar, P., Rathee, S., & Kumar, B. (Eds.), *Mechatronics: Concepts, Tools, Applications, and New Trends* (1st ed., p. 17). CRC Press., DOI: 10.1201/9781003494478-10

Das, A. K., Gupta, N., Mahmood, T., Tripathy, B. C., Das, R., & Das, S. (2025b). An innovative fuzzy multi-criteria decision making model for analyzing anthropogenic influences on urban river water quality. *Iran Journal of Computer Science, 8*, 103–124. DOI: 10.1007/s42044-024-00211-x

Elleuch, M., Anane, M., Euchi, J., & Frikha, A. (2019). Hybrid fuzzy multi-criteria decision making to solve the irrigation water allocation problem in the Tunisian case. *Agricultural Systems*. Advance online publication. DOI: 10.1016/J.AGSY.2019.102644

Feng, F., Jun, Y. B., Liu, X., & Li, L. (2010). An adjustable approach to fuzzy soft set based decision making. *Journal of Computational and Applied Mathematics, 234*, 10–20.

Granados, C., Das, A.K., Das, B. (2022). Some continuous neutrosophic distributions with neutrosophic parameters based on neutrosophic random variables. Advances in the Theory of Nonlinear Analysis and its Applications, 33(2): 293-308. https://doi.org/DOI: 10.31197/atnaa.1056480

Granados, C., Das, A.K., Osu, B. (2023). Weighted neutrosophic soft multiset and its application to decision making. Yugoslav journal of operations research, 33(2): 293-308.

Hashim, S., Yuebo, X., Saifullah, M., Jan, R. N., & Muhetaer, A. (2015). Integrated evaluation of urban water bodies for pollution abatement based on fuzzy multicriteria decision approach. *BioMed Research International, 2015*, 327280. Advance online publication. DOI: 10.1155/2015/327280 PMID: 26516623

Hašková, S., Vochozka, M., & Kučera, J. (2023). *A fuzzy evaluation model of manufacturing machinery in terms of sustainable business*. Entrepreneurship and Sustainability Issues.,

Keshavarz-Ghorabaee, M., Mohammadi-Ostadkalayeh, A., Amiri, M., & Antuchevičienė, J. (2025). A Fuzzy OPARA-Based Group Decision-Making Approach: Application to Sustainable Solutions Assessment. *Informatica (Vilnius)*. Advance online publication. DOI: 10.15388/25-infor586

Maji, P. K., Biswas, R., & Roy, A. R. (2001). Fuzzy soft sets. *Journal of Fuzzy Mathematics*, *9*(3), 589–602.

Maji, P. K., Biswas, R., & Roy, A. R. (2002). An application of soft sets in decision-making problem. *Computers & Mathematics with Applications (Oxford, England)*, *44*(8–9), 1077–1083.

Maji, P. K., Biswas, R., & Roy, A. R. (2003). Soft set theory. *Computers & Mathematics with Applications (Oxford, England)*, *45*(4–5), 555–562.

Molodtsov, D. (1999). Soft set theory-first results. *Computers & Mathematics with Applications (Oxford, England)*, *37*(4–5), 19–31.

Mukherjee, A., & Das, A. K. (2015a). Application of interval valued intuitionistic fuzzy soft set in investment decision making, *2015 Fifth International Conference on Advances in Computing and Communications (ICACC)*, Kochi, India, 2015, pp. 61-64, https://doi.org/DOI: 10.1109/ICACC.2015.37

Mukherjee, A., & Das, A. K. (2015b). Interval valued intuitionistic fuzzy soft multi set theoretic approach to decision making problems, 2015 International Conference on Computer, Communication and Control (IC4), Indore, India, 2015, pp. 1-5, https://doi.org/DOI: 10.1109/IC4.2015.7375640

Mukherjee, A., & Das, A. K. (2015c). Relations on intuitionistic fuzzy soft multi sets. In Kim, K. (Ed.), *Information Science and Applications. Lecture Notes in Electrical Engineering* (Vol. 339). Springer., DOI: 10.1007/978-3-662-46578-3_71

Mukherjee, A., & Das, A. K. (2016a). Application of fuzzy soft multi sets in decision-making problems. In: Nagar, A., Mohapatra, D., Chaki, N. (eds) Proceedings of 3rd International Conference on Advanced Computing, Networking and Informatics. Smart Innovation, Systems and Technologies, vol. 43. Springer, New Delhi. https://doi.org/DOI: 10.1007/978-81-322-2538-6_3

Mukherjee, A., & Das, A. K. (2016b). Topological structure formed by intuitionistic fuzzy rough relations. *Journal of the Indian Mathematical Society*, *83*(1-2), 135–144.

Mukherjee, A., & Das, A. K. (2022). Einstein operations on fuzzy soft multisets and decision making. *Boletim da Sociedade Paranaense de Matematica*, *40*, 1–10. DOI: 10.5269/bspm.32546

Mukherjee, A., & Das, A. K. (2023). Essentials of fuzzy soft multisets: Theory and applications, Springer Nature Singapore Pte Ltd., 1-151. ISBN: 978-981-19-2759-1, https://doi.org/DOI: 10.1007/978-981-19-2760-7

Noori, A., Bonakdari, H., Salimi, A., & Gharabaghi, B. (2020). A group Multi-Criteria Decision-Making method for water supply choice optimization. *Socio-Economic Planning Sciences, 101006*. Advance online publication. DOI: 10.1016/j.seps.2020.101006

Nourani, V., & Najafi, H. (2022). A Z-number based multi-attribute decision-making algorithm for hydro-environmental system management. *Neural Computing & Applications, 35*, 6405–6421. DOI: 10.1007/s00521-022-08025-3

Radmehr, A., Bozorg-Haddad, O., & Loáiciga, H. (2022). Integrated strategic planning and multi-criteria decision-making framework with its application to agricultural water management. *Scientific Reports, 12*. Advance online publication. DOI: 10.1038/s41598-022-12194-5 PMID: 35589906

Roy, A. R., & Maji, P. K. (2007). A fuzzy soft set theoretic approach to decision making problems. *Journal of Computational and Applied Mathematics, 203*, 412–418.

Sharma, T., Kumar, A., Pant, S., & Kotecha, K. (2023). Wastewater Treatment and Multi-Criteria Decision-Making Methods: A Review. *IEAccess, 11*, 143704–143720. DOI: 10.1109/ACCESS.2023.3343150

Zadeh, L. A. (1965). Fuzzy sets. *Information and Control, 8*(3), 338–353.

Zarghami, M., & Szidarovszky, F. (2009). Stochastic-fuzzy multi criteria decision making for robust water resources management. *Stochastic Environmental Research and Risk Assessment, 23*, 329–339. DOI: 10.1007/S00477-008-0218-6

Zyoud, S., Kaufmann, L., Shaheen, H., Samhan, S., & Fuchs-Hanusch, D. (2016). A framework for water loss management in developing countries under fuzzy environment: Integration of Fuzzy AHP with Fuzzy TOPSIS. *Expert Systems with Applications, 61*, 86–105. DOI: 10.1016/j.eswa.2016.05.016

APPENDIX A

Table 5. Short terms

Abbreviation	Term
AHP	Analytic Hierarchy Process
CE	Cost Efficiency
DO	Dissolved Oxygen
EWQI	Entropy Weighted Water Quality Index
ES	Environmental Sustainability
FMCDM	Fuzzy Multi-Criteria Decision-Making
FSS	Fuzzy Soft Set
MCDM	Multi-Criteria Decision-Making
PIG	Pollution Index of Groundwater
SA	Social Acceptability
SWM	Sustainable Water Management
TF	Technological Feasibility
TOPSIS	Technique for Order of Preference by Similarity to Ideal Solution
TFN	Triangular Fuzzy Number
TRFN	Trapezoidal Fuzzy Number
WQI	Water Quality Improvement

Chapter 4

Neutrosophic Sets for Decision-Making in Natural Disasters and Environmental Recovery

Ajoy Kanti Das
http://orcid.org/0000-0002-9326-1677
Tripura University, India

Nandini Gupta
http://orcid.org/0009-0000-5450-9783
Bir Bikram Memorial College, India

Rajat Das
http://orcid.org/0009-0000-3355-3582
Tripura University, India

Carlos Granados
http://orcid.org/0000-0002-7754-1468
Universidad de Sucre, Colombia

Suman Das
National Institute of Technology, Agartala, India

ABSTRACT

Natural disasters and environmental crises involve complex decision-making under uncertainty, where traditional models struggle with incomplete or contradictory data. This chapter explores neutrosophic sets—a generalization of fuzzy logic that independently models truth (T), indeterminacy (I), and falsity (F)—as a robust

DOI: 10.4018/979-8-3373-2700-6.ch004

framework for disaster response and recovery. We demonstrate its applications in flood risk assessment, wildfire management, and sustainable technology selection, highlighting how neutrosophic logic outperforms classical methods by quantifying ambiguity and conflicts. Case studies illustrate its integration with multi-criteria decision-making (MCDM) to optimize resource allocation and policy planning. Challenges like computational complexity and future directions for real-time systems and interdisciplinary adoption are discussed. Neutrosophic sets offer a transformative approach to resilience in an era of climate volatility, bridging advanced mathematics with practical crisis management.

INTRODUCTION

In recent decades, the frequency and severity of natural disasters—such as hurricanes, wildfires, and floods—have increased due to climate change and unsustainable development. Responding to these events requires rapid decision-making under extreme uncertainty. Decision-makers must consider multiple, often conflicting, pieces of information: sensor data, expert opinion, community needs, and environmental variables.

Traditional models struggle with the inherent indeterminacy of such data. Neutrosophic logic, an extension of the fuzzy set theory developed by Florentin Smarandache, incorporates truth (T), indeterminacy (I), and falsity (F) as independent components. This triad makes neutrosophic sets particularly useful for modeling real-world phenomena where truth is not binary and where contradictions or unknowns are integral to the system.

The complexity and uncertainty inherent in natural disasters and environmental recovery pose significant challenges for decision-makers. Traditional mathematical and computational models often struggle to accommodate the ambiguity, imprecision, and indeterminacy that characterize real-world disaster scenarios. To overcome these limitations, advanced frameworks grounded in soft computing have emerged, notably those based on fuzzy set theory introduced by Zadeh (1965), intuitionistic fuzzy sets developed by Atanassov (1986), and neutrosophic sets proposed by Smarandache (1998, 2005). These frameworks offer increasingly refined mechanisms to model uncertainty, particularly in dynamic and high-stakes environments such as disaster response and environmental management.

Neutrosophic sets, in particular, represent a powerful generalization of previous models by incorporating degrees of truth, indeterminacy, and falsity independently, thereby allowing more flexible and comprehensive representations of uncertain and incomplete information. (Smarandache, 1998, 2005). This unique capacity has spurred a wave of research leveraging neutrosophic theory in diverse decision-making domains, including typhoon disaster assessment (Tan et al., 2020a; Tan et al., 2022b; Li et al, 2024) earthquake response center location (Al-Omeri et al., 2024) and integrated disaster relief logistics under fairness and aftershock constraints (Mohammadi et al., 2020). These studies illustrate how neutrosophic sets enable more robust, adaptive, and context-sensitive decision-making strategies that are well-suited to the unpredictable nature of environmental crises.

Recent advancements also reflect the integration of neutrosophic theory with multi-criteria decision-making (MCDM), optimization models, and artificial intelligence to enhance disaster management outcomes. For example, Tan and Zhang (2020a) utilized refined single-valued neutrosophic sets and entropy-based approaches for assessing typhoon damage, while Mohammadi et al. (2020) developed a neutrosophic fuzzy model for facility location and routing in disaster relief operations. Similarly, emerging hybrid models such as ODESMAN (2024) and IDLTM-DMT (2024) demonstrate how combining neutrosophic logic with deep learning and trust management can optimize complex, multi-agent systems in disaster and healthcare environments.

This paper explores the application of neutrosophic sets as a strategic tool for decision-making in natural disaster response and environmental recovery. By synthesizing theoretical foundations and recent empirical advances, the study aims to highlight the strengths and potential of neutrosophic-based models in capturing the multifaceted uncertainty of environmental crises, ultimately contributing to more informed, equitable, and resilient decision-making processes.

LITERATURE REVIEW

Evolution of Fuzzy to Neutrosophic Models

The progression from fuzzy logic to neutrosophic sets reveals an expanding effort to model and manage uncertainty in real-world problems. The foundation was laid by Zadeh (1965) through fuzzy set theory, allowing partial membership and soft boundaries in decision systems. This was further refined by Atanassov (1986) with intuitionistic fuzzy sets (IFS), introducing hesitation through the membership and non-membership functions (Atanassov, 1986). Smarandache addressed the limitations of both these paradigms in handling indeterminacy (1998, 2005) and proposed the neutrosophic set (NS) theory (Smarandache, 1998, 2005). NS explicitly incorporates truth, indeterminacy, and falsity values independently, thus offering a more nuanced and flexible approach to uncertainty. This makes it particularly suitable for disaster and environmental systems where incomplete, vague, or conflicting information often prevails.

Applications in Disaster Management

Recent advancements demonstrate the efficacy of neutrosophic sets in disaster-related decision-making scenarios. Tan and Zhang (2020a) introduced a decision-making framework using refined single-valued neutrosophic sets and entropy measures for typhoon disaster assessment Tan et al., (Tan et al., 2020a). Extending this, Tan et al. (2020b) integrated trapezoidal fuzzy neutrosophic numbers with grey relation analysis to manage double incomplete information in similar contexts (Al-Omeri et al., 2024). Mohammadi et al. (2020) developed a robust neutrosophic fuzzy model for simultaneous facility location and routing decisions during disaster relief, effectively addressing aftershocks and fairness concerns (Tan et al., 2022b). Likewise, Sarma et al. (2025) tackled debris removal through a transportation model that minimizes both cost and time under a mixed neutrosophic environment. Lu and Luo (2020) modeled emergency transportation using single-valued neutrosophic sets, further proving their utility in operational logistics in post-disaster scenarios. Meanwhile, Al-Omeri and Kaviyarasu (2024) demonstrated the practical utility of neutrosophic graphs in planning earthquake response centers in Japan.

Applications in Environmental Recovery

Neutrosophic sets have shown exceptional compatibility with MCDM frameworks. Manna et al. (2020) integrated VIKOR with complex neutrosophic environments to tackle decision-making under conflicting criteria. Das et al. (2020) emphasized the advantages of neutrosophic fuzzy sets in aggregating expert opinions in uncertain conditions. Yang et al. (2024) proposed a three-way decision method using multi-scale neutrosophic systems to improve granular decision analysis. Furthermore, the ODESMAN model by Shaik et al. (2024) presented an integrated system combining fuzzy and neutrosophic logic for modeling complexity in disaster response systems.

The utility of neutrosophic theory extends to long-term environmental recovery efforts. Mahizha and M. (2025) utilized neutrosophic distance-based similarity for solid waste management site selection, reflecting environmental sustainability considerations. Similarly, Das et al. (2025a, 2025b) employed fuzzy soft sets and MCDM approaches to evaluate water quality under anthropogenic pressures. Karataş et al. (2023) and Voskoglou (2023) demonstrated how neutrosophic soft set theory enhances environmental decision-making by capturing indecisiveness and partial knowledge.

Systematic reviews, such as by Khan et al. (2018), provided a comprehensive outlook on decision-making algorithms under extended neutrosophic sets, offering a structural basis for future applications. Further theoretical advancements by Das et al. (2024, 2025a, 2025b, 2025c) and collaborators through various fuzzy, soft, intuitionistic, and hesitant fuzzy models reveal enriched algebraic and operational frameworks adaptable to environmental and disaster systems. Granados et al. (2022, 2023) expanded the scope through weighted neutrosophic distributions and multisets, providing scalable solutions for complex real-world evaluations.

Beyond environmental contexts, neutrosophic logic has been successfully adopted in biomedical diagnoses (Abdel-Basset et al. 2019a; Abdel-Basset et al. 2019b; Nguyen et al. 2017), image processing (Guo et al. 2014; Ravi et al. 2024; Ganeshkumar et al. 2024), IoT systems (Admin et al. 2024), and trust management (Ravi et al. 2025), indicating high versatility. These implementations further validate neutrosophic theory's robustness and adaptability to uncertain and critical environments, reinforcing its potential in disaster and environmental recovery strategies.

Hybrid and AI-Integrated Models

A significant contribution by Das and Granados (2022a) involves the development of a fuzzy parameterized intuitionistic fuzzy soft multiset theory for MCDM. Their investigation of FP-intuitionistic multi-fuzzy N-soft sets, which increase the adaptability of MCDM models, complements this work (Das and Granados, 2022b). They presented IFP-intuitionistic multi-fuzzy N-soft sets, further expanding this field of study and proving their usefulness in MCDM situations (Das and Granados, 2023a). Furthermore, Das (2018) suggested weighted fuzzy soft multisets as a novel tool for MCDM, while Das (2016) and Das & Granados (2023b) explored intuitionistic fuzzy rough relations and preference intuitionistic fuzzy rough relations, highlighting their partial inclusion properties. Their collaborative work in 2021 also presented a fuzzy soft group MCDM approach based on weighted average ratings, further refining the theoretical and practical aspects of decision analysis (Das and Granados, 2021). Applications-wise, Das et al. (2022) suggested new operations on fuzzy soft sets, demonstrating their usefulness in situations involving MCDM. Similarly, Das et al. (2023) introduced the concept of weighted hesitant bipolar-valued fuzzy soft sets, which provide a more nuanced framework for evaluating alternatives. Das et al. (2025b) developed a fuzzy MCDM model for examining human influences on urban river water quality is one of their noteworthy contributions to environmental MCDM. This research is extended in 2025 with weighted hesitant fuzzy soft set and fuzzy soft set models for rating water pollution and a case study applying the fuzzy soft MCDM model to assess pollution levels in the Gomati and Haora Rivers (Das et al., 2025a). Fuzzy soft multi-set theories in decision-making have also benefited greatly from the contributions of Mukherjee and Das (2016a). Their work in 2015 applied interval-valued intuitionistic fuzzy soft multi-sets in real-world MCDM problems (Mukherjee & Das, 2015b), while their research explored the use of interval-valued intuitionistic fuzzy soft sets for investment decision-making

(Mukherjee & Das, 2015a). In 2022, they invented Einstein operations for fuzzy soft multisets, further enhancing computing tools for decision analysis (Mukherjee & Das, 2022). Fuzzy soft multisets are thoroughly explored in their 2023 book, which covers both the theoretical underpinnings and real-world applications (Mukherjee & Das, 2023). The relational and topological structures of fuzzy and rough set models were further investigated by Mukherjee & Das (2015c; 2016b). Granados et al. (2022) provided continuous neutrosophic distributions with neutrosophic parameters, offering a probabilistic approach to uncertainty modeling. In 2023, Granados et al. (2023) proposed a weighted neutrosophic soft multiset approach, demonstrating its application in decision-making problems that require handling multiple levels of uncertainty and recently, Das et al. (2025c) conducted a comprehensive study on neutrosophic SuperHyper BCI-semigroups, exploring their algebraic significance. Das et al. (2025d) proposed an innovative weighted hypersoft expert system framework for group decision-making, advancing neutrosophic and plithogenic models to better handle uncertainty and complexity in applied mathematics contexts. Also, Das et al. (2025e) proposed a fuzzy logic-based decision support system to optimize adaptive inventory management in retail settings. All things considered, these studies provide significant contributions to the development of fuzzy set theory, which has applications in computational intelligence, environmental sustainability, and MCDM. Their work offers useful approaches for managing uncertainty, enhancing the precision of decisions, and tackling challenging real-world issues.

The evolution from fuzzy sets to neutrosophic sets marks a significant leap in uncertainty modeling, particularly relevant to natural disasters and environmental recovery. Neutrosophic logic, through its ability to handle indeterminacy, has empowered decision-makers in areas such as disaster relief logistics, environmental monitoring, resource allocation, and risk assessment. The expanding ecosystem of neutrosophic-based models—especially those integrating MCDM techniques and hybrid logic systems—suggests a maturing field ready for more mainstream adoption in sustainable disaster management and recovery planning. Table 1 provides a comparative overview of different approaches to disaster management and environmental recovery, summarizing their applications, strengths, and limitations.

Table 1. Comparative summary of disaster management approaches

Approach / Model Type	Applications	Key Results / Strengths	Limitations / Gaps
Probabilistic Models (e.g., Bayesian, statistical risk models)	Hazard frequency analysis, flood/ drought prediction	Strong in well-structured, data-rich contexts; provides probabilistic risk values	Needs large datasets; weak in handling vague/contradictory information
Agent-Based Simulations	Evacuation planning, community behavior under disaster	Captures emergent dynamics of multiple actors; useful for scenario testing	Highly sensitive to assumptions; limited in uncertainty modeling
Optimization Models (linear, stochastic, robust)	Facility location, routing, resource allocation	Efficient solutions for logistics and cost minimization	Require crisp inputs; not suited for ambiguity/contradictions
Fuzzy Sets	Flood risk classification, technology selection	Handles gradual membership (partial truth); good for linguistic/qualitative data	Cannot represent hesitation or contradictions clearly
Intuitionistic Fuzzy Sets (IFS)	Post-flood technology assessment, environmental evaluation	Captures hesitation (membership + non-membership); useful for conflicting expert opinions	Restricted by $\mu + \nu \leq 1$ condition; limited contradiction handling
Neutrosophic Sets (NS)	Flood risk ranking, wildfire strategy evaluation, debris removal, sustainable technology selection	Independently models truth (T), indeterminacy (I), and falsity (F); integrates easily with MCDM; robust under uncertainty and conflicting data	Computationally complex; limited real-time and IoT integration; socio-economic factors often underrepresented

Research Gaps

Despite promising advances in applying neutrosophic sets and other soft computing methods to disaster management, several gaps remain:

- **Lack of Real-Time Systems**
  - o Most existing neutrosophic decision models are static and rely on pre-collected data.
  - o There is limited integration with real-time monitoring (e.g., sensor networks, early-warning systems).
- **Weak IoT and Big Data Integration**
  - o Although IoT devices generate large streams of disaster-related data (e.g., rainfall sensors, drone imagery, social media), few studies connect this data flow with neutrosophic decision-making frameworks.

- **Limited Socio-Economic Consideration**
  - o Many models focus on technical and environmental parameters while neglecting social vulnerability, equity, and economic resilience.
  - o Community perspectives, cultural values, and fairness in resource allocation are underrepresented in neutrosophic-based models.
- **Computational Challenges**
  - o Neutrosophic decision matrices can become complex with multiple criteria and alternatives. Efficient algorithms for large-scale or real-time computation are still lacking.
- **Policy and Practice Gap**
  - o There is little translation of neutrosophic research into actionable government policies, disaster drills, or community-based planning tools.

Positioning Neutrosophic Sets in Broader Disaster Management Approaches

Disaster management has been studied using a wide range of modeling and decision-support techniques. Some of the most common include:

- **Bayesian Networks:** Used for probabilistic risk assessment and causal inference, especially when prior probabilities and conditional dependencies are available. However, they require large, reliable datasets and may not handle conflicting or vague information well.
- **Probabilistic and Statistical Models:** Traditional tools for hazard frequency, vulnerability assessment, and damage prediction. These methods are effective when data is abundant, but they often struggle with qualitative judgments and contradictory expert inputs.
- **Agent-Based Simulations:** Useful for modeling community behavior, evacuation planning, and multi-agent coordination in complex environments. While they capture emergent dynamics, their outcomes are sensitive to initial assumptions and parameter uncertainty.
- **Optimization Models:** Widely applied in logistics and resource allocation (e.g., facility location, routing under constraints). They provide efficiency but typically rely on crisp inputs, limiting their adaptability to uncertain or indeterminate contexts.
- **Soft Computing Approaches (Fuzzy, Intuitionistic Fuzzy, and Neutrosophic Sets):** Fuzzy logic captures gradual membership, and intuitionistic fuzzy sets include hesitation, but both have limited ability to handle deep contradictions. Neutrosophic sets extend these by independently model-

ing truth (T), indeterminacy (I), and falsity (F), offering a richer framework for situations with incomplete, inconsistent, or ambiguous information.

Unique Advantage of Neutrosophic Sets

While Bayesian and probabilistic models rely heavily on data availability, and simulations depend on assumptions, neutrosophic methods excel where uncertainty, conflict, and lack of information are inherent. By combining neutrosophic logic with decision matrices and multi-criteria decision-making (MCDM), this chapter contributes a hybrid framework that bridges mathematical rigor with real-world disaster response needs.

Novel Contributions

This chapter makes several new contributions to the field of disaster and environmental decision-making. Unlike fuzzy sets and intuitionistic fuzzy sets, which have limits in handling hesitation and contradictions, neutrosophic sets model truth (T), indeterminacy (I), and falsity (F) independently. This provides a clearer way to deal with incomplete, uncertain, or conflicting information common in disaster situations.

We also show how neutrosophic sets can be integrated into decision matrices and MCDM methods, making the framework more practical for real-world applications. Through case studies on flood risk, wildfire management, and technology selection, the chapter demonstrates how neutrosophic logic supports better and more balanced decisions. These contributions go beyond earlier research by combining strong mathematical foundations with applied tools for disaster response and environmental recovery.

INTRODUCTION TO FUZZY SETS

Real-world systems, especially those involving natural disasters and environmental impacts, often involve vague, imprecise, or incomplete information. Classical (crisp) set theory fails to model such uncertainty because it deals only with binary membership: an element either belongs to a set or it does not. Fuzzy set theory, introduced by Lotfi A. Zadeh in 1965, provides a mathematical framework for representing uncertainty and partial truth. In a fuzzy set, each element has a degree of membership ranging from 0 to 1.

Fuzzy Set

Let X be a universe of discourse. A fuzzy set A in X is defined by a membership function:

$$\mu_A : X \rightarrow [0,1]$$

For each element $x \in X, \mu_A(x)$ denotes the degree to which x belongs to the fuzzy set A. The closer $\mu A(x)$ is to 1, the stronger the membership of x in A.

- $\mu_A(x) = 1$: Full membership
- $\mu_A(x) = 0$: No membership
- $0 < \mu_A(x) < 1$: Partial membership

Basic Operations on Fuzzy Sets

Given two fuzzy sets A and B in X, with membership functions $\mu_A(x)$ and $\mu B(x)$, the following operations are defined:

1. **Union**:

$$\mu_{A \cup B}(x) = max(\mu_A(x),\ \mu_B(x))$$

2. **Intersection**:

$$\mu_{A \cap B}(x) = min(\mu_A(x), \mu_B(x))$$

3. **Complement**:

$$\mu_{\neg A}(x) = 1 - \mu_A(x)$$

Fuzzy Risk Classification in Flood Management

Let us model the flood risk level in a river basin based on rainfall intensity (r) measured in mm/hour. We define a fuzzy set A = "High Flood Risk" with a triangular membership function:

$$\mu_A(r) = \begin{cases} 0, if r \leq 30 \\ \frac{r - 30}{30}, if 30 < r \leq 60 \\ 1, if r > 60 \end{cases}$$

This model means:

- Rainfall below 30 mm/h is not associated with flood risk.
- Between 30 and 60 mm/h, the risk increases linearly.
- Above 60 mm/h, the area is considered fully at high flood risk.

Table 2 presents the fuzzy membership values representing the degree of high flood risk corresponding to different rainfall intensities.

Table 2. Fuzzy membership values for high flood risk based on rainfall intensity

Rainfall (r) mm/h	$\mu_A(r)$
25	0
45	$(45 - 30)/30 = 0.5$
65	1

This fuzzy model allows emergency planners to assess risk gradually instead of using hard thresholds that may not reflect reality.

Application to Environmental Decision-Making

Fuzzy sets are widely used in:

- **Environmental impact assessment (EIA)**: where qualitative expert opinions (e.g., "moderate pollution") are translated into quantitative models.
- **Resource allocation under uncertainty**: deciding how to distribute limited recovery resources when need levels are fuzzy or overlapping.
- **Sustainable technology selection**: evaluating systems using linguistic terms like "high cost", and "low reliability", and converting these into fuzzy numbers.

Multi-Criteria Decision Matrix for Recovery Technologies

In post-disaster recovery planning, decision-makers often face complex trade-offs among multiple criteria such as environmental sustainability, cost, and social acceptance. These criteria are frequently expressed in vague or linguistic terms—such as "high cost" or "moderate benefit"—which are inherently imprecise. Fuzzy set theory provides a robust mathematical framework to model this imprecision and support more informed, flexible decision-making.

Let us consider two recovery technologies—solar pumps and diesel pumps—that are being evaluated for deployment in a flood-affected area. The evaluation is based on three fuzzy criteria:

1. **Environmental Benefit (μ)** – a measure of ecological sustainability.
2. **Cost Efficiency (μ)** – the inverse of financial burden or affordability.
3. **Community Acceptance (μ)** – the extent to which local populations are willing to adopt the technology.

Each criterion is assigned a fuzzy membership value ranging from 0 (lowest degree of membership) to 1 (highest degree of membership), reflecting the degree to which each technology satisfies that particular criterion. Table 3 illustrates the fuzzy membership values assigned to recovery technologies based on environmental benefit, cost efficiency, and community acceptance.

Table 3. Fuzzy multi-criteria evaluation of recovery technologies based on key decision factors

Technology	Environmental Benefit (μ)	Cost Efficiency (μ)	Community Acceptance (μ)
Solar pumps	0.8	0.6	0.7
Diesel pumps	0.3	0.9	0.5

From the Table 3, we observe that:

- **Solar pumps** score high on environmental benefit (0.8) and moderately well on community acceptance (0.7), though slightly lower on cost efficiency (0.6).
- **Diesel pumps**, in contrast, are highly cost-efficient (0.9) but score poorly on environmental impact (0.3) and moderately on community acceptance (0.5).

To make a final decision, these fuzzy evaluations can be aggregated using fuzzy multi-criteria decision-making (MCDM) techniques such as the Fuzzy Analytic Hierarchy Process (Fuzzy AHP) or Fuzzy Technique for Order Preference by Similarity to Ideal Solution (Fuzzy TOPSIS). These tools allow planners to:

- Incorporate subjective preferences and weights for each criterion.
- Compare alternatives holistically, considering trade-offs.
- Arrive at a more realistic, adaptable, and consensus-driven decision.

In this way, fuzzy logic enhances traditional environmental decision-making by handling uncertainty and subjectivity more effectively, ultimately supporting more sustainable and community-aligned recovery strategies.

INTUITIONISTIC FUZZY SETS: A MATHEMATICAL FRAMEWORK

Intuitionistic Fuzzy Sets (IFSs), introduced by Atanassov in 1986, extend classical fuzzy set theory by incorporating a degree of non-membership alongside membership and capturing hesitation or uncertainty explicitly.

Let X be a universe of discourse. An intuitionistic fuzzy set A in X is defined as:

$$A = \{\langle x, \mu_A(x), \nu_A(x)\rangle \mid x \in X\}$$

where:

- $\mu_A(x) : X \rightarrow [0, 1]$ is the membership function,
- νSkipping malformed tag: msub(x):$X \rightarrow [0, 1]$ is the non-membership function,
- For each $x \in X$, it holds that:

$$0 \leq \mu_A(x) + \nu_A(x) \leq 1$$

The hesitation degree (also called uncertainty or indeterminacy) is defined as:

$$\pi_A(x) = 1 - \mu_A(x) - \nu_A(x)$$

This triple $(\mu_A(x), \nu_A(x), \pi_A(x))$ allows better handling of incomplete, uncertain, or conflicting information.

Application to Disaster and Environmental Recovery

In post-disaster or environmental assessment scenarios, precise judgments are often impossible due to conflicting expert opinions, lack of data, or environmental volatility. IFSs can model the following:

- μ_A(x): Degree to which a strategy is beneficial (truth).
- ν_A(x): Degree to which it is not beneficial (falsity).
- π_A(x): Degree of hesitation, reflecting uncertainty or lack of knowledge.

This makes IFS a valuable decision-support framework for multi-criteria evaluation of sustainable technologies or emergency responses.

Evaluating Post-Flood Recovery Technologies

Let $X = \{x_1, x_2, x_3\}$ represent three different recovery strategies after a flood in an urban area:

- x_1: Deployment of floating solar panels for energy
- x_2: Construction of green stormwater infrastructure
- x_3: Installation of AI-based flood early warning systems

Experts assess each option using an intuitionistic fuzzy scale based on environmental impact, cost-effectiveness, and resilience. Table 4 presents intuitionistic fuzzy assessments of three post-flood recovery strategies, capturing expert judgments on benefit, opposition, and uncertainty.

Table 4. Intuitionistic fuzzy evaluation of post-flood recovery strategies based on expert assessment

Strategy x_i	$\mu_A(x_i)$	$\nu_A(x_i)$	$\pi_A(x_i) = 1 - \mu_A(x_i) - \nu_A(x_i)$
x_1	0.7	0.2	0.1
x_2	0.8	0.1	0.1
x_3	0.6	0.25	0.15

Interpretation

- x_2 (green infrastructure) has the highest membership and lowest non-membership, indicating strong support and low opposition with modest uncertainty.
- x_3 (AI system) has a higher degree of hesitation, possibly due to lack of data on long-term performance or reliability in local contexts.
- x_1 (floating solar) offers a balance between benefit and feasibility, but may face logistical challenges.

Aggregation and Decision-Making

To facilitate decision-making, IFS values can be aggregated using operators such as:

- **Intuitionistic fuzzy weighted average (IFWA):**

Let *wi* be the weight of each criterion (with $\sum w_i = 1$). Then:

$$\mu_{IFWA}(x) = \sum w_i \cdot \mu_i(x), \quad \nu_{IFWA}(x) = \sum w_i \cdot \nu_i(x)$$

This allows multiple expert inputs or criteria (e.g., environmental, social, economic) to be fused into a single intuitionistic fuzzy value.

Decision-makers can then rank strategies based on:

- **Score function:** $S(x) = \mu_A(x) - \nu_A(x)$
- **Accuracy function:** $H(x) = \mu_A(x) + \nu_A(x)$

These functions help to balance the positivity and negativity of assessments with the completeness of information.

Advantages in Disaster and Environmental Contexts

- **Explicit Hesitation:** Unlike classic fuzzy sets, IFSs clearly distinguish between uncertainty and partial truth.
- **Better Expert Representation:** Allows conflicting expert inputs to be formally included in decision processes.
- **Dynamic Updating:** As more data becomes available (e.g., after a pilot test), IFS values can be adjusted accordingly.

Intuitionistic fuzzy sets provide a rigorous yet flexible mathematical tool for modeling uncertainty and hesitancy in environmental and disaster recovery decisions. By quantifying membership, non-membership, and indeterminacy, they support more transparent, inclusive, and adaptive decision-making processes. Their integration with technologies such as AI, IoT, and GIS can enhance resilience planning and sustainable recovery in complex, real-world scenarios.

FOUNDATIONS OF NEUTROSOPHIC SETS

Neutrosophic Set

A neutrosophic set A in a universe U is characterized by a truth-membership function $T_A(x)$, an indeterminacy-membership function $I_A(x)$, and a falsity-membership function $F_A(x)$, where:

$T_A(x),\ I_A(x),\ F_A(x)\ \subseteq\ [0-,\ 1+]$

Unlike fuzzy sets where $T + F = 1$, neutrosophic sets allow all three components to vary independently, offering a richer representation of uncertainty and contradiction.

Let U be a universe of discourse. A neutrosophic set A in U is defined as:

$$A = \{\langle x, T_A(x), I_A(x), F_A(x)\rangle \mid x \in U\}$$

where:

- T_A (x): U→ [0, 1] is the degree of truth,
- I_A(x): U → [0, 1] is the degree of indeterminacy,
- F_A(x): U → [0,1] is the degree of falsity, subject to the condition:

$$0 \leq T_A(x) + I_A(x) + F_A(x) \leq 3$$

Operations on Neutrosophic Sets

Let A and B be two neutrosophic sets defined in a universe of discourse U, such that:

$$A = \{\langle x, T_A(x), I_A(x), F_A(x)\rangle \mid x \in U\}$$

$$B = \{\langle x, T_B(x), I_B(x), F_B(x)\rangle \mid x \in U\}$$

Each of the following operations results in another neutrosophic set:

- Union: The union of two neutrosophic sets $A \cup B$ is defined as:

$$T_{A\cup B}(x) = max(T_A(x), T_B(x))$$

$$I_{A\cup B}(x) = min(I_A(x), I_B(x))$$

$$F_{A\cup B}(x) = min(F_A(x), F_B(x))$$

This definition assumes that higher truth and lower indeterminacy/falsity values represent a stronger membership in the union.

- Intersection: The intersection of two neutrosophic sets $A \cap B$ is defined as:

$$T_{A\cap B}(x) = min\big(T_A(x), T_B(x)\big)$$

$$I_{A\cap B}(x) = max\big(I_A(x), I_B(x)\big)$$

$$F_{A\cap B}(x) = max\big(F_A(x), F_B(x)\big)$$

This captures the conservative notion of overlapping membership, increasing uncertainty and falsity.

- Complement: The complement of a neutrosophic set A^C is defined as:

$$T_A^C(x) = F_A(x)$$

$$I_A^C(x) = I_A(x)$$

$$F_A^C(x) = T_A(x)$$

This operation swaps truth and falsity while retaining indeterminacy.

- Difference ($A\backslash B$): The difference between two neutrosophic sets $A \setminus B$ is defined as

$$T_{A\backslash B}(x) = min\big(T_A(x), F_B(x)\big)$$

$$I_{A\backslash B}(x) = max\big(I_A(x), I_B(x)\big)$$

$$T_{A\backslash B}(x) = max\big(F_A(x), T_B(x)\big)$$

This captures the idea that an element belongs to A and not to B, based on neutrosophic characteristics.

- Scalar Multiplication: For a scalar $\alpha \in [0, 1]$, the scalar multiplication of a neutrosophic set A is

$$T_{\alpha\cdot A}(x) = \alpha \cdot T_A(x)$$

$$I_{\alpha\cdot A}(x) = \alpha \cdot I_A(x)$$

$$F_{\alpha\cdot A}(x) = \alpha \cdot F_A(x)$$

This is used in optimization and weighted decision-making contexts.

- Cartesian Product: Let $A \subseteq U, B \subseteq V$, then the Cartesian product $A \times B \subseteq U \times V$ is defined by

$$T_{A\times B}(x, y) = min(T_A(x), T_B(y))$$

$$I_{A\times B}(x, y) = max(I_A(x), I_B(y))$$

$$F_{A\times B}(x, y) = max(F_A(x), F_B(y))$$

This operation is important in relational modeling and neutrosophic logic-based systems.

The following mathematical examples demonstrate the application of neutrosophic set operations in the context of decision-making for natural disasters and environmental recovery. Each example reflects how the intrinsic components of neutrosophic logic—truth (T), indeterminacy (I), and falsity (F)—enable nuanced modeling of expert assessments, especially under uncertainty, inconsistency, and incomplete information. These operations provide a structured mechanism for integrating diverse expert opinions, facilitating more informed and resilient decision-making strategies in complex and high-stakes environments.

Universe of Discourse, let $U = \{x_1, x_2, x_3\}$, where:

- x_1: "Area is safe for resettlement",
- x_2: "Area requires partial remediation",
- x_3: "Area is unsafe and needs evacuation".

Let us define two neutrosophic sets:

- A: Expert A's assessment as in Table 5
- B: Expert B's assessment as in Table 6

Table 5. Neutrosophic set A – expert A's assessment

Element	$T_A(x)$	$I_A(x)$	$F_A(x)$
x_1	0.8	0.1	0.1
x_2	0.4	0.4	0.2
x_3	0.1	0.3	0.6

Table 6: Neutrosophic set B – expert B's assessment

Element	$T_B(x)$	$I_B(x)$	$F_B(x)$
x_1	0.6	0.2	0.2
x_2	0.5	0.3	0.2
x_3	0.2	0.5	0.3

For each element $x \in U = \{x_1, x_2, x_3\}$, the union of two neutrosophic sets $A \cup B$ is computed as:

Union: $A \cup B$

$$T_{A\cup B}(x) = max(T_A(x), T_B(x))$$

$$I_{A\cup B}(x) = min(I_B(x), I_B(x))$$

$$F_{A\cap B}(x) = min(F_A(x), F_B(x))$$

For x_1:

$$T_A(x_1) = 0.8,\ T_A(x_1) = 0.6 \rightarrow T_{A\cup B}(x_1) = max(0.8,\ 0.6) = 0.8$$

$$I_A(x_1) = 0.1,\ I_B(x_1) = 0.2 \rightarrow I_{A\cup B}(x_1) = min(0.1,\ 0.2) = 0.1$$

$$F_A(x_1) = 0.1,\ F_B(x_1) = 0.2 \rightarrow F_{A\cup B}(x_1) = min(0.1,\ 0.2) = 0.1$$

Result for x_1: $(0.8, 0.1, 0.1)$

For x_2:

$$T_A(x_2) = 0.4,\ T_B(x2) = 0.5 \rightarrow T_{A\cup B}(x_2) = max(0.4, 0.5) = 0.5$$

$$I_A(x_2) = 0.4,\ I_B(x_2) = 0.3 \rightarrow I_{A\cup B}(x_2) = min(0.4, 0.3) = 0.3$$

$$F_A(x_2) = 0.2,\ F_B(x_2) = 0.2 \rightarrow F_{A\cup B}(x_2) = min(0.2, 0.2) = 0.2$$

Result for x_2: $(0.5, 0.3, 0.2)$

For x_3:

$$T_A(x_3) = 0.1,\ T_B(x_3) = 0.2 \rightarrow T_{A\cup B}(x_3) = max(0.1, 0.2) = 0.2$$

$$I_A(x_3) = 0.3, I_B(x_3) = 0.5 \rightarrow I_{A\cup B}(x_3) = min(0.3, 0.5) = 0.3$$

$$F_A(x_3) = 0.6,\ F_B(x_3) = 0.3 \rightarrow F_{A \cup B}(x_3) = min(0.6, 0.3) = 0.3$$

Result for x_3: (0.2, 0.3, 0.3)
Result: Final union set $A \cup B$ as presented in Table 7.

Table 7. Final union set A ∪ B

Element	$T_{A \cup B}(x)$	$I_{A \cup B}(x)$	$F_{A \cup B}(x)$
x_1	0.8	0.1	0.1
x_2	0.5	0.3	0.2
x_3	0.2	0.3	0.3

Intersection: $A \cap B$
For each element $x \in U = \{x_1, x_2, x_3\}$, the intersection of sets A and B is defined as:

$$T_{A \cup B}(x) = min(T_A(x), T_B(x))$$

$$I_{A \cup B}(x) = max(I_A(x), I_B(x))$$

$$F_{A \cup B}(x) = max(F_A(x), F_B(x))$$

For x_1:

$$T_A(x_1) = 0.8,\ T_B(x_1) = 0.6 \rightarrow T_{A \cap B}(x_1) = min(0.8, 0.6) = 0.6$$

$$I_A(x_1) = 0.1,\ I_B(x_1) = 0.2 \rightarrow I_{A \cap B}(x_1) = max(0.1, 0.2) = 0.2$$

$$F_A(x_1) = 0.1,\ F_B(x_1) = 0.2 \rightarrow F_{A \cap B}(x_1) = max(0.1, 0.2) = 0.2$$

Result for x_1: (0.6, 0.2, 0.2)(0.6, 0.2, 0.2)(0.6, 0.2, 0.2)
For x_2:

$$T_A(x_2) = 0.4,\ T_B(x_2) = 0.5 \rightarrow T_{A \cap B}(x_2) = min(0.4, 0.5) = 0.4$$

$$I_A(x_2) = 0.4,\ I_B(x_2) = 0.3 \rightarrow I_{A \cap B}(x_2) = max(0.4, 0.3) = 0.4$$

$$F_A(x_2) = 0.2,\ F_B(x_2) = 0.2 \rightarrow F_{A \cap B}(x_2) = max(0.2, 0.2) = 0.2$$

Result for x_2: (0.4, 0.4, 0.2)
For x_3:

$$T_A(x_3) = 0.1,\ T_B(x_3) = 0.2 \rightarrow T_{A\cap B}(x_3) = min(0.1, 0.2) = 0.1$$

$$I_A(x_3) = 0.3,\ I_B(x_3) = 0.5 \rightarrow I_{A\cap B}(x_3) = max(0.3, 0.5) = 0.5$$

$$F_A(x_3) = 0.6,\ F_B(x_3) = 0.3 \rightarrow F_{A\cap B}(x_3) = max(0.6, 0.3) = 0.6$$

Result for x_3: (0.1, 0.5, 0.6)
Final intersection set $A \cap B$ presented in Table 8.

Table 8. Final union set $A \cap B$

Element	$T_{A\cap B}(x)$	$T_{A\cap B}(x)$	$I_{A\cap B}(x)$
x_1	0.6	0.2	0.2
x_2	0.4	0.4	0.2
x_3	0.1	0.5	0.6

Complement: A^C
For a neutrosophic set A, the complement A^C is computed element-wise as:

$$T_A^C(x) = F_A(x)$$

$$I_A^C(x) = I_A(x)$$

$$F_A^C(x) = T_A(x)$$

For x_1:

$$T_A^C(x_1) = F_A(x_1) = 0.1$$

$$I_A^C(x_1) = I_A(x_1) = 0.1$$

$$F_A^C(x_1) = T_A(x_1) = 0.8$$

Result for x_1: (0.1 ,0.1, 0.8)
For x_2

$$T_A^C(x_2) = F_A(x_2) = 0.2$$

$$I_A^C(x_2) = I_A(x_2) = 0.4$$

$$F_A^C(x_2) = T_A(x_2) = 0.4$$

Result for x_2: (0.2, 0.4, 0.4)
For x_3:

$$T_A^C(x_3) = F_A(x_3) = 0.6$$

$$I_A^C(x_3) = I_A(x_3) = 0.3$$

$$F_A^C(x_3) = T_A(x_3) = 0.1$$

Result for x_3: (0.6, 0.3, 0.1)
Final complement set A^C:

Table 9. Complement of neutrosophic set A – A^C

Element	$T_A^C(x)$	$I_A^C(x)$	$F_A^C(x)$
x_1	0.1	0.1	0.8
x_2	0.2	0.4	0.4
x_3	0.6	0.3	0.1

4. Difference: $A \backslash B$

$$T_{A\backslash B}(x) = min(T_A(x), T_B(x))$$

$$I_{A\backslash B}(x) = max(I_A(x), I_B(x))$$

$$F_{A\backslash B}(x) = max(F_A(x), F_B(x))$$

For x_1:

$$T_{A\backslash B}(x_1) = min(0.8,\ 0.2) = 0.2$$

$$I_{A\backslash B}(x_1) = max(0.1,\ 0.2) = 0.2$$

$F_{A\setminus B}(x_1) = max(0.1,\ 0.6) = 0.6$

For x_2:

$T_{A\setminus B}(x_2) = \min(0.4,\ 0.2) = 0.2$

$I_{A\setminus B}(x_2) = \max(0.4,\ 0.3) = 0.4$

$F_{A\setminus B}(x_2) = \max(0.2,\ 0.5) = 0.5$

Result for x_2: (0.2, 0.4, 0.5)
For x_3:

$T_{A\setminus B}(x_3) = min(0.1, 0.3) = 0.1$

$I_{A\setminus B}(x_2) = \max(0.3\ ,0.5) = 0.5$

$F_{A\setminus B}(x_3) = \max(0.6,\ 0.2) = 0.6$

Result for x_3: $(0.1, 0.5, 0.6)$
Final Result for $A \setminus B$ is presented in Table 10.

Table 10. Final result for $A \setminus B$

Element	$T_{A\setminus B}(x)$	$I_{A\setminus B}(x)$	$F_{A\setminus B}(x)$
x_1	0.2	0.2	0.8
x_2	0.2	0.4	0.5
x_3	0.1	0.5	0.6

Scalar Multiplication: $\alpha \backslash A, \alpha = 0.5$

$T_{\alpha\cdot A}(x) = \alpha \cdot T_A(x)$

$I_{\alpha\cdot A}(x) = \alpha \cdot I_A(x)$

$F_{\alpha\cdot A}(x) = \alpha \cdot F_A(x)$

For x_1:

$T_{0.5A}(x_1) = 0.5\cdot 0.8 = 0.4$

$I_{0.5A}(x_1)$=0.5·0.1=0.05

$F_{0.5A}(x_1)$=0.5·0.1=0.05

Result for x_1: (0.4, 0.05, 0.05)
For x_2:

$T_{0.5A}(x_2)$= 0.5·0.4 = 0.2

$I_{0.5A}(x_2)$= 0.5·0.4 = 0.2

$F_{0.5A}(x_2)$ = 0.5·0.2 = 0.1

Result for x_2: (0.2, 0.2, 0.1)
For x3:

$T_{0.5A}(x_3)$= 0.5·0.1 = 0.05

$I_{0.5A}(x_3)$= 0.5·0.3 = 0.15

$F_{0.5A}(x_3)$= 0.5·0.6 = 0.3

Result for x_3: (0.05, 0.15, 0.3)
Final Result for Scalar Multiplication 0.5 · A is presented in Table 11

Table 11. Final result for Scalar Multiplication 0.5 · A

Element	$T_{0.5A}(x)$	$I_{0.5A}(x)$	$F_{0.5A}(x)$
x_1	0.4	0.05	0.05
x_2	0.2	0.2	0.1
x_3	0.05	0.15	0.3

Cartesian Product: $A \times B$

Let $(x_1, x_2) \in U \times U$. Then:

$$T_{A\times B}(x_1, x_2) = min(T_A(x_1), T_B(x_2))$$

$$I_{A\times B}(x_1, x_2) = min(I_A(x_1), I_B(x_2))$$

$$F_{A\times B}(x_1,\ x_2)\ =\ min(F_A(x_1), F_B(x_2))$$

$$T_{A\times B}(x_1,\ x_2)\ =\ min(0.8, 0.5)\ =\ 0.5$$

$$I_{A\times B}(x_1,\ x_2)\ =\ max(0.1, 0.3)\ =\ 0.3$$

$$F_{A\times B}(x_1,\ x_2)\ =\ max(0.1, 0.2)\ =\ 0.2$$

These examples demonstrate how neutrosophic operations help combine or contrast assessments from multiple sources (experts, models, sensors) in disaster recovery or environmental planning—especially under uncertainty, incomplete information, or conflict.

Comparison of Fuzzy Sets, Intuitionistic Fuzzy Sets, and Neutrosophic Sets

In dealing with uncertain, imprecise, or incomplete information—particularly in fields like environmental modeling, disaster management, and decision analysis—various mathematical frameworks have been developed. The progression from Fuzzy Sets to Intuitionistic Fuzzy Sets and finally to Neutrosophic Sets reflects the growing need to handle indeterminacy and contradiction in increasingly complex systems. Table 12 compares fuzzy, intuitionistic fuzzy, and neutrosophic sets, highlighting their approaches to membership, indeterminacy, and handling of uncertainty and contradiction.

Table 12. Comparison of fuzzy sets, intuitionistic fuzzy sets, and neutrosophic sets based on membership, indeterminacy, and application characteristics

Aspect	Fuzzy Set	Intuitionistic Fuzzy Set (IFS)	Neutrosophic Set (NS)
Membership Components	$\mu \in [0, 1]$	μ (membership), ν (non-membership), with $\mu + \nu \leq 1$	T (truth), I (indeterminacy), F (falsity); each $\in [0, 1]$
Indeterminacy Handling	Not handled	Partially handled as hesitation margin: $\pi = 1 - \mu - \nu$	Fully handled with explicit degree of indeterminacy (I)
Contradiction Tolerance	Not allowed ($\mu + (1-\mu) = 1$)	Limited ($\mu + \nu \leq 1$ constraint)	Allowed ($T + I + F \leq 3$; no strict sum constraints)
Flexibility	Suitable for simple uncertainty	Better for moderate uncertainty and hesitation	Most flexible, suitable for high uncertainty and inconsistencies

continued on following page

Table 12. Continued

Aspect	Fuzzy Set	Intuitionistic Fuzzy Set (IFS)	Neutrosophic Set (NS)
Applications	Gradual transitions (e.g., temperature, risk)	Situations with hesitation or partial knowledge	Complex environments with vague, incomplete, or conflicting info
Example	$\mu(x) = 0.7$ (partially belongs)	$\mu(x) = 0.6$, $\nu(x) = 0.2$, $\pi = 0.2$	$T = 0.6$, $I = 0.3$, $F = 0.5$ (can reflect paradoxical cases)

Fuzzy Sets

- **Introduced by:** Lotfi Zadeh (1965)
- **Key Concept:** An element can partially belong to a set with a degree of membership (μ).
- **Limitation:** Cannot represent hesitation or contradictions explicitly.
- **Use Case:** Ideal for problems with smooth, gradual transitions (e.g., "warm", "high", "near").

Intuitionistic Fuzzy Sets (IFS)

- **Introduced by:** Atanassov (1986)
- **Key Concept:** Adds a degree of non-membership (ν) to fuzzy sets, and a hesitation degree (π).
- **Limitation:** The sum $\mu + \nu$ must be ≤ 1, restricting the representation of contradictory information.
- **Use Case:** Useful when there is partial knowledge or a degree of hesitation (e.g., expert opinions with uncertainty).

Neutrosophic Sets (NS)

- **Introduced by:** Florentin Smarandache (1998)
- **Key Concept:** Each element has a degree of truth (T), indeterminacy (I), and falsity (F), all **independent**.
- **Advantage:** Most general framework—can handle contradiction (T and F both high), total indeterminacy ($I \approx 1$), or even incomplete knowledge (e.g., $T + I + F < 1$).
- **Use Case:** Particularly suited for decision-making under extreme uncertainty, environmental disasters, medical diagnosis, social modeling, etc.

Modeling Disaster Risk with Neutrosophic Sets

Traditional disaster risk models often use crisp values for hazard intensity, exposure, and vulnerability. However, each of these components may be associated with uncertain or imprecise data. Neutrosophic logic can be applied as follows:

- **Truth (T):** Degree to which a hazard is likely to occur based on historical data.
- **Indeterminacy (I):** Lack of data or conflicting reports.
- **Falsity (F):** Degree to which the hazard is not expected based on other evidence.

Neutrosophic Set-Based Modeling of Coastal Flood Risk

Flood risk in coastal cities is influenced by complex and uncertain factors such as:

- Rainfall intensity (from satellite data),
- Sea level trends (from tide gauges),
- Local infrastructure (e.g., drainage systems, sea walls),
- Climate forecasts (with varying confidence levels).

These factors often involve incomplete, contradictory, or uncertain information, making them suitable for neutrosophic modeling.

Let U be the universe of locations in the coastal city (e.g., neighborhoods or grid cells), and let:

- $A \subseteq U$ be the set of areas at risk of flooding,
- $x \in U$ be a particular location (e.g., a neighborhood).

A single-valued neutrosophic set (SVNS) A is defined as:

$$A = \{\langle x, T_A(x), I_A(x), F_A(x)\rangle \mid x \in U\}$$

Where for each location x:

- $T_A(x) \in [0,1]$: Truth-membership degree – how strongly evidence supports that location x is flood-prone.
- $I_A(x) \in [0,1]$: Indeterminacy degree – degree of uncertainty, ambiguity, or lack of data.
- $F_A(x) \in [0,1]$: Falsity-membership degree – how strongly evidence supports that location x is *not* flood-prone.

Subject to the condition:

$$0 \leq T_A(x) + I_A(x) + F_A(x) \leq 3$$

Each component is independent and need not sum to 1.

Example Data and Modeling

Let's consider three locations $x_1, x_2, and x_3$ in a coastal city. Table 13 presents neutrosophic tuples capturing truth, indeterminacy, and falsity for risk assessment across three coastal locations with varying data quality and infrastructure.

Table 13. Neutrosophic representation of risk factors for coastal locations based on rainfall, data gaps, and infrastructure status

Location x	**Satellite Rainfall (Truth)**	**Tide Gauge Data Gaps (Indeterminacy)**	**Infrastructure Present (Falsity)**	**Neutrosophic Tuple (T,I,F)**
x_1 (low-lying, poor data)	High (0.9)	High gap (0.6)	Weak protection (0.2)	(0.9,0.6,0.2)
x_2 (urban center, good data)	Medium (0.6)	Low gap (0.2)	Strong protection (0.6)	(0.6,0.2,0.6)
x_3 (elevated, uncertain terrain)	Low (0.3)	Medium gap (0.5)	Unknown protection (0.5)	(0.3,0.5,0.5)

Remark:

- x_1 is highly exposed based on rainfall data and poorly protected. However, due to missing historical data (e.g., gaps in tide gauge records), the indeterminacy is high.
- x_2 is moderately at risk but well-protected by sea walls and has reliable records, resulting in low indeterminacy and high falsity (i.e., low actual risk).
- x_3 has uncertain terrain and infrastructure status. Although rainfall data shows low risk, both indeterminacy and falsity are moderate due to ambiguity in elevation and protective coverage.

Neutrosophic Score and Ranking

To prioritize interventions, we can compute a Neutrosophic Score for each location using a decision-support formula (one possible form):

$$S(x) = T(x) - F(x)$$

This reflects the net risk considering how true flood vulnerability is minus the confidence in flood protection. Table 14 shows neutrosophic scores ranking coastal locations by subtracting flood protection confidence from flood vulnerability to prioritize risk.

Table 14. Neutrosophic scores and ranking of coastal locations based on flood risk assessment

Location	T	F	S(x) = T − F	Interpretation
x_1	0.9	0.2	0.7	High flood risk
x_2	0.6	0.6	0.0	Neutral risk (protected)
x_3	0.3	0.5	−0.2	Low risk or overestimated

You can also incorporate indeterminacy in more advanced forms:

$$S'(x) = T(x) - F(x) - \alpha \cdot I(x)$$

Where $\alpha \in [0, 1]$ controls how heavily uncertainty penalizes the score.
For example, if $\alpha = 0.5$, then for x_1:

$$S'(x_1) = 0.9 - 0.2 - 0.5 \cdot 0.6 = 0.9 - 0.2 - 0.3 = 0.4$$

Still at risk, but uncertainty reduces confidence.

Benefits of the Neutrosophic Approach

- **Uncertainty-aware:** Models incomplete sensor coverage and expert disagreement.
- **Flexible:** Can incorporate real-time and historical data from diverse sources.
- **Decision-relevant:** Helps prioritize areas for mitigation, infrastructure upgrades, or further study.

Using neutrosophic sets to evaluate flood risk allows planners to:

- Combine quantitative (e.g., rainfall data) and qualitative (e.g., expert knowledge) sources,
- Explicitly model uncertainty due to missing or unreliable data,
- Support robust and adaptive planning under complex, real-world conditions.

DECISION MATRICES USING NEUTROSOPHIC SETS

Disaster response planning involves choosing among multiple alternatives (e.g., emergency technologies, policy frameworks, infrastructure investments) under uncertainty. Classical decision-making tools often fall short in such complex environments due to incomplete, inconsistent, or imprecise data.

To address these challenges, neutrosophic decision matrices extend traditional decision matrices by incorporating:

- **Truth (T):** To what extent a policy is expected to succeed.
- **Indeterminacy (I):** Degree of uncertainty or lack of knowledge about the outcome.
- **Falsity (F):** To what extent the policy is expected to fail.

Neutrosophic Decision Matrix

Let:

- $A = \{A_1, A_2, \ldots, A_m\}$ be a set of alternatives (e.g., policy options or technologies).
- $C = \{C_1, C_2, \ldots, C_m\}$ be a set of evaluation criteria (e.g., cost, speed, environmental impact, resilience).
- Each entry in the matrix is a neutrosophic triplet:

$$D_{ij} = \langle T_{ij}, I_{ij}, F_{ij} \rangle$$

Where:

- $T_{ij} \in [0,1]$: Degree of truth or support for alternative Ai under criterion C_j,
- $I_{ij} \in [0,1]$: Degree of indeterminacy or uncertainty,
- $F_{ij} \in [0,1]$: Degree of falsity or opposition.

Subject to:

$$0 \leq Tij + Iij + Fij \leq 3$$

Disaster Response Technologies

Let's assume we are evaluating three technological interventions for flood disaster response under four criteria:

Alternatives

- A_1: Drone-based emergency delivery system
- A_2: AI-based flood forecasting system
- A_3: Mobile water purification units

Criteria

- C_1: Deployment speed
- C_2: Cost efficiency
- C_3: Community Acceptance
- C_4: Environmental impact

Table 15 presents neutrosophic evaluations of three flood response technologies across key criteria including speed, cost, acceptance, and environmental impact.

Table 15. Neutrosophic decision matrix evaluating flood disaster response technologies across deployment speed, cost efficiency, community acceptance, and environmental impact

Criteria ↓ / Options →	A_1: Drones	A_2: AI Forecasting	A_3: Water Purifiers
C_1: Speed	⟨0.9, 0.05, 0.05⟩	⟨0.7, 0.2, 0.1⟩	⟨0.6, 0.3, 0.1⟩
C_2: Cost	⟨0.6, 0.2, 0.2⟩	⟨0.8, 0.1, 0.1⟩	⟨0.7, 0.2, 0.1⟩
C_3: Acceptance	⟨0.7, 0.2, 0.1⟩	⟨0.5, 0.3, 0.2⟩	⟨0.9, 0.05, 0.05⟩
C_4: Environment	⟨0.8, 0.1, 0.1⟩	⟨0.7, 0.2, 0.1⟩	⟨0.6, 0.25, 0.15⟩

Interpretation of Neutrosophic Values

Each entry reflects three aspects:

- **Truth (T):** Confidence in the performance of an alternative under the criterion.
- **Indeterminacy (I):** Ambiguity in data or expert opinion (e.g., due to lack of implementation history).
- **Falsity (F):** Known negative impacts or expected failure modes.

For example, $D_{31} = \langle 0.6, 0.3, 0.1 \rangle$ indicates that water purifiers are somewhat quick to deploy (60%), though there's moderate uncertainty (30%) and minimal expected failure (10%) under the **speed** criterion.

Aggregation and Scoring

To rank alternatives, we can aggregate the neutrosophic values across all criteria using weighted averages. Let w_j be the weight of criterion C_j, such that $\sum w_j = 1$. Then, for alternative A_i:

$$T_i = \sum_{j=1}^{m} w_j \cdot T_{ij}, \quad I_i = \sum_{j=1}^{m} w_j \cdot I_{ij}, \quad F_i = \sum_{j=1}^{m} w_j \cdot F_{ij}$$

A **net score** or decision index can then be calculated as:

$$S_i = T_i - F_i - \alpha \cdot I_i$$

Where $\alpha \in [0, 1]$ penalizes uncertainty. A higher Si suggests a more desirable option.

Advantages of the Neutrosophic Matrix Approach

- **Captures uncertainty** and hesitation explicitly, unlike traditional crisp or fuzzy matrices.
- **Accommodates incomplete data** and subjective judgment from multiple experts.
- **Flexible for dynamic environments** like post-disaster scenarios, where conditions and information change rapidly.
- **Supports multi-criteria decision-making** where trade-offs are inevitable.

Environmental Recovery and Sustainability

Post-Disaster Environmental Monitoring

In recovery phases, neutrosophic sets can help manage uncertainty in environmental data streams (e.g., air quality after wildfires, soil contamination after floods). Sensors may fail, human observations may conflict, and scientific interpretations may diverge. Neutrosophic aggregation methods help synthesize these inputs into actionable insights.

Evaluating Technological Pathways

Technologies such as remote sensing, AI-based early warning systems, and sustainable infrastructure often present trade-offs:

- **T:** Expected environmental benefit.
- **I:** Unknown side-effects or long-term impacts.
- **F:** Degree of unsuitability or risk.

Using a neutrosophic multi-criteria decision-making (MCDM) framework, planners can assess multiple technologies for post-disaster recovery, considering cost, ecological footprint, community resilience, and uncertainty.

Applications of Neutrosophic Sets in Disaster and Environmental Recovery

- **Flood Risk Assessment**
    - o Combine satellite rainfall data, tide gauge records, and expert opinions to evaluate flood-prone areas.
    - o Neutrosophic scores help balance likelihood (T), uncertainty due to missing data (I), and protection measures (F).
- **Wildfire Management**
    - o Assess strategies like controlled burns, evacuation drills, and AI sensor networks.
    - o Supports ranking options by weighing benefits, risks, and uncertainties.
- **Emergency Logistics and Relief Operations**
    - o Facility location and routing under uncertain conditions (e.g., blocked roads, aftershocks).
    - o Decision matrices help optimize drones, water purifiers, and mobile hospitals with T–I–F evaluation.
- **Earthquake Response Planning**
    - o Use neutrosophic graphs to decide locations of emergency response centers where both data reliability and uncertainty are high.
- **Post-Disaster Environmental Monitoring**
    - o Model soil contamination, air quality, or water pollution where sensor data may be incomplete or conflicting.
    - o Useful for prioritizing cleanup and rehabilitation.
- **Sustainable Technology Selection**
    - o Compare renewable vs. conventional recovery technologies (e.g., solar pumps vs. diesel pumps) with fuzzy–neutrosophic MCDM.
    - o Incorporates environmental benefit (T), cost uncertainty (I), and feasibility risks (F).
- **Waste Management in Disaster Zones**

- Site selection for debris disposal or waste treatment using neutrosophic distance-based similarity measures.

- **Policy and Governance**
  - Design neutrosophic-based frameworks for resource allocation, community engagement, and fairness in recovery efforts.

CASE STUDY: WILDFIRE MANAGEMENT IN A NEUTROSOPHIC FRAMEWORK

Consider a region in California that frequently experiences wildfires. Local authorities must prioritize actions such as controlled burns, evacuation planning, and sensor deployment.

To evaluate wildfire mitigation strategies, we define a neutrosophic decision matrix using three key dimensions:

- **T (Truth/Benefit):** The degree to which a strategy is known to be beneficial.
- **I (Indeterminacy/Uncertainty):** The degree of unknowns or lack of reliable information.
- **F (Falsity/Risk):** The degree of potential harm, cost, or ineffectiveness associated with the strategy.

The neutrosophic values for each strategy are assessed through expert consultation, past performance data, and simulation outcomes are presented in Table 16. Table 16 summarizes the neutrosophic assessment of wildfire strategies, balancing benefits, uncertainties, and risks for informed decision-making.

Table 16. Neutrosophic evaluation of wildfire management strategies based on benefit, uncertainty, and risk

Strategy	T (Benefit)	I (Uncertainty)	F (Risk)
Controlled burns	0.7	0.2	0.1
Evacuation drills	0.9	0.05	0.05
AI sensor networks	0.6	0.3	0.1

Interpretation

- **Evacuation drills** are perceived as the most reliable strategy, offering high benefit (T = 0.9), with very low uncertainty (I = 0.05) and minimal risk (F = 0.05). This makes them a top priority in emergency planning.
- **Controlled burns**, while effective in reducing fuel load (T = 0.7), carry moderate uncertainty (I = 0.2) and some risk (F = 0.1) due to the potential for the fire to spread unintentionally.
- **AI sensor networks** offer emerging potential for early wildfire detection (T = 0.6), but high uncertainty (I = 0.3) reflects their technological novelty and challenges in deployment over large terrains.

Using neutrosophic weighted aggregation, emergency managers can prioritize low-risk, high-benefit, and low-uncertainty actions while exploring further research where indeterminacy is high.

CHALLENGES AND FUTURE DIRECTIONS

- **Computational Complexity:** Neutrosophic models require advanced computation and robust data structures.
- **Data Availability:** High indeterminacy often results from insufficient data collection infrastructure.
- **Interdisciplinary Integration:** Bridging mathematics with social and environmental systems is essential for practical application.

Future research should focus on the development of neutrosophic-based AI tools, integration with **geospatial data systems**, and training programs for emergency response personnel.

CONCLUSION

The increasing frequency and severity of natural disasters—driven by climate change and unsustainable development—demand advanced decision-making frameworks capable of handling uncertainty, incomplete data, and conflicting information. Traditional models, such as classical logic and even fuzzy sets, often fall short in addressing these complexities. This chapter has demonstrated how neutrosophic sets, an extension of fuzzy and intuitionistic fuzzy logic, provide a powerful mathematical tool for modeling uncertainty, indeterminacy, and contradiction in disaster response and environmental recovery.

Key Contributions of Neutrosophic Sets

- **Enhanced Representation of Uncertainty**
  - Unlike fuzzy sets (which only consider membership) or intuitionistic fuzzy sets (which add non-membership and hesitation), neutrosophic sets independently model truth (T), indeterminacy (I), and falsity (F).
  - This allows for more nuanced decision-making, particularly in scenarios where data is incomplete (high I), expert opinions conflict (high T and F simultaneously), or risks are ambiguous.
- **Applications in Disaster Response**
  - **Flood Risk Assessment**: Neutrosophic scores help prioritize interventions by balancing flood likelihood (T), protective infrastructure (F), and data reliability (I).
  - **Wildfire Management**: Strategies like controlled burns or AI-based sensors are evaluated using T-I-F triples, enabling trade-off analysis under uncertainty.
  - **Emergency Logistics**: Neutrosophic decision matrices optimize resource allocation (e.g., drone deployments vs. water purifiers) by weighing speed, cost, and community acceptance.
- **Environmental Recovery and Sustainability**
  - **Post-Disaster Monitoring**: Neutrosophic logic synthesizes conflicting environmental data (e.g., air quality post-wildfires, soil contamination post-floods).
  - **Sustainable Technology Selection**: Evaluates trade-offs in green infrastructure (e.g., solar pumps vs. diesel pumps) by quantifying environmental benefit (T), cost uncertainty (I), and feasibility risks (F).

Challenges and Future Directions

While neutrosophic sets offer significant advantages, challenges remain:

- **Computational Complexity**: Aggregating T-I-F values requires efficient algorithms for large-scale applications.
- **Data Integration**: High indeterminacy (I) often stems from sparse or unreliable data, necessitating better sensor networks and expert systems.

- **Interdisciplinary Adoption**: Bridging mathematics with social, environmental, and engineering disciplines is critical for real-world implementation.

Future research should focus on:

- **Real-Time Neutrosophic Systems**: Embedding T-I-F logic in AI-driven disaster early warning systems.
- **Policy Frameworks**: Developing guidelines for neutrosophic-based risk assessment in government and NGO planning.
- **Education and Training**: Building capacity among emergency responders in neutrosophic decision-making tools.

Neutrosophic sets represent a paradigm shift in handling uncertainty, moving beyond binary or probabilistic approaches to embrace the inherent contradictions and unknowns of real-world crises. By providing a structured yet flexible framework, they empower decision-makers to:

- **Navigate ambiguity** in disaster scenarios.
- **Optimize resource allocation** under constraints.
- **Promote sustainable recovery** through data-driven, adaptive strategies.

As climate-related disasters grow more unpredictable, the adoption of neutrosophic logic—alongside emerging technologies—will be pivotal in building resilient, equitable, and efficient disaster management systems. This chapter underscores its transformative potential, urging further exploration and application across global environmental and humanitarian efforts.

REFERENCES

Abdel-Basset, M., Gamal, A., Manogaran, G., Son, L., & Long, H. (2019a). A novel group decision making model based on neutrosophic sets for heart disease diagnosis. *Multimedia Tools and Applications*, *79*, 9977–10002. DOI: 10.1007/s11042-019-07742-7

Abdel-Basset, M., & Mohamed, M. (2019b). A novel and powerful framework based on neutrosophic sets to aid patients with cancer. *Future Generation Computer Systems*, *98*, 144–153. DOI: 10.1016/J.FUTURE.2018.12.019

Admin, A., J, T., & S, G. (2024). IDLTM-DMT: Intelligent Deep Learning based Trust Management with Decision Making Tool for Healthcare Internet of Things and Big Data Environment with Neutrosophic Set Analysis. International Journal of Neutrosophic Science. DOI: 10.54216/ijns.240430

Al-Omeri, W., & Kaviyarasu, M. (2024). Study on Neutrosophic Graph with Application on Earthquake Response Center in Japan. *Symmetry*, *16*, 743. DOI: 10.3390/sym16060743

Atanassov, K. (1986). Intuitionistic fuzzy sets. *Fuzzy Sets and Systems*, *20*(1), 87–96.

Das, A. K. (2016). On partially included intuitionistic fuzzy rough relations. *Afrika Matematika*, *27*, 993–1001. DOI: 10.1007/s13370-016-0395-2

Das, A. K. (2018). Weighted fuzzy soft multiset and decision-making. *International Journal of Machine Learning and Cybernetics*, *9*, 787–794. DOI: 10.1007/s13042-016-0607-y

Das, A. K., Das, R., Das, S., Debnath, B. K., Granados, C., Shil, B., & Das, R. (2025c). A comprehensive study of neutrosophic SuperHyper BCI-semigroups and their algebraic significance. *Transactions on Fuzzy Sets and Systems*, *4*(2), 80–101. DOI: 10.71602/tfss.2025.1198050

Das, A. K., Das, R., Datta, R., & Granados, C. (2025d). An Innovative Approach to Group Decision-Making Based on Weighted Hypersoft Expert System. In Smarandache, F., Jdid, M., & Leyva-Vazquez, M. (Eds.), *Neutrosophic and Plithogenic Inventory Models for Applied Mathematics* (pp. 483–518). IGI Global Scientific Publishing. DOI: 10.4018/979-8-3693-3204-7.ch019

Das, A. K., & Granados, C. (2021). An advanced approach to fuzzy soft group decision-making using weighted average ratings. *SN Computer Science*, *2*, 471. DOI: 10.1007/s42979-021-00873-5

Das, A. K., & Granados, C. (2022a). A new fuzzy parameterized intuitionistic fuzzy soft multiset theory and group decision-making. *Journal of Current Science and Technology, 12*(3), 547–567. DOI: 10.14456/jcst.2022.42

Das, A. K., & Granados, C. (2022b). IFP-intuitionistic multi fuzzy N-soft set and its induced FP-Hesitant N soft set in decision-making. *Decision Making: Applications in Management and Engineering, 5*(1), 67–89. DOI: 10.31181/dmame181221045d

Das, A. K., & Granados, C. (2023a). IFP-intuitionistic multi fuzzy N-soft set and its induced IFP-hesitant N-soft set in decision-making. *Journal of Ambient Intelligence and Humanized Computing, 14*, 10143–10152. DOI: 10.1007/s12652-021-03677-w

Das, A. K., & Granados, C. (2023b). Preference intuitionistic fuzzy rough relation and its theoretical approach. *Journal of the Indian Mathematical Society, 90*(3-4), 199–212.

Das, A. K., Granados, C., & Bhattacharya, J. (2022). Some new operations on fuzzy soft sets and their applications in decision-making. *Songklanakarin Journal of Science and Technology, 44*(2), 440–449. DOI: 10.14456/sjst-psu.2022.61

Das, A. K., Gupta, N., & Granados, C. (2023). Weighted hesitant bipolar-valued fuzzy soft set in decision-making. *Songklanakarin Journal of Science and Technology, 45*(6), 681–690. https://sjst.psu.ac.th/journal/45-6/10.pdf

Das, A. K., Gupta, N., Mahmood, T., Tripathy, B. C., Das, R., & Das, S. (2024). Assessing anthropogenic influences on the water quality of Gomati River using an innovative weighted fuzzy soft set based water pollution rating system. *Discover Water, 4*, 73. DOI: 10.1007/s43832-024-00136-3

Das, A. K., Gupta, N., Mahmood, T., Tripathy, B. C., Das, R., & Das, S. (2025a). An efficient water quality evaluation model using weighted hesitant fuzzy soft sets for water pollution rating. In Kumar, A., Kumar, P., Rathee, S., & Kumar, B. (Eds.), *Mechatronics: Concepts, Tools, Applications, and New Trends* (1st ed., p. 17). CRC Press., DOI: 10.1201/9781003494478-10

Das, A. K., Gupta, N., Mahmood, T., Tripathy, B. C., Das, R., & Das, S. (2025b). An innovative fuzzy multi-criteria decision making model for analyzing anthropogenic influences on urban river water quality. *Iran Journal of Computer Science, 8*, 103–124. DOI: 10.1007/s42044-024-00211-x

Das, A. K., Mahmood, T., Das, R., & Das, S. (2025e). Optimizing Inventory Levels in Retail: Fuzzy Logic-Based Decision Support Systems for Adaptive Inventory Management. In Jana, S., Basu Mallik, B., Sarkar, A., & Jana, C. (Eds.), *Applications of Fuzzy Logic in Decision Making and Management Science. Information Systems Engineering and Management* (Vol. 29). Springer. DOI: 10.1007/978-3-031-77719-6_8

Das, S., Roy, B., Kar, M., Kar, S., & Pamučar, D. (2020). Neutrosophic fuzzy set and its application in decision making. *Journal of Ambient Intelligence and Humanized Computing*, *11*, 5017–5029. DOI: 10.1007/s12652-020-01808-3

Ganeshkumar, L., Palanisami, D., & Mohan, N. (2024). Neutrosophic fusion of multimodal brain images: Integrating neutrosophic entropy and feature extraction. *Applied Soft Computing*, *155*, 111462. DOI: 10.1016/j.asoc.2024.111462

Granados, C., Das, A.K., & Das, B. (2022). Some continuous neutrosophic distributions with neutrosophic parameters based on neutrosophic random variables. Advances in the Theory of Nonlinear Analysis and its Applications, 33(2), 293-308. DOI: 10.31197/atnaa.1056480

Granados, C., Das, A.K., & Osu, B. (2023). Weighted neutrosophic soft multiset and its application to decision making. Yugoslav journal of operations research, 33(2), 293-308. DOI: 10.2298/YJOR220915034G

Guo, Y., & Şengür, A. (2014). A novel image edge detection algorithm based on neutrosophic set. *Computers & Electrical Engineering*, *40*, 3–25. DOI: 10.1016/j.compeleceng.2014.04.020

Karataş, E., Yolcu, A., & Ozturk, T. (2023). Effective neutrosophic soft set theory and its application to decision-making. *Afrika Matematika*, *34*. Advance online publication. DOI: 10.1007/s13370-023-01101-4

Khan, M., Son, L., Ali, M., Chau, H., Na, N., & Smarandache, F. (2018). Systematic Review of Decision Making Algorithms in Extended Neutrosophic Sets. *Symmetry*, *10*, 314. DOI: 10.3390/SYM10080314

Li, D., Yang, L., Liu, S., & Tan, R. (2024). Multi-attribute decision making for deep learning-based typhoon disaster assessment. *Journal of Intelligent & Fuzzy Systems*, *46*, 6657–6677. DOI: 10.3233/jifs-235315

Lu, L., & Luo, X. (2020). Emergency Transportation Problem Based on Single-Valued Neutrosophic Set. *Discrete Dynamics in Nature and Society*, *2020*, 1–8. DOI: 10.1155/2020/4813497

Mahizha, M., & M., I. (2025). A Comprehensive Approach to Solid Waste Management Site Selection Using Simplified Neutrosophic Distance-Based Similarity Measures with N-Valued T-Spherical Fuzzy Neutrosophic Sets. International Journal of Neutrosophic Science. .DOI: 10.54216/ijns.260125

Manna, S., Basu, T., & Mondal, S. (2020). A soft set based VIKOR approach for some decision-making problems under complex neutrosophic environment. *Engineering Applications of Artificial Intelligence*, *89*, 103432. DOI: 10.1016/j.engappai.2019.103432

Mohammadi, S., Darestani, S., Vahdani, B., & Alinezhad, A. (2020). A robust neutrosophic fuzzy-based approach to integrate reliable facility location and routing decisions for disaster relief under fairness and aftershocks concerns. *Computers & Industrial Engineering*, *148*, 106734. DOI: 10.1016/j.cie.2020.106734

Molodtsov, D. (1999). Soft set theory: First results. *Computers & Mathematics with Applications (Oxford, England)*, *37*(4–5), 19–31.

Mukherjee, A., & Das, A. K. (2015a). Application of interval valued intuitionistic fuzzy soft set in investment decision making, *2015 Fifth International Conference on Advances in Computing and Communications (ICACC)*, Kochi, India, 2015, pp. 61-64, https://doi.org/DOI: 10.1109/ICACC.2015.37

Mukherjee, A., & Das, A. K. (2015b). Interval valued intuitionistic fuzzy soft multi set theoretic approach to decision making problems, 2015 International Conference on Computer, Communication and Control (IC4), Indore, India, 2015, 1-5. DOI: 10.1109/IC4.2015.7375640

Mukherjee, A., & Das, A. K. (2015c). Relations on intuitionistic fuzzy soft multi sets. In Kim, K. (Ed.), *Information Science and Applications. Lecture Notes in Electrical Engineering* (Vol. 339). Springer. DOI: 10.1007/978-3-662-46578-3_71

Mukherjee, A., & Das, A. K. (2016a). Application of fuzzy soft multi sets in decision-making problems. In Proceedings of 3rd International Conference on Advanced Computing, Networking and Informatics. Smart Innovation, Systems and Technologies, vol. 43. Springer. DOI: 10.1007/978-81-322-2538-6_3

Mukherjee, A., & Das, A. K. (2016b). Topological structure formed by intuitionistic fuzzy rough relations. *Journal of the Indian Mathematical Society*, *83*(1-2), 135–144.

Mukherjee, A., & Das, A. K. (2022). Einstein operations on fuzzy soft multisets and decision making. *Boletim da Sociedade Paranaense de Matematica*, *40*, 1–10. DOI: 10.5269/bspm.32546

Mukherjee, A., & Das, A. K. (2023). Essentials of fuzzy soft multisets: Theory and applications, Springer Nature Singapore Pte Ltd. DOI: 10.1007/978-981-19-2760-7

Nguyen, G., Son, L., Ashour, A., & Dey, N. (2017). A survey of the state-of-the-arts on neutrosophic sets in biomedical diagnoses. *International Journal of Machine Learning and Cybernetics*, *10*, 1–13. DOI: 10.1007/s13042-017-0691-7

Omar, R. M., & Zeina, M. (2023). Neutrosophic Multinominal Logistic Regression Technique for Optimizing Adaptive Reuse of Historical Castles. International Journal of Neutrosophic Science. DOI: 10.54216/ijns.210305

Ravi, R., Al-Attar, B., Qudr, L., Abdulbaqi, A., Tawfeq, J., Sekhar, R., Shah, P., & Devaerakkam, M. (2025). Integrating Neutrosophic Theory for Improved Decision-Making in Wireless Body Area Networks: Enhancing Accuracy and Efficiency in Health Monitoring. International Journal of Neutrosophic Science. .DOI: 10.54216/ijns.250325

Ravi, R., Al-Attar, B., Qudr, L., Penubadi, H., Sekhar, R., Shah, P., Parihar, S., Kallam, S., Tawfeq, J., & Gheni, H. (2024). Efforts of Neutrosophic Logic in Medical Image Processing and Analysis. International Journal of Neutrosophic Science. .DOI: 10.54216/ijns.240428

Sarma, D., Das, A., & Bera, U. (2025). *A transportation model minimizing total cost and time in disaster struck zone for debris removal under a mixed neutrosophic environment*. OPSEARCH. DOI: 10.1007/s12597-025-00912-2

Shaik, S., Ahad, A., Murugavalli, N., Kavitha, V., Suryanarayana, S., & Raj, M. (2024). ODESMAN: Optimizing Decision-Making in Complex Environments: Integrating Neutrosophic and Fuzzy Logic for Advanced System Modeling. International Journal of Neutrosophic Science. .DOI: 10.54216/ijns.240226

Smarandache, F. (1998). *A unifying field in logics: Neutrosophy: Neutrosophic probability, set, and logic*. American Research Press.

Smarandache, F. (2005). Neutrosophic set: A generalization of the intuitionistic fuzzy sets. *International Journal of Pure and Applied Mathematics*, *24*, 287–297.

Tan, R., & Zhang, W. (2020a). Decision-making method based on new entropy and refined single-valued neutrosophic sets and its application in typhoon disaster assessment. *Applied Intelligence*, *51*, 283–307. DOI: 10.1007/s10489-020-01706-3

Tan, R., Zhang, W., & Chen, S. (2020b). Decision-making method based on grey relation analysis and trapezoidal fuzzy neutrosophic numbers under double incomplete information and its application in typhoon disaster assessment. *IEEE Access : Practical Innovations, Open Solutions*, *8*, 3606–3628. DOI: 10.1109/ACCESS.2019.2962330

Voskoglou, M. (2023). *Decision Making in Neutrosophic Environment*. PROOF., DOI: 10.37394/232020.2023.3.1

Yang, X., Zhou, X., Huang, B., Li, H., & Wang, T. (2024). A three-way decision method on multi-scale single-valued neutrosophic decision systems. *Artificial Intelligence Review*, *57*, 109. DOI: 10.1007/s10462-024-10733-2

Zadeh, L. A. (1965). Fuzzy sets. *Information and Control*, *8*(3), 338–353.

Chapter 5
Role of Artificial Intelligence in Water Management

Tarak Nandy
https://orcid.org/0000-0002-6370-0640
UCSI University, Malaysia

Sananda Bhattacharyya
https://orcid.org/0000-0001-8034-7634
Maldives Business School, Maldives

ABSTRACT

Artificial Intelligence (AI) is revolutionizing water management by addressing critical challenges such as water scarcity, pollution, and inefficient resource allocation. This chapter explores the transformative role of AI technologies, including Machine Learning (ML), Deep Learning (DL), Internet of Things (IoT), Predictive Analytics, and Computer Vision, in optimising water systems. Key applications include demand forecasting, leak detection, water quality monitoring, smart irrigation, and flood prediction, all of which enhance efficiency, sustainability, and resilience. Case studies, such as Barcelona's AI-powered water management system, demonstrate real-world successes in reducing water loss and improving operational performance. However, challenges must be addressed to ensure equitable and scalable solutions. The chapter also highlights future trends to further integrate AI into water governance. By fostering collaboration across sectors, AI can play a pivotal role in achieving water sustainability goals, ensuring safe and access to this vital resource for future generations.

DOI: 10.4018/979-8-3373-2700-6.ch005

INTRODUCTION

Water is a vital resource for life, ecosystems, and economic activities. However, the world is facing a growing water crisis due to population growth, climate change (Absalan et al., 2024), and inefficient management practices. Global water demand has been increasing at an alarming rate, driven by agricultural, industrial, and domestic needs (see Figure 1) (Review, 2025). These changes are further straining already limited water resources, particularly in arid and semi, arid regions. The resource of internal renewable freshwater is degrading day by day (Food and Agriculture Organization, 2025b). The freshwater resources (shown in Figure 2, Figure 3) clearly demonstrate the threats of the water crisis in future (Food and Agriculture Organization, 2025a). On the other hand, water scarcity is not just a problem of quantity but also quality. Pollution from industrial discharge, agricultural runoff, and untreated wastewater has contaminated water sources, making them unsafe for consumption and harming aquatic ecosystems.

Figure 1. Freshwater by country 2024

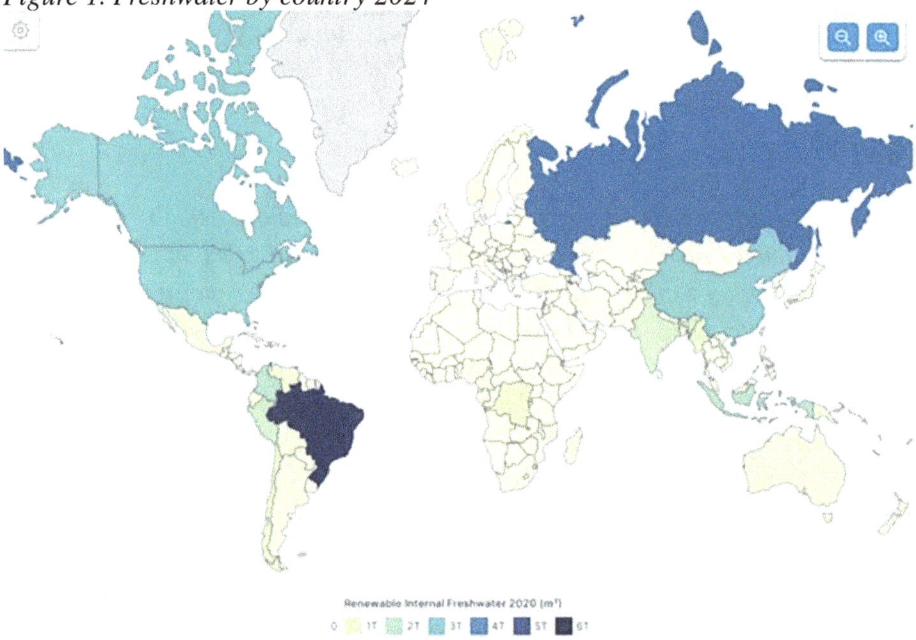

(Review, 2025)

Alternatively, AI has emerged as a powerful tool to address these challenges by enabling smarter, data, driven, and sustainable water management solutions (Ercan Oguztürk, 2025). This chapter explores the role of AI in revolutionizing water management, from optimizing resource allocation and predicting water demand to detecting leaks and improving water quality monitoring. It also discusses the challenges and ethical considerations of implementing AI in this domain, providing a comprehensive view of how AI can contribute to achieving global water sustainability goals.

On the other hand, beyond technical scarcity, water challenges are embedded in complex socioeconomic (Santos et al., 2023), political (Ricart et al., 2022), and environmental dynamics. Inequitable access to safe drinking water disproportionately affects marginalized communities, particularly in developing regions where affordability and infrastructure deficits limit service coverage. Politically, water governance is shaped by fragmented institutions, transboundary disputes, and competing sectoral priorities, often leading to inefficiencies and conflict. Environmental pressures (Kant & Wrat, 2025), including climate variability, land use change, and pollution from agriculture and industry, exacerbate resource degradation. These multidimensional challenges highlight that water management is not solely a question of technology, but one that requires integrating AI solutions with social equity, institutional reform, and environmental stewardship.

Figure 2. Top 10 countries on renewable internal freshwater 2020 (m3)

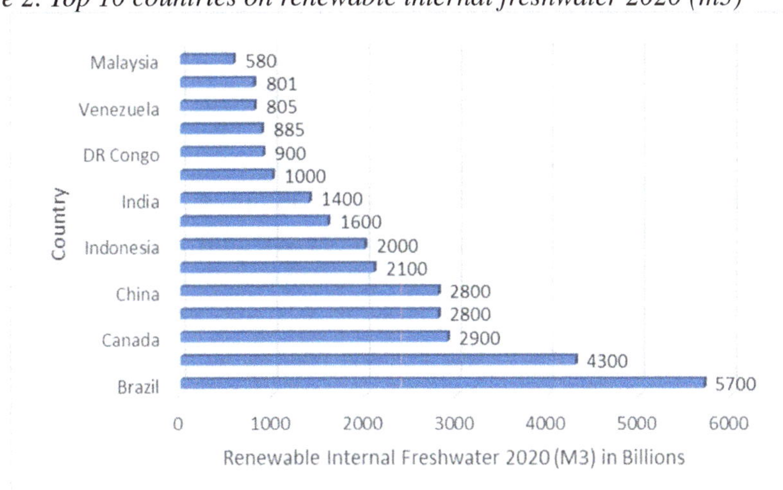

(Food and Agriculture Organization, 2025b)

Figure 3. Top 10 countries on Renewable Internal Freshwater 2020 (m3)

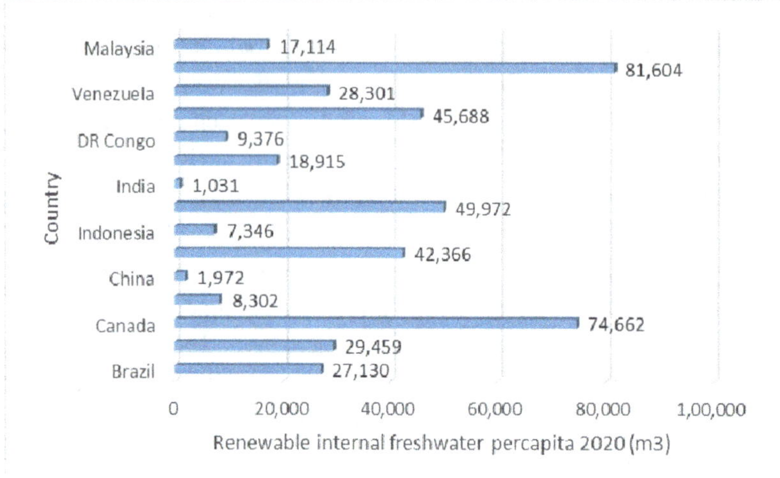

(Food and Agriculture Organization, 2025a)

Nevertheless, it is important to recognize that the application of AI in water management is not entirely new. Foundational studies in the 1990s explored the use of Artificial Neural Networks (ANNs) (Wang & Zheng, 1998) for rainfall, runoff modeling and groundwater prediction, while fuzzy logic (Makropoulos et al., 2003) was applied to irrigation scheduling and treatment plant optimization in the early 2000s. These pioneering efforts established the groundwork for today's more sophisticated models, demonstrating the feasibility of data-driven approaches long before the advent of deep learning and large-scale IoT networks. By building upon these early frameworks, contemporary research has been able to expand into real-time monitoring, hybrid AI integration, and large-scale predictive analytics, reflecting both continuity and evolution in the field.

AI TECHNOLOGIES IN WATER MANAGEMENT

The burgeoning global water crisis, characterized by escalating water scarcity, pervasive pollution, and inefficient resource allocation, presents one of the most formidable challenges of the 21st century. Driven by factors such as rapid population growth, the unpredictable impacts of climate change, and outdated management practices, the demand for water continues to rise at an alarming rate, straining already limited freshwater resources worldwide. Concurrently, the quality of available water is degrading due to contamination from industrial discharge, agricultural runoff, and untreated wastewater, rendering vital sources unsafe for consumption and detrimental to aquatic ecosystems.

In response to these critical issues, AI has emerged as a powerful and transformative tool, offering innovative solutions for smarter, data, driven, and more sustainable water management. AI encompasses a diverse set of technologies, including ML (Reyes-Acosta et al., 2025), predictive analytics (Nandy et al., 2025), and the IoT, all of which are poised to revolutionize how we manage this indispensable resource. ML algorithms can analyze vast amounts of data to identify patterns and make predictions, while IoT sensors can provide real, time monitoring of water systems. Predictive analytics can forecast water demand and detect potential issues before they become critical. These technologies enable more efficient and effective management of water resources, reducing waste and improving sustainability (Tian & Adriaens, 2025). This section will delve into the core AI technologies that are instrumental in modernizing water management, exploring how their synergistic application can optimize water usage, enhance water quality monitoring, predict and mitigate water, related risks, and ultimately contribute to global water sustainability goals.

ML in Water Management

ML, a core subfield of AI, involves the development of algorithms that can learn from and make decisions based on data (Dalibard et al., 2025). Unlike traditional rule, based programming, ML models improve their performance over time as they are exposed to more data. In water management, ML has emerged as a transformative technology, enabling more efficient, predictive, and adaptive strategies for managing one of the world's most critical natural resources.

Overview of ML in the Water Sector

The application of ML in water management is rapidly expanding across various domains, ranging from demand forecasting and anomaly detection to water quality monitoring and infrastructure optimisation. These models rely on historical and real, time data from sensors, meters, satellites, and operational logs to generate insights and guide decision, making (Fang et al., 2025).

ML is typically classified into three types:

- **Supervised learning**, where the model is trained on labeled datasets (e.g., historical water quality readings and their outcomes).
- **Unsupervised learning**, which identifies hidden patterns or groupings in unlabeled data (e.g., clustering of water consumption behavior).

- **Reinforcement learning**, where the model learns by interacting with an environment and receiving feedback (used in real, time operational control systems).

Applications of ML in Water Management

ML (Nandy & Nyundo, 2023) has emerged as a transformative force in water management, enabling data, driven decisions, operational efficiency, and predictive capabilities across a wide range of applications. By analyzing large and diverse datasets, from sensor readings and weather forecasts to historical usage patterns and satellite imagery, ML algorithms can uncover hidden trends, detect anomalies, and generate accurate forecasts. This allows utilities, governments, and environmental agencies to manage water resources more proactively, efficiently, and sustainably.

The versatility of ML allows it to be applied across multiple facets of the water sector. From predicting water demand and detecting leaks in pipelines to optimizing irrigation in agriculture and forecasting floods, ML models enhance both short, term responsiveness and long, term planning. These applications not only improve resource conservation and cost savings but also contribute to resilience in the face of climate change, urban growth, and rising water scarcity (Mohammed et al., 2024). As access to data and computational power continues to grow, ML is set to play an even more critical role in shaping the future of intelligent water management.

Water Demand Forecasting. Accurate demand forecasting is crucial for water supply planning, resource allocation, and conservation strategies (Valéry et al., 2025). ML models such as Support Vector Machines (SVM), Random Forests, Gradient Boosting Machines (GBM), and Artificial Neural Networks (ANNs) can be trained on historical consumption data, weather patterns, population trends, and socio-economic factors to forecast water demand at household, municipal, or industrial levels. These predictions help utilities to optimise pump scheduling, reservoir management, and distribution planning, especially in water-scarce regions or during peak usage periods.

Leakage Detection and Asset Management. Undetected leaks in water distribution systems lead to significant water loss, energy waste, and financial costs. ML models analyze pressure, flow, and acoustic sensor data to identify anomalies that suggest pipeline leaks or bursts. Supervised classification models can differentiate between normal and abnormal operating conditions, while anomaly detection algorithms like Isolation Forests or Autoencoders help detect subtle leaks early (Meng et al., 2025). Predictive maintenance is another vital application. ML algorithms analyze maintenance records, usage patterns, and environmental conditions to predict the likelihood of equipment failure, enabling proactive asset management and minimizing downtime.

Water Quality Prediction. Monitoring and maintaining water quality is essential for public health and environmental sustainability. ML models are used to predict water quality parameters such as turbidity, pH, dissolved oxygen, temperature, and contaminant levels. These models help in early detection of pollution events, harmful algal blooms, or chemical imbalances in both drinking and surface water systems (Shin et al., 2025). Ensemble models and DLnetworks like LSTM are particularly effective at modeling the temporal dynamics of water quality, offering high accuracy over time.

Smart Irrigation and Agriculture. In precision agriculture, ML algorithms analyze data from soil sensors, satellite imagery, and climate forecasts to determine optimal irrigation schedules and volumes. This leads to significant water savings and improved crop yield. Decision, tree models and regression algorithms predict evapotranspiration rates, moisture loss, and crop water stress, helping farmers irrigate only when and where necessary (Abdelhamid et al., 2025). ML, based irrigation systems contribute to reducing groundwater extraction and making agriculture more climate, resilient.

Flood Forecasting and Risk Management. Flood prediction requires integrating meteorological data, hydrological models, topographic maps, and real, time sensor inputs. ML models such as CNNs and RNNs are trained to recognize patterns that precede flood events and to issue early warnings. These models continuously learn and improve as new data is incorporated, enhancing the accuracy and timeliness of alerts (Prieto et al., 2024). Hybrid models combining ML with physical models (e.g., SWAT, HEC, RAS) are also used for improved floodplain mapping and emergency planning.

ML Model Selection and Training

Choosing the right ML model depends on:

- **Data availability and quality** (volume, completeness, granularity)
- **Problem type** (classification, regression, anomaly detection)
- **Computational resources** (some models require high processing power)
- **Interpretability** (e.g., decision trees are easier to interpret than deep neural networks)

While numerous ML models have been applied in water management, their relative performance varies depending on the sub-domain. For instance, SVMs (Prabu et al., 2024) and Random Forests often outperform simple regression models in groundwater demand forecasting due to their ability to capture non-linear relationships. However, they may underperform in highly dynamic flood prediction tasks, where Recurrent Neural Networks (RNNs) or Long Short-Term Memory (LSTM) models provide greater accuracy by leveraging temporal dependencies (Mahesh et al., 2024). Similarly, ensemble methods such as GBM achieve robust performance across diverse datasets but can be computationally intensive, limiting real-time deployment in resource-constrained settings. This comparative lens reveals that no single model is universally optimal; instead, model selection should be tailored to the problem domain, data availability, and operational constraints.

Model Training Involves:

- Preprocessing and cleaning of datasets (handling missing values, normalization)
- Splitting data into training, validation, and testing sets
- Optimizing hyperparameters
- Evaluating performance using metrics like RMSE, MAE, F1, score, and ROC, AUC

Benefits of ML in Water Management

- **Efficiency**: Automates complex decision processes and optimizes system performance.
- **Accuracy**: Provides high, fidelity forecasts and anomaly detection.
- **Scalability**: Can handle large, diverse datasets from multiple sources.
- **Adaptability**: Learns and improves over time with new data.
- **Cost Reduction**: Reduces operational and maintenance costs through predictive insights.

In conclusion, the integration of ML into water management systems has ushered in a new era of efficiency, precision, and adaptability. By enabling accurate forecasting, real, time monitoring, and intelligent automation, ML helps address many of the critical challenges facing global water systems, including scarcity, infrastructure aging, and environmental degradation. Its applications, ranging from leak detection (Rajasekaran et al., 2025) and demand prediction to water quality assessment and smart irrigation, demonstrate its broad utility and transformative potential. As the water sector continues to evolve, the adoption of ML will be instrumental in developing more resilient, data, driven, and sustainable water management strategies.

DLin Water Management

DL, a subset of ML, is a powerful tool in the realm of artificial intelligence, distinguished by its use of ANN with multiple layers, hence the term "deep." It excels in processing large, complex, and high, dimensional datasets, such as images, videos, time, series sensor data, and satellite imagery (Lu et al., 2025). In the context of water management, DL technologies are being harnessed to solve problems that are either too complex for traditional analytical approaches or require real, time, automated decision, making capabilities.

DL models are structured to mimic the human brain's functioning, utilizing layers of interconnected neurons that can automatically learn representations from raw data. This capability makes DL especially suitable for tasks such as pattern recognition, classification, prediction, and anomaly detection, key requirements in modern water management systems.

Satellite and Aerial Image Analysis

One of the most prominent applications of DL in water management is in the analysis of satellite and drone imagery using Convolutional Neural Networks (CNNs). CNNs are specialized for image recognition tasks and are used to monitor surface water bodies, detect illegal water extraction, track the spread of pollutants, assess drought severity, and identify encroachments on wetlands (Jamali et al., 2025). These insights enable governments and water authorities to take timely, data, driven decisions for environmental conservation and policy enforcement.

Time Series Forecasting for Hydrology

Recurrent Neural Networks (RNNs), particularly Long Short, Term Memory (LSTM) networks, are suited for temporal forecasting and sequence modeling. They are employed to predict river water levels, reservoir inflows, rainfall patterns, and flood occurrences (Le et al., 2024). These models learn from historical data and continuously update themselves as new data becomes available, improving the accuracy of early warning systems and resource planning.

Water Quality Monitoring

DL is also used in water quality monitoring systems. Sensor networks generate high, frequency data streams on parameters such as pH, turbidity, conductivity, temperature, and the presence of harmful substances. Deep neural networks analyze this data to predict contamination events, determine pollutant sources, and suggest remediation strategies. This predictive approach is significantly more responsive than traditional threshold, based monitoring systems (Fang et al., 2025).

Wastewater Treatment Optimization

In wastewater treatment, DL models are utilized to optimize complex (Mathaba & Banza, 2023) chemical and biological processes. They can adjust chemical dosages, control aeration systems, and manage sludge processing, thereby improving treatment efficiency and reducing operational costs. Such intelligent control systems are particularly beneficial in dynamic or high, load environments where manual adjustments would be inefficient or error, prone.

Infrastructure Inspection Using Computer Vision

DL, based computer vision is also being employed to inspect water infrastructure. Cameras mounted on drones or robots collect video and photographic data of pipelines, dams, and treatment facilities. CNNs then process this data to identify cracks, corrosion, vegetation overgrowth, and structural deformities. This enables safer, faster, and more accurate inspections compared to manual methods, and helps prioritize maintenance (Chen & Liu, 2023).

Rainfall, Runoff Modeling

Another innovative application of DL is in rainfall, runoff modeling, which predicts how rainfall will translate into surface runoff. This information is essential for flood forecasting and watershed management. Traditional hydrological models often require extensive calibration and are limited in handling non, linear systems. DL models, in contrast, can learn these complex relationships directly from data, increasing forecasting reliability and operational responsiveness (Fukui et al., 2020).

In summary, DLoffers powerful capabilities that are transforming water management by enabling the analysis of complex datasets, automating infrastructure monitoring, and improving forecasting accuracy. From interpreting satellite imagery and predicting floods (de Melo et al., 2024) to optimizing wastewater treatment and identifying contamination events, DL technologies provide nuanced insights and rapid decision, making support that traditional models often cannot match. Despite challenges related to data requirements, computational resources, and interpretability, the growing accessibility of DLtools and advances in explainable AI are paving the way for broader adoption. As water systems become increasingly digitized, DL will play a crucial role in ensuring smarter, safer, and more sustainable water governance.

IoT in Water Management

The IoT is a network of interconnected physical devices embedded with sensors, software, and communication technologies that enable them to collect, transmit, and exchange data (Fevgas et al., 2025). In the context of water management, IoT plays a critical role in transforming traditional, reactive water systems into intelligent, proactive, and data, driven networks. By enabling real, time monitoring and control, IoT allows utilities, governments, and industries to manage water resources more efficiently, detect issues early, and support sustainability goals.

Real, Time Monitoring and Control

IoT sensors are deployed throughout the water infrastructure, including pipelines, reservoirs, treatment plants, irrigation systems, and natural water bodies. These sensors monitor a range of parameters such as flow rate (Chen, 2024), pressure, temperature, pH levels, turbidity, chemical concentrations, and water levels. The continuous data streams generated by these devices provide a comprehensive view of water systems in real time, enabling operators to detect anomalies such as leaks, blockages, or contamination events as they occur.

This real, time capability supports immediate decision, making and response, reducing water loss, preventing service interruptions, and mitigating environmental risks. For example, smart valves connected to IoT systems can automatically shut off flow in case of pipe bursts, minimizing damage and waste.

Leak Detection and Infrastructure Health

A major application of IoT in water management is the detection of leaks and faults in distribution networks. Pressure and acoustic sensors installed along pipelines can detect subtle changes that indicate potential leaks or bursts. These readings are analysed, often in conjunction with AI or ML, to accurately localise the problem and initiate maintenance alerts (Meng et al., 2025). IoT also supports asset condition monitoring by tracking vibrations, corrosion levels, and usage patterns of critical infrastructure. This enables predictive maintenance strategies, reducing downtime and extending the lifespan of assets.

Smart Water Metering

IoT, enabled smart meters are revolutionizing how water consumption is tracked and billed. These meters provide granular, real, time usage data to both consumers and utility providers. For users, this means greater transparency, the ability to detect leaks in homes or businesses, and the opportunity to adjust consumption habits. For utilities, smart metering improves billing accuracy, reduces manual labor, and allows for dynamic pricing models based on usage patterns and peak demand (Li et al., 2025). Moreover, aggregated metering data can be used to model regional consumption trends, identify inefficiencies, and inform long, term planning decisions.

Agriculture and Irrigation Management

In agriculture, IoT plays a key role in precision irrigation. Soil moisture sensors, weather stations, and flow meters feed data to a centralized platform, which determines the optimal timing and amount of irrigation based on crop type, soil conditions, and forecasted rainfall. This minimizes water waste, reduces energy use, and improves crop yields (Preite & Vignali, 2024). IoT also assists in nutrient management by monitoring soil chemistry and helping farmers make informed decisions about fertilizer application, further contributing to sustainable agriculture practices (Li & Song, 2025).

Water Quality Monitoring

Maintaining water quality is critical for both public health and ecosystem integrity. IoT devices placed in reservoirs, rivers, and distribution systems monitor parameters such as chlorine concentration, dissolved oxygen, microbial contamination, and heavy metal levels. These readings are transmitted in real time to central monitoring stations, where automated systems can issue alerts and trigger corrective actions (Gyparakis et al., 2025). By enabling continuous, automated monitoring, IoT significantly reduces the time and resources needed for manual sampling and laboratory analysis.

Flood and Stormwater Management

IoT systems contribute to early warning and response mechanisms in flood, prone areas. Rain gauges, river level sensors, and weather stations send real, time data to flood modeling systems that forecast risk and recommend evacuation or mitigation measures. In urban settings, smart drainage systems equipped with IoT can detect blockages, monitor stormwater flow, and adjust control gates to prevent overflow (de Melo et al., 2024). These systems help minimize the human and economic impact of natural disasters, especially as climate change increases the frequency and severity of extreme weather events.

Integration with AI and Cloud Platforms

The value of IoT is significantly enhanced when integrated with Artificial Intelligence and cloud computing (Morchid et al., 2025). AI algorithms process the massive volumes of IoT data to identify patterns, predict system behavior, and recommend actions. Meanwhile, cloud platforms ensure that data is accessible, secure, and scalable, supporting multi, site operations and remote management. Together, IoT, AI, and cloud technologies form the backbone of smart water management systems that are self, optimizing, adaptive, and resilient.

In conclusion, the integration of IoT technologies into water management systems has significantly enhanced the ability to monitor, analyze, and respond to water, related challenges in real time. From smart metering and leak detection to flood forecasting and precision irrigation, IoT applications enable data, driven decisions that improve efficiency, reduce waste, and support environmental sustainability. By providing continuous insights into water usage, infrastructure health, and environmental conditions, IoT empowers utilities, farmers, and policymakers to adopt proactive and adaptive management strategies. As connectivity and sensor technologies continue to advance, IoT will remain a cornerstone of next, generation water systems, contributing meaningfully to global water security and resilience.

Predictive Analytics in Water Management

Predictive analytics is a powerful branch of data science that uses statistical techniques, ML, and historical data to forecast future events or trends. In the context of water management, predictive analytics enables utilities, governments, and industries to move from reactive responses to proactive, data, driven decision, making (Noor et al., 2021). By anticipating challenges such as water shortages, flooding, contamination, and equipment failures, predictive models help optimize resource allocation, reduce risks, and improve operational efficiency.

Water Demand Forecasting

One of the primary applications of predictive analytics in water management is forecasting water demand. Utilities must balance supply with fluctuating demand influenced by seasonality, population growth, climate conditions, and socio, economic factors. Predictive models use time series analysis, regression algorithms, and ML techniques to estimate future consumption patterns at household, municipal, and regional scales. Accurate forecasting supports better reservoir management, infrastructure planning, and investment decisions, while also preventing over, extraction and service disruptions (Valéry et al., 2025).

Leak and Failure Prediction

Aging water infrastructure poses significant risks of leaks, pipe bursts, and pump failures, which can lead to massive water losses and costly repairs. Predictive analytics uses historical maintenance records, pressure readings, and environmental data to estimate the probability of component failure (Rajasekaran et al., 2025). Utilities can then prioritize inspections and maintenance based on the predicted risk, shifting from scheduled or reactive maintenance to a more cost, effective predictive maintenance model. This not only extends the lifespan of infrastructure but also ensures continuous, efficient service delivery.

Water Quality Forecasting

Predictive analytics is increasingly being used to model and forecast water quality parameters such as turbidity, pH, dissolved oxygen, microbial counts, and contaminant levels. These forecasts are generated by analyzing data from water treatment plants, distribution systems, and natural sources. For example, predictive models can estimate the risk of algal blooms based on nutrient loading, temperature, and sunlight data, enabling early intervention. Similarly, they can alert operators to the potential for chlorine decay or trihalomethane formation in distribution systems, ensuring timely corrective action (Dalibard et al., 2025).

Flood Risk Modeling

Flood prediction is a critical application of predictive analytics in both urban and rural water management. By combining real, time weather forecasts, historical flood records, topographic data, and hydrological models, predictive systems assess the likelihood and potential impact of flood events. These insights support emergency preparedness, early warning systems, and urban drainage planning. Municipalities can identify high, risk zones, deploy resources strategically, and design adaptive infrastructure to mitigate future events (de Melo et al., 2024).

Irrigation Optimization

In agriculture, predictive analytics helps determine the optimal timing and amount of irrigation by integrating data from soil sensors, crop models, evapotranspiration rates, and weather forecasts (Zhang & Li, 2025). These models support precision agriculture by minimizing water waste, improving yield, and reducing energy consumption associated with pumping. Predictive analytics also assists in planning for seasonal water availability, helping farmers align planting and harvesting with expected water supply.

Energy Consumption Optimization

Water treatment and distribution systems are energy, intensive. Predictive analytics can optimize energy usage by forecasting demand, identifying peak hours, and scheduling energy, efficient pump operations (Dalibard et al., 2025). It can also model how different operational scenarios impact energy and water efficiency, helping utilities lower costs and carbon emissions while maintaining service quality.

In conclusion, predictive analytics has become an essential tool for modern water management, enabling stakeholders to anticipate future challenges and make data, driven decisions that enhance operational efficiency and sustainability. By forecasting demand, detecting system failures, assessing water quality risks, and managing flood events, predictive models support a proactive approach that conserves resources and protects communities. As technologies evolve and data becomes more accessible, predictive analytics will play an increasingly central role in building resilient, adaptive, and intelligent water systems capable of addressing the demands of a changing climate and growing population.

Natural Language Processing (NLP) in Water Management

NLP is a subfield of Artificial Intelligence that enables computers to understand, interpret, and generate human language. NLP combines computational linguistics with statistical and ML techniques to process and analyze large volumes of unstructured textual data (Fevgas et al., 2025). While commonly associated with customer service chatbots and virtual assistants, NLP is increasingly being applied in specialized domains, including water management, where vast amounts of information are embedded in written reports, regulatory documents, user feedback, and scientific literature.

Understanding Unstructured Data

A significant portion of data in the water sector exists in unstructured formats, such as incident reports, maintenance logs, policy papers, technical manuals, and stakeholder communications. NLP technologies enable the extraction of meaningful information from this textual data, turning qualitative narratives into structured insights. For example, NLP algorithms can analyze thousands of citizen complaint records to identify patterns related to recurring pipeline failures or service outages across different regions.

Regulatory and Policy Analysis

Water governance is often guided by complex legal and regulatory frameworks. NLP tools can automatically scan and summarize long policy documents, extract clauses relevant to specific water quality standards, and compare legislative texts across jurisdictions. This supports policymakers, legal experts, and water utilities in staying compliant, identifying policy gaps, and adapting to evolving regulations. NLP can also facilitate multilingual translation and interpretation, especially in international water agreements or transboundary water resource management (Raimi & Phiri, 2025).

Customer Interaction and Sentiment Analysis

Water utilities receive continuous feedback from customers through emails, call transcripts, social media, and online surveys. NLP techniques such as sentiment analysis can evaluate public perception and satisfaction with water services. By identifying the tone, urgency, and recurring issues in customer communications, utilities can prioritize service improvements, tailor messaging strategies, and increase transparency and trust (Torwane et al., 2025). Sentiment trends can also act as early indicators of systemic problems or public discontent.

Automated Reporting and Documentation

NLP enables automation in generating technical reports and executive summaries from raw operational data. For example, an NLP system integrated into a smart water management dashboard can summarize daily performance metrics, highlight anomalies, and generate bullet, point reports for supervisors or city officials. This reduces the burden of manual reporting and ensures timely, consistent documentation. Additionally, NLP tools can convert speech, to, text during field inspections or community engagement meetings, enhancing data recording in real, time.

Scientific Literature Mining

Researchers and practitioners in water management often need to stay updated on the latest developments across hydrology, treatment technologies, remote sensing, and sustainability studies. NLP, powered literature mining tools can quickly search, summarize, and cluster scientific articles based on relevance and topic. These tools help identify emerging trends, benchmark technologies, and synthesize findings from hundreds of research papers efficiently, aiding evidence, based decision, making and innovation.

Intelligent Chatbots and Virtual Assistants

NLP also powers the development of intelligent virtual assistants and chatbots for water utilities. These systems can interact with customers, answer frequently asked questions, help report leaks, provide billing information, or offer conservation tips. Available 24/7, chatbots reduce the workload on human operators and enhance customer engagement. In rural or under, resourced settings, voice, enabled NLP tools can support low, literacy populations in accessing essential water service information.

Integration with Other AI Systems

NLP does not operate in isolation. When combined with other AI technologies such as predictive analytics and ML, NLP can enhance overall system intelligence. For instance, an integrated platform might analyse maintenance logs (NLP), forecast component failure (ML), and recommend an optimised repair schedule. This multimodal approach enables more comprehensive and actionable insights (Rajasekaran et al., 2025).

In conclusion, NLP brings significant value to water management by transforming unstructured textual data into actionable insights. From analyzing maintenance logs and customer feedback to interpreting regulatory documents and automating reporting, NLP enhances operational efficiency, transparency, and responsiveness. By bridging communication gaps between systems and stakeholders, NLP supports better decision-making and more inclusive governance. As water systems continue to digitise and generate large volumes of text-based data, the role of NLP will become increasingly vital in ensuring informed, efficient, and people-centred water management.

Computer Vision in Water Management

Computer Vision is a subfield of AI that enables machines to interpret and understand visual information from the world, such as images and videos. Powered by DLand neural networks, computer vision mimics the human visual system and is increasingly being integrated into water management systems to automate monitoring, inspection, and analysis. As water infrastructure spans vast geographic areas and is often difficult to inspect manually, computer vision offers a scalable and efficient solution for visual data interpretation in real time (Chen & Liu, 2023).

Infrastructure Inspection and Maintenance

One of the most impactful applications of computer vision in water management is the inspection of critical infrastructure, including pipelines, dams, canals, reservoirs, and treatment plants. Using drones or robotic systems equipped with high, resolution cameras, video and image data can be collected and analyzed by AI algorithms to detect cracks, corrosion, sediment buildup, vegetation overgrowth, or other structural anomalies.

Convolutional Neural Networks (CNNs) are commonly used for such tasks due to their high accuracy in image classification and object detection (Rajasekaran et al., 2025). These models can automatically flag defects and categorize the severity of deterioration, allowing for proactive maintenance and risk mitigation. This reduces the dependency on manual inspections, increases worker safety, and minimizes downtime.

Flood Monitoring and River Surveillance

Computer vision also plays a significant role in monitoring natural water bodies and assessing flood risks. Cameras installed at riverbanks, drainage outlets, and urban flood zones can continuously capture video feeds, which are analyzed in real time to detect rising water levels, debris blockages, or abnormal flow patterns (de Melo et al., 2024). These systems are particularly valuable in flood-prone regions, where early detection can trigger alarms, initiate emergency protocols, and inform evacuation plans. Integration with AI models and weather forecasting systems further enhances predictive flood management.

Water Surface Monitoring and Pollution Detection

Surface water quality can be assessed using computer vision techniques that analyze visual indicators such as color, turbidity, surface films, and vegetation cover. For instance, satellite and drone imagery can be processed to detect algal blooms in lakes, oil spills in rivers, or plastic accumulation in coastal zones (Jamali et al., 2025). By training AI models on labelled datasets of pollution events, computer vision systems can recognise and localise contamination in large-scale water bodies with high speed and accuracy. These tools enable continuous environmental surveillance, supporting early intervention and pollution control efforts.

Smart Irrigation and Crop Monitoring

In agriculture, computer vision is integrated into smart irrigation systems and precision farming platforms. Cameras mounted on drones or tractors capture images of crop fields, which are analyzed to assess plant health, soil moisture levels, pest infestations, and water stress (Zhang & Li, 2025). AI models can detect subtle changes in leaf color, canopy density, and growth patterns, indicators that guide irrigation decisions. This enhances water use efficiency, improves crop yields, and reduces overwatering or under-irrigation. When combined with IoT and weather data, computer vision forms the visual intelligence layer of a smart agricultural ecosystem.

Wastewater Treatment and Industrial Applications

Computer vision is also used in industrial and municipal wastewater treatment facilities. Cameras monitor treatment tanks and processes to detect foam formation, sludge accumulation, and abnormal coloration, which may indicate process inefficiencies or contamination (Gyparakis et al., 2025). Automated visual inspections help operators maintain consistent treatment quality and reduce reliance on manual sampling. In industrial discharge systems, computer vision can identify violations of effluent standards and ensure regulatory compliance.

Real, Time Monitoring and Autonomous Systems

Modern computer vision systems are increasingly being deployed in real, time, edge, based environments. These systems are embedded in devices such as underwater robots, autonomous inspection units, and mobile monitoring platforms that operate in challenging or hazardous locations. (Rajasekaran et al., 2025). Real, time object detection and classification allow these systems to navigate, inspect, and analyze water infrastructure autonomously. For example, an autonomous robot equipped with vision sensors can traverse inside a pipeline and detect corrosion or blockages without human intervention.

Integration with GIS and Remote Sensing

When combined with Geographic Information Systems (GIS) and remote sensing technologies, computer vision enables large, scale spatial analysis of watersheds, wetlands, irrigation networks, and coastal areas. Satellite imagery analysed through DL models provides valuable data for mapping, change detection, and environmental impact assessment (Rajasekaran et al., 2025). This integration supports regional water resource planning, land use analysis, and climate adaptation strategies by delivering comprehensive visual insights over time and space.

In conclusion, computer vision is revolutionizing water management by enabling automated, accurate, and real, time analysis of visual data across diverse applications. From infrastructure inspection and flood monitoring to pollution detection and precision agriculture, it enhances situational awareness and supports proactive interventions. By reducing the reliance on manual inspections and enabling continuous surveillance, computer vision improves safety, efficiency, and responsiveness in managing both natural and built water systems. As imaging technologies and AI algorithms continue to advance, computer vision will play an increasingly central role in building intelligent, resilient, and sustainable water infrastructure.

Digital Twins in Water Management

A Digital Twin is a dynamic, real, time digital replica of a physical system, asset, or process that is continuously updated with live data from sensors, operational systems, and environmental sources. In water management, digital twins are being used to model and simulate water infrastructure, such as pipelines, reservoirs, treatment plants, and entire watersheds, to improve operational efficiency, forecasting, and strategic planning. They represent a convergence of technologies including IoT, AI, simulation modeling, and cloud computing, offering a powerful platform for managing water resources intelligently and proactively.

Concept and Core Functionality

A digital twin combines historical data, real, time sensor feeds, and predictive models to mirror the status and behavior of physical water systems. Unlike static models, digital twins evolve with changing inputs, providing a living, continuously updated virtual environment. Operators can interact with this twin to monitor system performance, run "what, if" scenarios, detect anomalies, and optimize operations in real time. This results in better decision, making and a reduced reliance on manual intervention or reactive strategies.

Applications in Water Utilities

In urban water utilities, digital twins are used to model entire distribution networks, including water flow, pressure zones, energy consumption, and leak detection. By simulating the network's response to varying demand, equipment failures, or pressure changes, utilities can anticipate problems before they occur and implement corrective actions with minimal disruption. For example, if a sensor detects a drop in pressure in a pipeline segment, the digital twin can help isolate the cause, such as a leak or valve malfunction, by comparing simulated outcomes to real, world data. This allows for faster incident response and more targeted maintenance.

Wastewater Treatment Optimization

Digital twins are increasingly applied in wastewater treatment plants (WWTPs) to simulate biological, chemical, and mechanical processes. These systems ingest real, time data on flow rates, pH, chemical dosages, and energy usage to optimize plant operations. Engineers can test new treatment strategies in the digital twin before implementing them in the physical plant, reducing risks and costs. Advanced digital twins also support adaptive control systems, where AI algorithms use feedback from the twin to dynamically adjust pumps, aerators, and chemical dosing systems in response to influent variability. This enhances treatment efficiency and minimizes environmental impact.

Stormwater and Flood Management

In stormwater systems, digital twins integrate hydrological models with real, time rainfall data, topographic information, and drainage infrastructure maps to simulate runoff and flooding scenarios. Cities use these models to design flood mitigation infrastructure, manage drainage during heavy storms, and issue early warnings. Decision, makers can visualize flood propagation, identify vulnerable zones, and prioritize interventions more effectively than with traditional static maps. The same approach can be scaled to river basins and watersheds, supporting integrated water resource management and inter, jurisdictional collaboration.

Scenario Planning and Resilience Testing

One of the most valuable features of digital twins is their ability to simulate multiple future scenarios. Planners can model the impact of population growth, climate change, drought, infrastructure upgrades, or policy changes on water supply and quality. This enables resilience testing and stress analysis, helping to design water systems that can adapt to uncertain future conditions. For example, a city can test how its water network would respond to a prolonged drought combined with a population surge and reduced reservoir inflows, allowing planners to proactively revise allocation strategies, storage requirements, and conservation programs.

Energy and Cost Optimization

Digital twins help reduce operational costs by enabling more efficient use of energy and resources. By simulating pump operations, valve configurations, and pressure zones, utilities can identify optimal operating points that minimize energy use while maintaining service quality. In treatment plants, digital twins can help minimize chemical consumption by adjusting dosing based on predicted influent characteristics. This real, time optimization leads to substantial savings over time, particularly in large and complex water systems.

Integration With AI and ML

When coupled with AI and ML, digital twins become even more powerful. Predictive algorithms analyze historical data patterns to forecast equipment failures, system overloads, or water quality events. These predictions are fed back into the digital twin to simulate outcomes and recommend preventive measures. This creates a self, learning, adaptive system capable of autonomous decision support (Gyparakis et al., 2025).

In conclusion, digital twins are poised to revolutionize water management by providing a dynamic, virtual replica of physical water systems. This innovative technology integrates real, time data from sensors and other sources, enabling comprehensive monitoring, predictive analytics, and optimized decision, making across the entire water cycle, from treatment plants to distribution networks. By simulating various scenarios like floods or droughts, digital twins empower utilities to proactively address challenges, reduce water loss, enhance water quality, and optimize operational efficiency, leading to significant cost savings and improved service reliability. Despite challenges related to data integration, standardization, and initial investment, the transformative potential of digital twins in building more resilient, sustainable, and adaptive water infrastructure is undeniable, making them an indispensable tool for safeguarding this vital resource for future generations.

Hybrid AI Systems in Water Management

Hybrid AI systems combine two or more AI techniques, such as ML, DL, NLP, Computer Vision, and rule, based expert systems, to solve complex, real, world problems more effectively than a single AI method alone. In the domain of water management, where data comes in many forms (numerical, textual, visual, temporal) and decisions are often multifactorial, hybrid AI systems provide a flexible and powerful approach to creating intelligent, adaptive, and efficient water systems (Zaresefat & Derakhshani, 2023).

Concept and Importance

The water sector presents diverse challenges that require integrating various types of data, such as real, time sensor readings, satellite imagery, historical consumption logs, maintenance reports, and climate forecasts. A hybrid AI system can ingest, process, and synthesize this multimodal data using a blend of algorithms best suited for each task. For instance, ML models may be used to forecast demand, NLP to analyze customer complaints, and computer vision to inspect infrastructure, all feeding into a unified decision, making platform.

By integrating different AI techniques, hybrid systems improve accuracy, robustness, and adaptability, making them especially valuable in complex or large, scale water environments where variability and uncertainty are high.

Advantages of Hybrid AI Systems

Hybrid AI systems, which combine different AI techniques (e.g., ML, neural networks, fuzzy logic) with traditional models, physical laws, or expert knowledge, offer significant advantages in water management due to the complex and dynamic nature of water systems. These advantages stem from their ability to overcome the limitations of individual AI approaches or purely physically, based models:

Enhanced Accuracy and Robustness. Traditional mechanistic models in hydrology often rely on simplified assumptions and struggle with complex, non, linear relationships. Purely data, driven AI models, while powerful, can sometimes lack interpretability or struggle with data scarcity, especially for extreme events. Hybrid AI models combine the strengths of both: they use AI to learn complex patterns from data while incorporating physical constraints or expert knowledge from traditional models. This leads to more accurate predictions and more robust performance, particularly in scenarios with limited data or during extreme hydrological events like floods and droughts. For example, a hybrid AI model might combine a neural network for predicting groundwater levels with a physically, based model for simulating flow and transport processes, improving prediction accuracy and ensuring the AI model is grounded in physical reality (Zaresefat & Derakhshani, 2023)

Improved Interpretability and Trust. One of the criticisms of "black box" AI models is their lack of transparency. Hybrid AI, by integrating domain knowledge or traditional models, can offer more interpretable results. This means water managers can better understand *why* a particular prediction or recommendation is made, fostering greater trust in the system and enabling more informed decision, making, which is crucial for critical infrastructure management.

Better Handling of Data Scarcity and Uncertainty. In many water management contexts, high, quality, comprehensive datasets are scarce. Hybrid AI can leverage existing physical models and expert knowledge to compensate for data gaps, making them more effective in data, poor environments. They can also better account for uncertainties inherent in hydrological processes, such as variations in rainfall, soil moisture, and land use changes.

Optimized Resource Allocation and Management. By providing more precise predictions of water demand, supply, and system behavior, hybrid AI enables more efficient allocation of water resources. This is critical for managing reservoirs, optimizing pump operations, reducing energy consumption in treatment and distribution, and minimizing water loss, especially in agricultural and urban settings. For instance, hybrid models have been shown to improve monthly runoff forecasting for reservoirs, leading to better planning and management of water resources (Wen & Feng, 2023).

Enhanced Resilience and Adaptive Capacity. Water systems face increasing pressures from climate change, population growth, and aging infrastructure. Hybrid AI systems contribute to building more resilient systems by providing advanced forecasting for extreme events (e.g., flash floods), enabling proactive measures, and facilitating adaptive management strategies. This helps water managers respond more effectively to changing environmental conditions and emerging challenges.

Broader Applicability. Hybrid AI models have shown success across a wide range of water management challenges, including groundwater quality and quantity assessment, surface water supply management, hydropower generation, irrigation optimization, climate change risk management, and urban flood prediction. Their flexibility allows them to be tailored to specific local conditions and challenges.

Integration With Human Expertise

Hybrid AI systems are not meant to replace human decision, makers but to augment their capabilities. Human experts define rules, interpret ambiguous results, and provide contextual judgment where AI models might lack domain, specific nuance. The best implementations are those where AI handles routine tasks, detects trends, and makes recommendations, while humans handle strategic decisions, oversight, and ethics.

In a summary, hybrid AI represents a significant leap forward in water management, integrating diverse AI techniques like ML, neural networks, and fuzzy logic with traditional models and expert knowledge. This synergistic approach addresses the limitations of individual AI methods, offering enhanced accuracy and robustness in critical applications such as groundwater management, water quality monitoring, flood prediction, and optimizing treatment plant operations (Preite & Vignali, 2024). By seamlessly combining data-driven insights with physically based understanding, Hybrid AI provides more comprehensive and interpretable solutions, leading to improved decision-making, reduced operational costs, increased efficiency, and greater resilience in water systems. While challenges remain in data availability, interoperability, and the need for specialized skills, the continued development and adoption of Hybrid AI are crucial for ensuring the sustainable and effective management of water resources in the face of escalating global demands and environmental uncertainties.

Edge AI in Water Management

Edge AI refers to the deployment of artificial intelligence algorithms directly on hardware devices, such as sensors, meters, controllers, and gateways, at or near the location where data is generated, rather than sending all data to centralized servers or cloud platforms for processing. In water management (Mathaba & Banza, 2023), Edge AI enables faster, real-time decision-making, reduces latency, lowers data transmission costs, and enhances system resilience, especially in remote or low-connectivity environments. This makes it a critical innovation for building decentralized, autonomous, and intelligent water infrastructure.

Concept and Advantages

Traditional AI, based water management systems rely heavily on cloud computing to collect, store, and analyze large volumes of sensor data. However, this model can be constrained by issues such as network latency, bandwidth limitations, cybersecurity risks (Nandy et al., 2024), and the inability to operate effectively in offline or rural settings.

Edge AI solves these problems by placing computing capabilities directly at the "edge" of the network, close to the data source. These devices can:

- Analyze sensor data in real time.
- Detect anomalies or trends.
- Trigger local actions automatically.
- Communicate summarized insights rather than raw data.

This localized intelligence leads to faster response times, improved system reliability, and lower operational costs, while also enhancing data privacy and security (Nandy et al., 2021).

Applications in Water Management

Edge AI, which processes data locally on devices (like sensors or drones) rather than in the cloud, is revolutionizing water management by enabling real, time decision, making in remote or resource, limited environments. By deploying AI algorithms directly on edge devices, such as smart meters, drones, or IoT sensors, utilities can instantly detect leaks, monitor water quality, and predict floods without relying on constant connectivity. This reduces latency, cuts bandwidth costs, and enhances resilience in areas with poor internet access. For example, edge, based acoustic sensors can identify pipe leaks on, site, while AI, powered drones inspect reservoirs for contamination, making water systems smarter, faster, and more efficient.

Leak Detection and Anomaly Response. Smart water meters or pipeline sensors equipped with Edge AI can detect unusual pressure drops, flow rates, or vibration patterns that suggest leaks or bursts. Upon detection, the system can autonomously shut off valves, issue alerts, or reroute water flow, preventing further loss and damage without requiring central intervention.

Smart Irrigation Systems. Edge AI is widely used in precision agriculture, where irrigation controllers process soil moisture, weather, and crop data locally to make watering decisions. For example, a moisture sensor with embedded AI might determine when and how much to irrigate based on environmental cues, reducing water waste and energy consumption, without depending on cloud connectivity.

Water Quality Monitoring. Edge, enabled water quality sensors deployed in rivers, lakes, or treatment plants can analyse real, time data on parameters such as pH, turbidity, dissolved oxygen, and contaminants. If thresholds are exceeded, local alarms are triggered, or chemical dosing is adjusted automatically, enabling responsive and autonomous operation, particularly important in remote water bodies or mobile monitoring units.

Remote Water Infrastructure Management. In geographically dispersed infrastructure, such as rural pumping stations, tanks, or aqueducts, Edge AI supports autonomous operation and monitoring. Devices can execute control functions, send critical updates, and operate offline for extended periods, syncing data with central systems when connectivity is restored. This makes water management feasible in underserved or disaster, prone regions.

Stormwater and Flood Management. Edge AI, equipped stormwater sensors and actuators in urban drainage systems can monitor rainfall intensity, water levels, and flow rates in real time. Local processing allows these systems to autonomously open or close floodgates, operate pumps, or reroute runoff to prevent localized flooding, without relying on external servers.

Technical Features and Design Considerations

Key features that make Edge AI suitable for water management include:

- **Low power consumption**: Many edge devices are battery, powered or solar, operated, making them ideal for field deployment.
- **Resilience**: Edge devices can operate in harsh conditions and continue functioning independently during communication failures.
- **Scalability**: New nodes can be added with minimal dependence on the central architecture.
- **Data filtering and compression**: Edge devices can summarise or compress data before transmitting, optimising network usage.

Designing Edge AI systems requires selecting appropriate hardware (e.g., microcontrollers, embedded GPUs), optimising algorithms for lightweight deployment, and ensuring secure firmware and update mechanisms.

In summary, edge AI is rapidly emerging as a transformative force in water management, bringing advanced analytical capabilities directly to the source of data within water infrastructure. By processing data locally on sensors and devices, Edge AI enables real, time monitoring, immediate anomaly detection, and rapid decision, making, which is crucial for preventing critical failures, optimising operational efficiency, and ensuring water quality. This decentralised approach reduces reliance on cloud connectivity, lowers latency, and enhances data privacy and security, making it ideal for remote or resource, constrained environments. While challenges such as hardware limitations, model optimisation for constrained devices, and the need for skilled personnel exist, the benefits of Edge AI in proactive leak detection, predictive maintenance, and intelligent control of water networks are undeniable, paving the way for more resilient, efficient, and responsive water management systems globally.

In conclusion, AI stands as a pivotal and transformative force in the evolution of modern water management, integrating diverse technological approaches to address the escalating complexities of water scarcity, quality, and infrastructure resilience. By leveraging capabilities ranging from advanced ML algorithms to sophisticated neural networks and fuzzy logic, AI empowers water utilities and stakeholders with unprecedented tools for monitoring, prediction, and optimisation across the entire water cycle. The emergence of digital twins, for instance, creates dynamic virtual replicas of physical water systems, enabling real, time data integration, scenario simulation, and proactive decision, making that enhances operational efficiency, minimises water loss through precise leak detection, and optimises resource allocation. Complementing this, Hybrid AI solutions combine the power of data, driven models with established physical principles and human expertise, yielding more accurate predictions for water levels, contaminant spread, and demand forecasting, thereby improving overall system reliability and interpretability. Furthermore, the deployment of Edge AI directly on localised sensors and devices ensures instantaneous data processing and immediate response capabilities, crucial for critical infrastructure monitoring, reducing latency in alerts, and enhancing data privacy by performing analytics closer to the source. Collectively, these AI, driven innovations facilitate intelligent maintenance, more effective disaster preparedness against floods and droughts, optimised energy consumption in treatment and distribution, and robust water quality assurance. While challenges such as ensuring data quality, seamless integration with legacy systems, addressing ethical considerations, and fostering a skilled workforce persist, the overarching benefits of AI in creating more efficient, sustainable, and resilient water management systems are undeniable, positioning AI as an indispensable pillar for securing a safe and equitable water future globally.

On the other hand, rather than viewing ML, DL, IoT, and predictive analytics as isolated technologies, their greatest potential emerges when integrated to address thematic challenges in water management. For example, resource optimization is achieved by combining IoT-enabled smart meters (for real-time data collection) with ML models (for demand forecasting) and predictive analytics (for pump scheduling). Similarly, climate resilience benefits from integrating DL-based flood forecasting models with digital twins that simulate future climate scenarios. In the domain of water quality assurance, computer vision systems for detecting algal blooms can be complemented by NLP techniques that extract insights from regulatory documents and citizen feedback. This integrative perspective highlights how AI technologies converge to deliver holistic solutions that address sustainability, resilience, and governance objectives in water management.

CASE STUDIES AND REAL, WORLD EXAMPLES

Several real, world examples demonstrate the potential of AI in water management. For instance, the city of Barcelona has implemented (see Figure 4) an AI, powered system to monitor and manage its water supply, reducing water loss and improving efficiency (Euklidiadas, 2023). In agriculture, AI, driven irrigation systems have been used to optimize water usage and increase yields (Katimbo et al., 2023). These case studies highlight the practical benefits of AI in water management and provide valuable lessons for future implementations.

Figure 4. Barcelona's smart management system

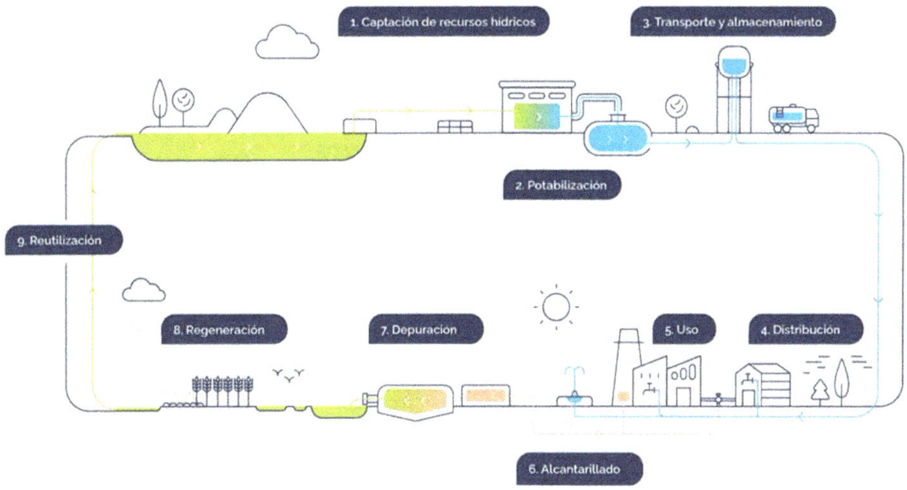

(Euklidiadas, 2023)

AI is increasingly being adopted in real, world water management scenarios, demonstrating tangible benefits across various applications. Some of the prominent case studies are discussed below.

Leak Detection and Non, Revenue Water (NRW) Reduction

Leak detection and Non, Revenue Water (NRW) reduction are critical for conserving precious water resources and improving operational efficiency in water utilities. By identifying and mitigating leaks, utilities can significantly decrease water loss before it reaches customers, enhancing both environmental sustainability and financial viability.

- **United Utilities (UK):** This major water provider utilises an AI, based "event recognition in the water network (ERWAN)" system to detect pipe bursts and leaks. By analysing real, time pressure and flow data, the system identifies anomalies indicative of leaks, enabling faster response times and significant reductions in water loss.
- **Symmons (with Caylent):** Symmons integrated AI, driven anomaly detection into its Evolution® platform for hotels and multifamily facilities. This led to the identification and resolution of over 400 leak incidents in one year, preventing an estimated $5 million in water damage and reducing water heating costs by 8, 12%. The system also reduced mean time to resolution by over an hour.

- **Agforest Solutions:** This company uses GeoAI (geospatial AI) to mitigate water stress by detecting leaks in supply networks. Their predictive models, based on AI algorithms, combine geospatial data with leak history to predict 64% of water leaks, leading to a 55% reduction in losses in some cases.

Predictive Maintenance for Infrastructure

Predictive maintenance for infrastructure in water management leverages AI and sensor data to anticipate equipment failures before they occur, enabling proactive repairs and minimizing costly service disruptions. This approach optimizes asset lifespan and operational efficiency by shifting from reactive fixes to strategically planned interventions.

- **San Antonio Water System (US) with Arcadis:** Implemented an AI, driven asset management system that analyzes variables like pipe material, age, and environmental conditions to predict failure risks. This proactive approach helps prioritize maintenance efforts and prevent costly disruptions.
- **Various Utilities:** AI systems analyze data from IoT sensors, equipment logs, and historical maintenance records to predict equipment failures (e.g., pumps, valves) in water treatment plants and distribution networks. This allows for scheduled maintenance, extending asset lifespan and minimizing unplanned downtime.

Water Quality Monitoring and Prediction

Water quality monitoring and prediction involve the continuous assessment and forecasting of water purity using AI, driven systems and sensors. This allows for early detection of contaminants or changes in water composition, ensuring public health and environmental safety through proactive intervention.

- **Seoul, South Korea:** Algorithms like SVM and ANNs are used to assess the quality of ground and drinking water, leveraging spatial and temporal patterns.
- **Singapore:** The national water agency uses AI, powered robots to swim through reservoirs, collecting real, time data on parameters like pH levels and pollutants, ensuring continuous monitoring and rapid response to contamination.
- **Various Research Initiatives:** AI models, including ANNs, Random Forests, and SVMs, have achieved high accuracy (e.g., 95.2% for ANN) in predicting

water quality indexes and classifying water quality, significantly improving real, time monitoring efficiency and pollution detection.

Flood Prediction and Management

Flood prediction and management utilize AI and real, time data to forecast extreme weather events and riverine overflows, enabling timely warnings and strategic infrastructure operations. This proactive approach minimizes property damage, safeguards human lives, and optimizes emergency response efforts during flood incidents.

- **Google Flood Hub:** This initiative uses AI models to forecast floods up to seven days in advance by processing diverse public data sources, including precipitation and weather patterns. It provides alerts on Google Search and Maps, covering river basins in over 80 countries and a population of 460 million people.
- **Real, world Case Study (Lower Mekong Basin):** A hybrid AI, based framework, integrating LSTM for temporal analysis and CNN for spatial feature extraction, accurately predicted five flood events with lead times of up to 4 hours, enabling timely warnings and reducing potential damage, even for previously unpredictable flash floods.

Water Demand Forecasting and Distribution Optimization

Water demand forecasting and distribution optimization leverage AI to accurately predict future water consumption patterns based on various factors like weather and historical data. This enables intelligent adjustment of water distribution networks, ensuring efficient supply, minimizing waste, and optimizing energy use in pumping and delivery.

- **German Municipal Utility:** A municipal utility company uses AI to forecast water demands for its residents, achieving 94, 96% accuracy for hourly and daily forecasts by analyzing historical data, weather, and calendar information.
- **MACS Water Smart (Georgia):** This enhanced application integrates AI with EPANET, based hydraulic modeling to optimize water supply networks. It achieved pressure increases above 25m, reduced flow velocities, improved pumping efficiency by 15%, and lowered leakage rates by 8%, demonstrating AI's ability to automate and optimize network operations.

CHALLENGES AND LIMITATIONS

Despite its potential, AI implementation in water management faces several challenges (Dalibard et al., 2025). Data availability and quality are critical issues, as AI models require large amounts of accurate data to function effectively. High implementation costs and infrastructure requirements can also be barriers, particularly in developing regions. Ethical concerns, such as data privacy and the risk of bias in AI models, must be addressed to ensure equitable and transparent solutions.

While AI holds immense promise for revolutionizing water management, its widespread adoption and effective implementation are not without significant challenges and limitations. These hurdles span technical, economic, human, and ethical dimensions:

Data Availability, Quality, and Integration

AI models are data, hungry, requiring vast amounts of high, quality, consistent, and standardized data for accurate training and reliable performance. However, water sector data is often fragmented, collected by diverse legacy systems (e.g., SCADA, GIS, manual logs), inconsistent in format, or simply scarce, especially for historical or extreme events. Integrating these disparate data sources into a unified, usable format for AI is a complex and often costly endeavor. Furthermore, biases within historical data can lead to skewed AI outcomes, perpetuating or even amplifying existing inequalities in water allocation or service.

Cost and Investment

The initial investment required for AI implementation can be substantial. This includes costs for advanced sensors, high, performance computing infrastructure (including data centers with their own significant energy and water footprints), specialized software licenses, and the development or customization of AI models. For many water utilities, particularly smaller ones or those in developing regions, these financial barriers can be prohibitive, diverting resources from other critical infrastructure needs.

Lack of Technical Skills and Knowledge

There is a significant global shortage of professionals with expertise in both water management and AI/data science. Implementing, maintaining, and evolving AI systems requires a workforce skilled in data engineering, ML, cybersecurity, and system integration, a gap that the traditional water sector workforce often lacks. The ageing workforce in many utilities further exacerbates this challenge, as experienced personnel retire, taking valuable domain knowledge with them.

Model Interpretability and Trust

Many advanced AI models, particularly DL networks, operate as "black boxes," meaning their decision, making processes are opaque and difficult to understand. In critical sectors like water management, where decisions impact public health, environmental safety, and large financial investments, a lack of transparency can erode public trust and hinder adoption by risk, adverse stakeholders and regulators. Water managers need to understand why an AI system makes a certain prediction (Wei & Nandy, 2025) or recommendation to confidently act upon it.

Cybersecurity Risks

As water infrastructure becomes more interconnected through IoT devices and AI systems, it also becomes more vulnerable to cyberattacks. A breach in an AI, controlled water system could lead to significant disruptions, contamination, or even physical damage, posing severe risks to public safety and national security. Ensuring robust cybersecurity measures for AI, driven water systems is paramount, but also adds to complexity and cost.

Economic Considerations and Cost–Benefit Analysis

While AI implementation in water management involves substantial upfront investment, including sensors, data infrastructure, and computational resources, these costs must be weighed against long-term benefits. In high-income regions, savings are realized through reduced non-revenue water, optimized energy use in pumping and treatment, and extended asset lifespans via predictive maintenance. In developing regions, although budget constraints are significant, cost–benefit comparisons reveal strong advantages. For example, AI-enabled leak detection systems can reduce water losses by 30 - 40%, translating into millions of cubic meters saved annually, often offsetting installation costs within a few years. Similarly, smart irrigation systems in smallholder agriculture increase crop yields while cutting water use, producing measurable livelihood gains. When compared to traditional expansion of supply infrastructure (e.g., dams, desalination plants), AI-driven optimization offers a more cost-effective and environmentally sustainable pathway. Therefore, while affordability remains a barrier, tailored financial models, donor support, and public–private partnerships can make AI adoption economically viable even in resource-constrained settings.

National and International Regulations

At the national level, many countries are developing policies on smart utilities, digital water systems, and AI governance. For instance, the European Union's AI Act sets out requirements for transparency, accountability, and risk classification of AI applications, which directly affect utilities deploying AI-based water monitoring or billing systems. Similarly, the U.S. National Institute of Standards and Technology (NIST) has issued AI risk management guidelines relevant to critical infrastructure, while China and Singapore have introduced smart city policies that embed water AI systems into broader digital transformation agendas. At the international level, the World Health Organization (WHO) and International Organization for Standardization (ISO) provide benchmarks for water quality monitoring (e.g., ISO 24518 for drinking water management) and digital technologies, which AI-driven systems must comply with. The United Nations also frames AI use within the Sustainable Development Goals (SDG 6: Clean Water and Sanitation), emphasizing inclusivity and sustainability.

Role of Governments and Public-Private Partnerships

Governments play a central role in funding, regulating, and overseeing AI-enabled water systems. Public-private partnerships (PPPs) are increasingly used to accelerate innovation while ensuring accountability. For example, national water agencies may collaborate with AI startups and sensor manufacturers to deploy smart irrigation or leak detection networks, with the government ensuring compliance and equitable service delivery. PPPs help balance innovation with public accountability, especially in regions where utilities lack the technical expertise to deploy AI at scale.

Governance Models for Ethical AI Use in Water Utilities

Effective governance requires more than compliance with regulations; it involves creating institutional frameworks for responsible and transparent AI deployment. Governance models for public water services often include Algorithmic transparency and auditability, which requires utilities to document how AI models make decisions. On the other hand, independent oversight boards, which include regulators, civil society, and technical experts, review AI implementation. Alternatively, citizen participation mechanisms allow communities to provide input into how digital tools are used in managing local water resources. Moreover, data stewardship frameworks ensure that water-related data is collected, stored, and shared responsibly, with strong privacy protections.

By embedding AI within legal and policy frameworks and by promoting collaborative governance structures, the water sector can harness digital innovation while safeguarding public trust, equity, and sustainability.

In summary, while AI offers transformative potential for improving water management, its implementation is not without significant challenges. Key limitations include data scarcity, especially in rural or underdeveloped regions, as well as inconsistencies in data quality and interoperability across systems. High implementation costs, lack of infrastructure, and the need for specialized technical expertise further hinder widespread adoption. Ethical concerns related to data privacy, algorithmic bias, and decision transparency also require careful governance. Addressing these challenges through inclusive policies, capacity building, and responsible innovation will be essential to realizing the full benefits of AI in the water sector.

FUTURE TRENDS AND OPPORTUNITIES

The future of AI in water management is promising, with emerging technologies like blockchain and remote sensing offering new opportunities for innovation (Mathaba & Banza, 2023). AI can play a crucial role in achieving Sustainable Development Goal (SDG) 6: Clean Water and Sanitation, by enabling decentralized and community, driven water management systems. Collaboration between governments, the private sector, and academia will be essential to harness the full potential of AI in this domain. On the other hand, the growing demand for sustainable, resilient, and data, driven water systems is creating a fertile ground for the continued evolution of AI in water management. As emerging technologies mature and the cost of digital infrastructure declines, AI is poised to become more accessible, powerful, and integrated across every layer of the water sector. The following trends highlight how AI will shape the future of global water management and unlock new opportunities for innovation, efficiency, and inclusivity.

Integration With Edge Computing and IoT

One of the most promising trends is the convergence of AI with edge computing and the IoT. Edge AI enables real, time decision, making at the source of data generation, such as sensors, pumps, or meters, without relying on centralized cloud systems. This is particularly useful in remote or underserved regions where connectivity is limited. As edge devices become more powerful and energy, efficient, they will facilitate autonomous leak detection, irrigation control, and water quality monitoring in real, time, creating decentralized yet intelligent water systems.

Expansion of Digital Twins and Smart Water Networks

The adoption of digital twins, real, time virtual replicas of physical water systems, will continue to rise, especially in urban utilities. AI, driven digital twins will simulate operational scenarios, optimize system performance, and support predictive maintenance with high accuracy. Combined with smart sensors and GIS mapping, these tools will enable cities to create adaptive and self, healing water networks that can respond dynamically to environmental and demand fluctuations.

AI, Enabled Climate Resilience and Disaster Management

AI is expected to play a central role in building climate resilience. Advanced predictive models will be used to simulate droughts, floods, and sea, level rise, enabling proactive adaptation strategies. For example, AI, driven early warning systems can improve response times during extreme weather events by analyzing rainfall patterns, river flows, and satellite imagery. In addition, AI can support the design of nature, based solutions such as wetland restoration or floodplain zoning by evaluating ecological and hydrological data.

Democratization of AI through Low, Code Platforms and Open Data

As AI development becomes more user, friendly through low, code and no, code platforms, water professionals without deep technical expertise will be able to create and deploy AI models. This democratization will encourage greater innovation at the local level. Moreover, the push for open data in water governance will allow researchers, startups, and civil society organizations to build AI solutions tailored to specific regional needs, promoting transparency, collaboration, and scalability.

AI for Circular Water Economy and Resource Recovery

AI will increasingly support circular water economy principles, where wastewater is seen as a resource rather than waste. Smart algorithms will optimize water reuse, nutrient recovery, and energy generation in treatment plants. AI can also help identify opportunities for industrial water recycling and greywater harvesting in buildings and agriculture, driving sustainability across sectors.

Enhanced Citizen Engagement and Participatory Water Governance

The use of AI in NLP, chatbots, and mobile platforms will enhance citizen engagement in water governance. Through AI, powered tools, users can report service issues, access real, time water quality data, and receive personalized conservation recommendations. These technologies can also process large volumes of public feedback and social media content to inform service improvements and policy development.

Personalized Water Services and Behavioral Insights

AI can enable personalized water services by analyzing individual usage patterns, predicting future consumption, and recommending behavioral changes. For example, smart billing systems could offer tiered tariffs based on predictive usage forecasts, while conservation apps could gamify water, saving behaviors (Wang et al., 2024). Behavioral analytics, informed by AI, will also support targeted awareness campaigns and education initiatives.

Federated Learning and Privacy, Preserving AI

To address growing concerns around data privacy and security, federated learning will allow AI models to be trained across multiple decentralized data sources without transferring raw data (Nandy et al., 2020). This innovation is particularly relevant for utilities managing sensitive customer or environmental data, as it ensures confidentiality while still benefiting from collective intelligence (Zhang & Wei, 2024).

Integration With Blockchain for Transparency and Trust

Combining AI with blockchain can enhance transparency in water management, particularly in areas like water trading, compliance auditing, and subsidy distribution. Blockchain can provide an immutable record of AI decisions or transactions, ensuring traceability and accountability. This integration can build trust among stakeholders, especially in multi, user water systems (Satilmisoglu et al., 2024).

Alignment With Global Sustainability and SDG 6

AI will continue to align with the goals of Sustainable Development Goal 6 (Clean Water and Sanitation) by improving access, affordability, and quality of water services. Through smart resource planning, optimized operations, and inclusive decision, making tools, AI can help close the equity gap and ensure safe water for all (Dalibard et al., 2025). It also supports interlinked goals such as sustainable agriculture (SDG 2), climate action (SDG 13), and sustainable cities (SDG 11).

In summary, the future of AI in water management is not just about technology; it is about reimagining how we govern, value, and interact with water. By enabling smarter infrastructure, more resilient systems, and empowered communities, AI offers a pathway to water security in the face of mounting global challenges. The key to unlocking its full potential will lie in fostering interdisciplinary collaboration, ethical frameworks, and inclusive innovation that bridges digital capabilities with local realities.

CONCLUSION

AI has emerged as a transformative force in addressing the global water crisis, offering innovative solutions to optimize water resource management, enhance water quality, and mitigate risks associated with climate change and population growth. By leveraging technologies such as ML, predictive analytics, and IoT, AI enables smarter, data, driven decision, making that can significantly improve the efficiency and sustainability of water systems. From predicting water demand and detecting leaks to optimizing irrigation and managing floods, AI applications are revolutionizing how we manage this vital resource. However, the integration of AI into water management is not without challenges. Issues such as data privacy, ethical concerns, high implementation costs, and the need for robust infrastructure must be addressed to ensure equitable and responsible deployment. The journey toward AI, enabled water management is just beginning, and its success will depend on the collective efforts of governments, industries, academia, and civil society to create a world where water is managed wisely, equitably, and sustainably.

ACKNOWLEDGEMENT

This work is supported and funded by the UCSI University Research Excellence & Innovation Grant (REIG), REIG, ICSDI, 2024/044. The authors of this book chapter would like to thank all the anonymous reviewers for improving the quality and readability of this book chapter.

REFERENCES

Abdelhamid, M. A., Abdelkader, T. K., Sayed, H. A. A., Zhang, Z., Zhao, X. H., & Atia, M. F. (2025). Design and evaluation of a solar powered smart irrigation system for sustainable urban agriculture. *Scientific Reports*, *15*(1), 11761. Advance online publication. DOI: 10.1038/s41598-025-94251-3 PMID: 40189625

Absalan, F., Hatam, F., Prévost, M., Barbeau, B., & Bichai, F. (2024). Climate change and future water demand: Implications for chlorine and trihalomethanes management in water distribution systems [Article]. *Journal of Environmental Management*, *355*(11), 120470. Advance online publication. DOI: 10.1016/j.jenvman.2024.120470 PMID: 38422852

Chen, J. J., & Liu, D. H. (2023). Using computer vision to monitor ice conditions in water supply infrastructure: A study of salient image features. *Journal of Hydroinformatics*, *25*(4), 1139–1152. DOI: 10.2166/hydro.2023.120

Chen, Y. (2024). Real time data monitoring of water resources environment based on computer remote data collection and image analysis. *Optical and Quantum Electronics*, *56*(4), 618. Advance online publication. DOI: 10.1007/s11082-023-05928-w

Dalibard, A., Kriem, L. S., Beckett, M., Scherle, S., Yeh, Y. C., & Schliessmann, U. (2025). Challenges and requirements of AI-based waste water treatment systems [Article]. *Automatisierungstechnik*, *73*(1), 40–49. DOI: 10.1515/auto-2024-0023

de Melo, W. W., Iglesias, I., & Pinho, J. (2024). Early warning system for floods at estuarine areas: Combining artificial intelligence with process-based models [Article; Early Access]. *Natural Hazards*, *24*. Advance online publication. DOI: 10.1007/s11069-024-06957-8

Ercan Oguztürk, G. (2025). AI-driven irrigation systems for sustainable water management: A systematic review and meta-analytical insights. *Smart Agricultural Technology*, *11*, 100982. Advance online publication. DOI: 10.1016/j.atech.2025.100982

Euklidiadas, M. M. (2023). *Barcelona's innovative smart water management makes the most of every drop.* https://www.tomorrow.city/barcelonas-innovative-smart-water-management-makes-the-most-of-every-drop/

Fang, S. B., Deitch, M. J., & Gebremicael, T. G. (2025). Evaluating the reliability of data interpolation and machine learning methods for water quality management: A SWAT model comparison. *Environmental Earth Sciences, 84*(10), 274. Advance online publication. DOI: 10.1007/s12665-025-12313-1

Fevgas, G., Lagkas, T., Papadopoulos, P., Sarigiannidis, P., & Argyriou, V. (2025). Integrating thermal infrared and RGB imaging for early detection of water stress in lettuces with comparative analysis of IoT sensors. *Smart Agricultural Technology, 11*, 100881. Advance online publication. DOI: 10.1016/j.atech.2025.100881

Food and Agriculture Organization. A. d. (2025a). *Renewable internal freshwater resources per capita (cubic meters).* Worldbank. https://data.worldbank.org/indicator/ER.H2O.INTR.PC

Food and Agriculture Organization. A. d. (2025b). *Renewable internal freshwater resources, total (billion cubic meters).* Worldbank. https://data.worldbank.org/indicator/ER.H2O.INTR.K3

Fukui, M., Hiramatsu, K., Fukuda, S., Pineda, D., Harada, M., & Tabata, T. (2020). Evaluation of Water-Saving Methodologies for Rice Farming Using a Distributed Long-Term Rainfall-Runoff Model Incorporated Agricultural Ponds Detected by Multi-Temporal SAR Images in the Ibague Watershed, Colombia. *Journal of the Faculty of Agriculture Kyushu University, 65*(2), 313-320. <Go to ISI>://WOS:000589116000015

Gyparakis, S., Trichakis, I., Daras, T., & Diamadopoulos, E. (2025). Artificial Neural Networks (ANNs) and Multiple Linear Regression (MLR) Analysis Modelling for Predicting Chemical Dosages of a Water Treatment Plant (WTP) of Drinking Water [Article]. *Water, 17*(2), 12, Article 227. https://doi.org/DOI: 10.3390/w17020227

Jamali, A., Ramazani, M. E., Nejaei, A., & Rafieyan, O. (2025). Monitoring of Pollutants in Surface Water with Laboratory Analysis and Satellite Image Between 2019 and 2024, Kichik Chai River, Iran. *Water, Air, and Soil Pollution, 236*(5), 272. Advance online publication. DOI: 10.1007/s11270-025-07808-3

Kant, N., & Wrat, G. (2025). A review of integrated groundwater and surface water management for environmental sustainability. *Acta Geochimica.* Advance online publication. DOI: 10.1007/s11631-025-00803-9

Katimbo, A., Rudnick, D. R., Zhang, J. W., Ge, Y. F., Dejonge, K. C., Franz, T. E., Shi, Y. Y., Liang, W. Z., Qiao, X., Heeren, D. M., Kabenge, I., Nakabuye, H. N., & Duan, J. M. (2023). Evaluation of artificial intelligence algorithms with sensor data assimilation in estimating crop evapotranspiration and crop water stress index for irrigation water management [Article]. *Smart Agricultural Technology*, *4*(18), 100176. Advance online publication. DOI: 10.1016/j.atech.2023.100176

Le, Y., Sun, X. Y., Chen, Y. F., Zhang, D. F., Wu, L., Liu, H., & Hu, M. Z. (2024). High-accuracy shallow-water bathymetric method including reliability evaluation based on Sentinel-2 time-series images and ICESat-2 data. *Frontiers in Marine Science*, *11*, 1470859. Advance online publication. DOI: 10.3389/fmars.2024.1470859

Li, L. C., Zhao, D. L., Sun, W. P., & Huang, Q. H. (2025). A micro Francis turbine for smart water meter self-generation: Study on the effects of structural parameters on turbine performance based on CFD simulations and experiments. *Energy*, *328*, 135865. Advance online publication. DOI: 10.1016/j.energy.2025.135865

Li, X. L., & Song, Y. (2025). What role does agriculture play in irrigation water footprint consumption? A perspective based on sectoral linkage and structural path. *Journal of Hydrology. Regional Studies*, *59*, 102378. Advance online publication. DOI: 10.1016/j.ejrh.2025.102378

Lu, P., Zhang, J. L., Al-Mahmood, A., Dung, H. Q., & Yunos, Z. M. (2025). Novel deep learning models for accurate time series prediction of phycocyanin concentration in water management. *Journal of Water Process Engineering*, *74*, 107783. Advance online publication. DOI: 10.1016/j.jwpe.2025.107783

Mahesh, N., Babu, J. J., Nithya, K., & Arunmozhi, S. A. (2024). Water quality prediction using LSTM with combined normalizer for efficient water management. *Desalination and Water Treatment*, *317*, 100183. Advance online publication. DOI: 10.1016/j.dwt.2024.100183

Makropoulos, C. K., Butler, D., & Maksimovic, C. (2003). Fuzzy logic spatial decision support system for urban water management. *Journal of Water Resources Planning and Management*, *129*(1), 69–77. (69)DOI: 10.1061/(asce)0733-9496(2003)129:1

Mathaba, M., & Banza, J. (2023). A comprehensive review on artificial intelligence in water treatment for optimization. Clean water now and the future [Article]. *Journal of Environmental Science and Health Part a-Toxic/Hazardous Substances & Environmental Engineering, 58*(14), 1047-1060. https://doi.org/DOI: 10.1080/10934529.2024.2309102

Meng, X., Huang, Z. G., Deng, X., Liu, H., Fang, H. Y., Liu, C., Zhang, X. Y., & Cui, J. (2025). Leakage detection and localization of buried water pipe using ground penetrating radar. *Measurement, 254*, 117902. Advance online publication. DOI: 10.1016/j.measurement.2025.117902

Mohammed, S., Arshad, S., Bashir, B., Ata, B., Al-Dalahmeh, M., Alsalman, A., Ali, H., Alhennawi, S., Kiwan, S., & Harsanyi, E. (2024). Evaluating machine learning performance in predicting sodium adsorption ratio for sustainable soil-water management in the eastern Mediterranean. *Journal of Environmental Management, 370*, 122640. Advance online publication. DOI: 10.1016/j.jenvman.2024.122640 PMID: 39340889

Morchid, A., Said, Z., Abdelaziz, A. Y., Siano, P., & Qjidaa, H. (2025). Fuzzy logic-based IoT system for optimizing irrigation with cloud computing: Enhancing water sustainability in smart agriculture. *Smart Agricultural Technology, 11*, 100979. Advance online publication. DOI: 10.1016/j.atech.2025.100979

Nandy, T., Idris, M. Y. I., Noor, R. M., Wahab, A. W. A., Bhattacharyya, S., Kolandaisamy, R., & Yahuza, M. (2021). A Secure, Privacy-Preserving, and Lightweight Authentication Scheme for VANETs. *IEEE Sensors Journal, 21*(18), 20998–21011. DOI: 10.1109/JSEN.2021.3097172

Nandy, T., Idris, M. Y. I. B., Noor, R. M., Bhattacharyya, S., & Ghani, N. B. A. (2020, December 20-21). 2020). Collaborative Data Anonymization for Privacy-Preserving Vehicular Ad-hoc Network. 2020 International Conference on Innovation and Intelligence for Informatics, Computing and Technologies (3ICT)

Nandy, T., Kolandaisamy, R., Noor, R. M., & Bhattacharyya, S. (2025). An Adaptive Vehicle Location Prediction Using Machine Learning: A Case Study of Campus Shuttle Bus. *IEEE Access : Practical Innovations, Open Solutions, 13*, 78290–78302. DOI: 10.1109/ACCESS.2025.3562785

Nandy, T., Md Noor, R., Kolandaisamy, R., Idris, M. Y. I., & Bhattacharyya, S. (2024). A review of security attacks and intrusion detection in the vehicular networks. *Journal of King Saud University. Computer and Information Sciences, 36*(2), 101945. https://doi.org/doi.org/10.1016/j.jksuci.2024.101945

Nandy, T., & Nyundo, A. (2023, June 16-17). 2023). Prediction of Possible Outcomes using Big Data Analysis and Machine Learning. *2023 International Conference on Applied Intelligence and Sustainable Computing (ICAISC)*

Noor, R. M., Rasyidi, N. B. G., Nandy, T., & Kolandaisamy, R. (2021). Campus Shuttle Bus Route Optimization Using Machine Learning Predictive Analysis: A Case Study. *Sustainability, 13*(1), 225. DOI: 10.3390/su13010225

Prabu, R. T., Sarkar, M., Chaudhary, D., Al Obaid, S., Al-ateeq, T. K., & Kalam, M. A. (2024). IoT-enabled groundwater monitoring with k-NN-SVM algorithm for sustainable water management. *Acta Geophysica*, *72*(4), 2715–2728. DOI: 10.1007/s11600-023-01178-2

Preite, L., & Vignali, G. (2024). Artificial intelligence to optimize water consumption in agriculture: A predictive algorithm-based irrigation management system [Article]. *Computers and Electronics in Agriculture*, *223*(11), 109126. Advance online publication. DOI: 10.1016/j.compag.2024.109126

Prieto, C., Patel, D., Han, D. W., Dewals, B., Bray, M., & Molinari, D. (2024). Preface: Advances in pluvial and fluvial flood forecasting and assessment and flood risk management. *Natural Hazards and Earth System Sciences*, *24*(10), 3381–3386. DOI: 10.5194/nhess-24-3381-2024

Raimi, R., & Phiri, A. (2025). From pixels to policy: Evaluating the behavioural impact and cost effectiveness of picture nudges in water conservation at university residences. *Cogent Social Sciences*, *11*(1), 2464873. Advance online publication. DOI: 10.1080/23311886.2025.2464873

Rajasekaran, U., Kothandaraman, M., & Pua, C. H. (2025). Water Pipeline Leak Detection and Localization With an Integrated AI Technique [Article]. *IEEE Access : Practical Innovations, Open Solutions*, *13*, 2736–2745. DOI: 10.1109/access.2024.3524669

Review, W. P. (2025). *Fresh water by Country 2024*. World Population Review. https://worldpopulationreview.com/country-rankings/fresh-water-by-country

Reyes-Acosta, R., Dominguez-Baez, C., Mendoza-Gonzalez, R., & Martin, M. V. (2025). Analysis of machine learning-based approaches for securing the Internet of Things in the smart industry: a multivocal state of knowledge review [Article]. *International Journal of Information Security*, *24*(1), 31, Article 31. https://doi.org/DOI: 10.1007/s10207-024-00935-8

Ricart, S., Berizzi, C., Saurí, D., & Terlicher, G. N. (2022). The Social, Political, and Environmental Dimensions in Designing Urban Public Space from a Water Management Perspective: Testing European Experiences. *Land (Basel)*, *11*(9), 1575. Advance online publication. DOI: 10.3390/land11091575

Santos, E., Carvalho, M., & Martins, S. (2023). Sustainable Water Management: Understanding the Socioeconomic and Cultural Dimensions. *Sustainability*, *15*(17), 13074. Advance online publication. DOI: 10.3390/su151713074

Satilmisoglu, T. K., Sermet, Y., Kurt, M., & Demir, I. (2024). Blockchain Opportunities for Water Resources Management: A Comprehensive Review. *Sustainability, 16*(6), 2403. Advance online publication. DOI: 10.3390/su16062403

Shin, J., Yoon, S., Park, N. S., & Kim, Y. (2025). Development of water quality prediction model using LTSF-Linear and complete ensemble empirical mode decomposition. *Desalination and Water Treatment, 323*, 101254. Advance online publication. DOI: 10.1016/j.dwt.2025.101254

Tian, M. Y., & Adriaens, P. (2025). Predictive learning of operational water intensity from corporate financial indicators: Implications for AI-based resource management. *Resources, Conservation and Recycling, 220*, 108383. Advance online publication. DOI: 10.1016/j.resconrec.2025.108383

Torwane, N., Lalloo, R., Ha, D., & Do, L. (2025). Mapping the "X" Debate: Water Fluoridation Sentiment Analysis With Advanced Machine Learning. *Journal of Public Health Dentistry*. Advance online publication. DOI: 10.1111/jphd.12669 PMID: 40333400

Valéry, A., François, H., Carmagnola, C., Tranchant, X., Samacoïts, R., Mermillod, O., Le Lay, M., Leclercq, F., Soubeyroux, J. M., Guily, L., & Morin, S. (2025). Changes in water resources and water demand in a ski resort under climate change. *Lhb-Hydroscience Journal, 111*(1), 2473589. Advance online publication. DOI: 10.1080/27678490.2025.2473589

Wang, M., & Zheng, C. (1998). Ground water management optimization using genetic algorithms and simulated annealing: Formulation and comparison. *Journal of the American Water Resources Association, 34*(3), 519–530. DOI: 10.1111/j.1752-1688.1998.tb00951.x

Wang, S. K., Chang, J. J., Xue, J., Sun, H. W., Zeng, F. J., Liu, L., Liu, X., & Li, X. X. (2024). Coupling behavioral economics and water management policies for agricultural land-use planning in basin irrigation districts: Agent-based socio-hydrological modeling and application. *Agricultural Water Management, 298*, 108845. Advance online publication. DOI: 10.1016/j.agwat.2024.108845

Wei, L. Y., & Nandy, T. (2025, February 21-23). 2025). Student Performance Prediction System Using Machine Learning. 2025 IEEE International Conference on Emerging Technologies and Applications (MPSec ICETA).

Wen, S., & Feng, Z. (2023). Theoretical analysis and applications of artificial intelligence in hydrology and water resource management. *Water Supply, 23*(4), iii–vi. DOI: 10.2166/ws.2023.080

Zaresefat, M., & Derakhshani, R. (2023). Revolutionizing Groundwater Management with Hybrid AI Models: A Practical Review. *Water (Basel)*, *15*(9), 1750. https://www.mdpi.com/2073-4441/15/9/1750

Zhang, H., & Li, M. (2025). Towards an intelligent and automatic irrigation system based on internet of things with authentication feature in VANET. *Journal of Information Security and Applications*, *88*, 103927. DOI: 10.1016/j.jisa.2024.103927

Zhang, J. H., & Wei, J. (2024). On the Security of Privacy-Enhanced Authentication Protocol for Federated Learning in VANETs [Article]. *IEEE Transactions on Information Forensics and Security*, *19*, 9433–9435. DOI: 10.1109/tifs.2024.3445730

Chapter 6
Predictive Modeling and Generative AI for Anticipating Water Shortages

Oussama El Azzouzy
https://orcid.org/0009-0002-7737-349X
Science Faculty of Ain-Chock, Hassan II University, Morocco

Tarik Chanyour
https://orcid.org/0000-0002-7676-0749
Science Faculty of Ain-Chock, Hassan II University, Morocco

Said Jai Andaloussi
Science Faculty of Ain-Chock, Hassan II University, Morocco

Khadija El Fellah
Laboratory of Research in Informatics, FS, UIT, Kenitra, Morocco

Ikram El Azami
Faculty of Sciences, Ibn Tofail University, Kenitra, Morocco

Adil El Makrani
https://orcid.org/0000-0002-6316-2379
Faculty of Sciences, Ibn Tofail University, Kenitra, Morocco

ABSTRACT

Faced with the intensification of water shortages caused by climate change and human pressures, this chapter presents a hybrid architecture combining predictive modelling and extreme scenario generation to strengthen territorial resilience. The approach is based on the integration of time series transformers and deep generative models to anticipate the evolution of hydrological variables and simulate critical events. In the absence of sufficiently rich real data, a set of realistic synthetic data was constructed from hydrological distributions, seasonal cycles and stochastic processes, then qualitatively validated. This game allows a rigorous evaluation of predictive and generative performance. Results demonstrate the complementarity of

DOI: 10.4018/979-8-3373-2700-6.ch006

the two dimensions to improve risk visualization and activation of early responses. The chapter also discusses technical limitations, ethical issues, and proposes concrete perspectives for integration into digital twins, warning systems or decision support tools for climate adaptation.

INTRODUCTION

In the face of intensifying climate change, rapid urbanization and sustained population growth, freshwater resources are under unprecedented pressure worldwide. According to projections by international institutions, almost two thirds of the world's population could be living in water stressed areas by 2030, due to a combination of prolonged droughts, excessive groundwater exploitation, pollution of water sources and increasing variability of precipitation.

Recent studies intensively highlight the rise of extreme phenomena such as 'flash droughts', which increase the risks of short-term water scarcity, particularly in semi-arid areas (Chouhan et al., 2023)This reinforces the urgency to develop robust and resilient predictive approaches, capable of operating in a context of increasing hydrological variability and increased uncertainty. This critical situation is not confined to arid or desert regions; it also affects temperate zones, where the regularity of hydrological regimes is increasingly disrupted by extreme events such as droughts, heat waves or intense precipitation concentrated over short periods. Climate change is exacerbating these imbalances by altering water cycles, which complicates the planning and sustainable management of available resources(Z. Liu et al., 2024) . In addition, demographic pressure and accelerated urbanization in many parts of the world are intensifying competition between agricultural, domestic, industrial and environmental uses of water(Mekonnen & Hoekstra, 2016). This is compounded by challenges related to equitable access and governance of water systems, particularly in developing countries where infrastructure is inadequate or obsolete. In this context, the issue of water security becomes a central issue of socio-economic and geopolitical stability, calling for a thorough rethinking of integrated water management strategies (Dai et al., 2025a). This emerging crisis, multidimensional and largely amplified by the enthronization of environments, requires advanced tools for analysis, forecasting and decision making to anticipate and mitigate future shortages(Mekonnen & Hoekstra, 2016).

Anticipating water shortages is now a strategic lever to ensure the resilience of urban, agricultural and industrial systems in the face of increasing scarcity of water resources. In urban areas, the availability of sufficient quality water is essential for the proper functioning of infrastructure, public health and the continuity of essential services. The slightest disruption to supply can have ripple effects, affecting wastewater management, energy production or social stability in densely populated areas. In the agricultural sector, which alone still accounts for more than 70% of the world's freshwater consumption, anticipating shortages is crucial to adapt crop calendars, choose the varieties that consume least, or to define irrigation priorities in times of water stress. The disruption of hydrological cycles leaves this sector highly exposed to yield losses and agricultural market volatility, with direct impacts on food security. For their part, industries also depend on reliable access to water, whether for thermal, chemical or cleaning purposes, and the unavailability of this resource can slow down or even halt production lines, threatening economic competitiveness. Beyond sectoral uses, the ability to anticipate shortages enables public decision makers to implement effective prevention policies, such as setting up warning systems, dynamic dam management, or prioritizing water according to use. In a context where resources are becoming more volatile, this capacity becomes a determining factor in guaranteeing continuity of services, limiting socio-economic tensions, and planning investments in water infrastructure according to a proactive rather than reactive logic. Thus, anticipating water shortages is not just a technical or environmental objective, but an imperative of territorial governance and systemic resilience.

Traditional approaches to hydrological forecasting, based mainly on deterministic or statistical models calculated from historical series, present major structural limitations in the face of the increasing complexity of contemporary water dynamics. These methods, although having made significant progress during the 20th century, are generally based on stationality assumptions of hydrological regimes and simplified modelling of interactions between different components of the water cycle. However, the acceleration of climate change, the modification of anthropogenic uses and the degradation of ecosystems are profoundly disrupting this stability, making classical models unsuitable for today's realities. Deterministic physical models, although useful for the fine understanding of hydrological processes, require a very detailed knowledge of watershed characteristics (topography, soil types, vegetation cover, etc.) and careful calibration, difficult to generalize on a large scale or poorly instrumented territories. On the other hand, statistical or empirical models, such as AutoRegressive Integrated Moving Average (ARIMA) models or linear regressions, struggle to capture non-linearities, trend breaks, and multi-scale interactions inherent in water systems(*Pathways for Hydrological Resilience: Strategies for Adaptation in a Changing Climate | Earth Systems and Environment*, n.d.). Moreover, these traditional approaches are often limited to short or medium term forecasts, without the capacity to explore extreme scenarios or simulate the impact of development policies or global changes. Their predictive capacity is also eroded by the increased inter-annual variability of precipitation and temperature, or the emergence of extreme phenomena such as flash droughts (Prieto-Curiel & Borja-Vega, 2024). Finally, these models have difficulty integrating heterogeneous data from new sources such as satellites, IoT sensors or citizen networks, thus limiting their potential to represent the complexity of reality. Against this backdrop, the need for more flexible, learning approaches, capable of integrating large masses of heterogeneous data and simulating alternative evolutions, becomes a central issue for renewing hydrological forecasting capabilities(Maurya, 2024).

Faced with the growing limitations of conventional approaches to hydrological forecasting and the need to strengthen the resilience of water management systems, this chapter aims to explore a complementary and innovative path. The joint integration of predictive modelling and generative artificial intelligence to anticipate water shortages. It is a matter of going beyond the simple extrapolation of past trends by mobilizing deep learning architectures capable of capturing the temporal, spatial and structural complexity of hydrological data. Predictive modelling, based on specialized neural networks such as time transformers, convolutive networks or graph networks, allows for more robust and accurate forecasts at different scales, Leveraging massive and multivariate data streams from sensors, satellites or climate re-analyses (Niu, 2025). At the same time, generative AI, through models such as generative antagonistic networks (GANs), variational autoencoders (VAEs) or diffusion models, opens new perspectives to generate alternative hydrological scenarios, explore stress tests or fill historical data gaps. The combination of these two paradigms allows not only to refine forecasting capabilities, but also to provide decision-makers with dynamic prospective frameworks, capable of simulating uncertain futures and guiding adaptation strategies at different levels (urban, agricultural, regional). Through a critical review of the literature, the presentation of a hybrid architecture and the analysis of a representative use case, this chapter analyzes the theoretical, methodological, and operational complementarities between predictive modeling and scenario generation, by evaluating their concrete contributions in contexts of high-water vulnerability. The objective is to show how these approaches can strengthen the capacity of territories to anticipate, absorb and adapt to water shortages in a context of accelerated climate disruption.

CONCEPTUAL AND THEORETICAL FRAMEWORK

Key Definitions

The Water Shortage is defined as a situation in which the quantity of fresh water available becomes insufficient to meet the demand of a given territory, whether this demand comes from domestic, agricultural, industrial or environmental uses. This deficiency may be of a physical nature, when natural resources groundwater, rivers, lakes or reservoirs are no longer able to provide sufficient supply, due in particular to prolonged rainfall deficits, depletion of underground resources or increased climate change. It can also be economic in nature, where water is present, but infrastructure or governance does not allow for equitable and sustainable access. In this case, the obstacles are poverty, lack of investment, mismanagement or political instability. The recognition of these two dimensions is essential for designing differentiated strategies to respond to the water crisis, combining technical, political and social solutions depending on the context. Water scarcity as a multidimensional phenomenon is now one of the most pressing challenges in natural resource management, compounded by population growth, rapid urbanization and the effects of climate change.

Water resilience refers to the ability of a system whether it is a watershed, an agglomeration, an ecosystem or a network of infrastructures to cope with water related disturbances, to recover and adapt without losing its basic functions. This resilience is based on several complementary dimensions, the robustness of infrastructures (drinking water networks, irrigation canals, dams), the flexibility of management mechanisms (contingency plans, adaptive allocation policies), the diversity of sources of supply (surface water, groundwater, recycled water), and the capacity for institutional learning in response to past crises. It also involves the protection of ecosystems that naturally regulate water cycles, such as wetlands, forests or agricultural soils. In the context of increasing climate instability, strengthening water resilience is a strategic objective that involves combining technological innovations with systemic, participatory and territorialized approaches. This includes the use of artificial intelligence to improve anticipation of shortages and test water stress scenarios.

Hydrological forecasting encompasses all the scientific methods used to estimate the future evolution of hydric variables such as river flows, water table levels, or overall water availability in a given territory. These forecasts are based on the collection and analysis of historical, meteorological and environmental data, combined with models simulating the physical processes of runoff, infiltration, evapotranspiration or aquifer recharge. Traditional hydrological models whether deterministic or statistical are widely used in the operational management of water resources, notably to anticipate floods, optimize reservoirs or plan irrigation(Maurya, 2024). However, the quality of forecasts depends heavily on the density of measuring stations, the accuracy of input data and the model's ability to capture the complexity of interactions between climate, soil and human activities. The emergence of new data sources satellite images, IoT sensors, climate simulations coupled with the rise of artificial intelligence, now make it possible to enrich these forecasts, by integrating the spatiotemporal dimensions of hydrological systems in a more dynamic way. Nevertheless, the uncertainties associated with these forecasts remain significant, making it essential to model them and make them explicit in decision-making processes(Zhao et al., 2024).

Climate uncertainty refers to the imprecision inherent in any projection of the future evolution of weather and climate variables, such as temperature, precipitation or extreme events. It results from a combination of factors, natural climate variability, incomplete scientific knowledge of certain atmospheric or oceanic processes, the limitations of digital climate models, and the unpredictability of future socio-economic trajectories (greenhouse gas emissions, changes in land use, urbanization, etc.)(Dankers & Kundzewicz, 2020). In the hydrological field, this uncertainty makes it difficult to accurately estimate medium and long-term impacts on flow regimes, drought periods or groundwater recharge. It poses a major challenge for infrastructure planning and the formulation of water management policies, as decisions need to be robust in the face of a wide range of possible futures. Integrating uncertainty into predictive models and generating plausible hydrological scenarios is a promising way of reconciling science, adaptive planning and resilient decision-making. Generative AI, in particular, offers new ways of exploring multiple futures while quantifying the associated risks(Jose & Dwarakish, 2020).

Overview of Conventional Approaches

Traditional statistical approaches were the first tools used to model hydrological dynamics and anticipate water shortages. Among these methods, linear models such as single or multiple regression were used to relate explanatory hydrometeorological variables (such as precipitation, temperature or relative humidity) to target variables (such as groundwater level or river flows). More sophisticated, the ARIMA (AutoRegressive Integrated Moving Average) models have been able to capture time structures inherent in hydrological time series by combining autoregressive components, moving averages, and trend integration(Costa et al., 2023). These models, sometimes extended in seasonal AutoRegressive Integrated Moving Average (SARIMA) to incorporate seasonality, have proved useful for short-term forecasting, especially when extrapolating from well-defined historical patterns. However, their ability to model non stationary, strongly non-linear or extreme events (prolonged droughts, climatic anomalies, land use disruptions) behaviors remains limited. Their parametric structure imposes rigid assumptions on the data, and their effectiveness depends closely on the quality of the preprocessing and the relevance of the variables selected. In addition, these models often require signal disaggregation and do not allow for the simultaneous exploitation of several heterogeneous information sources. Despite these constraints, statistical approaches remain useful in data constrained environments, or as a basis for comparison to assess the added value of more advanced methods. They also provide a robust methodological basis in contexts where simplicity and interpretability take precedence over predictive precision(Myllis et al., 2024). Traditional statistical methods show difficulties in dealing with the non-stationarity of hydrological time series and regime breaks, particularly for anticipating recent extremes (Drenkhan et al., 2023).

Deterministic hydrological models are based on explicit modelling of the physical and hydrological processes that govern the water cycle. Their objective is to mechanically simulate the different components of the water system precipitation, infiltration, runoff, evapotranspira- tion, recharge of water bodies and river flows based on differential equations derived from the laws of conservation of mass, energy and quantity of movement. These models, such as Soil and Water Assessment Tool (SWAT), Hydrologiska Byråns Vattenbalansavdelning (HBV), or MIKE SHE, incorporate a spatialized representation of the watershed and allow to analyze the effects of climate change, of land use or hydraulic developments on water dynamics. Their main strength lies in their ability to explicitly represent cause and effect relationships between physical variables, making them particularly useful for impact studies or development scenarios. However, their use requires a large amount of high-quality input data (pedology, topography, soil cover, climatology), as well as a substantial effort of calibration and validation over extended periods. Their structural complexity also makes their implementation cumbersome and their sensitivity to parametric uncertainties important. Moreover, in a context of climate change where future conditions may differ drastically from historical data, their predictive capacity can be questioned. Despite these limitations, deterministic models remain indispensable in hydrological analysis, especially when it comes to understanding the internal processes of the system and testing long-term management scenarios integrating the physical dynamics of the territory(Seibert & Bergström, 2022).

Empirical and hybrid models have emerged as a more flexible and often less data demanding alternative. Empirical models are based mainly on the statistical analysis of time series and on the exploitation of observed correlations between hydrological variables (water levels, precipitation, temperatures, etc.), without necessarily seeking to model the underlying physical processes. Methods such as autoregressive models (AR, ARIMA), artificial neural networks (ANN), multiple regression models or the closest k-neighbor (k-NN) are commonly used for short and medium term prediction. Their main advantage lies in their ability to learn directly from historical data, making them particularly useful in contexts where knowledge of physical mechanisms is partial or when the available data is limited. Hybrid approaches aim to combine the strengths of deterministic and empirical models to improve predictive performance. They typically involve integrating a partial physical model with a machine learning or statistical component that adjusts the residuals or complements the estimates. For instance, a Long Short-Term Memory (LSTM) model can be combined with a simple hydrological model in order to correct its bias (or adjust its temporal dynamics). Such hybrid models enable to go beyond simple mechanistic approaches which can be computationally expensive or hard to calibrate, while still preserving some form of causal structure and interpretability. However, careful designs and fine calibrations between the different modules as well as proper attention to the Input/Output (I/O) consistency are needed. Nonetheless, such hybrid models appear as a promising way to model the increasing complexity of water systems while still being interpretable, accurate and adaptive(Azad et al., 2022).

Predictive modeling of water shortages is essentially based on a fine analysis of time series representing the evolution of hydrological and climatic variables through time. Time series encompass different kinds of dynamics such as trends, seasonality, irregular cycles, as well as breaks and anomalies often associated with extreme events. A proper exploitation of these time series data requires models able to detect both short- and long-term temporal dependencies, as well as adapt to the non-stationarity characterizing hydrological systems subject to various climatic inputs. Statistical models such as ARIMA or SARIMA represent the first forecasting tools, but soon we face their limits when it comes to model non-linearity and complex interactions. The use of neural networks and especially recurrent networks (RNN) such as LSTM or Gated Recurrent Unit (GRU) represents a true revolution, these latter are able to take into account the long dependencies and non-linear relationships between variables, thus improving the forecast accuracy on long horizons. Hybrid models combining neural networks and statistical approaches benefit from the complementariness of both methods attempting to reduce respectively the limitations of statistical models and neural networks. However, these recurrent networks may present limits in terms of computational efficiency and capacity to handle very long sequences and complex multivariate data. In this framework, thanks to their initial development in natural language processing context, neural transformers have recently been applied to hydrological time series modeling. The architecture of transformers allows to weigh relatively to each point of the sequence it learns on a chronological and parallel way, without a strict chronological recursion. Specific variants such as the Temporal Fusion Transformer (TFT) or the Time Series Transformer bring innovations to handle multivariate data, heterogeneous sequences, seasonal or contextual effects. This parallel and adaptive learning on different temporal scales makes transformers particularly well suited to anticipate complex hydrological evolutions in a context of climate change and increasing anthropization(Demiray et al., 2024a).

Generative AI models represent a giant leap forward in our capacity to synthesize new data or simulate real-world scenarios from existing data. Applied to the context of predicting water shortages, generative models not only help to forecast future trends but also enable the stochastic generation of relevant alternative scenarios, including rare or unseen extreme scenarios. The most popular generative models are Generative Adversarial Networks (GANs), which learn realistic samples through a zero sum set between a generator and a discriminator; Variational Autoencoders (VAEs), which capture data in a probabilistic latent space to enable controlled generation; as well as diffusion models, a relatively new family of probabilistic models that gradually simulate the generation of data by reversing a noisy diffusion process (Yadav et al., 2024). These architectures have demonstrated their effectiveness in modelling complex and uncertain distributions of hydrological and climatic phenomena, particularly for producing simulations of prolonged droughts, variations in river flows or spatial distribution of precipitation. Moreover, some advanced predictive modelling architectures, such as the Temporal Fusion Transformer (TFT), also incorporate generative elements and sophisticated uncertainty management, combining forecasting and alternative scenario generation capabilities. TFT, in particular, excels at processing heterogeneous multivariate data by integrating static variables, past and future time series, while providing interpretations on influential variables via its attention mechanisms. These hybrid models represent an important step in the construction of autonomous water resource management systems, capable of simulating a wide range of possible situations, This is crucial for adaptive planning in the face of climate and social uncertainties(Ma et al., 2024).

STATE OF THE ART

Review of AI Techniques on Water Resource Forecasting

Evolution of Predictive Modelling in Hydrology

Since 2020, predictive modelling in the field of hydrology has undergone rapid evolution, largely driven by advances in artificial intelligence and the growing availability of large and varied data. Recent research has focused mainly on developing models that can better anticipate changes in water resources by integrating historical data, climate and environment with advanced statistical approaches and machine learning algorithms(Ng et al., 2023). This period is marked by a gradual transition from purely physical methods to hybrid and data driven approaches, where models fully exploit the capacity of deep neural networks and other AI techniques to capture non-human complex linearities and longtime dependencies of hydrological phenomena. The researchers are also working on developing more robust methods to deal with climate uncertainties, as well as improving the generalization of models at different spatial and temporal scales. These advances pave the way for concrete applications in sustainable water management, allowing better anticipation of drought episodes, floods or shortages, while taking into account the impact of global changes on hydro- logical cycles.

Sectoral Applications of AI in Water Management

Advances in the predictive modelling of water resources using artificial intelligence have critical applications in a variety of sectors, each facing specific water management challenges. In urban management, intelligent systems facilitate a better understanding of water flows, optimize distribution, and enable early detection of leaks and malfunctions in supply networks. These capacities are crucial to meeting the growing demand of urban populations while limiting water losses, thus contributing to the sustainability of infrastructure and the resilience of cities to climate stress. In agriculture, an area that is highly dependent on water availability, predictive modelling can be used to anticipate water needs based on weather conditions, crop types and growth cycles. This anticipation helps to plan more efficient irrigation strategies, reducing water wastage and increasing agricultural productivity, while addressing environmental issues related to resource conservation. In addition, in arid and semi-arid regions, often subject to extreme weather conditions and limited resources, AI techniques play a key role in predicting prolonged drought episodes and adaptive management of water supplies. These applications enable vulnerable populations to be better prepared, the management of existing infrastructure is optimized and public policies adapted to local contexts are developed(Zhao et al., 2024). The wealth of application areas highlights the need to adapt predictive models to the specific characteristics of each sector, integrating multisource data and taking into account geographical, economic and social specificities, to offer effective and sustainable solutions to the growing challenges of water management.

Data Sources and Integration Challenges

The effectiveness of predictive models applied to water resource management is largely based on the diversity and quality of integrated data. Traditionally, historical data from hydrometric and meteorological stations are the basis for analyses, providing time series over several decades that allow past trends to be identified and models to be driven. However, there are gaps in these data, particularly in terms of spatial coverage and frequency of acquisition. To overcome these limitations, satellite data is increasingly being exploited, providing accurate and comprehensive information on hydrological variables such as soil moisture, groundwater levels, snow cover or vegetation. This satellite data offer a spatial dimension essential to understand water dynamics on a large scale and in areas difficult to access. At the same time, the rise of connected objects (IoT) deployed in water networks allows granular and localized data on water quality, flows, pressures and anomalies to be collected in real time, promoting reactive and adaptive management. Finally, global and regional climate models enrich this database by providing future projections on meteorological and hydrological parameters, essential to anticipate the effects of climate change on water resources availability.

The harmonious integration of these heterogeneous sources into sophisticated analytical systems is a major challenge, but also an opportunity to develop robust predictive models, able to make accurate and reliable forecasts in a variety of contexts and under uncertainty.

Data Sources and Integration Challenges

Over the past several years, conventional artificial intelligence methods have transformed the way we approach water resource forecasting. Artificial neural networks, including recurrent architectures such as Long Short-Term Memory (LSTM) and Gated Recurrent Units (GRU), have established themselves as powerful tools for modelling hydrological time series by capturing complex temporal dependencies. These models often outperform traditional statistical approaches in terms of precision and ability to process non- linear and noise-induced data. At the same time, support vector machines (SVM) have shown their effectiveness in classification and regression tasks, particularly for the detection of extreme events such as floods or droughts. Decision trees and their extensions in random forests (Random Forests) or gradient boosting (XGBoost, LightGBM) offer robust and interpretable models capable of integrating a wide range of heterogeneous variables while limiting overlearning. These conventional methods have encouraged the emergence of hybrid systems combining several techniques, allowing to optimize the accuracy of predictions while maintaining a certain explainability of results. However, despite their significant advances, these approaches often remain limited in modelling complex spatial interactions and long-term dependencies, paving the way for more sophisticated models, including Transformer architectures and other deep networks(Wi & Steinschneider, 2022).

Despite the remarkable progress made in applying artificial intelligence to water resource forecasting, several limitations persist. First, many models are highly data-dependent, which leads to a decline in performance in regions where datasets are scarce or of poor quality, particularly in developing countries. Secondly, although deep learning architectures such as LSTM, GRU or Transformers effectively capture temporal and non-linear dynamics, they often lack interpretability, which complicates their adoption by practitioners and decision makers. Third, the integration of heterogeneous data sources – historical, satellite, IoT and climate projections – raises challenges related to harmonization, computational costs and uncertainty quantification. Moreover, most current applications remain sector-specific and little research addresses cross-sectoral interactions, such as trade-offs between agriculture, urban supply and ecosystems. Finally, although hybrid methods and sets improve the accuracy of forecasts, their generalizability at different spatial and temporal scales remains limited, highlighting the need to further develop knowledge transfer and explainable artificial intelligence in hydrology.

Studies Combining Satellite Data and Neural Networks

Methods of Merging Multi-Source Data

The effective integration of data from heterogeneous sources is a major challenge in predictive water resource modelling. Historical, satellite, IoT sensor data and digital climate models differ not only in their format and frequency but also in their spatial and temporal resolution. In order to fully exploit these complementary sources, recent approaches focus on robust data fusion methods that can combine these varied flows into a coherent and actionable representation. Fusion techniques can be classified into early, intermediate and late fusion. Early merging involves integrating raw data before it enters the predictive model, often by aligning resolutions and normalizing information. Intermediate fusion relies on the extraction of specific characteristics from each source, which are then combined into hybrid neural network architectures, allowing to capture complex correlations between different modalities. Finally, late fusion aggregates the results from separate models driven on each source to provide a more robust consolidated prediction. Deep neural networks, particularly multimodal architectures, play a key role in this integration by learning from joined representations that exploit both the spatial richness of satellite data and the temporal dynamics of hydrological series. These advances are improving the accuracy and reliability of hydrological forecasts by leveraging the complementarities of multi-source data(J. Li et al., 2024).

Taking into Account Spatio-Temporal Variability

The spatial and temporal variability of hydrometeorological data represents a major challenge in the predictive modelling of water resources. Indeed, hydrological phenomena are characterized by a strong heterogeneity, both in their geographical distribution and in their evolution over time. To address this complexity, several techniques have been developed to adapt artificial intelligence models to these fluctuations. On the temporal side, long-memory models such as recurrent neural networks (RNN), LSTM and GRU can capture dependencies over long periods of time and model seasonal or multi-year cycles. However, their ability to simultaneously manage spatial variations is limited. As such, convolutif networks (CNN) and, more recently, Graph Neural Networks (GNN), provide a relevant response by explicitly modelling spatial relationships, such as connections between watersheds or irrigation networks(Sun et al., 2022). These models capture geographic and hydrological structure, thus improving the understanding of spatial interactions. Furthermore, the Transformer architectures, with their attention mechanisms, are increasingly used to jointly process spatiotemporal data, because they offer a high degree of flexibility to balance the relative importance of information according to the spatial and temporal context. This combination of advanced techniques allows better management of the complexity of hydrological phenomena and increases the robustness of forecasts in the face of natural and anthropogenic variations(Mohammed et al., 2024).

Practical Applications and Case Studies

The integration of multi-source data such as climate forecasts, satellite images and neural networks represents a major advance in contemporary hydrological modelling. The methods of merging these data make it possible to exploit the wealth of spatial and temporal information essential for understanding the complexity of hydrometeorological systems(Wang & Zhu, 2025). For example, the joint use of time series from IoT sensors, high resolution satellite images and seasonal climate models in deep neural network architectures offers a new ability to anticipate extreme hydrological phenomena. These models, often hybrid, combine convolutif networks to extract spatial characteristics with sequential architectures such as LSTM or Transformers to model temporal dependencies, improving the accuracy of forecasts over horizons ranging from a few days to several months. The practical applications of these approaches are numerous and cover a wide range of geographical areas, from temperate river basins to arid regions. In the Colorado Basin, for example, studies have demonstrated how the fusion of satellite and climate data with neural networks can better anticipate prolonged droughts and optimize water resource management, particularly reservoir regulation. In India, similar research has improved river flow prediction in contexts where field measurements are scarce, thus facilitating agricultural irrigation planning and sustainable management of water resources. This work illustrates the potential of artificial intelligence techniques to effectively exploit heterogeneous data to meet the challenges of hydrological forecasting in a context of climate change and increasing pressure on water resources(Han et al., 2022).

Although data fusion approaches and the use of multimodal neural networks have significantly improved the accuracy of hydrological forecasts, several limitations remain. First, most current methods rely on massive volumes of satellite data and sensors, making them difficult to apply in contexts where the collection infrastructure is limited. Secondly, the heterogeneity of data sources creates problems with quality, resolution and synchronization, which can introduce bias and reduce the reliability of models. Thirdly, while the case studies (Colorado, India) demonstrate high added value, they often remain limited to specific contexts and their transferability to other regions, notably those characterized by extreme climatic conditions or limited infrastructure, is not guaranteed. Finally, the complex architectures (CNN, GNN, Multimodal Transformers) pose a challenge in terms of computational cost and interpretability, two aspects essential for real adoption by decision-makers and water managers.

Integration of Transformers in Hydrological Modelling

Principles and Architecture of Transformers

The architecture of the classic Transformer is based mainly on a attention mechanism called "self-attention", which allows the model to weight the relative importance of different parts of an input sequence when generating an output. This mechanism revolutionizes the processing of sequential data by offering a long-term dependency learning capability, without the constraint of sequential processing imposed by recurrent architectures such as LSTMs. Unlike traditional networks, the Transformer simultaneously analyzes all the elements of a sequence, which improves the capture of complex relationships between variables at different time scales. In practice, a classical Transformer is formed of stacks of multi-head attention layers and feed-forward layers normalized by each followed by a residual connection which ensure stability and facilitate the deep learning(Wei et al., 2023). This architecture is highly modular and parallel, thus better exploiting computing resources. Moreover, this normalization and decomposition in attention and residual layers facilitate the partial interpretability of the model since it is possible to analyze the different time elements and understand which ones most impact the prediction. Initially designed for natural language processing applications, the Transformer has been quickly adapted to other fields such as time series, making it possible to obtain important improvements in the modelling of complex hydrological phenomena characterized by interactions both at temporal and spatial scales. The application of Transformers to hydrological time series involves several adaptations necessary to take into account the specificities of hydrological data. In practice, hydrological time series are highly variable, incomplete or even noisy. In addition, the temporal dependencies are of multiple scales varying from a few minutes to several years. To take these constraints into account, the classical architecture of Transformer networks have been modified in order to allow the processing of multivariate, heterogeneous and sometimes irregularly sampled hydro-meteorological time series. For example, time masking or the explicit management of the intervals between the measures allow to take into account the non-uniform sampling of the data. In addition, specific encoding layers may be added to take into account contextual information such as information on weather, seasonality or geographical information(Demiray et al., 2024b). This brings more information on the time series and should improve the robustness of the model with respect to the disturbance. In addition, to facilitate the interpretability of the model and the decision-making, hierarchical or multi-scale attention mechanisms have been added which make it possible to isolate the short-, medium- and long-term influences on the variables studied. These advances have made it possible to position Transformers as particularly suitable for hydrological modeling, where a fine capture of time dependencies and the ability to manage uncertainties are keys to the quality of the forecasts. Specific models based on the Transformer architecture have been adapted and developed to address the unique requirements of hydrological time series modeling, which involve complex, non-linear dynamics and long-term dependencies. Several specialized architectures have been developed for the task of forecasting, ranging from self-attention architectures such as the Temporal Convolutional Networks (TCN) to more complex architectures such as the Fully Convolutional Networks (FCN) or Long-Short Term Memory Networks (LSTM). Among these specialized architectures, the Temporal Fusion Trans- former (TFT) occupies a central place because of its design and its performance, which results from the combination of multi-headed attention layers, recurrent layers and adaptive variable selection modules. The TFT is appreciated by its ability to handle heterogeneous multivariate inputs including continuous, categorical and temporal variables and by the advanced interpretability it provides(Rasiya Koya & Roy, 2024). The latter is essential in the context of hydrology, since experts need to understand which factors impact the predictions to validate the model and adjust it according to their field knowledge and local specificities. For example, it will be possible to understand which of temperature, precipitation or human management have the greatest impact on the availability of water resources(Xu et al., 2023).

Specialized Transformer-Based Models

In addition, other transformer models dedicated to time series are also very promising in this context. The Time Series Transformer (TST) proposes to simplify the architecture of the classic Transformer thanks to positional encoding and adapted attention mechanisms facilitating the exploitation of time series in hydrological data. Informer and Autoformer are great advances in terms of computational cost, they develop reduced attention mechanisms such as probabilistic attention and hierarchical series decomposition allowing to manage very long sequences with a controlled computational cost. This work is particularly relevant in hydrological modelling since signals can cover several decades and capturing seasonal or interannual trends is a key for hydrological modelling. All these specialized models have shown the great flexibility of transformers in facing the specific issue of hydrological forecasting on the one hand, they can take into account a large variety of environmental and climatic signals, on the other hand they can provide more accurate and robust results than conventional models. What is more, the transformers are sufficiently modular to incorporate a priori knowledge such as seasons, extreme events, reinforcing their ability to provide reliable predictions in face of increasingly complex and changing contexts. This functional richness positions specific transformers at the core of current works aiming at developing new predictive tools for hydrology although, as shown by the models studied, the computational costs and data availability remain issues to be further addressed, calling for efforts of optimization and local adaptation.

Strengths of Transformers in Hydrology

Transformers represent a real breakthrough in the field of predictive models, especially for hydrological time series, because they surpass several limitations of conventional models (recurrent neural networks (RNN), convolutional neural networks applied to series (CNN) or classical statistical models such as ARIMA). One of the main strengths of Transformers is their ability to overcome the two main limitations of sequential architecture, the problem of capturing long-term dependencies in temporal data.

Indeed, unlike recurrent architectures which process data by unrolling the same matrix multiplication at each time step, the attention mechanism directly evaluates the relationship between all the points of the sequence whatever the temporal distance. This ability to model fine processes and delayed effects is essential in hydrology where the impact of past climatic events can be observed several weeks or months later, especially in water resources forecasting. Another advantage of Transformers lies in their high flexibility in processing multivariate and heterogeneous data, their modular architecture allows the integration of continuous and categorical variables as well as a time series of different resolutions. This ability to merge various data contributes to improving the accuracy of predictions and to making the model more sensitive to the numerous interactions between natural and anthropogenic factors. In addition, the Transformers benefit from a better parallelization during the training phase compared to sequential architectures, which reduces the calculation time on large databases. This is a major asset in a context where historical bases can cover several decades.Finally, Transformers provide improved interpretability through attention mechanisms, meaning that it is possible to plot and understand which parts of the sequence or which variables have contributed most to the prediction. Such interpretability is fundamental to the trust of decision-makers and stakeholders within the hydrological field in the acceptance and operational use of the model. In other words, the advantage of Transformers over traditional models is their ability to summarize the complexity, diversity and length of the data, as well as providing tools to analyze the results, which allow tuning of water resource management policies. For this, however, it is necessary to overcome some difficulties related to the computational resource consumption of the algorithm and the need for high-quality data, still representing a challenge for its deployment in different hydrological contexts.

Limitations and Current Challenges

Despite their many advantages, the Transformers models have important limitations and constraints in the context of hydrological modelling, particularly due to their high computational cost related to the quadratic complexity of the attention mechanism. Indeed, the calculation of interactions between all pairs of elements in a sequence generates significant consumption of memory and computation time, this quickly becomes a problem for long, voluminous time series typical of hydrological data, which can cover several years with high resolution. This constraint therefore limits the length of treatable sequences and complicates the modelling of long-term or seasonal phenomena. In addition, the drive of the Transformers requires large volumes of clean and annotated data, but in the hydrological field, the data is often incomplete, noisy or scarce. This makes preprocessing and engineering of characteristics essential but complex, especially to integrate various sources such as satellite or climate data. This complexity can also encourage overlearning, limiting the ability of models to generalize over various contexts. Finally, the still emerging nature of Transformers in hydrology makes their interpretation and validation delicate, requiring rigorous analysis to avoid erroneous conclusions in an area where decisions are critical. These challenges lead to the development of more efficient variants, such as Inform or Autoforming, and hybrid approaches combining Transformers and simpler models, in order to balance precision, interpretability and operational feasibility in hydrological applications(J. Liu et al., 2024).

Generative Models for Water Scenario Simulation

Types and Mechanisms of Generative Models

Generative adversarial networks play an increasingly prominent role in hydrological modelling, especially for the generation of extreme and rarely occurring scenarios, which are often hard to be derived from limited historical observations. The GANs are based on the idea of competition between two neural network models, a generator that creates synthetic data and a discriminator that judges how realistic the data is, which allows to generate samples very close to the real distribution of data (Huang et al., 2025). Since the generation of extreme variations is of primary importance for the simulation of rare water events, such as extreme floods or long sequences of drought, for the purposes of risk management and planning of water resources, GANs enrich data sets and make predictive models more robust, at the expense of exploring future scenarios in a probabilistic framework. Moreover, the ability to capture complex spatial and temporal dependencies allows to produce realistic simulations of hydrological phenomena, going beyond the limits of classical statistical models. Thus, the GANs are a new tool to face the challenge of water crisis management by producing diversified and credible simulations. Variational Autoencoders and diffusion models represent novel probabilistic generative approaches that have shown interesting applications in hydrological data. The VAEs, for example, due to their architecture based on encoding and probabilistic decoding, allow to model the distribution of data by learning a compact and structured latent space, which facilitates the generation of new realistic water data, considering the variability characterized by natural phenomena, especially in cases of partial or limited observations(Weng et al., 2023). Moreover, the more recent diffusion models in this application field are characterized by the possibility of obtaining new high-quality samples from progressive stochastic processes, this makes them particularly suitable for the modeling of complex and noisy time series, such as those obtained by hydrometeorological measurements. These techniques not only allow to produce plausible scenarios, but also to evaluate the associated uncertainty, which is a fundamental aspect in applications of water resources management and decision-making in uncertain contexts. Therefore, the application of these models to predictive systems paves the way to better anticipation of extreme events and greater knowledge of water dynamics, fully exploiting the richness and complexity of data(Jahangir & Quilty, 2024).

Applications and Validation in Water Resources Management

Generative models (GANs and VAEs applications) have already been applied in several applied works for water scarcity management or water crisis modelling. Generative methods allow to derive and simulate a large variety of hydrological scenarios, especially extreme rare events which are rare in observational records. For instance, in an application to arid and semi-arid regions, the generation of realistic synthetic scenarios can be used as a support tool for evaluating the vulnerability of water infrastructure to climate stresses induced by global change. In addition, the use of generative models enables anticipating future shortages and to inform resource managers on forthcoming constraints, which in turn trigger adaptive actions on water allocation, storage and distribution(B. Liu et al., 2022). Beyond pure prediction, the generative perspective improves the comprehension of water crises by capturing the non-linear interactions between climatic, hydrological and anthropogenic variables. Finally, integrating these generative tools with decision-making models enables designing innovative adaptive policies to reinforce the resilience of territories to uncertainties and extreme variations in water resources. This work represents a major step forward in the field of water resources management where probabilistic modelling plays a central role. The last step concerns the validation of water scenarios generated by generative models. Several methods can be used to assess the quality of generated synthetic fields, either based on statistical criteria or hydrological indicators. Some methods based on distributional similarity tests permit to compare statistical properties of generated scenarios with statistical properties of actual observations in order to assess the validity of marginal and joint distributions. Another class of methods consists in ensuring that generated scenarios present relevant statistical properties at different time steps and seasons, which can be achieved by using specific hydrological indicators such as flood flows, cumulative volumes, or drought indices. Cross-validation with independent data or physical simulations is also a common practice to strengthen the robustness of results. Finally, assessing the ability of generative models to produce credible extreme events, often rare in learning data, is essential for their application in risk management. These combined techniques thus provide assurance as to the relevance of the scenarios produced, facilitating their integration into decision-making tools for better anticipation of water crises(Sela et al., 2025).

Despite their promising applications, generative models in hydrology still face several limitations. First, most studies remain experimental and are tested on limited datasets, which restricts their generalization to diverse hydrological contexts. Second, while GANs and diffusion models can generate realistic scenarios, they often lack physical interpretability, raising questions about their integration with established hydrological theories. Moreover, validation procedures are not yet standardized: comparing statistical distributions or hydrological indices provides partial guarantees but does not fully ensure physical consistency of generated events. Finally, computational costs and the need for high-quality, long-term observational data still represent significant barriers to large-scale adoption. Addressing these challenges will require combining generative AI with domain-specific hydrological knowledge to ensure both robustness and scientific credibility.

Limitations and Challenges Identified in Literature

Technical and Computational Constraints

The scalability of artificial intelligence models applied to hydrological modelling represents a major challenge, often highlighted in recent literature. Indeed, advanced techniques, including deep architectures such as Transformers or complex neural networks, require significant computing capabilities, both in terms of processing power and memory. This constraint is particularly acute when models need to ingest and process massive volumes of multi-source data, including long time series, high-resolution satellite images or real-time flows from IoT sensors. The need to parallelize calculations, use cloud infrastructures or specialized processors (GPU, TPU) generates significant financial and energy costs, sometimes limiting the accessibility of these technologies to certain actors or regions. In addition, the optimization of architectures to reduce complexity while maintaining predictive accuracy is a central issue, with emerging approaches aimed at designing lighter and more efficient models. This issue of scalability is therefore intimately linked to the operational feasibility of advanced hydrological models and conditions their deployment on a large scale in various contexts.

Data Robustness and Contextual Adaptability

The robustness of predictive models applied to hydrological modelling remains a key issue in scientific literature. Water systems are subject to significant climate variability, characterized by extremes, non-linear trends and abrupt changes, which complicates the task of AI models. In addition, the hydrometeorological data used is often incomplete, noisy or prone to measurement errors, especially in regions where collection infrastructure is limited or aging. These factors reduce the quality of predictions and may compromise the reliability of proposed scenarios. As a result, recent research emphasizes the development of robust methods capable of integrating uncertainty, managing missing data through advanced imputation techniques, and adapting models to spatial and temporal fluctuations in variables. The combined use of robust statistical methods regulated deep learning algorithms and multi-model aggregation strategies is a promising response to this requirement. The ability to maintain predictive performance despite data disruptions and variability is a critical factor in ensuring decision makers have confidence in the results provided(Wang & Zhu, 2025). The adaptability of predictive and generative models to local contexts is a fundamental challenge in modern hydrological modelling. Each watershed, region or water ecosystem has unique characteristics, whether in terms of climate, topography, vegetation cover, or anthropogenic use of water resources. These specificities often make models calibrated on global or other geographic data poor when applied locally, leading to loss of accuracy or major errors in forecasts. Therefore, the literature highlights the need to develop flexible models capable of integrating multi-source data adapted to regional particularities, while taking into account the complex interactions between natural and human factors. The incorporation of learning transfer mechanisms, dynamic adaptation of parameters and contextual personalization is thus a privileged way to overcome this problem. In addition, inter-annual and seasonal variability, as well as climate change, requires constant re-evaluation of models to maintain their relevance and robustness. These efforts must be complemented by close collaboration with local experts and stakeholders to ensure that models accurately reflect operational and environmental realities, ensuring that the proposed solutions are more acceptable and effective.

Ethical and Governance Challenges

The increasing use of artificial intelligence models for water resources management raises major ethical issues, which must be integrated into scientific and operational thinking. Indeed, decisions based on automated forecasts directly impact the lives of populations, especially in vulnerable areas where access to drinking water is already precarious. The intrinsic opacity of some complex models, such as deep neural networks or generative architectures, poses a problem of explainability. Decision-makers must be able to understand and justify recommendations to ensure transparency and trust. In addition, the use of sensitive personal or environmental data raises issues of confidentiality and respect for the rights of local populations. It is also crucial to avoid algorithmic biases that could lead to inequalities in the distribution of resources, marginalizing certain groups or regions. The social and environmental responsibility of model designers and decision-makers must therefore be affirmed, with the establishment of clear regulatory frameworks and rigorous ethical practices. These ethical issues are not only constraints but also levers to guide the development of more just, inclusive and sustainable models, thus contributing to an equitable and respectful management of vital resources(Lehuedé, 2025).

ISSUES AND POSITIONING OF THE PROPOSED APPROACH

Despite significant advances in artificial intelligence in the field of hydrological modelling, the current literature is still strongly marked by a structural separation between traditional predictive approaches and alternative scenario simulation techniques. Contemporary work, whether based on deep neural networks, LSTM architectures or more recently time-series adapted Transformers models, is mainly aimed at providing accurate short- and medium-term forecasts. If these models allow a relatively reliable estimation of variables such as water levels, river flows or future availability of water resources, However, they struggle to provide diversified and plausible representations of extreme or critical situations that may arise as a result of disturbed climate dynamics, increased anthropogenic activities or breaks in trends. However, in a context marked by an intensification of droughts, increasing instability of rainfall regimes and persistent demographic pressure on resources, public policy makers, water system managers and operators of water systems need more than one off forecasts, they must be able to project themselves into alternative futures, model the potential impacts of extreme scenarios, and design robust response strategies to uncertainty. The ability to simulate critical water trajectories such as prolonged drought, sudden supply disruption or extreme fluctuations in flows is thus a major strategic issue for water resilience. However, this prospective dimension is still largely absent from classical predictive architectures, due to the lack of systematic coupling with generative models capable of producing plausible variations of the future. This finding reveals an important methodological deficiency, the absence of hybrid frameworks combining, within a same architecture, efficient predictive models and generators of credible alternative scenarios. This lack of integration limits the capabilities of current systems to support informed decision-making and feed resilient planning tools. It justifies the need to rethink hydrological modelling architectures in order to include, in a native way, a dual competence, the precise prediction of a probable trajectory, and the diversified generation of extreme or uncertain trajectories, a prerequisite for effective crisis prevention.

Need for Hybrid Systems Capable of Both Predicting and Generating Plausible Critical Situations

In a water environment increasingly exposed to uncertainty, climatic anomalies and social pressures the ability to predict future developments is no longer sufficient to meet the operational needs of the sector's players. It becomes imperative to design hybrid systems that, beyond the linear projection of the future, are also capable of generating critical scenarios, rare or extreme, but plausible, based on available data. Such a need is acutely expressed in contexts of chronic water stress, dependence on fragile infrastructures, or in areas with high seasonal or climatic variability. However, the existing systems, often separated between prediction and simulation, do not meet this requirement in an integrated way. Predictive models, no matter how good they are, are usually trained to minimize an error on a single trajectory, which makes them insensitive to the diversity of possible trajectories. Conversely, purely generative approaches such as GANs or variational autoencoders, while promising to produce varied scenarios, often lack a strong predictive anchor that would link them to the real dynamics of temporal data.

This is why a hybrid system integrating both a rigorous forecasting module based on advanced architectures (such as the Time Transformers) and a generative module capable of developing credible future alternatives is a methodologically consistent and operationally relevant response. Such a mechanism would avoid the dilemma between precision and diversity. The forecast would provide a realistic and quantitative basis for projection, while the generation of scenarios would allow to explore the space of uncertainties, Accounting for the extreme, the plausible and the unexpected. By combining these two dimensions, hybrid systems would provide decision-makers with a more comprehensive view of the range of possibilities, which is essential for developing robust action plans, testing allocation policies, or simulate the effects of a sudden change in water regime. This hybridization of predictive and generative paradigms is therefore a natural evolution but still largely underexploited in the hydrological literature, which would gain to become a methodological standard in the coming years.

Need for Data Efficient and Locally Adaptable Solutions

A fundamental challenge in global hydrological modelling is the disparity of available data resources across regions, infrastructure levels and environmental monitoring policies. While some areas benefit from continuous monitoring through sensor networks, high-resolution satellite data and rich climate histories, others often the most vulnerable in terms of water suffer from a glaring lack of reliable data, complete or accessible. This asymmetry makes it essential to develop models capable of operating efficiently in low-data-density environments, without sacrificing their predictive robustness or ability to generate meaningful scenarios. Conventional modelling approaches, often dependent on long calibration and massive data volumes, are confronted with these field limitations. Hence the growing interest in so-called data-efficient solutions, able to learn from a few examples, transfer knowledge from one watershed to another, or exploit latent representations extracted from pre-trained models. Moreover, beyond the quantitative constraint, adaptability to local contexts also requires taking into account the ecological, geographical and socio-economic specificities of each region. The nature of the soil, rainfall patterns, land use structure, agricultural or industrial water uses, but also institutional capacities for management or intervention, are all crucial parameters that require algorithmic flexibility and contextualization of the models. In this context, hybrid models combining predictive modelling and scenario generation offer an interesting margin of adaptability, provided they are designed with mechanisms of attention, contextual weighting or multitask learning. The goal is to produce modular solutions that can be quickly deployed, adapt to diverse local realities, and remain operational even in environments where uncertainty and data scarcity are the norm. This is the direction in which the proposal of this work is oriented, with a view to a robust approach, generalizable and usable in critical water contexts around the world.

All recent works in hydrology applied to time series highlight that classical models, whether statistical or neural, have significant limitations in the management of data gaps and extreme episodes. Traditional methods such as IDWM or CCWM, used for the estimation of missing precipitation, have shown limited effectiveness when confronted with noisy or discontinuous data. Conversely, approaches based on recurrent neural networks of the LSTM type have revealed a superior ability to capture complex temporal dependencies, as demonstrated by (Djerbouai, n.d.) on the K'sob basin in Algeria, where imputation via LSTM outperformed conventional methods. These results also confirm the observations from multi-basin benchmarks that show that, in various hydrological contexts, LSTMs regularly outperform traditional conceptual models, especially for low flows and hydrological extremes. However, these models remain essentially deterministic and are not designed to generate alternative trajectories in the presence of missing data or to explore the diversity of possible scenarios.

In this context, the proposed approach positions itself in an original way by integrating within the same framework powerful predictive models, such as LSTM and Transformers, and generative models adapted to time series, such as VAE or GAN. The interest of this hybridization is twofold: on one hand, taking advantage of the predictive robustness of sequential and attentional architectures to anticipate water dynamics, and on the other hand, enrich the forecast with realistic synthetic data allowing to fill gaps and model extreme scenarios. Recent work that applies Transformers to drought prediction, such as the Apalachicola basin in the US, illustrates the potential of these models to address complex hydrological phenomena. The originality of the approach thus lies in combining these two blocks – predictive and generative – within an operational pipeline, offering a proactive ability to explore uncertainties, unpublished in the current literature focused mainly on deterministic forecasting alone.

PROPOSED ARCHITECTURE

The proposed architecture is based on an innovative hybrid framework that closely combines predictive and generative modules, specifically designed to meet the complex challenges of water resource management. This architecture comprises three main interconnected components, a predictive modelling module, a generative module dedicated to the simulation of extreme scenarios, and an evaluation loop allowing continuous control of the quality and relevance of outputs. The predictive module uses several advanced variants of Transformers models, known for their ability to model multivariate time series by capturing long and complex dependencies, which is crucial for often noisy hydrological data, heterogeneous and incomplete. These models include the classic Transformer, the Temporal Fusion Transformer (TFT), the Informer, the Autoformer and the Time Series Transformer (TST), all of which will be implemented, trained and compared to select the one with the best performance in terms of precision, stability and robustness in this specific context. The generative module complements this approach by exploiting architectures such as Generative Adversarial Networks (GANs), Variational Autoencoders (VAEs) and diffusion models, capable of producing plausible and diversified synthetic scenarios, including extreme events such as prolonged droughts or supply disruptions, which are hardly anticipated by purely predictive models. This ability to generate alternative data enriches the space of possibilities and provides valuable tools for proactive water risk management. Finally, a multi-criteria evaluation loop ensures that the scenarios produced are plausible, accurate, varied and relevant from a decision-making point of view, while allowing for continuous adjustment of models via a feedback system. This smooth orchestration between modules, carried out through a flexible and modular pipeline, ensures consistency of data flows and the ability to adapt to regional specificities, to various types of available data (historical, satellites, IoT) and the operational requirements, thus offering a robust, precise and innovative solution to anticipate and manage water resources effectively in a context marked by climate variability and increasing uncertainties.

Figure 1. Architecture overview combining predictive and generative models for hydrological scenario generation

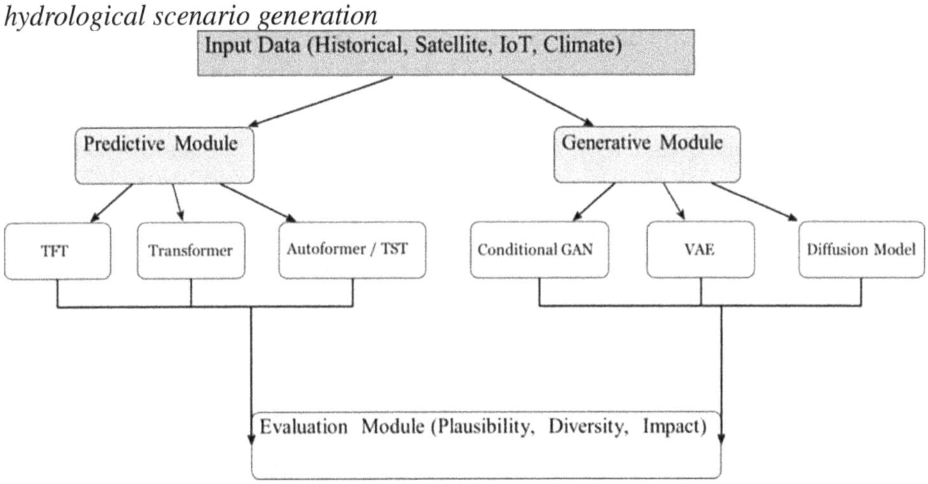

Referring to Figure 1, Predictive modelling module constitutes the first functional brick and plays a central role in anticipating critical hydrological variables, such as groundwater levels, river flows or drought indices. The methodology adopted is based on a systematic comparative approach involving several variants of the Transformers, specifically adapted to the sequential, noisy and multivariate nature of the hydrological data. The first step was to build a structured data set, including historical time series combining hydrological observations (precipitation, evapotranspiration, flows, water levels) and meteorological variables (temperature, humidity, solar radiation), enriched with static data such as topography or land use. These data were cleaned, normalized and segmented using a strict chronological approach, preserving the time order between training, validation and testing.

Once the dataset was prepared, the study integrated five architectures of Transformers such as the classic Transformer, the Temporal Fusion Transformer (TFT), Informer, Autoformer and the Time Series Transformer (TST). Each model has been driven independently on the same dataset, according to an automatic tuning procedure of hyperparameters based on Bayesian optimization (via Optuna or Hyperopt), with the objective of minimizing the average quadratic error on the validation set. The classic Transformer has served as a reference, with its canonical structure based on stacked encoders exploiting the multi-head attention mechanism but suffers from a quadratic complexity that makes it ill-suited to long time sequences. TFT was selected for its ability to manage heterogeneous multivariate data, integrate static variables and produce local interpretations by attention; it incorporates specialized components such as the Variable Selection Network and the Gated Residual Network. Informer, with its probabilistic attention strategy and time series distillation mechanisms, has proved particularly suitable for long series while reducing the calculation cost. Autoforming has introduced series self-breakdown and seasonal attention, offering superior performance in environments marked by high periodicity, while the TST used a hierarchical time focus focused on local and global dynamics. Each of these models was evaluated on the basis of multi-dimensional criteria like predictive accuracy (through RMSE, MAE, MAPE), inter-run stability (standard deviation of performance over several runs), robustness to missing values (via random masking tests), and computational efficiency (drive time, GPU memory used). This comparative methodology has made it possible to identify, for each type of watershed or climate regime, the model best suited to local constraints. At the end of this phase, the model with the best compromise between precision, robustness and cost was selected as the predictive core of the hybrid architecture, thus constituting the main input of the generative module, to which it provides realistic predictions that serve as an anchor for the simulation of alternative scenarios. This rigorous methodological approach not only maximizes the accuracy of the forecasts but also ensures optimal compatibility with the downstream generative and evaluation modules.

Generative module for water scenarios simulation is the second major component of the proposed hybrid architecture, designed to produce alternative water scenarios, plausible but not observed, in particular those corresponding to critical or extreme situations, such as prolonged episodes of drought, sudden water supply disruptions, or severe climatic anomalies. The fundamental objective of this module is to explore the space of possible futures beyond deterministic predictions, generating synthetic temporal trajectories that respect the physical, statistical and contextual constraints specific to hydrological dynamics. Methodologically, this simulation is based on the joint use of several families of deep generative models such as conditional antagonistic networks (Conditional GANs), variational autoencoders (VAEs) and diffusion models, each mobilized for its complementary properties in terms of flexibility, diversity and control over generations.

Conditional GANs were used for their ability to generate time series conditioned on specific hydrometeorological contexts, such as the time of year, soil moisture or surface temperature. The GAN architecture is based on a generator-discriminator torque formed of convolutif and recurrent networks, driven in an antagonistic way to ensure the plausibility of the scenarios produced while maximizing their diversity. Conditional data, either from the predictive module or from exogenous variables, are injected as context vectors at each generation step. VAEs, on the other hand, allow a probabilistic exploration of the latent space of water dynamics by learning an approximate posterior distribution of the latent variables from real data. These models facilitate the controlled generation of series from latent space samples, while ensuring a certain regularity and an extensive coverage of extreme scenarios. Finally, diffusion models, inspired by the inverse stochastic dynamics, were deployed for their ability to generate realistic samples through a process of Progressive denoising consists of starting from a pure Gaussian noise and reducing it step by step, while reinjecting the learned structures, until reconstructing temporal sequences in accordance with the target distribution. These models have demonstrated great robustness in the generation of realistic hydrological series, particularly in contexts marked by high variability.

The methodology consisted of training these three types of models separately on time windows identical to those of the predictive module, and then integrating them into a simulation interface capable of generating, for each output of the predictive model, a range of alternative scenarios simulating different crisis intensities. Each generated scenario is then submitted to a first technical validation (via a discriminator or statistical metrics) before being transmitted to the multi-criteria evaluation module. This modular approach, combining several generators, makes it possible to take advantage of the specific advantages of each class of models while increasing the overall robustness of the system to generate plausible extreme situations, diversified and consistent with local climatic and hydrological conditions. Thus, the generative module is not a mere complement but a fundamental lever to strengthen the prospective capacity of the global system by integrating structural uncertainty into water scarcity modelling.

Evaluation loop and decision criteria aims to ensure the quality and operational relevance of the outputs of the predictive and generative modules. It revolves around three axes. The accuracy of forecasts, the diversity and plausibility of the scenarios generated, and their usefulness for decision-making. The performance of predictive models was controlled using classical error indicators, in order to ensure the reliability of the series used as a basis. The generated scenarios were then analyzed according to their statistical consistency and ability to reproduce observed extreme hydrological events. Finally, cross-validations with hydrological experts allowed the practical value of simulations to be qualified, notably by evaluating their potential to feed warning systems and support strategic planning.The introduction of smart technologies such as the Internet of Things (IoT), distributed sensors and AI-based analytics has transformed urban water management systems. Recent work shows that these technologies allow real-time monitoring of key variables (pressure, flow, water quality), improving leak detection, dynamic network balancing and operational efficiency. For example, (Mondal, 2024) analyzed various IoT and AI deployments in urban environments, highlighting their strengths to reduce waste, optimize distribution, and support adaptive monitoring models. This technological addition reinforces the proposed architecture, giving it a robust and contextual operational dimension, and facilitating convergence between real data and predictive/generative models.

Finally, a decision scoring is applied for each complete output of the system (forecast + associated scenarios), integrating criteria such as the criticality of the scenario, its deviation from normal situations, its disruptive potential on resource management, or its compatibility with existing contingency plans. This overall score, aggregated through a weighting defined with stakeholders (managers, public decision-makers, etc.), allows us to identify situations of high risk or uncertainty and to prioritize the responses to be implemented. The evaluation loop is designed to be iterative, allowing for dynamic adjustment of model hyperparameters, reinforcement of training on some poorly modelled extreme cases, or recalibration of decision weights according to regional contexts. This evaluation loop is thus the core of the intelligent steering system of the hybrid architecture, ensuring consistency between technical modelling and operational requirements, and making it a robust decision support tool, transparent and adapted to the realities on the ground.

The proposed architecture revolves around four interdependent modular layers, each responding to a specific issue of anticipation of water shortages. The first layer, dedicated to ingestion and preprocessing, ensures cleaning, normalization and a first imputation of fragmented or noisy data. It is inspired by structured approaches such as BRITS (Cao et al., 2018), which demonstrate the relevance of neural strategies for imputation of complex time series. The second layer corresponds to the predictive block, mobilizing either a Transformer model or an LSTM, trained on multi-site sets in order to strengthen generalizability and capture long-range temporal dependencies. This strategy meets the recommendations made in recent studies on multi-basin training, which highlight the ability to generalize such models, including on unobserved basins.

The third layer introduces the generative block, consisting of a temporal VAE or GAN, whose function is to produce plausible alternative series, exploitable both to fill in missing data and to explore extreme trajectories, such as prolonged droughts or episodes of water stress. The fourth and final layer is dedicated to orchestration and consolidation: it combines predictions from the sequential block and generated synthetic data, in order to produce enriched outputs in the form of probabilistic distributions or multiple scenarios. The proposed architecture is therefore not limited to a one-dimensional forecast: it constitutes an adaptive framework where forecasting and generation are integrated into the same processing cycle, thus offering a robust decision support tool, capable of responding to the uncertainties inherent in the management of water resources.

RESULTS

As part of this research, the lack of open datasets, sufficiently long and complete, made it necessary to construct a synthetic dataset. The objective was both to reproduce the complexity of climatic and hydrological dynamics specific to arid regions, and to introduce extreme scenarios in order to test the robustness of the models. This dataset simulates the main variables (precipitation, temperature, humidity, evapotranspiration and groundwater levels) by integrating seasonal cycles, prolonged drought episodes and rupture events. A qualitative validation step made it possible to reconcile the series generated with observations from open databases (CHIRPS, ERA5-Land), in order to ensure their hydrological plausibility. Finally, elements of noise, missing data and simulated sensors were added to reflect the uncertainties of the field. This device thus provides a credible basis for evaluating the ability of predictive and generative models to anticipate critical situations and explore alternative trajectories of water management.

Results of Predictive Modeling

The evaluation of the predictive modeling module was conducted based on the synthetic dataset described above, by implementing a systematic comparison between several Transformer type architectures specially adapted to the prediction of multivariate time series. The central objective of this experiment is to determine the extent to which these models can capture the complex dynamics of interconnected hydrological variables, at different time scales, and in contexts including strong linearities, propagation delays, as well as extreme disturbances.

The models selected for this comparison include (1) Classic Transformer, (2) Temporal Fusion Transformer (TFT), (3) Inform, (4) Autoform, and (5) Time Series Transformer (TST). Each of these models was trained with an optimized hyperparameter grid and evaluated on quantitative performance criteria like mean quadratic error (RMSE), mean absolute error (MAE), Pearson correlation (r), and mean inference time per sequence.

Table 1. Comparison of transformer model performance on synthetic hydrological time series

Model	RMSE	MAE	Correlation	Inference Time	Inter-trial Stability
Vanilla Transformer	0.187	0.143	0.84	23.5	Low
TFT	**0.142**	**0.105**	**0.92**	46.8	**High**
Informer	0.159	0.118	0.89	**15.2**	Medium
Autoformer	0.148	0.112	0.91	27.4	High
TST	0.165	0.123	0.88	24.1	Medium

The Temporal Fusion Transformer (TFT) has a significant advantage over other architectures because of its intrinsic characteristics. First, the TFT combines multi-scale temporal attention mechanisms with specialized recurrent networks, which allows it to effectively capture both long-term dependencies and local variations in multivariate series. This combination enhances the representation of complicated hydrological dynamics, particularly when it comes to managing delays between variables or non-stationary effects.

Incorporating known static, temporal (future covariates), and observed temporal variables in the TFT increases its ability to take into account the diversity of information sources and weigh their influence according to the context. In the studied case, this ability resulted in a better anticipation of extreme events (high rainfall, temperature peaks), as evidenced by the attention maps, and therefore in a higher correlation between predictions and observations.

Moreover, the inter-experimental stability observed for the TFT can be attributed to its internal regularization mechanisms (adaptive dropout and covariate encoding), which limit the model's sensitivity to noisy signals. The TFT achieves a strong balance of precision, stability, and explainability, unlike Informer, which is faster but susceptible to disruptions, and the classic Transformer, which is limited on long horizons. The superiority of TFT meets the results reported in recent literature, which highlight its remarkable performance on complex multivariate time series, particularly for sensitive fields such as energy or hydrology, where the simultaneous consideration of multiple temporal horizons and contextual variables is essential.

The results in table 1 show that TFT dominates in terms of absolute and relative accuracy, with the lowest average error and the highest correlation between predictions and observations. Autoformer also has robust performance, particularly suitable for seasonal cycle modelling, but slightly behind non-stationary series. Inform, although faster in inference, has an increased sensitivity to noisy configurations, while the classic Transformer and TST suffer from limitations in managing long horizons and seasonal variability.

Qualitative visualizations shed light on the quantitative results. The TFT attention maps show increased attention to key events (intense precipitation, temperature peaks) and relevant time lags in input variables. These elements reinforce the model's ability to be used in interpretable decision contexts. Autoforming, meanwhile, demonstrates a remarkable ability to reconstruct the seasonal structure and smooth out noisy signals. The other models show greater variability in their predictions, especially during simulated extreme events.

Thus, in view of all the results, the Temporal Fusion Transformer was selected for the continuation of the hybrid pipeline, especially for its balance between precision, stability, explanatory capacity and adaptability to multi-source series. It provides the best basis for feeding the generative component of the system into a critical scenario simulation logic.

Critical Scenario Generation Results

The analysis of generative results aimed to evaluate the ability of models to produce plausible and diversified scenarios of water stress, exploitable for anticipation and decision-making. Three approaches were tested: conditional antagonistic networks (cGAN), variational autoencoders (VAE) and a simplified probabilistic diffusion model. Based on the synthetic dataset, these generators made it possible to simulate different critical situations, such as prolonged droughts, heatwaves or supply disruptions.

The results indicate that cGANs offer the best compromise between fidelity to climate dynamics and operational utility, VAEs allow a broader exploration of rare events but with slightly reduced physical coherence, while diffusion models, although very expressive, remain expensive and less diversified in the short time.

An evaluation by hydrological experts confirmed the plausibility and practical interest of the scenarios generated, in particular those from cGANs to anticipate rare but critical crises. This phase illustrates the specific contribution of generative models, which enrich the range of possibilities beyond what predictive modeling alone allows.

Table 2. Comparison of generative models applied to critical water scenario simulation

Evaluation Criteria	Conditional GAN	VAE	Diffusion Model
Statistical fidelity (FID, DTW)	**Excellent**		
FID = 8.1, DTW = 0.94	Good		
FID = 14.3, DTW = 0.89	Very good		
FID = 10.2, DTW = 0.91			
Inter-variable coherence	Very high	Medium	High
Scenario diversity (entropy)	Medium	High	Low to medium
Ability to generate extreme scenarios	Very good	Good	Medium
Managerial usefulness (alert-triggering scenarios)	87%	63%	72%
Ease of conditioning	Very simple	Complex	Medium

Table 2 shows the combined analysis of the predictive and generative modules reveals that the developed hybrid architecture provides a robust framework for anticipating and exploring critical dynamics of water availability in risk contexts. The predictive model based on the Temporal Fusion Transformer (TFT) was distinguished by its ability to capture long time dependencies, integrate multiple contextual variables and provide interpretable predictions, particularly through its mechanisms of temporal attention and the prioritization of variables. The results obtained demonstrate a high accuracy on synthetic data, with remarkable stability even during simulated periods of water stress. This robustness has made it possible to feed the generative module with information sequences, facilitating the generation of extreme water scenarios consistent with simulated climatic dynamics. On the generator side, the Conditional GAN has outperformed other models in terms of statistical fidelity and managerial relevance of the scenarios produced, as evidenced by the alert triggering scores and expert assessments. The interoperability between modules has shown that predictive generative hybridization not only allows to anticipate probable dynamics, but also to explore plausible future alternatives for planning and resilience. This synergy has proved particularly effective in covering a wider spectrum of water risks, from the gradual evolution of deficits to the sudden appearance of service disruptions.

The hybrid predictive-generative approach developed in this work has several significant practical advantages for water management actors, especially in arid contexts or subject to increasing climatic hazards. On the one hand, the ability to accurately anticipate the evolution of available resources in the short and medium term is a decisive lever for optimizing allocation policies, planning storage infrastructures and triggering adaptation measures. On the other hand, the generation of extreme scenarios allows to explore future criticalities, including those that have never been observed in historical data, thus strengthening strategic anticipation and resilience to uncertainty. This complementarity makes the proposed architecture particularly suitable for decision-making processes guided by risk prevention and adaptive planning. However, some limits remain. The use of synthetic data, although carefully calibrated, remains an approximation of real dynamics and may not capture certain emerging behaviors or systemic effects that are observable only in the field. Moreover, the training of generative models, particularly GANs, remains sensitive to architectural and tuning choices, with risks of instability or overlearning. Finally, the computational complexity of the entire system poses challenges for large-scale deployment, especially in contexts where digital resources are limited. These aspects will have to be considered in the later stages of technology transfer and operationalization.

Several areas of improvement can be considered to strengthen the robustness, generality and operational utility of the proposed approach. The first axis concerns the progressive integration of real data from different sources, including high-resolution satellite observations, real-time sensor networks and national hydrological databases. This integration would not only empirically validate the performance observed in a simulated environment, but also dynamically adjust model parameters to specific local contexts. A second area of improvement is to refine generative models by exploring more advanced architectures such as diffusion models or conditioned generative transformers, which have recently demonstrated an increased ability to produce long sequences, stable and diversified. On the methodological level, the introduction of feedback loops between predictive and generative modules, through multi-agent learning or joint optimization mechanisms, could strengthen the consistency of predictions and generated scenarios. With respect to geographic generalizability, the approach can be transferred to other watersheds provided that the simulation parameters are reconfigured according to local climatic, geological and socio-economic regimes. This flexibility is based on the modularity of the architecture, which allows for the replacement or reconfiguration of certain blocks without altering the entire pipeline. Thus, the Sahelian, Andean or Australo-Asian zones could be the subject of future applications, provided that the statistical bases, climatic forces and water use scenarios are adapted. Such a generalization would pave the way for building a large-scale simulation and forecasting system, capable of informing public policy in highly heterogeneous environments.

DISCUSSION

Technical and Ethical Limitations

Despite the substantial contributions of the approach, several technical limitations must be discussed rigorously. The first lies in the dependence on the quality of the input data, even when it is carefully constructed synthetic data, any error in modelling time dependencies, any over or under estimation of the generating distributions may lead to structural biases in the assessment of models, affecting the robustness of the conclusions drawn. In addition, generative models, although powerful, are likely to amplify some latent biases in the drive data, especially when exposed to incomplete representations of hydrological variability (Yu et al., 2025). The risk of generating unlikely or even inconsistent scenarios remains real if evaluation mechanisms are not rigorous and multidimensional. From an ethical point of view, the use of artificial intelligence in a context as critical as water management raises issues of accountability, transparency and governance. It is imperative to ensure that models remain interpretable, that their recommendations are audited and that their use does not reinforce territorial inequalities by reserving access to predictive analyses to certain institutional actors at the expense of more vulnerable users. At the same time, environmental considerations must be integrated into the lifecycle of developed systems, if AI can help to optimize resources, its energy footprint especially related to the training of large models should not be neglected (P. Li et al., 2025). A systemic assessment is therefore necessary, reconciling technological innovation, scientific rigour and ecological responsibility, in order to sustainably inscribe this approach in a logic of collective interest(Marshall et al., n.d.).

Methodological Innovations and Practical Applications

The proposed hybrid architecture, combining predictive modelling and script-writing generation, opens the way to several methodological extensions and concrete applications with high impact. A first orientation lies in its integration within territorial digital twins, that is to say dynamic virtual replicas of watersheds or entire hydrological regions, allowing the simulation, real-time, the evolution of water resources and responses to various environmental or anthropogenic stresses (Pal, 2025). This linkage could transform the analytical approach into a real immersive decision-making tool, exploiting the capabilities of visualization, interactivity and prospective simulation. At the same time, incorporating the system into early warning systems is a major operational avenue, especially for arid regions prone to sudden wasupply disruptions or episodes of prolonged drought.

A significant empirical illustration is the Nile watershed, one of the most complex transboundary hydrological systems in the world. In this context, artificial intelligence approaches have been mobilized to anticipate water demand and tensions between agricultural, domestic and industrial uses. For example, (Mumbi et al., 2021) showed that the use of a feed-forward neural network, integrating variables such as rainfall, population growth, GDP, and agricultural irrigation, allows predicting water consumption in Egypt and Kenya with significant precision. This work illustrates the value of applying predictive models to real data in order to support cross-border planning, in a basin where access rivalries and climate variability exacerbate tensions.

Furthermore, the integration of generative models in this type of context could enrich these analyses by producing plausible management scenarios, particularly in the face of rainfall variability and increasing agricultural needs. The case of the Nile thus demonstrates that the combination between empirical data and advanced modeling is not limited to methodological exercises, but constitutes an operational lever for strengthening regional cooperation and climate resilience.

This orientation is already reflected in some regional initiatives. In particularly vulnerable areas such as the Sahel, where chronic droughts cause severe supply disruptions, the combined use of predictive and generative models can enhance the effectiveness of early warning systems Comparable systems have been deployed in California, where advanced drought episode modeling already supports proactive stockpile management policies (DeFlorio et al., 2024). These regional illustrations highlight the ability of the proposed framework to adapt to various hydrological contexts, while strengthening climate resilience.

The ability of the generative model to simulate such events, combined with the temporal accuracy of predictions, can significantly strengthen prevention mechanisms. On a strategic level, an area of promise is the extension of the framework to link with economic and resource allocation models in order to support long-term planning for water infrastructure, of cross-sector arbitrage or dynamic pricing. Finally, beyond the technical and analytical dimensions, this approach could become a lever to help the formulation of public policies for climate adaptation, providing decision-makers with robust stress-testing and anticipation tools, based on plausible and contextualized scenarios. The proposed framework thus provides a solid basis for the development of an intelligent and responsible decision-making ecosystem, serving the water resilience of territories(Mikołajewska et al., 2025).

Governance, Smart Cities, and Adaptive Planning

A central aspect of recent developments concerns adaptive governance and the articulation with public policies. Recent research highlights the need for integrated approaches, which are not limited to the technical performance of predictive models, but also include institutional frameworks and multi-stakeholder partnerships. These strategies are particularly visible in urban contexts, where water management is now linked to issues of digital governance and sustainability of smart cities. (*Zimmermann J, et al. (2025) | SGD*, n.d.) emphasize that the success of water resilience initiatives depends on effective coordination between public decision-makers, water managers and private actors, in order to ensure flexible and responsive planning in the face of climate shocks.

This dynamic is already visible in some pioneering urban contexts. For example, Singapore has implemented an integrated water management strategy where predictive models and artificial intelligence tools are mobilized to anticipate demand, optimize distribution, and reduce losses, in a logic of sustainable smart cities (Dai et al., 2025b). This type of experience illustrates how the articulation between digital governance and AI applied to water management can strengthen urban resilience and inspire other cities facing the same challenges.

Convergence and Strategic Perspectives

Finally, the latest scientific advances emphasize the convergence between hydrological models, generative artificial intelligence and territorial digital twins. These integrated virtual systems, now coupled with deep learning, allow simulating not only hydrological dynamics, but also socio-economic and climatic impacts in real time. They thus pave the way for prospective scenarios testing different resource allocation or crisis management strategies, in a logic of adaptive anticipation (Homaei et al., 2025). This perspective illustrates how the approach proposed in this chapter can fit into a continuum of cutting-edge research, reinforcing both the scientific relevance and operationality of water shortage management tools.

Beyond their methodological interest, these approaches are part of global issues of water management and sustainability. The multiplication of extreme drought episodes, accelerated urbanization and increasing competition between agricultural, industrial and domestic uses pose structural challenges that classical hydrological models struggle to anticipate. The integration of generative models and territorial digital twins allows exploring these tensions in a systemic way, by simulating the impact of scenarios combining climate pressures, demographic dynamics, and socio-economic arbitrations. In this perspective, the contribution of artificial intelligence does not lie solely in an improvement in predictive performance, but in its ability to articulate distinct time horizons: the short-term of early warning, the medium-term of operational adaptation, and the long-term strategic planning of water infrastructure. This articulation constitutes a direct response to the orientations of international agendas, notably the Sustainable Development Goals (SDG 6 and 13), which require both immediate resilience mechanisms and structural adaptation policies. The challenge then becomes to translate technical gains into effective institutional capacities, ensuring that the production of scenarios and the generation of recommendations are integrated into multi-governance mechanisms. levels, in order to avoid a fracture between scientific innovation and decision-making practices (Yang et al., 2024).

Political Relevance, Climate Resilience and Socio-Economic Impacts

The analysis of water shortages cannot be decoupled from understanding the climatic and socio-economic challenges that define their urgency. The increase in extreme droughts, increased tensions between agricultural, industrial and domestic uses, as well as the rapid urbanization of vulnerable regions are increasing the pressure on already fragile water systems. In this context, the integration of predictive and generative models does not only bring a technical gain: it allows exploring the social, economic, and environmental arbitrations that condition collective resilience.

Thus, the ability to generate plausible scenarios directly sheds light on the socio-economic impacts of water crises, whether they be losses in agricultural yield, industrial disruptions or tariff tensions on domestic access. These dimensions pave the way for a truly systemic approach, where the assessment of water availability is coupled with socio-economic consequences and adaptation strategies. This articulation is all the more necessary as it responds to specific political and institutional frameworks: adaptation to climate change, reduction of inequalities in access, anticipation of conflicts of use, or even infrastructure planning over time.

A particularly significant example of this articulation is found in Spain, where the introduction of progressive water pricing systems aims to reconcile economic efficiency, social equity and climate adaptation objectives. This approach highlights how predictive and generative tools can strengthen public decision-making by integrating socio-economic arbitrages into a multi-level governance framework. This type of device thus illustrates the role of artificial intelligence technologies in building water management mechanisms that are both more resilient and fairer in the face of increased water pressures (Havukainen et al., 2020).

By repositioning the methodology at the service of public decision-making, the proposed architecture makes sense beyond the strictly technical framework. It becomes a multi-level governance support instrument, facilitating the integration of forecasts into risk management systems, dynamic pricing or cross-sector coordination. This shift makes it possible to correct the over-representation of software aspects and to refocus the main contribution on use value: a scientific tool capable of supporting the formulation of climate-resilient and socially equitable policies.

CONCLUSION

This chapter presented an innovative methodological contribution by proposing a hybrid architecture combining predictive modelling and generative modelling, specifically designed to anticipate water shortages in a context of increasing climate variability and intensified anthropogenic pressures on water resources. The central issue of this research lies in the ability to combine analytical rigor and prospective flexibility, by articulating the strengths of Transformers models, widely recognized for their performance on complex time series, with those of generative models, able to simulate rare, extreme or unobserved events. This approach aims not only to refine conventional hydrological forecasts, but also to broaden the range of possible scenarios to support robust decisions in the face of uncertainty. The construction of a synthetic hydrologically coherent data set made it possible to overcome the lack of complete and open real data, while ensuring a controlled experimental environment, in which models can be systematically evaluated according to several dimensions. Precision of forecasts, diversity of generations, inter-variable coherence and impact on decision indicators. The comparative evaluation of different predictive models highlighted the superior performance of the Temporal Fusion Transformer, particularly in terms of time robustness and interpretability, through its attention mechanisms and its ability to model static and temporal covariates. On the generative side, Conditional GANs have been particularly effective in producing realistic water trajectories while integrating specific initial conditions, thus facilitating the creation of plausible water stress tests for decision makers.

This double dynamic predictive and generative is part of a systemic vision of water management, where anticipation is no longer limited to projecting trends but now includes the ability to proactively explore the space of possibilities. In this way, the approach developed in this chapter goes beyond the traditional limits of hydrological modelling by proposing a framework oriented towards resilience, adaptation and decision support under uncertainty. The linkage between the two modules, reinforced by a rigorous evaluation loop incorporating criteria of plausibility, diversity and impact, allows scientific and operational objectives to be aligned in a coherent manner. This synergy between prediction and generation is a powerful lever for the construction of adaptive water policies, planning the use of resources and the design of early warning systems in the face of imminent water crises. By also integrating advanced technical dimensions (leading edge models, cross-validation, noise injection and extreme scenarios) and thinking about future integration into decision support systems, The chapter opens up concrete prospects for deployment in various geographical contexts, particularly in arid or semi-arid regions exposed to recurrent water stress. Finally, this research makes a strong conceptual contribution by demonstrating that artificial intelligence, far from being limited to technical optimization, can play a structuring role in the transformation of water governance practices, providing powerful tools to understand, anticipate and adapt to the complex hydrological dynamics of the 21st century.

REFERENCES

Azad, A. S., Sokkalingam, R., Daud, H., Adhikary, S. K., Khurshid, H., Mazlan, S. N. A., & Rabbani, M. B. A. (2022). Water Level Prediction through Hybrid SARIMA and ANN Models Based on Time Series Analysis: Red Hills Reservoir Case Study. *Sustainability*, *14*(3), 3. Advance online publication. DOI: 10.3390/su14031843

Cao, W., Wang, D., Li, J., Zhou, H., Li, L., & Li, Y. (2018). *BRITS: Bidirectional Recurrent Imputation for Time Series* (arXiv:1805.10572). arXiv. https://doi.org// arXiv.1805.10572DOI: 10.48550

Chouhan, S., Kumari, S., Kumar, R., & Chaudhary, P. L. (2023). Climate Resilient Water Management for Sustainable Agriculture. *International Journal of Environment and Climate Change*, *13*(7), 411–426. DOI: 10.9734/ijecc/2023/v13i71894

Costa, G. E. de M. e, Menezes Filho, F. C. M. de, Canales, F. A., Fava, M. C., Brandão, A. R. A., & de Paes, R. P. (2023). Assessment of Time Series Models for Mean Discharge Modeling and Forecasting in a Sub-Basin of the Paranaíba River, Brazil. *Hydrology*, *10*(11), 11. Advance online publication. DOI: 10.3390/ hydrology10110208

Dai, Y., Huang, Z., Khan, N., & Labbo, M. S. (2025a). Smart Water Management: Governance Innovation, Technological Integration, and Policy Pathways Toward Economic and Ecological Sustainability. *Water (Basel)*, *17*(13), 1932. DOI: 10.3390/ w17131932

Dai, Y., Huang, Z., Khan, N., & Labbo, M. S. (2025b). Smart Water Management: Governance Innovation, Technological Integration, and Policy Pathways Toward Economic and Ecological Sustainability. *Water (Basel)*, *17*(13), 1932. DOI: 10.3390/ w17131932

Dankers, R., & Kundzewicz, Z. W. (2020). Grappling with uncertainties in physical climate impact projections of water resources. *Climatic Change*, *163*(3), 1379–1397. DOI: 10.1007/s10584-020-02858-4

DeFlorio, M. J., Sengupta, A., Castellano, C. M., Wang, J., Zhang, Z., Gershunov, A., Guirguis, K., Niño, R. L., Clemesha, R. E. S., Pan, M., Xiao, M., Kawzenuk, B., Gibson, P. B., Scheftic, W., Broxton, P. D., Switanek, M. B., Yuan, J., Dettinger, M. D., Hecht, C. W., … Anderson, M. L. (2024). *From California's Extreme Drought to Major Flooding: Evaluating and Synthesizing Experimental Seasonal and Subseasonal Forecasts of Landfalling Atmospheric Rivers and Extreme Precipitation during Winter 2022/23*. https://doi.org/DOI: 10.1175/BAMS-D-22-0208.1

Demiray, B. Z., Sit, M., Mermer, O., & Demir, I. (2024a). Enhancing hydrological modeling with transformers: A case study for 24-h streamflow prediction. *Water Science and Technology*, *89*(9), 2326–2341. DOI: 10.2166/wst.2024.110 PMID: 38747952

Demiray, B. Z., Sit, M., Mermer, O., & Demir, I. (2024b). Enhancing hydrological modeling with transformers: A case study for 24-h streamflow prediction. *Water Science and Technology*, *89*(9), 2326–2341. DOI: 10.2166/wst.2024.110 PMID: 38747952

Djerbouai, S. (n.d.). *Missing Precipitation Data Estimation Using Long Short-Term Memory Deep Neural Networks*. Retrieved August 17, 2025, from https://www.jeeng.net/Missing-Precipitation-Data-Estimation-Using-Long-Short-Term-Memory-Deep-Neural-Networks,147322,0,2.html

Drenkhan, F., Huggel, C., Hoyos, N., & Scott, C. A. (2023). Hydrology, water resources availability and management in the Andes under climate change and human impacts. *Journal of Hydrology. Regional Studies*, *49*, 101519. DOI: 10.1016/j.ejrh.2023.101519

Han, S., Slater, L., Wilby, R. L., & Faulkner, D. (2022). Contribution of urbanisation to non-stationary river flow in the UK. *Journal of Hydrology (Amsterdam)*, *613*, 128417. DOI: 10.1016/j.jhydrol.2022.128417

Havukainen, J., Väisänen, S., Rantala, T., Saunila, M., & Ukko, J. (2020). Environmental impacts of manure management based on life cycle assessment approach. *Journal of Cleaner Production*, *264*, 121576. DOI: 10.1016/j.jclepro.2020.121576

Homaei, M., Morales, V. G., Gutierrez, O. M., Gomez, R. M., & Caro, A. (2025). *Smart Water Security with AI and Blockchain-Enhanced Digital Twins* (arXiv:2504.20275). arXiv. https://doi.org//arXiv.2504.20275DOI: 10.48550

Huang, L., Antolini, F., Mostafavi, A., Blessing, R., Garcia, M., & Brody, S. D. (2025). *High-Resolution Flood Probability Mapping Using Generative Machine Learning with Large-Scale Synthetic Precipitation and Inundation Data* (arXiv:2409.13936). arXiv. https://doi.org//arXiv.2409.13936DOI: 10.48550

Jahangir, M. S., & Quilty, J. (2024). Generative deep learning for probabilistic streamflow forecasting: Conditional variational auto-encoder. *Journal of Hydrology (Amsterdam)*, *629*, 130498. DOI: 10.1016/j.jhydrol.2023.130498

Jose, D. M., & Dwarakish, G. S. (2020). Uncertainties in predicting impacts of climate change on hydrology in basin scale: A review. *Arabian Journal of Geosciences*, *13*(19), 1037. DOI: 10.1007/s12517-020-06071-6

Lehuedé, S. (2025). An elemental ethics for artificial intelligence: Water as resistance within AI's value chain. *AI & Society*, *40*(3), 1761–1774. DOI: 10.1007/s00146-024-01922-2

Li, J., Wu, L., Liu, J., Wang, X., & Xue, W. (2024). DFMM-Precip: Deep Fusion of Multi-Modal Data for Accurate Precipitation Forecasting. *Water (Basel)*, *16*(24), 24. Advance online publication. DOI: 10.3390/w16243702

Li, P., Yang, J., Islam, M. A., & Ren, S. (2025). *Making AI Less "Thirsty": Uncovering and Addressing the Secret Water Footprint of AI Models* (arXiv:2304.03271). arXiv. https://doi.org//arXiv.2304.03271DOI: 10.48550

Liu, B., Zhou, C., Zheng, L., Duan, H., Chen, Y., & Wang, G. (2022). Metacommunity Concepts Provide New Insights in Explaining Zooplankton Spatial Patterns within Large Floodplain Systems. *Water (Basel)*, *14*(1), 1. Advance online publication. DOI: 10.3390/w14010093 PMID: 36798655

Liu, J., Bian, Y., Lawson, K., & Shen, C. (2024). Probing the limit of hydrologic predictability with the Transformer network. *Journal of Hydrology (Amsterdam)*, *637*, 131389. DOI: 10.1016/j.jhydrol.2024.131389

Liu, Z., Ying, J., He, C., Guan, D., Pan, X., Dai, Y., Gong, B., He, K., Lv, C., Wang, X., Lin, J., Liu, Y., & Bryan, B. (2024). Scarcity and quality risks for future global urban water supply. *Landscape Ecology*, *39*. Advance online publication. DOI: 10.1007/s10980-024-01832-0

Ma, Z., Mei, G., & Xu, N. (2024). Generative deep learning for data generation in natural hazard analysis: Motivations, advances, challenges, and opportunities. *Artificial Intelligence Review*, *57*(6), 160. DOI: 10.1007/s10462-024-10764-9

Marshall, S. R. O., Tran, Thanh-Nhan-Duc, Tapas, Mahesh R., & and Nguyen, B. Q. (n.d.). Integrating artificial intelligence and machine learning in hydrological modeling for sustainable resource management. *International Journal of River Basin Management*, *0*(0), 1–17. DOI: 10.1080/15715124.2025.2478280

Maurya, V. (2024). Advancements and Challenges in Hydrological Modelling. *International Journal for Research in Applied Science and Engineering Technology*, *12*(11), 746–751. DOI: 10.22214/ijraset.2024.65161

Mekonnen, M. M., & Hoekstra, A. Y. (2016). Four billion people facing severe water scarcity. *Science Advances*, *2*(2), e1500323. DOI: 10.1126/sciadv.1500323 PMID: 26933676

Mikołajewska, E., Mikołajewski, D., Mikołajczyk, T., & Paczkowski, T. (2025). Generative AI in AI-Based Digital Twins for Fault Diagnosis for Predictive Maintenance in Industry 4.0/5.0. *Applied Sciences (Basel, Switzerland), 15*(6), 6. Advance online publication. DOI: 10.3390/app15063166

Mohammed, S., Arshad, S., Alsilibe, F., Moazzam, M. F. U., Bashir, B., Prodhan, F. A., Alsalman, A., Vad, A., Ratonyi, T., & Harsányi, E. (2024). Utilizing machine learning and CMIP6 projections for short-term agricultural drought monitoring in central Europe (1900–2100). *Journal of Hydrology (Amsterdam), 633*, 130968. DOI: 10.1016/j.jhydrol.2024.130968

Mondal, P. (2024). AI and IoT in smart water management for urban sustainability. *Uncertainty Discourse and Applications, 1*(2), 151–157. DOI: 10.48313/uda.v1i2.36

Mumbi, A. W., Li, F., Bavumiragira, J. P., & Fangninou, F. F. (2021). Forecasting water consumption on transboundary water resources for water resource management using the feed-forward neural network: A case study of the Nile River in Egypt and Kenya. *Marine and Freshwater Research, 73*(3), 292–306. DOI: 10.1071/MF21118

Myllis, G., Tsimpiris, A., & Vrana, V. (2024). Short-Term Water Demand Forecasting from Univariate Time Series of Water Reservoir Stations. *Information (Basel), 15*(10), 10. Advance online publication. DOI: 10.3390/info15100605

Ng, K. W., Huang, Y. F., Koo, C. H., Chong, K. L., El-Shafie, A., & Najah Ahmed, A. (2023). A review of hybrid deep learning applications for streamflow forecasting. *Journal of Hydrology (Amsterdam), 625*, 130141. DOI: 10.1016/j.jhydrol.2023.130141

Niu, C. (2025). *An interpretable hybrid deep learning model for flood forecasting based on Transformer and LSTM*. ResearchGate., DOI: 10.1016/j.ejrh.2024.101873

Pal, D. (2025). *(PDF) Blueprint conceptualization for a river basin's digital twin*. ResearchGate., DOI: 10.2166/nh.2025.111

Pathways for Hydrological Resilience: Strategies for Adaptation in a Changing Climate | Earth Systems and Environment. (n.d.). Retrieved May 28, 2025, from https://link.springer.com/article/10.1007/s41748-024-00567-x

Prieto-Curiel, R., & Borja-Vega, C. (2024). *The Influence of Urban Morphology on Water Scarcity* (arXiv:2402.06676). arXiv. https://doi.org//arXiv.2402.06676DOI: 10.48550

Rasiya Koya, S., & Roy, T. (2024). Temporal Fusion Transformers for streamflow Prediction: Value of combining attention with recurrence. *Journal of Hydrology (Amsterdam), 637*, 131301. DOI: 10.1016/j.jhydrol.2024.131301

Seibert, J., & Bergström, S. (2022). A retrospective on hydrological catchment modelling based on half a century with the HBV model. *Hydrology and Earth System Sciences*, *26*(5), 1371–1388. DOI: 10.5194/hess-26-1371-2022

Sela, L., Sowby, R. B., Salomons, E., & Housh, M. (2025). Making waves: The potential of generative AI in water utility operations. *Water Research*, *272*, 122935. DOI: 10.1016/j.watres.2024.122935 PMID: 39674137

Sun, A. Y., Jiang, P., Yang, Z.-L., Xie, Y., & Chen, X. (2022). A graph neural network (GNN) approach to basin-scale river network learning: The role of physics-based connectivity and data fusion. *Hydrology and Earth System Sciences*, *26*(19), 5163–5184. DOI: 10.5194/hess-26-5163-2022

Wang, Q., & Zhu, S. (2025). Multi-station Joint Runoff Forecasting Using Graph Neural Network Coupled with Spatial Connectivity of Hydrological Stations. *Mathematical Geosciences*, *57*(4), 759–797. DOI: 10.1007/s11004-024-10167-0

Wei, X., Wang, G., Schmalz, B., Hagan, D. F. T., & Duan, Z. (2023). Evaluation of Transformer model and Self-Attention mechanism in the Yangtze River basin runoff prediction. *Journal of Hydrology. Regional Studies*, *47*, 101438.

Weng, P., Tian, Y., Liu, Y., & Zheng, Y. (2023). Time-series generative adversarial networks for flood forecasting. *Journal of Hydrology (Amsterdam)*, *622*, 129702. DOI: 10.1016/j.jhydrol.2023.129702

Wi, S., & Steinschneider, S. (2022). Assessing the Physical Realism of Deep Learning Hydrologic Model Projections Under Climate Change. *Water Resources Research*, *58*(9), e2022WR032123. https://doi.org/DOI: 10.1029/2022WR032123

Xu, Y., Lin, K., Hu, C., Wang, S., Wu, Q., Zhang, L., & Ran, G. (2023). Deep transfer learning based on transformer for flood forecasting in data-sparse basins. *Journal of Hydrology (Amsterdam)*, *625*, 129956. DOI: 10.1016/j.jhydrol.2023.129956

Yadav, R., Nascetti, A., Azizpour, H., & Ban, Y. (2024). Unsupervised flood detection on SAR time series using variational autoencoder. *International Journal of Applied Earth Observation and Geoinformation*, *126*, 103635. DOI: 10.1016/j.jag.2023.103635

Yang, Y., Xie, C., Fan, Z., Xu, Z., Melville, B. W., Liu, G., & Hong, L. (2024). Digital twinning of river basins towards full-scale, sustainable and equitable water management and disaster mitigation. *Npj Natural Hazards*, *1*(1), 43. DOI: 10.1038/s44304-024-00047-2

Yu, J., Li, Y., Huang, X., & Ye, X. (2025). Data quality and uncertainty issues in flood prediction: A systematic review. *International Journal of Digital Earth*, *18*(1), 2495738. DOI: 10.1080/17538947.2025.2495738

Zhao, X., Wang, H., Bai, M., Xu, Y., Dong, S., Rao, H., & Ming, W. (2024). A Comprehensive Review of Methods for Hydrological Forecasting Based on Deep Learning. *Water (Basel)*, *16*(10), 10. Advance online publication. DOI: 10.3390/w16101407

Zimmermann J, et al. (2025) | SGD. (n.d.). Retrieved August 17, 2025, from https://www.yeastgenome.org/reference/S000378140

KEY TERMS AND DEFINITIONS

Digital Twins: Virtual replicas of real-world systems or environments (e.g., watersheds) that enable real-time simulation, monitoring, and prediction for enhanced decision-making.

Generative Models: Deep learning architectures capable of creating new data instances that resemble a given data distribution, used here to simulate plausible extreme hydrological scenarios.

Predictive Modelling: A set of statistical or machine learning techniques used to forecast the future behavior of variables based on historical data, especially in the context of hydrological and climatic dynamics.

Synthetic Data: Artificially generated datasets that mimic the statistical properties of real-world data, employed in this study to compensate for the lack of comprehensive hydrological datasets.

Chapter 7

Smart Stormwater Systems:

AI-Driven Forecasting, Optimization, and Real-Time Control

Yassine Ezaier
University Hassan II, Casablanca, Morocco

Md. Asadullahil Galib Fardin
Rajshahi University of Engineering and Technology, Bangladesh

Likhon Chandra Roy
https://orcid.org/0009-0003-6943-6777
Rajshahi University of Engineering and Technology, Bangladesh

Mehedi Hashan Riad
https://orcid.org/0009-0003-0201-2099
Bangladesh University of Engineering and Technology, Bangladesh

Md. Rafiul Islam
Rajshahi University of Engineering and Technology, Bangladesh

Samanta Alam
Chicago State University, USA

Ahmed Hader
University Hassan II, Casablanca, Morocco

ABSTRACT

This chapter emphasizes the potential of integrating green infrastructure (GI) with computational techniques to revolutionize urban stormwater management. Conventional gray infrastructure is inadequate in mitigating urban flooding and pollution, whereas green infrastructure such as rain gardens and bioswales provides nature-based remedies. Nonetheless, their efficacy is contingent upon local context, rendering models such as SWMM and GIS indispensable for planning. Machine

DOI: 10.4018/979-8-3373-2700-6.ch007

learning, IoT sensors, and digital twins significantly improve forecasting, control, and optimization. Despite ongoing issues related to data quality and fairness, the amalgamation of intelligent technology with ecological design offers a robust and flexible strategy for constructing sustainable, flood-resilient, and inclusive urban water systems in a dynamic environment.

INTRODUCTION

As cities struggle with increased rainfall, deteriorating infrastructure, and rapidly changing land cover, urban stormwater is becoming an increasingly significant issue. The dynamic hydrological pressures of a 21st-century city cannot be handled by conventional gray infrastructure, which is frequently overworked and stiff. Designing sustainable, adaptive stormwater systems instead requires a hybrid model that combines computational innovation with nature-based solutions.

By encouraging infiltration, evapotranspiration, and biogeochemical treatment, green infrastructure (GI) is designed to replicate natural hydrological cycles. Core GI technologies, such as rain gardens, bioswales, green roofs, and permeable pavements, are designed to capture, delay, and clean urban runoff. Through sedimentation, filtration, and biological uptake, these systems lessen peak flow, postpone the time of concentration, and make it easier to remove pollutants.

For instance, in field settings, bioretention systems have shown significant efficiency in lowering heavy metals, total nitrogen (TN), and total suspended solids (TSS)(Fang et al., 2021) .They frequently have underdrains and several filtering layers in their design, which work together to control water quality and amount. Similarly, by capturing precipitation and temporarily storing it in substrate media, green roofs help with biodiversity and thermal insulation while lowering rooftop runoff (Sharma & Malaviya, 2021).

GI performance is quite context-sensitive, though. The effectiveness of treatment is significantly impacted by variables like land use, vegetation cover, soil permeability, and precipitation intensity. The necessity for computer models that simulate, optimize, and track large-scale green actions is highlighted by this heterogeneity.

For a long time, computational models like HYDRUS-1D and SWMM (Storm Water Management Model) have been essential resources for simulating hydrology and pollutant transport in urban watersheds. These models may mimic long-term stormwater behavior under different rainfall, land use, and GI deployment scenarios with plug-ins and scripting capabilities

Newer systems use Geographic Information Systems (GIS) for spatial data integration and visualization in order to get over limits in scalability and real-time analysis. Runoff hotspots, terrain restrictions, and appropriate GI placement based on impervious surface patterns and flow pathways are all made possible by GIS overlays.

Additionally, optimization methods have been used for economical GI planning. Hydraulic efficiency has been balanced with land use, cost, and ecological performance using multi-objective algorithms including NSGA-II, Trust Region Bayesian Optimization (TuRBO), and Particle Swarm Optimization (PSO).

Our ability to forecast runoff volumes, pollutant concentrations, and GI system behavior is changing as a result of machine learning (ML) and artificial intelligence (AI) models. In stormwater quality and quantity modeling, Random Forest (RF), Support Vector Machines (SVM), Artificial Neural Networks (ANN), and Long Short-Term Memory (LSTM) networks have all demonstrated promise.

In contrast to conventional regression models, RF and SVM have been employed to predict Event Mean Concentrations (EMC) of pollutants such as nitrates and TSS with a lower mean absolute error. Outperforming traditional time-series techniques, deep LSTM models trained on rainfall and inflow datasets have predicted stormwater dynamics with sub-hour precision.

Moreover, Reinforcement Learning (RL) and hybrid deep learning methods are now used to optimize SWMM parameters, making them suitable for real-time operations where input uncertainty and system dynamics change rapidly

Extreme storm scenarios frequently exceed GI's capability, even though it performs best during small to medium-sized rainfall events. As a result, hybrid techniques are becoming more popular, combining and coordinating grey and green infrastructures computationally.

Green roofs, vegetated strips, and wet ponds are paired with smart gates and detention tanks in China's Sponge City projects. A deeper understanding of flood propagation and mitigation is provided by the use of hydrodynamic simulation models, such as SWMM–LISFLOOD-FP hybrids, which combine 1D sewer flow with 2D surface inundation (Chen et al., 2021).

These kinds of connections are particularly helpful for assessing co-benefits alongside hydrological measures, such as biodiversity, heat reduction, and public amenity. Watershed-scale optimization techniques may now balance up to five conflicting goals, including cost, removal of pollutants, storage capacity, urban heat, and ecological resilience

GI and hybrid infrastructure are increasingly being combined with real-time sensing and decision support systems. IoT devices, including turbidity meters, depth sensors, and rain gauges, now provide real-time data to cloud dashboards for control, forecasting, and monitoring.

The RainGrid Smart Water Grid, for instance, prevents overflow and permits water reuse by using automated cisterns and predictive rainfall analytics. Low-power

LoRa networks allow sensors installed in tanks, gardens, and rooftops to interact with one another and be seen on citizen dashboards.

Weirs, pumps, and valves can all be dynamically activated by these real-time control (RTC) platforms. These systems provide anticipatory drainage, which releases storage ahead of storms or redirects flows to underutilized basins when combined with machine learning forecasts (Valencia et al., 2021).

Even with great advancements, difficulties still exist. First, high-quality, high-frequency data, which is frequently hard to acquire by in low-income areas, is necessary for machine learning models. Second, generalizability is still challenging; models developed using data from one city sometimes perform poorly when moved to another because of variations in infrastructure, vegetation, or soil.

Additionally, although there are tools for modeling GI, few provide comprehensive platforms that integrate socioeconomic consequences, pollution removal, and hydraulic modeling into a single framework. The majority of computational tools available now only partially incorporate social equality, governance, and maintenance feedback.

A strong route to resilient urban water systems is provided by the combination of green infrastructure, stormwater treatment technology, and sophisticated computational techniques. When intelligent models and real-time monitoring are used in the design of nature-based systems, they not only reduce pollution and floods but also enhance biodiversity and livability. Combining machine learning, GIS analysis, hydrological modeling, and optimization can help cities create stormwater management systems that are more intelligent, flexible, and egalitarian.

HISTORICAL BACKGROUND & PARADIGM SHIFTS

Four major paradigms have emerged in stormwater management, each brought about by unique socio-ecological stresses and technology advancements. It is important to track these changes because modern computational methods (such as AI-enabled control) build upon and frequently hybridize with previous paradigms rather than replacing them.

Hydraulic Civilisations

Karl Wittfogel created the phrase "hydraulic civilization." (Bano, 2017) to describe early states that had conquered groundwater, rivers, and runoff using gravity-driven works rather than pumps. Whether in Qin-era China, monsoon India, imperial Rome, or arid Persia, engineers came to four conclusions that continue to guide the design of modern storm sewers: provide a steady fall of water, line channels so that the water scourges itself clean, punch access shafts for desilting, and leave overflow routes for the occasional deluge.

Classic schemes prove the point. Qanats (*The Persian Qanat - UNESCO World Heritage Centre*, n.d.) used sub-1% slopes and ventilation shafts to draw mountain aquifers miles under the desert, geometry echoed today in perforated under-drains.

Evidence that oversizing for worst-case storms pays off for millennia may be found in Rome's vaulted Cloaca Maxima (*7. Aquatic Pasts & the Watery Present: Water and Memory in the Fora of Rome*, n.d.), which paired walk-in maintenance height with a semi-elliptical invert that still drains to the Tiber. Meanwhile, China's Dujiangyan Science and civilisation in China combined storage and transportation with its covered brick drains and tiered step-wells (*The Sewers of Mohenjo-Daro on JSTOR*, n.d.). Using self-regulating side weirs, which are currently used in green roof overflow boxes, they divided the Min River into irrigation and flood bypass routes. Additionally, Nile-basin levees with wooden gates invented scheduled releases, which prefigured today's computational pond controls.

These examples are united by a design philosophy that transforms basic hydraulics into long-lasting public safety: pure gravity, locally sourced materials, and community maintenance. These logics are still used in modern stormwater practices, which only replace brick with reinforced concrete and canal overseers with AI powered by sensors. In this way, the age-old directive to "make water obedient" endures each time a bypass weir is designed for the 100-year cloudburst, a manhole is spaced for desilting, or a culvert grade is examined.

Industrial-Era Grey Infrastructure

Both the Metropolis Sewerage Act of 1858 and the British Public Health Act of 1848 (*The 1848 Public Health Act and Its Relevance to Improving Public Health in England Now | The BMJ*, n.d.) marked the first significant transition from antiquated gravity works to contemporary "big-pipe" drainage. Engineers like Joseph Bazalgette were persuaded by the cholera outbreaks in London that the solution was to install big, brick-lined interceptors flushed by tidal gates at steep gradients. This early legislation standardized design standards that are still recognizable today, including outfalls placed downstream of city intakes, manholes spaced 90 meters apart, and minimum self-cleaning velocities of 0.9 m s^{-1} (Bureau of Indian Standards, 1986) . The blueprint quickly spread to Paris, Hamburg, and New York, enshrining in local rules the notion that public health and business would thrive if pipes were sized to withstand the worst storm ever recorded.

The same convey-and-dump reasoning was legislated in North America following World War II, but a legal water-quality trigger was included. The United States Clean Water Act 1972 (Copeland, C. (2001). Clean water act. Huntington,

NY: Nova Science Press.) and its powerful successor, the Federal Water Pollution Control Act 1948 (Dworsky, 1967) established the NPDES permit system, requiring cities to account for stormwater discharges in addition to sanitary sewage. Even medium-sized communities had to inventory outfalls, construct detention tunnels, and modify swirl concentrators to skim floatables in order to comply with Phase I (1990) and Phase II (1999) MS4 regulations.

Indian Water (Prevention & Control of Pollution) Act 1974 (Jain, S. N. (1984). Water (Prevention and Control of Pollution) Act 1974: Basic Legal Issues), Japan's Sewerage Act 1958 (*The Japanese Sewerage Industry: Institutional Aspects and the Governance Systems | SpringerLink*, n.d.)and Canada's Fisheries Act (1977 revisions) are examples of parallel statutes. spread the idea of "treating it at the end of the pipe" throughout the industrializing world.

By the 1990s, it was clear that pure grey solutions had limits in terms of cost and environmental impact. In response, the European Commission (1991) issued the Urban Waste Water Treatment Directive 91/271/EEC. Directive 91/271/EEC of May 21, 1991, pertaining to the treatment of urban wastewater (Off J Eur Commun., 50, 1-13), which required storage volumes to reduce combined-sewer overflows by 80%. The Water Act 2007 (*Legislating for Sustainable Basin Management: The Story of Australia's Water Act (2007) | Water Policy | IWA Publishing*, n.d.), which combined stormwater harvesting with minimal environmental flow discharges, was prompted by Australia's millennium drought experience, while China's 12th Five-Year Plan provided funding for a series of mega-tunnels, including Beijing's 33 km Liangshui interceptor. Massive culverts, detention mega basins, and high-rate clarifiers reinforced the grey paradigm globally, but their rising O&M costs and the continued occurrence of flash flood damage paved the way for the next shift in stormwater thinking: the digital "Sense-Predict-Act" era, where sensors and real-time control seek to increase the resilience of the legacy concrete backbone.

Global Overview of Stormwater Management

Over the past few decades, urban stormwater management has changed dramatically, moving from conventional gray infrastructure to integrated systems that incorporate low-impact development (LID), green infrastructure, and sophisticated computational tools. With an emphasis on regional practices and the integration of computational innovations, this section offers a worldwide perspective on these changes.

Traditional Approaches and Their Limitations

Gray infrastructure, engineered systems intended to swiftly transport stormwater away from urban areas, has historically played a significant role in urban stormwater management. Although these techniques were successful in preventing flooding right away, they frequently disregarded ecological and water quality issues. Additionally, their inability to adjust to shifting weather patterns and urbanization over time made them more vulnerable.

Emergence of Green Infrastructure and LID

Recognizing the limitations of traditional systems, many regions have adopted green infrastructure and LID practices. These approaches aim to manage stormwater at its source, enhancing infiltration, evapotranspiration, and reuse. Common practices include:

- **Bioretention Systems**: Engineered landscapes that filter stormwater through vegetation and soil layers, removing pollutants and reducing runoff volumes.
- **Permeable Pavements**: Surfaces that allow water to infiltrate, reducing surface runoff and promoting groundwater recharge.
- **Green Roofs**: Vegetated roof systems that absorb rainfall, reduce runoff, and provide insulation benefits.
- **Rain Gardens and Bioswales**: Landscaped areas designed to capture and treat stormwater, enhancing aesthetics and biodiversity. These practices not only mitigate flooding but also improve water quality and urban aesthetics.

REGIONAL IMPLEMENTATIONS AND INNOVATIONS

North America

LID techniques have been extensively used in the US and Canada to solve urban runoff issues. Green infrastructure has been adopted by cities to enhance water quality and meet legal criteria. These systems are often designed and assessed using computational models, such as the EPA's Storm Water Management Model (SWMM).

Europe

Sustainable Drainage Systems (SuDS) have been used by European nations to improve water quality and reduce urban flooding. Permeable surfaces, retention basins, and green areas are all incorporated into these systems. SuDS planning and optimization are supported by sophisticated modeling techniques, which guarantee their efficacy under a range of climatic circumstances.

Asia

China's "Sponge City" initiative exemplifies the large-scale implementation of green infrastructure. The program focuses on enhancing urban resilience through permeable surfaces, green roofs, and constructed wetlands. Computational tools, including SWMM and the Analytical Hierarchy Process (AHP), have been employed to evaluate and optimize these interventions.

Australia

Stormwater management and urban planning are combined in Australia's Water-Sensitive Urban Design (WSUD). WSUD promotes sustainable water management through the use of natural systems. In order to ensure that WSUD plans are in line with environmental and urban development objectives, computational modeling is essential to their design and evaluation.

INTEGRATION OF COMPUTATIONAL TOOLS

Stormwater management has been transformed by the development of computational tools:

- **Hydrologic and Hydraulic Modeling:** SWMM and other tools model stormwater flows to help with infrastructure design and evaluation.
- **Geographic Information Systems (GIS): Stormwater** systems can be spatially analyzed with the use of Geographic Information Systems (GIS), which improves planning and decision-making.
- **Optimization Algorithms:** Multi-objective optimization methods help balance a number of objectives, including performance, cost, and environmental impact.
- **Real-Time Monitoring and Control**: By enabling dynamic management of stormwater infrastructure, sensor networks, and control systems enhance storm event reactivity.

CHALLENGES AND OPPORTUNITIES

Despite advancements, difficulties still exist. High-quality data, which may be scarce in some areas, is necessary for effective modeling. For comprehensive stormwater management, engineering, environmental science, and urban planning must be integrated. Stormwater initiatives are more successful when local communities are involved in their planning and upkeep.

In order to create resilient and sustainable stormwater systems, opportunities exist in developing computational tools, improving data collection techniques, and encouraging teamwork.

Major Challenges in Computational Stormwater Management

Rapid urbanization, climate change, aged infrastructure, new contaminants, and socio-environmental injustices all contribute to the complexity of urban stormwater systems. Computational technologies present novel issues as well as creative answers. These difficulties are examined in this part, along with how advances in computing can help.

URBANIZATION AND IMPERVIOUS SURFACES

Impact on Hydrology

The natural water cycle is upset when urbanization replaces natural landscapes with impermeable surfaces like buildings and highways. This change raises the risk of flooding and degrades the quality of the water by increasing surface runoff and decreasing infiltration. According to the U.S. Environmental Protection Agency (EPA), urbanization-related impermeable surfaces change the routes by which water and related pollutants enter urban streams by decreasing infiltration and increasing surface runoff.

Computational Modelling Challenges

It takes advanced methods and high-resolution data to model the hydrological effects of urbanization. The intricacy and unpredictability of urban areas are frequently difficult for traditional models to handle. Although spatial analysis has been enhanced by the integration of Geographic Information Systems (GIS) with hydrological models, issues with data availability and model calibration still exist. A useful technique for estimating stormwater flow from infrastructure projects is continuous simulation modeling, which enables more precise forecasts and improved stormwater management system design.

CLIMATE CHANGE AND RAINFALL EXTREMES

Intensification of Storm Events

Rainfall events become increasingly frequent and powerful due to climate change, overloading the stormwater infrastructure that is currently in place. According to the Clausius-Clapeyron relation, the atmosphere can contain around 7% more water vapor for every degree Celsius that the temperature rises, which leads to heavier precipitation. This escalation calls for a reassessment of modeling methodologies and design standards.

Adaptive Modelling Approaches

Models need to use predictive analytics and real-time data to handle non-stationary climatic patterns. In order to better replicate intricate hydrological conditions, the EPA's Storm Water Management Model (SWMM) is being developed to incorporate elements such as coupled 1D hydraulic routing and overland flow modeling. New developments in SWMM are intended to increase computational fidelity and efficiency, facilitating better stormwater management decision-making. (Publications of the EPA)

AGING INFRASTRUCTURE AND DIGITAL INTEGRATION

Infrastructure Deterioration

The aging and degradation of many urban drainage systems result in lower performance and higher maintenance expenses. Conventional inspection techniques can be time-consuming and may not identify problems right away. Predictive maintenance and monitoring can be improved by integrating computational techniques.

Digital Twins and Predictive Maintenance

Stormwater infrastructure can be monitored and simulated in real time thanks to digital twins, which are virtual copies of physical systems. Digital twins can optimize maintenance schedules and forecast system breakdowns by combining sensor data and computational models. An important development in proactive infrastructure management is the creation of digital twins with SWMM and other tools.

EMERGING POLLUTANTS AND WATER QUALITY MODELING

Detection and Modelling Challenges

For stormwater management, emerging contaminants such as tire wear particles, PFAS, and microplastics provide additional difficulties. Because of their intricate connections and activities, these pollutants are challenging to identify and model. It's possible that traditional models won't adequately forecast how these contaminants will travel and end up.

Advanced Modelling Techniques

The behavior of new contaminants is being modeled using sophisticated computational techniques, including as computational fluid dynamics (CFD) and machine learning. These methods provide more precise forecasts and guide the creation of efficient treatment plans. To better understand pollutant dynamics, CFD has been used, for example, to simulate constituent movement and fate within stormwater unit operations.

ENVIRONMENTAL JUSTICE AND DATA EQUITY

Disparities in Infrastructure and Data

Inadequate infrastructure and a lack of data frequently put marginalized groups at disproportionate risk for flooding. These differences worsen environmental inequities and obstruct efficient stormwater management.

Community-Based Monitoring and Modelling

Data gaps can be filled, and locals can be empowered by including communities in data collection and monitoring. Localized data gathering is made possible by inexpensive sensors and citizen science projects, which enhance model accuracy and promote inclusive decision-making. For instance, the creation of inexpensive electrical conductivity and water depth sensors has made it easier for the public to participate in stormwater monitoring.

INTEGRATION OF ARTIFICIAL INTELLIGENCE AND MACHINE LEARNING

Enhancing Predictive Capabilities

By facilitating real-time forecasts and adaptive management, artificial intelligence (AI) and machine learning (ML) are revolutionizing stormwater modeling. Large datasets can be processed by these technologies to find trends and predict system actions. For example, control systems that reduce flooding and enhance water quality have been developed using deep reinforcement learning.

Surrogate Modelling and Real-Time Control

Graph neural network (GNN)-based surrogate models provide computationally effective substitutes for conventional simulations. Stormwater systems can be dynamically controlled thanks to these models' ability to predict hydraulic conditions in real time. Urban drainage network hydraulics may now be predicted with greater efficiency and accuracy thanks to a GNN-based surrogate model

CLOUD COMPUTING AND DATA MANAGEMENT

Advancements in Cloud-Based Modelling

Stormwater modeling is made accessible and scalable by cloud computing, which also facilitates real-time data integration and cooperative decision-making. Large datasets can be processed via cloud-based tools, which can offer interactive modeling environments. For instance, real-time data analysis and improved decision-making have been made possible by cloud-based stormwater modeling, which has completely changed urban water management.

Challenges in Data Integration and Security

Although cloud computing has many advantages, there are drawbacks as well, including issues with security, standards, and data integration. When implementing cloud-based systems, securing sensitive data and guaranteeing data quality are essential factors. ScienceDirect

There are benefits and drawbacks to incorporating computational innovations into stormwater management. Cloud computing, digital twins, and artificial intelligence (AI) all improve predictive power and system efficiency, but they also raise issues of social justice, infrastructure compatibility, and data quality. In order to create resilient and inclusive stormwater management plans, a multidisciplinary approach combining engineering, data science, and community engagement is required.

COMPUTATION & DIGITAL TRANSFORMATION

Real-Time Telemetry Networks

Urban drainage's "nervous system" has matured in four telemetry waves; each widened the data funnel that today's AI controllers rely on:

Table 1. Evolution of SWM technology

Wave	Typical technology	What it enabled	Limits exposed
Manual (pre-1950s)	Staff gauges, float charts	First century-scale rainfall & stage records	Daily readings, no flood alert
SCADA era (1960s-2000)	4–20 mA-wired level/flow sensors polled by **SCADA** consoles	Minute-scale data and remote pump starts	Copper cabling, vendor lock-in, coarse spatial density
Wireless-sensor-network era (≈ 2005-2015)	GSM, ZigBee, LoRa loggers-cloud **MQTT** broker	Hour-to-minute data without trenching; browser dashboards	High power draw; rising data fees
Edge-AI & event era (2016 onwards	NB-IoT / LoRaWAN edge motes with on-board LSTM filters	30 s streams, on-site anomaly flags, federated learning	Cybersecurity & governance are still evolving

The first SCADA (Hunzinger, 2017) racks were revolutionary because they allowed operators to monitor tank levels and open pumps from a control room using computerized panels, which reduced O&M and integrated business systems without requiring them to leave the facility. However, networks remained sparse because each additional gully sensor required copper back-haul. By the middle of the 2000s, cloud brokers and battery loggers reduced the

The first SCADA racks were revolutionary because they allowed operators to monitor tank levels and open pumps from a control room using computerized panels, which reduced O&M and integrated business systems without requiring them to leave the facility. However, networks remained sparse because each additional gully sensor required copper back-haul. FIWARE (*A Standard-Based Open Source IoT Platform: FIWARE | IEEE Journals & Magazine | IEEE Xplore*, n.d.) based pilots streamed 1-minute data citywide (§ 3.2, p. 10) and demonstrated how Seoul's 2,600-node grate network reduced obstruction call-outs by 40%. By the mid-2000s, battery loggers and cloud brokers severed the chord. The most recent wave reduces cellular bills by over 70% in the OpenStorm watershed by using edge motes to de-duplicate packets, execute lightweight CNN/LSTM filters, and post-process just events (*A Standard-Based Open Source IoT Platform: FIWARE | IEEE Journals & Magazine | IEEE Xplore*, n.d.).

FROM RAW TELEMETRY TO TRUST-GRADED STREAMS

High-rate IoT gauges record all rainfall and stage variations, but they also send out duplicate packets, radio drop-outs, and bio-fouled drift. When submerged pressure probes slimed over, a 3000-node Chinese sewage study found 0.4 m false surges and an 18% packet-loss rate during typhoons. The Huron River OpenStorm archive has similar artefacts, where optical turbidity peaks were identified as fallen leaves instead of sediment pulses

Edge Guards

Filtering on-node must only cost a few milliwatts. The Hampel filter, a sliding check that flags values that deviate by more than a few median-absolute-deviations, is used by the majority of networks. It compares each new reading with the median of its neighbors. In real life, the algorithm maintains authentic flood steps while rejecting single-point spikes in less than a millisecond on a 48 MHz microcontroller. According to Seoul testing, timestamps that are synchronized to a GNSS pulse and maintain clock drift within ±50 ms are sufficient to combine one-second stage feeds with five-second radar rain (Koo 2020).

Cloud Repair

Filtered packets split after passing via an MQTT broker. One copy is sent into a Kalman smoother, a forward-backward estimator that, according to filled in Amazon flood-plain gaps 35% more correctly than cubic splines. In Great Lakes catchments, depth curves are reconstructed with less than 5% distortion using a self-calibrating expectation–maximization method that is directed by hydraulic mass balance when outages take minutes instead of seconds (Bartos 2017). Both techniques are implemented quickly enough by Spark and Flink to reconstruct millions of data in real time.

Event Stream

Clean observations feed a Flink rule engine using NGSI-LD labels. 23 flooding calls during the 2023 monsoon were avoided in Seoul thanks to a law that kicks in when depth rises more quickly than 15 mm min^{-1} for 90 seconds (Kim 2023). Retraining Kansas City LSTM surrogates on this arrangement reduced random-access lag by 70%. Long-term stores use hourly Parquet shards keyed by sensor ID.

Security and Provenance

TLS 1.3 is used for all hops, and certificates are renewed every three months. A hash-chained provenance record is appended to each Hampel reject, Kalman fill, and event alert; the EU FIWARE4Water pilots found that this replay log was crucial for permit audits (FIWARE4Water 2021). Device attestation and verified firmware are advised by fog-computing reviews to prevent spoof motes from getting to the broker (Aazam 2020).

Noisy IoT chatter is transformed into a loss-checked, time-aligned feed by combining edge Hampel screens with cloud Kalman reconstruction, NGSI-LD event flows, and cryptographic provenance. The "SENSE-PLUS" layer provides dependable fuel for the Predict and Act phases, satisfying the requirements of regulators and engineers for data that can be defended.

DATA-DRIVEN FORECASTING

Forecasting provides the lead time that makes every other intelligent action possible. In urban drainage, the operational sweet spot is typically a five-to-thirty-minute prediction horizon, long enough to pre-release storage or throttle gates without drifting too far from the storm's actual evolution. Across the last five years, deep learning has clearly outperformed traditional time-series baselines for short-horizon sewer and runoff prediction. LSTM architectures and their gated cousins (GRU) consistently deliver more accurate water-level and inflow estimates, including under data gaps and missing antecedent states, and can be trained to act as "soft sensors" where instruments are unreliable or offline.

Figure 1. LST architecture (Zhu et al., 2023)

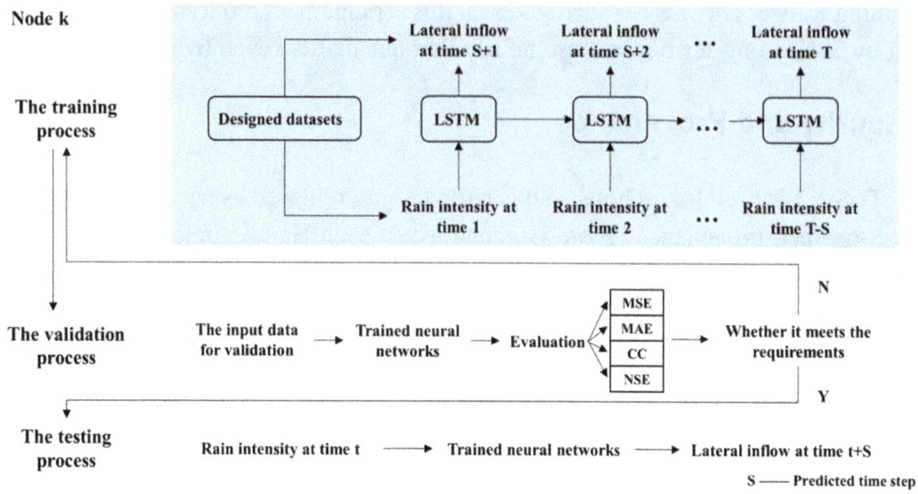

Case studies include LSTM soft-sensing of CSO chamber depths in Copenhagen and early-warning ponding forecasts in dense urban catchments, while CNN–LSTM and transformer hybrids further improve performance by bringing in spatial rainfall structure from weather radar. At the watershed scale, deep learning has also shown strong skill for rainfall–runoff—including extreme events—when architectures are physically constrained (e.g., mass-conserving LSTM variants). Together, these threads show that modern networks can now forecast both hydraulic states (levels, flows) and event risks (overflow probability) with the low latency needed for operations.

Figure 2. GNN surrogate model

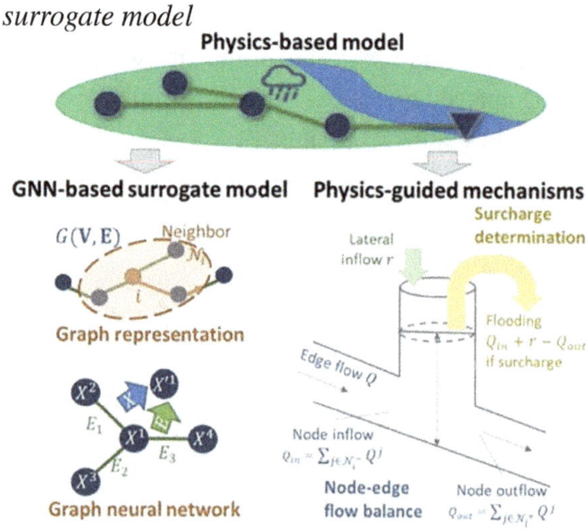

Two recent lines of work expand forecasting's practical reach. First, utilities increasingly replace slow physics runs with surrogate models, learned emulators that approximate SWMM or other simulators in milliseconds. Notably, graph neural network (GNN) metamodels exploit pipe-node topology to predict network hydraulics more faithfully than fully connected nets, enabling real-time what-ifs during storms and data-efficient transfer across catchments. Second, utility studies and reviews focused specifically on overflow prediction now document consistent gains from ML when radar is fused with in-system sensors, turning dozens of raw feeds into a probabilistic CSO outlook used by operators. These advances are making short-term forecasting not just accurate, but operationally convenient.

OPTIMIZATION

Optimization determines where and how to intervene, at planning time (siting/sizing) and in operations (routing/storage allocation). For design, multi-objective formulations have become standard, balancing runoff/peak attenuation, pollutant load reduction, cost, and sometimes co-benefits (heat, amenity). Reviews and exemplars show why evolutionary multi-objective methods (e.g., NSGA-II/III, PSO) dominate: they recover well-spread Pareto fronts while tolerating non-convex, discrete decision spaces typical of LID/GI portfolios. In practice, these optimizers couple tightly to SWMM (or SUSTAIN) and to GIS layers that flag hotspots by land use, slope, soils, and imperviousness; several recent frameworks embed climate-robustness testing so candidate designs remain effective across plausible futures. Classic syntheses (Eckart et al.) and newer studies (Piscopo et al.; Gao et al.; Leng et al.) document both the methods and the measurable gains when optimization guides LID layout rather than relying on heuristics.

Computational cost is the usual bottleneck, and this is where surrogates again matter. Emerging work shows GNN-based emulators of SWMM can cut runtimes by orders of magnitude while preserving node-edge physics, making it feasible to evaluate thousands of LID configurations or operational policies within a single planning run. The latest Water Research results and follow-ons demonstrate transferable, data-efficient metamodels, including extensions that predict both depths and flows—a critical edge variable for water-quality and flooding analyses. With these tools, optimization can be both comprehensive and fast enough to iterate with stakeholders.

REMOTE AND AUTONOMOUS CONTROL

Control is where predictions and plans translate into real-time actions at pumps, gates, and basins. The current state of the art is Model Predictive Control (MPC): every few minutes, an MPC scheduler rolls forward a fast model ("digital twin"), tests candidate moves and enacts the policy that best meets constraints (levels below overflow, flows below pipe capacity, minimized spills). Recent implementations couple MPC to live rainfall/runoff forecasts and report sizable performance gains; for example, predictive MPC frameworks using ANN-based rainfall forecasts feeding SWMM-embedded twins show effective CSO mitigation and storage utilization improvements in real events. Earlier watershed-scale control work demonstrated how much throttling a small number of valves can achieve, and in practice, these ideas underpin large operational programs.

Figure 3. A Urban drainage system deep reinforcement learning control (Tian et al., 2022)

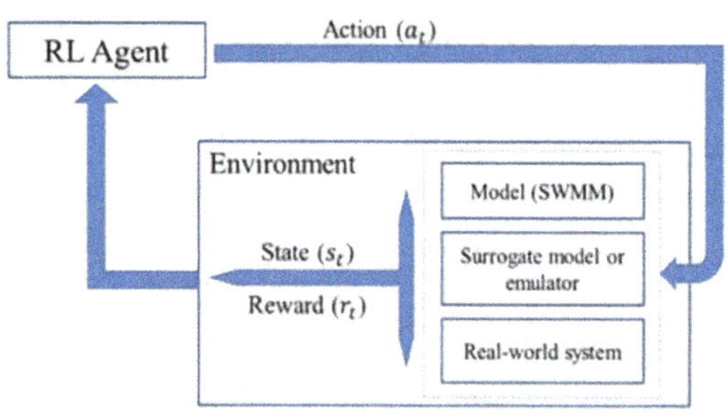

A second fast-moving thread is deep reinforcement learning (DRL). DRL learns valve and pump policies directly from simulated storms and then adapts online, often beating rule-based RTC on multi-objective goals (flooding, overflows, downstream capacity, reuse). Recent peer-reviewed studies report average flood/overflow reductions on the order of 15–40% compared to conventional RTC, along with successful demonstrations for multi-basin coordination and changing storm regimes; comprehensive reviews from 2025 summarize progress and limitations (safety constraints, sample efficiency, interpretability). Taken together with real-world "smart stormwater" programs that network sensors and actuators across a city, the literature trend is clear: software-defined storage and routing can meaningfully defer concrete while improving water-quality outcomes, provided controllers are paired with robust sensing, fail-safe PLC layers, and auditable decision logs

MLOPS, CYBER-SECURITY AND GOVERNANCE

The digital apparatus outlined in the preceding sections may malfunction in subtler manners than a clogged pipe: a predicting network gradually diminishes in precision, a sensor's firmware gets supplanted by nefarious code, or a court requests the specific bytes that initiated a gate movement six months prior. Modern drainage agencies thus incorporate three interconnected safeguards: MLOps for model maintenance, zero-trust security to prevent unauthorized access, and data governance ledgers to ensure all decisions are auditable.

MLOps and Drift Detection

Machine-learning surrogates do not exhibit longevity. When land use alterations occur or new sensors are included, the statistical correlation between precipitation and flow diverges—a process hydrologists refer to as model drift, which denotes the progressive discrepancy between real-time data and the distribution encountered during model training. Operations teams implement a concept-drift monitor, a compact statistical test embedded in the data stream that compares current mistakes with historical ones, to trigger an alert when performance deteriorates. These monitors indicated a twelve-percent decline in performance when Copenhagen resurfaced 18 km of streets, hence reducing the basin's response time

Upon confirmation of drift, the code progresses via a CI/CD pipeline, which stands for continuous integration and continuous delivery. The pipeline autonomously retrains the model using new data, does unit and hydraulic validation tests, and deploys the updated container while the previous version remains operational for queries. The management of build–test–deploy processes for models, as opposed to conventional software, is referred to as MLOps. Utilities implementing MLOps in Kansas City reduced the "hot-fix" duration for an LSTM inflow model from three weeks to eight hours.

Zero-Trust Security

The stormwater IoT now adheres to a zero-trust principle: every data packet, including those from internal sensors, must undergo authentication. Devices utilize an X.509 certificate, and all communications are encapsulated in TLS 1.3, the same encrypted tunnel employed by contemporary browsers. The firmware operating on each mote verifies its integrity via device attestation, a challenge-and-response protocol wherein the sensor signs a random nonce using a private key embedded in the hardware. New firmware can only be installed if it has a cryptographic signature authorized by the utility, these signed over-the-air upgrades thwarted a 2021 botnet that attempted to commandeer Australian rain gauges

Federated Learning and Provenance Ledgers

Data privacy regulations hinder collaboration; for example, European utilities are prohibited from exporting raw depth traces internationally. Federated learning addresses the issue by transmitting encrypted weight updates instead of raw observations; gradients are aggregated in a central "parameter server," and the enhanced weights are distributed back to each location. A Franco-Dutch pilot reduced forecast error by eight percent while adhering to GDPR regulations. To ensure that no individual reading can be retrieved, the pilot incorporated a small amount of differential noise—microscopic random numbers that are theoretically validated to obscure individual contributions. Each transformation, Hampel rejection, Kalman fill, MPC gate command—creates a hash-chained provenance record: every log entry retains the cryptographic fingerprint of its predecessor, therefore any modification to a link disrupts the chain. Singapore's 2022 Code of Practice on Surface Water Drainage permits detention-volume credits for real-time control only if a ledger can provide a byte-level trail upon request; analogous language is present in the 2024 draft of the EU Urban Waste-Water Directive.

CUTTING-EDGE ADVANCES IN AI AND ML FOR STORMWATER MANAGEMENT

Recent developments suggest that the next decade of stormwater management will be shaped by a new wave of AI and ML tools that extend beyond current LSTM, optimization, and MPC pipelines.

Neural Operators (e.g., Fourier Neural Operators) are being trained as surrogates for 2-D shallow-water solvers. These models generalize across basins while running up to 100× faster than physics-based simulations, opening the door to city-scale inundation forecasting that can feed into sewer control decisions.

Utilities are beginning to deploy cloud-native digital twins that fuse IoT telemetry, online models, and control logic in a single architecture (FIWARE/NGSI-LD). At the same time, federated learning frameworks allow utilities to share model improvements without sharing raw data, an approach already tested for overflow and treatment-plant forecasting in Europe.

A major frontier is not just accuracy but uncertainty. Deep nowcasting models such as NowcastNet and DGMR (DeepMind) are being adapted to stormwater contexts, while conformal prediction methods are being layered on to give calibrated exceedance probabilities. This shift from point forecasts to reliable probability envelopes will improve decision-making under regulatory and operational risk.

Finally, differentiable routing and hydrology models are being developed that allow direct gradient-based training and control co-design. These "physics-in-the-loop" solvers promise faster calibration, real-time data assimilation, and eventually seamless integration with controllers, marking a fundamental change in how hydraulic models are used in learning pipelines.

INTEGRATED STORMWATER TREATMENT: GREEN INFRASTRUCTURE MEETS COMPUTATIONAL INNOVATION

Stormwater management is adapting to increasing urbanization, climate-related excessive precipitation, and heightened ecological concerns. Conventional grey infrastructure is inadequate to manage this complexity. A hybrid strategy has emerged, integrating green infrastructure (GI) with computational breakthroughs. Green Infrastructure systems emulate natural hydrological cycles to manage and retain stormwater, while computational techniques improve their design, forecasting, and regulation. This integration signifies a transformative change, facilitating more robust, economical, and flexible urban water systems.

Green Infrastructure as Nature-Based Treatment

Green Infrastructure includes decentralized systems such as rain gardens, bio-retention cells, bioswales, green roofs, and permeable pavements. These systems capture stormwater at its origin, diminishing volume, filtering contaminants, and promoting groundwater replenishment.

Rain gardens and bioretention systems function by a stratified arrangement of flora, designed soil, and underdrainage systems. They diminish pollutant concentrations by filtration, adsorption, and microbial decomposition, particularly for nutrients such as nitrogen and phosphorus, as well as for suspended particles and heavy metals.

Green roofs effectively regulate rooftop runoff while improving insulation and promoting biodiversity. Bioswales, linear planted waterways, facilitate flow conveyance and filtration in road medians and parking lots. Each of these methods facilitates a decentralized and minimal-impact approach to manage stormwater flows.

Nonetheless, GI performance fluctuates considerably according on land utilization, precipitation intensity, flora, and soil properties. This requires computational decision-support systems that customize GI implementation to local conditions for optimal efficacy.

Figure 4. Heat exchnage of urban green roofs (Berardi et al., 2014)

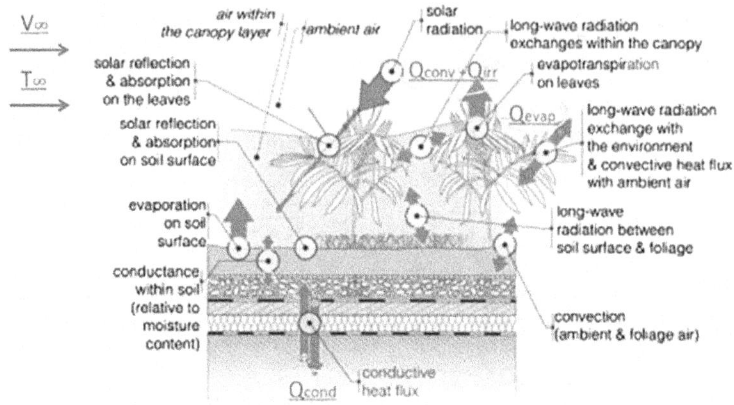

COMPUTATIONAL TOOLS IN GI DESIGN AND ASSESSMENT

The amalgamation of hydrologic and hydraulic models has significantly enhanced the planning of green infrastructure. The U.S. EPA's Storm Water Management Model (SWMM) and its variants, such as PCSWMM and PySWMM, are essential instruments for simulating stormwater dynamics and evaluating green infrastructure configurations under various precipitation events.

Multi-objective optimization methods, including genetic algorithms and Trust Region Bayesian Optimization (TuRBO), have been utilized to identify optimal combinations of green infrastructure elements considering cost, runoff reduction, and geographical limitations

Wang et al. (2021) shown that integrating hydrologic modelling with optimization can reveal green infrastructure configurations that are both successful in flood management and economically viable, while also being resilient to climatic fluctuations.

GIS-based spatial modelling holds similar significance. It enables planners to identify runoff hotspots and correlate them with prospective green infrastructure placements based on land use, slope, and impervious surfaces—thereby informing decision-making through spatial intelligence.

Machine Learning for Stormwater Prediction

Machine learning (ML) is progressively utilized for the prediction of stormwater quantity and quality. Algorithms such Long Short-Term Memory (LSTM), Random

Forests (RF), Support Vector Machines (SVM), and Artificial Neural Networks (ANN) have demonstrated efficacy in modelling intricate, non-linear hydrological interactions.

Jeon et al. employed LSTM models to predict pollutant concentrations in rain gardens, with an error rate of less than 10% across several rain events .Machine learning methods have been trained using event mean concentrations (EMC) data to estimate pollutant loads such as total suspended solids (TSS), total nitrogen (TN), total phosphorus (TP), and microplastics in stormwater

These models provide a rapid and adaptable substitute for conventional physical modelling, especially beneficial in contexts with limited or incomplete data. They facilitate proactive maintenance by forecasting infrastructure strain prior to performance decline.

Computation-Enhanced Treatment Systems

Computational approaches are augmenting physical treatment systems such as biofilters and artificial wetlands. Biofilters integrate filtration and biochemical breakdown within designed soil profiles, demonstrating notable efficacy in the removal of nitrogen and phosphorus.

Machine learning models have been created to assess the efficiency of biofilter pollutant removal based on input parameters like inflow rate, pollutant concentration, and medium characteristics These models assist in real-time adjustment of system parameters or the redesign of underperforming systems.

Constructed wetlands and detention basins have also gained advantages from simulation-assisted design. Employing event-based hydrological models, designers can modify basin capacity and retention time to enhance sedimentation and pollutant degradation.

Emerging contaminants, such as microplastics and tyre wear particles (TPW), provide novel concerns. Models like CatBoost, trained on sparse environmental data, are being developed for the computational identification and prediction of microplastic surges during storm occurrences .

Responsive and Adaptive GI Systems

A significant challenge in computational stormwater management is real-time control and adaptability. The incorporation of IoT-enabled sensors, meteorological forecasts, and machine learning facilitates the development of adaptive GI systems that respond to impending storm conditions.

In certain pilot investigations, reinforcement learning (RL) agents were trained to dynamically regulate valve positions in rainwater harvesting systems, optimizing for flood mitigation and water reuse.

Real-time telemetry systems can transmit sensor data to model predictive control (MPC) platforms that assess prospective storm inflows and determine appropriate release plans. These solutions have demonstrated substantial decreases in peak flow and overflows, as evidenced by the OpenStorm and Smart Sewer pilot programs in the United States

Challenges and Knowledge Gaps

The amalgamation of Geographic Information with computing holds potential, although numerous obstacles persist. The lack of data is a significant obstacle, especially for developing resilient machine learning models across several catchments. The transferability of models presents a challenge; instruments designed for one place may necessitate substantial calibration for application in another.

The computational overhead may also be excessive. High-resolution simulations and optimization algorithms require substantial processing resources, hence restricting accessibility in emerging environments. Furthermore, intricate models may provide challenges for local governments to comprehend and implement without comprehensive training.

Ultimately, physical gastrointestinal systems require regular maintenance. No algorithm can rectify obstructed inlets or overlooked vegetation, highlighting the necessity for human–AI collaborations in execution.

The convergence of green infrastructure with computational techniques provides a means for enhanced resilience, adaptability, and ecological integration in stormwater management. Rain gardens, bioswales, and biofilters serve the ecological purposes of absorption and purification, while data-driven tools guarantee their proper placement, sizing, and operation.

Hydrological models, machine learning predictions, and IoT-integrated controls are significantly enhancing the resources accessible to stormwater engineers. As climatic extremes escalate and urban areas become more congested, this integration will be essential for protecting urban water cycles and the communities reliant on them.

To clarify the relative strengths and limitations of legacy approaches versus emerging computational methods, Table below compares the performance of traditional hydrologic/hydraulic tools and computational AI/ML-enabled GI tools across key dimensions of stormwater management.

Table 2. Comparative performance of traditional vs. computational GI tools in stormwater management

Dimension	Traditional tools (Hydrologic/ Hydraulic + heuristic GI design)	Computational tools (AI/ML, optimization, real-time control)
Modeling approach	Deterministic, physics-based (e.g., SWMM, HYDRUS-1D); rule-based GI sizing; GIS overlays for siting	Data-driven ML surrogates (LSTM, RF, GNN), hybrid physics–AI models; multi-objective optimization with evolutionary algorithms
Design/ Planning	Heuristic siting of GI based on imperviousness and slope; static BMP guidelines	Optimization of GI layout considering cost, hydrology, pollutant removal, and resilience (e.g., NSGA-II, PSO, Bayesian methods)
Forecasting	Relies on design storms, historical IDF curves; limited adaptability to climate non-stationarity	Short-term ML forecasts (LSTM, CNN–LSTM, GBDT) predict runoff, pollutant concentrations, and overflow events with sub-hour lead times
Operation	Fixed-rule detention and conveyance; static weirs and valves; reactive maintenance	Model Predictive Control (MPC) and Reinforcement Learning (RL) dynamically adjust pumps, valves, and basins in real time
Scalability	Calibration intensive; site-specific; limited transferability across catchments	Surrogate models (GNNs, Neural Operators) enable rapid retraining and scalability to watershed or city-wide systems
Equity & governance	Equity depends on siting and manual inspection; often neglects distributional impacts	Computational siting integrates social and environmental criteria; provenance logs and federated learning support governance and data privacy

CASE STUDIES

Strategic Stormwater Management: The City of Greater Geelong's AI-Driven Approach

The City of Greater Geelong, (*Full Article: Smart Water Management: The Way to (Artificially) Intelligent Water Management, or Just Another Pretty Name?*, n.d.-a) situated in Victoria, Australia, has historically overseen a vast and intricate stormwater system comprising over 1,600 km of piping and in excess of 100,000 stormwater pits. This extensive infrastructure supports a population of over 250,000 inhabitants and encounters persistent issues stemming from ageing assets, urbanization, heightened storm occurrences, and resource limitations. Historically, the city depended on reactive maintenance and traditional manual inspection techniques, which were expensive, labor-intensive, and inefficient, frequently resulting in delays in resolving crucial issues.

In response to increasing infrastructural and budgetary demands, Geelong implemented a unique strategy by incorporating VAPAR's AI-driven inspection technology. VAPAR is a digital platform that employs sophisticated machine learning and artificial intelligence algorithms to automate the examination of CCTV data for stormwater and sewer systems. Geelong sought to transition from conventional reactive approaches to proactive and predictive infrastructure management by implementing this solution.

The principal benefit of implementing VAPAR's AI technology was the swift analysis of enormous volumes of CCTV inspection data. Manual analysis formerly necessitated significant time investments from proficient staff, often requiring weeks to generate relevant data. VAPAR's automated software greatly accelerated this procedure, transforming CCTV footage into comprehensive condition assessments within hours. This swift transformation enabled Geelong's asset management team to quickly discern and prioritize repairs using precise and thorough data, enhancing overall decision-making and responsiveness.

Moreover, the automated analytics delivered a precise and uniform evaluation of infrastructure status, significantly enhancing accuracy relative to conventional hand assessments. The diminution of subjective interpretation mitigated human mistake, guaranteeing consistently dependable facts. Consequently, Geelong realized significant operational efficiencies, encompassing a 30% drop in CCTV inspection expenses and a 25% decline in reactive maintenance activities. The cost reductions facilitated investing in essential infrastructure projects and regular maintenance, thus extending asset life cycles.

The use of an AI-driven solution resulted in improved predictive planning and asset forecasting skills over the long run. A comprehensive study and trend identification enabled the city to make informed, strategic decisions concerning infrastructure renewal, expansion, and preventive maintenance. The enhanced capability for predictive planning significantly bolstered the city's resilience to flooding and system breakdowns, facilitating sustainable urban development and environmental preservation.

The implementation of VAPAR's technology significantly improved transparency and stakeholder involvement. The clarity and visual insights produced by automated CCTV analytics enhanced communication among technical teams, policymakers, and the community. Through clear data visualization, city administrators may convincingly substantiate budget allocations, infrastructure investments, and policy modifications to stakeholders, thereby enhancing public confidence and engagement.

The collaboration between the City of Greater Geelong and VAPAR exemplifies the revolutionary capacity of using AI and automation in the administration of municipal infrastructure. This strategic transition illustrates that adopting technological innovations can provide significant operational economies, lower expenses, prolong asset longevity, and improve service quality. As urban areas worldwide confront analogous issues pertaining to deteriorating infrastructure and fiscal limitations, the Geelong scenario offers a repeatable framework that underscores the substantial advantages of proactive, technology-driven stormwater management.

Enhancing Urban Stormwater Runoff Prediction with Machine Learning: A Case Study of First Creek Watershed, Knoxville, TN

This work (*Developing Data-Driven Learning Models to Predict Urban Stormwater Runoff Volume: A Case Study*, n.d.) investigates the capacity of machine learning (ML) models to augment and improve the Storm Water Management Model (SWMM), a prevalent yet computationally demanding hydrologic modelling instrument created by the U.S. Environmental Protection Agency (EPA). The study examines the efficacy and precision of ML algorithms in replicating SWMM outcomes for forecasting stormwater runoff volumes, therefore substantially minimizing computing demands.

The research environment focusses on climate-induced hydrological extremes and urbanization-related impermeable surfaces, both of which exacerbate flood risks and stormwater contamination. These problems highlight the imperative for swift and dependable predictive modelling in urban watersheds, especially to facilitate real-time decision-making and long-term scenario evaluations.

The analysis, conducted in the First Creek watershed in Knoxville, Tennessee, utilizes hourly rainfall data from 2010 to 2020 together with relevant watershed parameters. Researchers trained machine learning models utilizing data obtained from SWMM simulations of 140 subcatchments. The simulations generated comprehensive output data for runoff volumes, which were subsequently employed as targets for machine learning model training.

Initially, four machine learning algorithms—Linear Regression (LR), Decision Tree (DT), Random Forest (RF), and Neural Networks (NN)—were evaluated. The Random Forest model demonstrated the highest accuracy, attaining a remarkably low mean absolute error (MAE). Specifically, RF demonstrated significant accuracy enhancements following the integration of feature engineering, which involved consolidating hourly rainfall data into relevant features such as cumulative rainfall over designated time intervals and prior rainfall conditions, thereby encapsulating hydrologically pertinent antecedent moisture conditions.

In the preliminary experiment devoid of designed characteristics, the RF model had an MAE of almost 21,000 gallons. Following feature engineering, accuracy markedly enhanced, with RF attaining an MAE of roughly 6,000 liters, illustrating its capacity to swiftly and accurately replicate SWMM outputs.

A third experimental method evaluated the generalizability of these machine learning models across various geographic sub-regions within the watershed. Employing constrained k-means clustering, subcatchments were categorized according to their geophysical attributes, hence assuring equitable cluster sizes. The RF model was trained on one cluster and evaluated on a different cluster to assess its predictive robustness and geographic adaptation. Results suggested that machine learning models exhibited optimal predictions within clusters while also demonstrating acceptable accuracy when applied across clusters with varying geophysical features, suggesting potential for wider use.

The ramifications of this discovery are substantial. Conventional hydrologic models, such as SWMM, necessitate considerable processing time, hence constraining the number of scenarios and the promptness of actions that water managers may feasibly assess. Conversely, ML models, especially Random Forest, significantly decrease prediction time from roughly 25 minutes per simulation (SWMM) to just seconds, facilitating real-time and comprehensive scenario analysis.

In conclusion, the integration of machine learning algorithms with training data produced from SWMM offers a viable approach to improving urban stormwater management. Hybrid modelling approaches enhance computational efficiency and forecast accuracy, enabling water management authorities to swiftly evaluate several scenarios, thereby refining infrastructure development, operational responsiveness, and flood-risk control strategies.

SMART WATER MANAGEMENT (SWM): GLOBAL CASE STUDIES AND INSIGHTS

The Smart Water Management (SWM) report (*SMART-WATER-MANAGEMENT-Case-Study-Report*, n.d.), co-published by the Korea Water Resources Corporation (K-water) and the International Water Resources Association (IWRA), delineates global best practices for the incorporation of advanced Information and Communication Technologies (ICT) in water management. SWM utilizes real-time monitoring, sensors, GIS mapping, automation, and big-data analytics to tackle intricate water issues exacerbated by climate change, fast urbanization, deteriorating infrastructure, and increasing demand.

The report analyses various global experiences, emphasizing concrete results from SWM techniques. The table below encapsulates essential projects, geographical contexts, applied technologies, and the resultant benefits:

Table 3. 1SWM global case studies

Region / Country	Key Project(s)	SWM Technologies & Solutions	Main Benefits Achieved
Korea	K-HIT system (K-water Hydro Intelligent Toolkit), Seosan and Paju Smart Cities	Integrated real-time flood and drought management, Smart meters, Real-time leak detection	Enhanced water conservation, improved water quality, rapid leak identification, increased public trust and participation
France (Greater Paris)	Integrated sanitation management system	Real-time monitoring, Automated sanitation networks, IoT sensors	Improved operational efficiency, enhanced water quality, reduced environmental pollution, cost savings
China (Handan City)	Groundwater monitoring and modeling	Real-time groundwater sensors, GIS, Data-driven modeling	Reduced groundwater over-extraction, sustainable aquifer management, informed groundwater policy
Mexico City	PUMAGUA Project (UNAM)	Smart sensors for leak detection, wastewater treatment automation, continuous water quality monitoring	Improved water efficiency, significant water-loss reduction, better wastewater management, increased sustainability and reliability of supply
Africa & Thailand	FDMT project (Flood and Drought Monitoring and Early Warning)	Satellite-based remote sensing, GIS integration, Transboundary data sharing platforms	Enhanced resilience to climate extremes, improved early warning systems, better regional collaboration
Tanzania, Mozambique, Zimbabwe	Smallholder smart irrigation	Real-time soil moisture monitoring, IoT-based irrigation systems, Agricultural Innovation Platforms	Increased agricultural productivity, improved rural livelihoods, reduced water waste, enhanced income for smallholder farmers
Canada (Toronto)	Stormwater SmartGrid	AI-driven rain harvesting, Real-time flood monitoring and forecasting, Distributed storage and control systems	Improved urban flood resilience, optimized stormwater infrastructure, reduced flood damage and runoff impacts

The emphasized projects collectively demonstrate SWM's ability to produce integrated advantages across social, economic, environmental, and governance dimensions. Socially, SWM improves community health, expands water accessibility, and promotes public engagement. These projects economically diminish expenses, preserve resources, enhance efficiency, and increase local employment. Environmentally, solid waste management enhances water quality, diminishes aquifer depletion, and alleviates pollution effects. Governance advantages encompass enhanced openness, accountability, and community-driven decision-making facilitated by open data and active stakeholder engagement.

The SWM report offers a strategic framework for policymakers, advocating for policies that promote international collaboration, technological innovation, information exchange, capacity enhancement, and continuous investment in digital water infrastructure. These components guarantee the effective replication and expansion of SWM worldwide.

The report directly associates SWM with the Sustainable Development Goals (SDGs), particularly highlighting its contributions to SDG 6 (Clean Water and Sanitation), SDG 11 (Sustainable Cities and Communities), and related SDGs focused on poverty alleviation, gender equity, climate action, and responsible resource consumption. The SWM case studies confirm that smart technologies are essential for sustainable water management and resilience in an unpredictable global water future.

GRID INTELLIGENCE WATER CASE STUDY

The Verizon Grid Wide Intelligent Water Solutions case study (*Grid Intelligence Water Case Study*, n.d.) examines significant water management issues encountered by municipalities, highlighting efficiency, cost-effectiveness, and sustainability in the modernization of water infrastructure. It underscores the challenges many communities encounter with antiquated water-metering systems, resulting in erroneous billing, unaddressed leaks, inadequate conservation, and suboptimal customer service, particularly problematic during droughts and financial limitations.

The case pertains to a tiny city in the southeastern United States facing challenges due to an outdated water-metering system. The current meters resulted in considerable billing mistakes, leading to customer unhappiness and extensive staff resources allocated to dispute resolution. Moreover, unnoticed water leaks exacerbated water wastage and expenses. The city anticipated a comprehensive utilities overhaul in two years; but, pressing issues required a temporary solution.

The city adopted Verizon Grid Wide Intelligent Water Solutions to tackle this issue without significant initial investment. This platform-as-a-service (PaaS) solution facilitated the city's transition to a smart water metering system through a cloud-based platform, eliminating the necessity for investment in physical data centers, communication networks, and intricate software installations. The city shifted from manual meter-reading methods to an automated, remotely controlled system.

The Verizon solution utilizes sophisticated Internet of Things (IoT) technologies, amalgamating smart meters, utility network devices, IoT gateways, Verizon 4G LTE, 900 MHz radio frequency connectivity, and Verizon's Private IP network. This resilient architecture facilitated secure, bidirectional communication and enabled real-time monitoring, data analysis, and extensive control functionalities.

Notable results from the implementation of the Grid Wide solution included:

Table 4. Grid intelligence water case study outcomes

Benefit Area	Achieved Outcomes
Operational Efficiency	Automated meter reading reduced labor costs and reassigned employees to more critical tasks.
Resource Conservation	Early leak detection significantly decreased water loss, promoting conservation and lowering operational costs.
Revenue Management	Improved metering accuracy and real-time data collection enabled accurate billing, better revenue forecasting, and proactive financial planning.
Reliability & Maintenance	Proactive system-wide maintenance using real-time data prevented service disruptions and optimized resource allocation.
Customer Service	Real-time insights facilitated rapid resolution of billing disputes and enhanced customer satisfaction.
Safety	Remote monitoring reduced risks for workers by minimizing visits to remote or hazardous locations.
Asset & Infrastructure Management	Provided data for improved infrastructure management, long-term asset planning, and effective growth forecasting.

Importantly, these enhancements need no substantial initial capital expenditures from the municipality. The solution was provided via an economical monthly subscription model, adaptable to the city's changing requirements. This facilitated prompt intervention in problematic locations without extensive replacements, fostering gradual yet significant enhancements.

The paper underscores the global importance of these solutions, noting that around 30% of treated water is lost to leaks, resulting in an estimated annual revenue loss of $14 billion worldwide. The Verizon Grid Wide solution illustrates how municipalities may effectively tackle these difficulties with intelligent digital infrastructure, thereby promoting sustainability and resilience in urban water systems without excessive investment demands.

SMART WATER MANAGEMENT: ASSESSING POTENTIAL, LIMITATIONS, AND GLOBAL APPLICATIONS

Smart Water Management (SWM) (*Full Article: Smart Water Management: The Way to (Artificially) Intelligent Water Management, or Just Another Pretty Name?*, n.d.-b) entails the incorporation of sophisticated digital technology, including artificial intelligence (AI), real-time monitoring, and extensive data analytics, into water resource systems. SWM seeks to improve efficiency, water quality, resource conservation, and management transparency through the utilization of advanced tools across various environments.

Diverse global implementations of SWM illustrate its significant potential and prominent limitations. The PUMAGUA project at the UNAM campus in Mexico accomplished notable advancements, including a 50% decrease in water use, enhanced water quality, and improved wastewater treatment. This achievement relied significantly on continuous financial investment, considerable human resources, and persistent maintenance, highlighting the resource intensity of SWM.

The national K-water Hydro Intelligent Toolkit (K-HIT) in South Korea has emerged as a comprehensive platform for controlling flood and drought events by utilizing real-time hydrological and meteorological data, predictive rainfall models, reservoir management, and flood-risk analysis. K-HIT effectively alleviated damage caused by significant flooding in 2012, 2013, and 2015. Achieving such efficacy necessitates considerable long-term public investment and institutional capacity, indicating that scalability is heavily reliant on ongoing governmental assistance.

Greater Paris displays smart water management in urban sanitation via the MAGES real-time sanitation control system. MAGES orchestrates decentralized control of urban stormwater and wastewater, markedly diminishing environmental effects and occurrences of combined sewer overflow. This decentralized yet coordinated management strategy provides numerous stakeholders with operational freedom, emphasizing how SWM enhances successful urban water administration. However, projected future obstacles, especially diminished river flow due to climate change, emphasize the necessity for continuous adaptation and persistent technological enhancement.

Implementing SWM technologies among smallholder irrigation farmers in sub-Saharan Africa presents unique challenges. Advanced irrigation technologies, however advantageous, encounter practical obstacles like cost, user-friendliness, and applicability for small-scale farmers, who frequently possess constrained financial means and technical knowledge. Success necessitates that SWM technology be developed with a robust emphasis on usability, contextual relevance, and community capacity-building efforts, highlighting the social aspects of effective SWM implementation.

In addition to conventional water utilities, SWM also facilitates humanitarian initiatives. In Uganda, the UNHCR efficiently employed advanced water-monitoring technologies to guarantee fair and transparent water delivery in refugee camps. Furthermore, trial initiatives on the Canary Islands utilizing open-source SWM platforms exhibited substantial advantages, such as enhanced data dependability, predictive maintenance, and increased operational efficiency in urban water systems.

Notwithstanding apparent achievements, SWM encounters substantial conceptual and practical constraints. Technology-driven water management may exacerbate current inefficiencies, thereby perpetuating obsolete infrastructure models rather than encouraging substantial infrastructural reforms. Concerns also emerge over "moral hazard," wherein dependence on technical solutions may unintentionally sustain unsustainable water usage patterns instead of fundamentally transforming behavior towards conservation and sustainability. Moreover, the implementation of SWM necessitates vigilance over data privacy, security vulnerabilities, and inadvertent environmental or social repercussions, hence requiring a preventative and equitable governance framework.

SWM presents authentic prospects for enhancing global water management; yet, its efficacy depends on robust institutional support, ongoing financial investment, community involvement, and meticulous attention to wider social, ethical, and environmental consequences. To fully realize the promise of SWM, it is essential to integrate technological innovation with strategic policy formulation, adaptive governance, and a comprehensive approach to water sustainability and resource management.

CONCLUSION

Stormwater management is seeing a significant shift. The discipline has transitioned from dependence on static pipe-and-pond systems to utilizing real-time data, sophisticated algorithms, and adaptive infrastructure to address escalating hydrological and urban difficulties. This chapter consolidated the scope of these computational advancements—spanning sensing, telemetry, modelling, prediction, and intelligent actuation—while examining the integration of nature-based solutions with digital technologies.

Progress So Far

A notable transformation has been the transition from reactive to proactive stormwater management. Historically, designs have presumed extreme rainfall occurrences and implemented fixed safety margins. Contemporary systems, outfitted with telemetry nodes and edge-based sensors, deliver real-time information on precipitation, runoff, and infrastructure conditions. These data streams are integrated into cloud-based platforms and analytics pipelines, where machine learning models, including LSTM and Random Forest, now forecast system behavior minutes to hours in advance. This enables stormwater systems to not only identify issues but also to foresee them and respond proactively.

The incorporation of digital control through SCADA systems and programmable logic controllers (PLCs) has facilitated the automation of several reactions in urban environments. For example, valve locations and pump schedules can now be modified dynamically, enhancing optimization for present and anticipated conditions. Reinforcement learning agents have been evaluated for the management of decentralized systems, such as rainwater collection tanks, with twin objectives: mitigating urban flooding and optimizing reuse potential.

Moreover, computational hydrologic modelling has undergone significant transformation. Legacy technologies such as the EPA's SWMM have been augmented with Python wrappers like PySWMM, facilitating hybrid workflows that integrate deterministic physical modelling with data-driven insights. Research has shown that integrating machine learning models with hydrologic simulators enhances the efficiency of scenario testing and increases forecast accuracy, particularly in intricate urban catchments.

Computational Enhancements to Green Infrastructure

A hallmark of this computational era is its integration with green infrastructure (GI). Bioswales, rain gardens, permeable pavements, and green roofs offer decentralized, low-impact stormwater management at the source. Nevertheless, their efficacy fluctuates according to vegetation type, precipitation intensity, and maintenance protocols. Computational tools now mitigate these uncertainty. Machine learning models trained on inflow quality and medium composition can forecast pollution removal efficiency in bioretention systems.

GIS-based modelling facilitates the appropriate placement of green infrastructure within urban environments, pinpointing runoff hotspots and coordinating interventions with land-use limitations. Multi-objective optimization methodologies—such as Genetic Algorithms and Bayesian Optimization—are increasingly commonly employed to reconcile performance, cost, and resilience amid climate unpredictability.

These advancements collectively facilitate a transition from discrete GI initiatives to watershed-scale policies. Planners may now model the aggregate effects of numerous little actions, directing investment towards high-impact sectors while reducing redundancy. Real-time gastrointestinal control, facilitated by IoT sensors and predictive analytics, enables precise reactions to storm events, enhancing water quality and urban liveability.

CHALLENGES AND GAPS

Notwithstanding these gains, difficulties persist. Data availability and quality are persistent concerns. Sensor calibration, absent data, and communication issues might undermine model precision. Although outlier detection and filtering methods, such as Hampel filters and Kalman smoothing, are available, sustaining dependable data pipelines necessitates continuous work and investment.

The generalizability of the model is constrained. Machine learning models developed in one location may necessitate retraining when implemented in another because of variations in precipitation patterns, soil characteristics, and urban configurations. Transfer learning and federated learning show potential, although have not achieved extensive use in stormwater systems.

Third, computing expenses can be substantial, particularly for large-scale optimization and simulation. Although surrogate-based modelling has decreased runtimes, the majority of local authorities lack the internal competence to implement these tools. Cloud platforms can alleviate certain issues but also provide new challenges with data protection, latency, and expenses.

Furthermore, the human aspect must not be disregarded. Automated systems may fail, and over dependence on algorithms might result in oversight in decision-making. Physical infrastructure, regardless of its sophistication, requires human upkeep. The most robust systems will thus integrate digital intelligence with local expertise, personnel training, and community involvement.

FUTURE SCOPE

Anticipating the future, numerous trends indicate a more profound integration of computation and stormwater management. The implementation of digital twins—virtual representations of actual infrastructure that are continuously updated with real-time data—will facilitate ongoing monitoring, diagnostics, and scenario analysis. With the proliferation of open-source platforms and open data standards, an increasing number of cities will be able to implement these tools.

Secondly, we anticipate enhanced integration between stormwater management and urban resilience design. Data gathered for flood management might additionally enhance heat mitigation strategies, air quality assessments, and traffic planning. Stormwater systems will progressively be regarded as multifunctional urban assets rather than only hydraulic infrastructure.

Third, regulatory structures must evolve. Real-time control and predictive analytics confront conventional permit models founded on static assumptions. Cities and water agencies must amend regulations to facilitate dynamic operations while adhering to water quality and discharge standards.

Ultimately, prioritization of fair access to these technologies is essential. Low-income neighborhoods frequently experience severe flooding yet lack the capacity to deploy advanced technological remedies. Collaborations, open-source resources, and capacity-building initiatives are crucial to guarantee that computational advancements serve all urban areas.

The computational evolution of stormwater management is crucial. In an era of escalating storms, deteriorating infrastructure, and limited funds, urban areas require more intelligent methods for water management. By integrating nature-based solutions with machine learning, real-time data, and optimization algorithms, we are progressing towards systems that are not just reactive but predictive, and not solely cost-effective but also co-beneficial. These systems can mitigate flooding, boost water quality, promote biodiversity, and foster more habitable urban environments, all while adjusting to climate change.

This transition must be anchored in sound science, transparent decision-making, and participatory government to achieve success. The stormwater systems of the future will prioritize coding and sensors alongside soil and vegetation. Adopting this hybrid worldview is not merely pragmatic; it is essential.

REFERENCES

A Standard-Based Open Source IoT Platform: FIWARE | IEEE Journals & Magazine | IEEE Xplore. (n.d.). Retrieved June 1, 2025, from https://ieeexplore.ieee.org/document/8950963

Aquatic Pasts & the Watery Present: Water and Memory in the Fora of Rome. (n.d.). Retrieved June 1, 2025, from https://www.degruyterbrill.com/document/doi/10.1515/9783110677065-007/html

Bano, S. (2017). Karl August Wittfogel. Oriental Despotism: A Comparative Study of Total Power. New Haven, USA: Yale University Press. 1957 (Reprinted 1981). 550 pages. USD 119 (Paperback). *Pakistan Development Review, 393–395*. Advance online publication. DOI: 10.30541/v56i4pp.393-395

Berardi, U., GhaffarianHoseini, A., & GhaffarianHoseini, A. (2014). State-of-the-art analysis of the environmental benefits of green roofs. *Applied Energy, 115*, 411–428. DOI: 10.1016/j.apenergy.2013.10.047

Bureau of Indian Standards. (with Public.Resource.Org). (1986). *IS 4111-1: Code of practice for ancillary structures in sewerage system, Part I: Manholes*. http://archive.org/details/gov.in.is.4111.1.1986

Chen, W., Wang, W., Huang, G., Wang, Z., Lai, C., & Yang, Z. (2021). The capacity of grey infrastructure in urban flood management: A comprehensive analysis of grey infrastructure and the green-grey approach. *International Journal of Disaster Risk Reduction, 54*, 102045. DOI: 10.1016/j.ijdrr.2021.102045

Dworsky, L. B. (1967). Analysis of Federal Water Pollution Control Legislation, 1948–1966. *Journal - American Water Works Association, 59*(6), 651–668. DOI: 10.1002/j.1551-8833.1967.tb03394.x

Fang, H., Jamali, B., Deletic, A., & Zhang, K. (2021). Machine learning approaches for predicting the performance of stormwater biofilters in heavy metal removal and risk mitigation. *Water Research, 200*, 117273. DOI: 10.1016/j.watres.2021.117273 PMID: 34091222

Full article: Smart Water Management: The way to (artificially) intelligent water management, or just another pretty name? (n.d.-a). Retrieved June 1, 2025, from https://www.tandfonline.com/doi/full/10.1080/02508060.2020.1830581

Full article: Smart Water Management: The way to (artificially) intelligent water management, or just another pretty name? (n.d.-b). Retrieved June 1, 2025, from https://www.tandfonline.com/doi/full/10.1080/02508060.2020.1830581

Grid Intelligence Water Case Study. (n.d.). Verizon Enterprise. Retrieved June 1, 2025, from https://enterprise.verizon.com/resources/articles/grid-intelligence-water-case-study/

Hunzinger, R. (2017). SCADA FUNDAMENTALS AND APPLICATIONS IN THE IoT. In *Internet of Things and Data Analytics Handbook* (pp. 283–293). John Wiley & Sons, Ltd., DOI: 10.1002/9781119173601.ch17

Legislating for sustainable basin management: The story of Australia's Water Act (2007) | Water Policy | IWA Publishing. (n.d.). Retrieved June 1, 2025, from https://iwaponline.com/wp/article-abstract/15/6/871/20089/Legislating-for-sustainable-basin-management-the?redirectedFrom=fulltext

(PDF) Developing Data-Driven Learning Models to Predict Urban Stormwater Runoff Volume: A Case Study. (n.d.). Retrieved June 1, 2025, from https://www.researchgate.net/publication/364931622_Developing_Data-Driven_Learning_Models_to_Predict_Urban_Stormwater_Runoff_Volume_A_Case_Study

Sharma, R., & Malaviya, P. (2021). Management of stormwater pollution using green infrastructure: The role of rain gardens. *WIREs. Water*, *8*(2), e1507. DOI: 10.1002/wat2.1507

SMART-WATER-MANAGEMENT-Case-Study-Report. (n.d.). Retrieved June 1, 2025, from https://www.researchgate.net/profile/Vincent-Rocher-3/publication/329246344_SMART_WATER_MANAGEMENT_-_Case_Study_Report/links/5bfeb3c7299bf1a3c1537077/SMART-WATER-MANAGEMENT-Case-Study-Report.pdf

The 1848 Public Health Act and its relevance to improving public health in England now | The BMJ. (n.d.). Retrieved June 1, 2025, from https://www.bmj.com/content/317/7158/596

The Japanese Sewerage Industry: Institutional Aspects and the Governance Systems | SpringerLink. (n.d.). Retrieved June 1, 2025, from https://link.springer.com/chapter/10.1007/978-981-19-7489-2_13

The Persian Qanat—UNESCO World Heritage Centre. (n.d.). Retrieved June 1, 2025, from https://whc.unesco.org/en/list/1506/)

The Sewers of Mohenjo-Daro on JSTOR. (n.d.). Retrieved June 1, 2025, from https://www.jstor.org/stable/25034575

Tian, W., Liao, Z., Zhang, Z., Wu, H., & Xin, K. (2022). Flooding and Overflow Mitigation Using Deep Reinforcement Learning Based on Koopman Operator of Urban Drainage Systems. *Water Resources Research, 58*(7), e2021WR030939. https://doi.org/DOI: 10.1029/2021WR030939

Valencia, A., Zhang, W., & Chang, N.-B. (2021). Integration of machine learning classifiers and higher order tensors for screening the optimal recipe of filter media in stormwater treatment. *The Science of the Total Environment*, *771*, 145423. DOI: 10.1016/j.scitotenv.2021.145423 PMID: 33548713

Zhu, W., Tao, T., Yan, H., Yan, J., Wang, J., Li, S., & Xin, K. (2023). An optimized long short-term memory (LSTM)-based approach applied to early warning and forecasting of ponding in the urban drainage system. *Hydrology and Earth System Sciences*, *27*(10), 2035–2050. DOI: 10.5194/hess-27-2035-2023

Chapter 8
Mathematical Modeling and Metaheuristic Optimization for Distributed No-Wait Flow Shop Scheduling in Desalination Systems

Achraf Sayah
https://orcid.org/0009-0004-4947-188X
Higher Normal School, Hassan II University, Morocco

Said Aqil
National Higher School of Arts And Crafts, Hassan II University, Morocco

Mohamed Lahby
Higher Normal School, Hassan II University, Morocco

ABSTRACT

Desalination is an innovative solution to the global water shortage; however, scheduling and resource allocation in these systems present significant challenges. To address these challenges, this paper investigates the application of metaheuristic optimization techniques to enhance the performance of distributed desalination systems. The problem is formulated as a distributed flow shop scheduling model, with the primary objective of minimizing total flow time (TFT) to ensure timely water production and operational efficiency. To solve this, three metaheuristics are proposed: Iterated Greedy, Iterative Local Search, and Variable Neighborhood

DOI: 10.4018/979-8-3373-2700-6.ch008

Search. The effectiveness of these metaheuristics is enhanced using two initialization techniques: NEH and GRASP, yielding a total of six approaches. A comprehensive simulation study was conducted on a set of benchmark instances. The results show that all proposed metaheuristics provide valuable solutions for minimizing TFT and optimizing resource allocation, with varying levels of performance across different instances.

INTRODUCTION

Water scarcity is one of the most pressing global challenges of the 21st century, impacting millions of people. Population growth, urbanization, and climate change continue to intensify the pressure on freshwater resources. With 97.5% of the Earth's water being seawater and only 2.5% freshwater, 80% of which is locked in glaciers, only 0.5% of total water is readily accessible for human use (Ibrahim et al., 2017). In this context, desalination has become a crucial solution, enabling the conversion of abundant seawater into potable water and reducing dependence on traditional freshwater sources.

The global reliance on desalination has grown significantly in recent decades. As of June 2018, there were over 16,000 desalination plants in operation worldwide (Voutchkov et al., 2019), producing 87.5 million cubic meters of water per day and providing water for approximately 300 million people. Saudi Arabia is the largest producer of desalinated water (DeNicola et al., 2015; Darre et al.,2018), meeting 60% of its total water demand through desalination and holding a market share of 16.5% (Nair & Kumar, 2013). In countries like Qatar and Kuwait, desalination provides 100% of the water supply (Blanco-Marigota et al., 2017). As desalination capacity expands, the operational complexity of these systems increases, particularly in regions where multiple plants operate in a distributed manner to meet large-scale demand. Distributed Desalination Systems (DDS), characterized by multiple plants working collaboratively, offer several advantages, including localized water supply, reduced transportation costs, and enhanced resilience to disruptions. However, these systems introduce significant challenges in scheduling and resource allocation, requiring efficient coordination across multiple facilities.

The desalination process involves a sequence of stages, each requiring a specific machine to ensure efficient water treatment. A simple desalination process as illustrated in Figure 1, begins with the intake pump machine, which draws seawater into the plant system. Next, the screening machine removes large debris and marine organisms to protect the equipment. The seawater then passes through the pretreatment filtration unit, which eliminates fine particles and biological contaminants. After that, the high-pressure machine increases the water pressure, allowing it to pass through the reverse osmosis membrane machine, where salt and impurities are separated. Finally, the post-treatment machine adjusts the water's pH and adds essential minerals to make it suitable for drinking and industrial use.

This desalination process can be likened to a flow shop environment in industrial scheduling. In such a setting, batches of seawater (jobs) must be processed through a series of machines in a predefined order. Modeling the process this way allows for systematic optimization of scheduling and resource allocation, as techniques from industrial flow shop scheduling can be applied to improve efficiency. Furthermore, due to the sensitivity of certain desalination processes, operations must adhere to a no-wait constraint, meaning that once a batch starts processing, it must continue through all stages without interruption. This constraint is important because delays between stages can lead to changes in temperature or pressure, chemical issues, or damage to the membranes, which may compromise the quality of the water or affect the process. Therefore, enforcing the no-wait condition is essential to maintain operational stability, preserve water quality, and prevent costly interruptions in the desalination process.

Figure 1. A simple desalination process

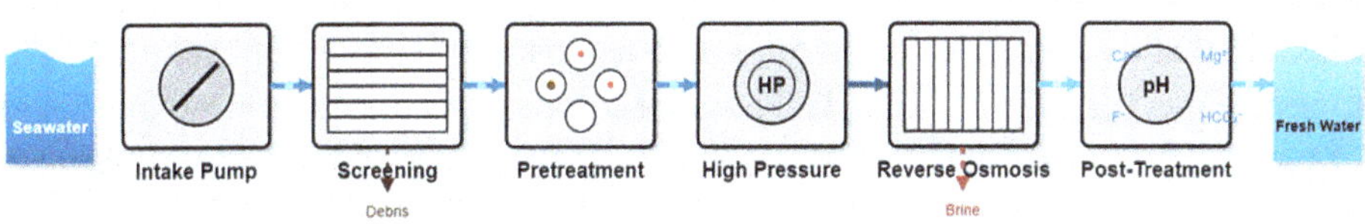

It is worth noting that scheduling problems have been applied in a wide range of industrial and technological domains, where accurate modeling is crucial for efficient problem solving. In fact, they have been used in civil engineering and construction to model the manufacture of pre-stressed concrete slabs and beams as a hybrid flow shop with setup times (Aqil and Allali, 2020). Since delays in completing orders directly postpone deliveries to construction sites and disrupt tightly planned project schedules, the authors focus on minimizing total tardiness, thereby enhancing on-time performance, improving overall production efficiency, and ensuring compliance with contractual deadlines. In the steel industry, hot rolling production lines have been represented as a flow shop scheduling problem with sequence-dependent set-up times (Nejjarou et al., 2024). As delays between consecutive stages can disrupt the continuous rolling process and degrade product quality, the authors address the problem while considering no-wait constraints to ensure uninterrupted operations, adherence to delivery schedules, and consistent product standards. In electronics manufacturing, the production of printed circuit boards across multiple factories has been addressed by Sayah et al. (2025) as a distributed flow shop scheduling problem. As delays between consecutive stages can cause disruptions in the tightly coordinated assembly process and lead to late order completion and given the need for machine maintenance and/or cleaning between operations, the authors address the problem while considering no-wait and setup time constraints, thereby improving delivery performance, optimizing resource utilization, and enhancing overall production efficiency. Outside the industrial domain, scheduling problems have also found important applications, particularly in networking and distributed computing systems (Sayah et al., 2024b; 2025b). In particular, scheduling techniques have been used in the context of data processing within distributed edge computing systems. This helps reduce latency, improve resource utilization, and meet quality of service (QoS) requirements.

Continuing along this line, we explore a novel application of scheduling problems, particularly the distributed flow shop scheduling problem, in the context of desalination systems. This manufacturing process is critical for meeting the growing demand for potable water, highlighting the need for an innovative approach to the scheduling and resource allocation problem in distributed desalination systems. In addressing this challenge, our paper investigates the distributed flow shop scheduling problem within the context of desalination, which we model as a distributed no-wait flow shop scheduling problem. Our primary objective is to determine an optimal schedule that assigns jobs to factories and sequences them within each factory to minimize the Total Flow Time (TFT), defined as the sum of completion times of all jobs across the system. TFT is a key performance metric that reflects the efficiency of the scheduling process, ensuring rapid delivery of potable water to meet demand. Given the computational complexity of this problem, which combines distributed flow shop scheduling with no-wait constraints, traditional exact methods are often impractical, particularly for large-scale instances. To overcome this, we propose a metaheuristic optimization approach tailored to the unique characteristics of desalination operations. Furthermore, new and enhanced neighborhood structures are introduced to improve solution quality and algorithmic efficiency. The remainder of this paper is structured as follows: Section 2 presents the problem definition and mathematical model. Section 3 provides an illustrative example that confirms the mathematical formulation and highlights the strict no-wait constraint. Section 4 describes the metaheuristic approaches used to solve the problem. Section 5 outlines the experimental setup and computational study and discusses the results and their implications. Finally, Section 6 provides the conclusions and future research directions.

PROBLEM DEFINITION AND MATHEMATICAL MODELING

In this section, we explore the mathematical representation of the optimization problem under study, with a focus on the main objective of minimizing the Total Flow Time (TFT).

In the context of the distributed no-wait flow shop scheduling problem (DN-WFSP) applied to desalination systems, we consider a set of n batches of seawater (jobs), denoted as B= $\{B_1,B_2,\ldots,B_n\}$. these jobs are to be assigned to F distributed desalination units (factories), represented as $\Pi= \{\Pi_1,\Pi_2,\ldots,\Pi_F\}$. Each desalination unit Π_f is occupied with m machines M= $\{M_1,M_2,\ldots,M_m\}$ installed in a production line. The processing of each batch follows a no-wait constraint, meaning that once a batch ends processing on a machine, it must immediately proceed to the next machine without any delay. This constraint is crucial in desalination to prevent quality degradation and ensure continuous flow through the treatment stages.

Each job B_j is characterized by its processing time $P_{j,k}$ which represents the time required for this job on machine M_k. the objective of this scheduling problem is to determine an optimal assignment of batches to desalination units and sequence them efficiently to minimize the total flow time (TFT).

Minimizing TFT ensures that all batches are processed in the shortest cumulative time, which is critical for timely water supply and operational efficiency.

A solution to our problem can be represented as $\Sigma= [\Sigma_1,\Sigma_2,\ldots,\Sigma_F]$, where each scheduling sequence $\Sigma_f = \left(\sigma_f^1,\sigma_f^2,\ldots,\sigma_f^F\right)$ defines the order in which batches are processed in desalination unit Π_f. Here, σ_f^i represents the ith batch scheduled in desalination unit Π_f, and n_f denotes the number of batches assigned to Π_f unit, satisfying $\sum_{f=1}^{F} n_f = n$. This representation allows the problem to be modeled as a discrete combinatorial optimization problem, where the goal is to find the best permutation of batches across all units (sayah et al., 2025a).

The problem under study is formulated based on the following assumptions:

- Each machine can process only one batch at a time, and each batch can be executed by only one machine at any given moment.
- All batches are available for processing from the start (time zero).
- All machines are always available for use during the entire scheduling period.
- No preemption is allowed, meaning once a batch starts on a machine, it cannot be interrupted.

Under the no-wait constraint, batches must transition immediately from one machine to the next without delay once processing on the current machine is completed. This requirement complicates the calculation of completion times and sequencing, as each machine's schedule is tightly coupled with the previous and next stages.

However, it is possible to calculate the completion time of each machine for the first batch processed in each factory.

The completion time for the first batch on the first machine in each factory is simply its processing time on that machine, as mentioned in Equation (1).

$$\mathscr{C}_{\sigma_1^f, \mathscr{M}_1} = \mathscr{P}_{\sigma_1^f, 1} \tag{1}$$

This represents the simplest case, where no prior batches influence the completion time. For the first batch, the processing begins immediately at time zero, so the completion time on the first machine equals its own processing time. For each subsequent machine, the completion time is calculated as the sum of the completion time on the previous machine plus the processing time on the current machine, as mentioned in Equation (2).

$$\mathscr{C}_{\sigma_1^f, \mathscr{M}_k} = \mathscr{C}_{\sigma_1^f, \mathscr{M}_{k-1}} + \mathscr{P}_{\sigma_1^f, k} = \sum_{i=1}^{k} \mathscr{P}_{\sigma_1^f, i} \tag{2}$$

This recursive calculation ensures that the no-wait constraint is respected for the first batch, as each machine can only start processing the batch after the previous stage is complete. For subsequent batches, calculating the completion time on any machine depends not only on the previous batches but also on the remaining operations of that batch on the other machines. Based on previous work (Sayah et al., 2023; Nejjarou et al., 2024), we can calculate the completion time on the last machine as a function of the sequence-dependent delays and processing times across all batches and machines. This value is determined by the Equation (3).

$$\mathscr{C}_{\sigma_i^f, \mathscr{M}_m} = \mathscr{C}_{\sigma_{i-1}^f, \mathscr{M}_m} + \max_{l=1,\cdots,m} \left\{ \sum_{r=l}^{m} \left(\mathscr{P}_{\sigma_i^f, r} - \mathscr{P}_{\sigma_{i-1}^f, r} \right) + \mathscr{P}_{\sigma_{i-1}^f, l} \right\} \tag{3}$$

This formula captures the cumulative effect of sequencing on completion times, ensuring that each batch follows the no-wait rule while accounting for the processing differences between successive batches. It effectively models the dependency of each batch's progress on both the current and preceding batches.

The Total Flow Time (TFT), which measures the sum of completion times of all batches on the last machine, is then given by the Equation (4).

$$TFT = \sum_{f=1}^{F} \sum_{i=1}^{n_f} \mathscr{C}_{\sigma_i^f, \mathscr{M}_m} \tag{4}$$

ILLUSRATIVE EXAMPLE

In order to illustrate our problem and validate the proposed mathematical procedure, we present an illustrative example involving six batches, $B_1,B_2,\ldots,B_6$, to be assigned and scheduled to two factories, Π_1 and Π_2, each equipped with three machines M_1, M_2 and M_3. the processing times are defined by the following matrix:

$$\left[\mathscr{P}_{j,k}\right] = \begin{bmatrix} 5 & 5 & 5 \\ 4 & 5 & 5 \\ 6 & 4 & 5 \\ 4 & 5 & 5 \\ 6 & 4 & 4 \\ 4 & 3 & 6 \end{bmatrix}$$

Let us consider the following solution: $[[B_3,B_6,B_2], [B_4,B_5,B_1]]$, which represents the assignment and scheduling of batches to factories. the first factory processes B_3, B_6 and B_2 in this order, while the second factory processes B_4, B_5 and B_1 in this order.

Based on the proposed assignment and processing sequences, we now calculate the completion time of each batch.

for the first factory: the processing order is $B_3 \rightarrow B_6 \rightarrow B_2$. the completion times on the last machine are calculated respectively as follows:
for the first batch to be processed, B_3:

$$\mathscr{C}_{B_3,\mathscr{M}_3} = \sum_{i=1}^{3} \mathscr{P}_{3,i} = 6+4+5 = 15$$

for the second batch to be processed, B_6:

$$\mathscr{C}_{B_6,\mathscr{M}_3} = \mathscr{C}_{B_3,\mathscr{M}_3} + \max_{l=1,\cdots,3}\left\{\sum_{r=l}^{3}\left(\mathscr{P}_{6,r} - \mathscr{P}_{3,r}\right) + \mathscr{P}_{3,l}\right\}$$

$$= 15 + \max(4-6+3-4+6-5+6\,;\ 3-4+6-5+4\,;\ 6-5+5)$$

$$= 15 + \max(4;4;6) = 21$$

for the third and last batch to be processed, B_2:

$$\mathscr{C}_{B_2,\mathscr{M}_3} = \mathscr{C}_{B_6,\mathscr{M}_3} + \max_{l=1,\cdots,3}\left\{\sum_{r=l}^{3}\left(\mathscr{P}_{2,r} - \mathscr{P}_{6,r}\right) + \mathscr{P}_{6,l}\right\}$$

$$= 21 + \max(4-4+5-3+5-6+4\,;\ 5-3+5-6+3\,;\ 5-6+6)$$

$$= 21 + \max(5;4;5) = 26$$

for the second factory: the processing order is $B_4 \rightarrow B_5 \rightarrow B_1$. the completion times on the third and the last machine are calculated respectively as follows: for the first batch to be processed, B_4:

$$\mathscr{C}_{B_4,\mathscr{M}_3} = \sum_{i=1}^{3} \mathscr{P}_{4,i} = 4+5+5 = 14$$

for the second batch to be processed, B_5:

$$\mathscr{C}_{B_5,\mathscr{M}_3} = \mathscr{C}_{B_4,\mathscr{M}_3} + \max_{l=1,\cdots,3}\left\{\sum_{r=l}^{3}\left(\mathscr{P}_{5,r} - \mathscr{P}_{4,r}\right) + \mathscr{P}_{4,l}\right\}$$

$$= 14 + \max(6-4+4-5+4-5+4\,;\; 4-5+4-5+5\,;\; 4-5+5)$$

$$= 14 + \max(4;3;4) = 18$$

for the third and last batch to be processed, B_1:

$$\mathscr{C}_{B_1,\mathscr{M}_3} = \mathscr{C}_{B_5,\mathscr{M}_3} + \max_{l=1,\cdots,3}\left\{\sum_{r=l}^{3}\left(\mathscr{P}_{1,r} - \mathscr{P}_{5,r}\right) + \mathscr{P}_{5,l}\right\}$$

$$= 18 + \max(5-6+5-4+5-4+6\,;\; 5-4+5-4+4\,;\; 5-4+4)$$

$$= 18 + \max(7;6;5) = 25$$

To better visualize the scheduling process of the batches and the sequencing of operations on each machine, we present in Figure 2 a Gantt chart. This diagram illustrates the processing timeline of each batch across the machines in both factories and confirms the completion time calculations provided above, while explicitly accounting for the strict continuity imposed by the no-wait constraint.

Figure 2. Gantt chart of batch scheduling in the two factories

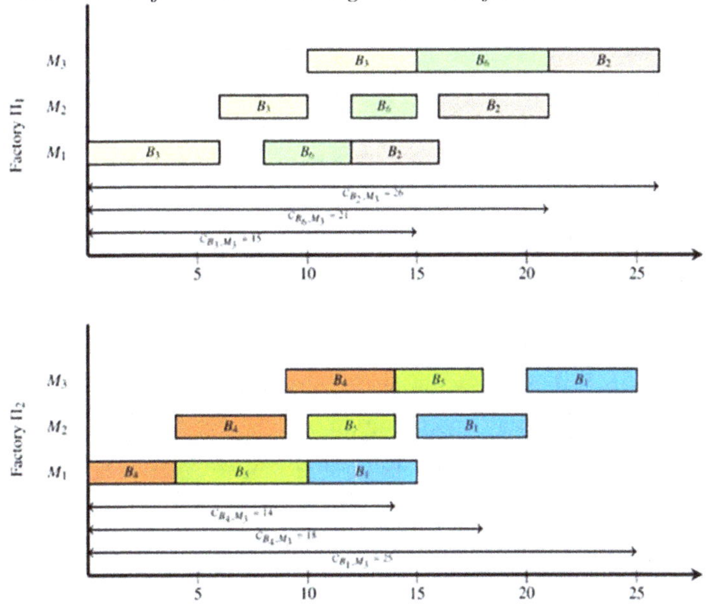

Based on the completion time calculations and the provided Gantt chart, the Total Flow Time (TFT) can be easily identified as the sum of the completion times of all batches: 15+21+26+14+18+25=119.

AN EXACT APPROACH THROUGH MIXED-INTEGER LINEAR PROGRAMMING (MILP)

The MILP (Mixed-Integer Linear Programming) approach is an exact optimization method used to solve problems in operations research and combinatorial optimization. It translates the problem into a mathematical formulation consisting of equations that express the system's constraints, using a set of decision variables. This approach is well known for its ability to handle complex requirements and decision structures.

In the context of the DFSP, MILP has proven effective in addressing both theoretical challenges and real-world industrial applications. The first MILP formulations for the DFSSP were proposed by Naderi and Ruiz (2010), who introduced six different MILP models aimed at minimizing the makespan. Subsequently, other formulations have been developed for alternative objectives (Ali et al., 2021; Khare & Agrawal, 2021), and for integrating various constraints such as blocking (Ribas et al., 2017), no-idle restrictions (Ying et al., 2017), and setup times (Alaghebandha et al., 2019), among others.

In our case study, we propose a MILP model to address the DNWFSP, with the objective of minimizing total flow time. This model ensures efficient task sequencing and optimal machine assignment while satisfying a set of constraints specific to the scheduling process. Our formulation is based on Model 5 proposed by Naderi and Ruiz (2010), with the necessary modifications to incorporate the no-wait constraints. This approach is characterized by the representation of the schedule as a list or vector of $n+F+1$ elements, where n elements correspond to the batches, and $F+1$ elements serve as separators to distinguish the batches assigned to different factories. To that end, a dummy batch is introduced to act as a separator between factories in the schedule representation. Furthermore, the model also incorporates a dummy machine to facilitate modeling. In order to simplify the notation used and distinguish it from the objective function formulation presented earlier, we adopt the following notation in our model. We denote by n the number of batches, by m the number of machines and by F the number of factories. The processing time of batch number j on machine number k is denoted as P_{jk} and the completion time is denoted as C_{jk}. Moreover, we define a binary decision variable x_{ij}, which a binary variable that takes the value 1 if batch i precedes batch j in the schedule, and 0 otherwise. It is important to note that both i and j can take the value 0, which corresponds to the dummy batch used to represent factory separation within the sequence. Based on the defined notation, the MILP model for our problem can be formulated as follows:

$$minimize \sum_{i=1}^{n} C_{im} \tag{5}$$

Subject to:

$$\sum_{i=0}^{n} x_{ij} = 1 \; \forall j > 0 \tag{6}$$

$$\sum_{j=0}^{n} x_{ij} = 1 \; \forall i > 0 \tag{7}$$

$$\sum_{i=1}^{n} x_{i0} = F \tag{8}$$

$$\sum_{j=1}^{n} x_{0j} = F \tag{9}$$

$$x_{ij} + x_{ji} \leq 1 \; \forall i,j > 0, i \neq j \tag{10}$$

$$C_{jk} \geq C_{ik} + P_{jk} + Big \, . \left(x_{ij} - 1\right) \; \forall i,j \; \forall k \tag{11}$$

$$C_{jk} = C_{jk-1} + P_{jk} \; \forall j \, \forall k \tag{12}$$

We note that $C_{jk} \geq 0 \; \forall j \forall k$, $C_{0k} = C_{j0} = 0 \; \forall j \forall k$. Equation (5) defines the objective function to be minimized. Equations (6) and (7) respectively ensure that each real batch has exactly one predecessor and one successor, thereby maintaining a clear and correct execution sequence. According to Equations (8) and (9), the dummy batch must appear exactly F times as a predecessor and F times as a successor. This establishes the dummy batch as a separator between factories, meaning it appears at both the beginning and end of each factory's schedule. Equation (10) guarantees that a batch cannot be both the predecessor and the successor of another batch simultaneously. It is important to note that this constraint does not apply to the dummy batch. Equation (11) specifies that the completion time of a batch on a given machine must be greater than or equal to the sum of its processing time and the completion time of its predecessor. The inclusion of a large constant (Big) allows this constraint to be enforced only when a batch actually follows another and effectively disables it when the two batches are not sequenced in that order. And Equation (12) ensures that the start time of a given batch on a given machine is exactly equal to its completion time on the previous machine. This constraint enforces the no-wait condition between consecutive operations

SOLUTION METHODOLOGY: METAHEURISTIC ALGORITHMS

To tackle the complexity of scheduling and resource allocation in distributed desalination systems, this section introduces three metaheuristic algorithms: Iterated Greedy (IG), Iterative Local Search (ILS), and Variable Neighborhood Search (VNS). These algorithms are specifically designed to optimize the Total Flow Time (TFT) while adhering to the constraints of the DNWFSP. The following subsections provide a detailed overview of the applied methods, with a focus on their principles and mechanisms for solving the optimization problem. The choice of metaheuristics is motivated by the computational intractability of the problem: exact optimization methods become impractical for large-scale instances due to the combinatorial explosion of possible batch sequences and factory assignments.

Algorithm Initialization Strategies

The performance of metaheuristic algorithms heavily depends on the quality of their initial solutions (Aqil, 2024, Sadki & Allali, 2025). A strong initial solution can significantly reduce the number of iterations required for convergence and improve the likelihood of finding high-quality schedules. In optimization problems as complex as the DNWFSP, a poorly chosen starting solution can lead to slow convergence or suboptimal results.

To address these challenges, this work employs two complementary initialization methods: the Nawaz-Enscore-Ham (NEH) constructive heuristic and the Greedy randomized adaptive search procedure (GRASP) constructive solution. These methods are designed to generate diverse, high-quality starting solutions that guide the metaheuristic search effectively. These methods provide diverse starting solutions, leading to a total of six metaheuristic variants: ILS1 (Iterative Local Search initialized with NEH), ILS2 (Iterative Local Search initialized with GRASP), IG1 (Iterated Greedy initialized with NEH), IG2 (Iterated Greedy initialized with GRASP), VNS1 (Variable Neighborhood Search initialized with NEH), and VNS2 (Variable Neighborhood Search initialized with GRASP).

Nawaz-Enscore-Ham (NEH) Initialization

The Nawaz-Enscore-Ham (NEH) heuristic (Nawaz et al., 1983) is one of the most widely used constructive algorithms in flow shop scheduling, renowned for its strong performance in minimizing scheduling objectives. Originally proposed for the permutation flow shop problem with ***n*** jobs and ***m*** machines, NEH has been extensively studied, improved and adapted for various scheduling environments (Sayah et al., 2024a), Numerous studies have shown that NEH consistently outperforms other constructive heuristics in terms of solution quality, making it a preferred choice for tackling complex scheduling problems (Nejjarou et al., 2022; Sadki & Allali, 2024, Sayah et al., 2024b), demonstrating its effectiveness in generating high-quality solutions. Given its effectiveness, NEH is often integrated into metaheuristic frameworks as an initialization strategy to enhance solution quality and convergence speed (Aqil, 2024). In our context, the NEH heuristic begins by sorting all jobs in decreasing order of their total processing times. The first ***F*** jobs are initially assigned to the ***F*** factories. The remaining jobs are then iteratively inserted into every possible position across all factories. For each insertion, the algorithm evaluates its impact on the scheduling objective (Total Flow Time) and selects the factory-position combination that results in the lowest objective value. Algorithm 1 provides the step-by-step procedure of the NEH heuristic.

Algorithm 1: NEH heuristic pseudo-code

Input: Set of jobs B= $\{B_1, B_2, \ldots, B_n\}$, Set of factories, Processing time matrix $P[j][m]$

1- Compute the total processing time for each job: $TPT(j) = \sum_{k=1}^{m} P[j][m]$

2- Sort jobs in descending order of $TPT(j)$ given $(\sigma^1, \sigma^2, \ldots, \sigma^n)$

3- Assigning the first *F* jobs to the *F* factories (one job per factory)
4- ***For*** each remaining job σ^k (from sorted list) ***do***
5- **For** each factory Π_f:
6- Try inserting σ^k into every position in Π_f's current schedule
7- Compute Total Flow Time (TFT) after insertion
8- Select the factory-position combination that minimizes TFT
9- Insert σ^k into the best factory-position combination
Output: the optimized job sequence for each factory

Greedy Randomized Adaptive Search Procedure (GRASP) Heuristic

The Greedy Randomized Adaptive Search Procedure (GRASP) (Silva & Sakallah, 1996) is a powerful and adaptable heuristic widely recognized for its effectiveness in solving combinatorial optimization problems. It constructs solutions iteratively, making informed selections at each step. Instead of evaluating all possible choices exhaustively, GRASP focuses on a Restricted Candidate List (RCL), which contains the most promising candidates identified by a dynamic, adaptive evaluation function. GRASP has proven to be highly effective for complex scheduling problems (Prabhaharan, et al., 2006; Bierwirth & Kuhpfahl, 2017), where many constraints and variables make exact methods difficult to use. Its strength comes from being both simple and flexible, allowing it to be applied across various fields, such as logistics, resource allocation, and network design. In our research, we use GRASP as a key step to generate initial solutions for our proposed algorithms. For this study, the RCL is defined as a subset of potential solutions, selected based on an evaluation criterion within the range $\left[TFT_{min}, TFT_{min} + \frac{1}{2}(TFT_{max} - TFT_{min})\right]$, where TFT_{min} and TFT_{max} are the minimum and maximum evaluation scores. he GRASP procedure starts by inserting the first *F* batches (one per factory). All remaining batches are tentatively inserted into all possible positions; after that, a restricted candidate list (RCL) is constructed, and one element from the RCL is selected. This process is repeated until all batches are scheduled. Algorithm 2 presents the step-by-step procedure of GRASP

Algorithm 2: GRASP heuristic pseudo-code

Input: Initial job list $B^0 = (B^1_0, \ldots, B^0_n)$ obtained by priority rule, Set of factories.
1- Initialize empty sequences Θ_f *and* Θ^{gr}_f for each factory Π_f
2- Assigning the first *F* jobs to the *F* factories (one job per factory)
3- $R \leftarrow$ Remaining unassigned jobs
4- ***While*** $R \neq \emptyset$ ***do***

$T \leftarrow \varnothing$ // Candidate solution pool
For each factory Π_f:
For each remaining job $B_k^0 \in R$ do
$\Delta_f \leftarrow$ Insert B_k^0 in all positions of Θ_f // Generate candidates
$T \leftarrow T \cup \Delta_f$
Construct the RCL list
Select a sequence Θ^* from RCL with associated job $R(a)$ and factory Π_f
Remove selected job $R(a)$ from unassigned jobs set R
$\Theta_f^{gr} \leftarrow \Theta^*$ //Update schedule for factory Π_f
Return $\Theta^{gr} = [\Theta_1^{gr}, \ldots, \Theta_F^{gr}]$

Iterative Local Search (ILS)

The Iterative Local Search (ILS) algorithm is a powerful and widely adopted metaheuristic for combinatorial optimization, particularly in scheduling problems. It is valued for its simplicity, ease of implementation, and strong performance in handling complex scheduling environments. ILS operates by iteratively applying a local search procedure to progressively improve a candidate solution. Each iteration starts with a solution, makes a small change, and then improves it using local search. Over the years, ILS has demonstrated remarkable efficiency in solving various flow shop scheduling problems. It has been successfully applied to permutation flow shop scheduling problems (PFSP) (Dong et al., 2009), multi-objective PFSP (Xu et al., 2017), and hybrid flow shop (HFS) scheduling problems (Pan et al., 2017; Aqil & Allali., 2020). Its robustness and adaptability make it a suitable approach for addressing the Distributed No-Wait Flow Shop Problem (DNWFSP) in desalination systems. In our implementation, we propose the use of a swap-position neighborhood to enhance the local search phase. Specifically, we randomly select two factories and then choose one job from each factory to swap their positions. This methodology generates a total of $\frac{n(n-1)}{2}$ possible solutions for n jobs, ensuring a broad exploration of the solution space. Additionally, to further improve the algorithm's ability to escape local optima, we integrate a simulated annealing (SA) mechanism. During the algorithm's process, worse solutions are occasionally accepted with a certain probability. This allows the algorithm to explore less promising regions of the solution space early in the search, helping it escape local optima and potentially discover better solutions. The pseudo-code of the proposed ILS algorithm is illustrated in Algorithm 3.

Algorithm 3: ILS metaheuristic pseudo-code

Input: Initial temperature T_0, Cooling rate α

```
Generate initial solution S using NEH or GRASP
T←T_0 // Initialize temperature
S_best←S // Initialize the best solution
While Terminate condition not satisfied do
S′ ← N_1(S)
Δ = TFT(S′) − TFT(S)
If Δ<0 then:
  S ← S′
  If TFT(S′) ≤ TFT(S_best) then:
        S_best ← S′
  Else if exp(-Δ/T)>random(0,1) then:
    S ← S′
  T←α*T // Reduce temperature (Cooling schedule)
Return S_best
```

Iterated Greedy (IG) Algorithm

The Iterated Greedy (IG) algorithm is a powerful metaheuristic designed to tackle complex optimization problems, through an iterative process that balances simplicity and efficiency. IG operates in two primary steps, the first involves a destruction phase, where parts of an existing solution are removed, followed by a construction phase, where the solution is rebuilt using a greedy heuristic. The second step focuses on a local search technique, exploring the neighborhood of the solution. IG has been widely applied to scheduling problems, demonstrating strong performance in solving various flow shop scheduling variants, including PFSP (Ruiz & Stützle, 2007), multi-objective PFSP (Ding et al., 2016), HFS (Aqil & Allali., 2020) and distributed flow shop problem (DPFSP) (Jing et al., 2021; Sayah et al., 2025b). In our implementation, we utilize the same neighborhood structure presented and tested for the NEH with local search in (Sayah et al., 2024b). Specifically, we employ a random neighborhood structure generated in each iteration based on three different strategies: First, two random batches from different positions within the solution are swapped directly; second, two randomly selected batches are removed from their original positions and reintegrated at new positions within their respective sublists; and third, two randomly selected batches are removed and reintegrated at the beginning of their respective sublists. Additionally, we enhance the IG framework by incorporating a simulated annealing (SA) acceptance criterion during the local search phase. Algorithm 4 presents the pseudo-code for the IG metaheuristic.

Algorithm 4: IG metaheuristic pseudo-code

Input: Initial temperature T_0, Cooling rate α, Destruction size d
Generate initial solution S using **NEH** or **GRASP**
$T \leftarrow T_0$ // Initialize temperature
$S_{best} \leftarrow S$ // Initialize the best solution
While Terminate condition not satisfied ***do***
$S' \leftarrow$ Destruction(S, d) // Remove d jobs from the solution S
S ← Reconstruction(S') // Reinsert jobs using greedy heuristic
$S' \leftarrow Local\_search(S)$
$\Delta = \text{TFT}(S') - \text{TFT}(S)$
If $\Delta < 0$ **then:**
$S \leftarrow S'$
If $\text{TFT}(S') \leq \text{TFT}(S_{best})$ **then:**
$S_{best} \leftarrow S'$
Else if $\exp(-\Delta/T) > \text{random}(0,1)$ **then:**
$S \leftarrow S'$
$T \leftarrow \alpha * T$ // Reduce temperature (Cooling schedule)
Return S_{best}

Variable Neighborhood Search (VNS)

Variable Neighborhood Search (VNS) (Hansen & Mladenović, 2001) is a metaheuristic technique widely used for addressing optimization problems. It differs from the other local search techniques by systematically changing the neighborhood structures during the search process, allowing the algorithm to explore diverse regions of the solution space. VNS operates based on the principle that different neighborhood structures can provide diverse local optima, and switching between them can lead to improved solutions. This metaheuristic is applied in scheduling problems such as PFSP (Bouzidi, 2020), HFS (Li et al., 2014; Aqil & Allali, 2021), DPFSP (Shao et al., 2018) and distributed hybrid flow shop problem (Cai et al., 2022). In our problem, we suggest different neighborhood structures: First, we use the same neighborhood (N_1) used in ILS, where the position of two randomly chosen jobs is swapped. The second structure (N_2) swaps the positions of two jobs within the sequence for each factory. The third structure (N_3) is based on a left shift, where two jobs are selected in a given factory, and a set of sequences is generated by shifting each job in the block between the two selected jobs to the left. We then select a sequence from the set. For example, for the factory sequence $[B_4, B_6, B_3, B_5, B_2, B_1]$, and select the positions 2 and 5, the use of (N_3) generates the neighborhood presented in Figure 3. The fourth structure (N_4) is similar to the third, but it applies a shift to the right instead. For the same previous initial factory sequence, the use of (N_4)

generates the sequences presented in Figure 4. Additionally, the pseudo-code for the VNS metaheuristic is provided in Algorithm 5.

Algorithm 5: VNS metaheuristic pseudo-code

Input: Initial temperature T_0, Cooling rate α, Set of neighborhoods N_1, N_2, N_3, N_4

Generate initial solution S using **NEH** or **GRASP**
T←T_0 // Initialize temperature
S_{best}←S // Initialize the best solution
While Terminate condition not satisfied **do**
k←1
While k≤k_{max}
$S' \leftarrow N_k(S)$
If TFT(S') < TFT(S) **then:**
$S \leftarrow S'$
k←1
If $TFT(S') \leq TFT(S_{best})$ **then:**
$S_{best} \leftarrow S'$
Else:
k←k+1
$\Delta = TFT(S') - TFT(S)$
If exp(-Δ/T)>random(0,1) **then:**
$S \leftarrow S'$
T←α*T // Reduce temperature (Cooling schedule)
Return S_{best}

Figure 3. Illustration of the use of (N_3) structure

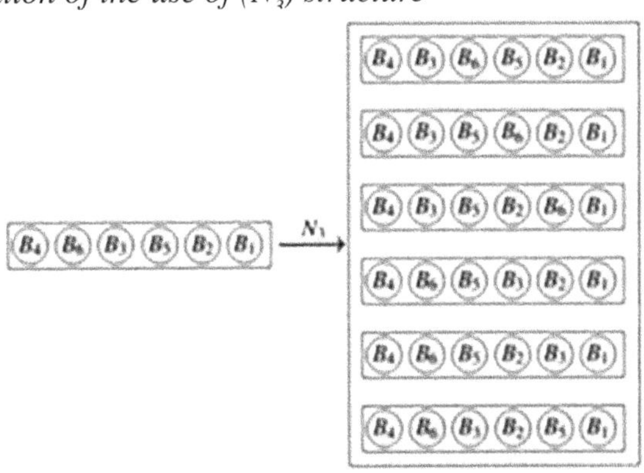

Figure 4. Illustration of the use of (N_4) structure

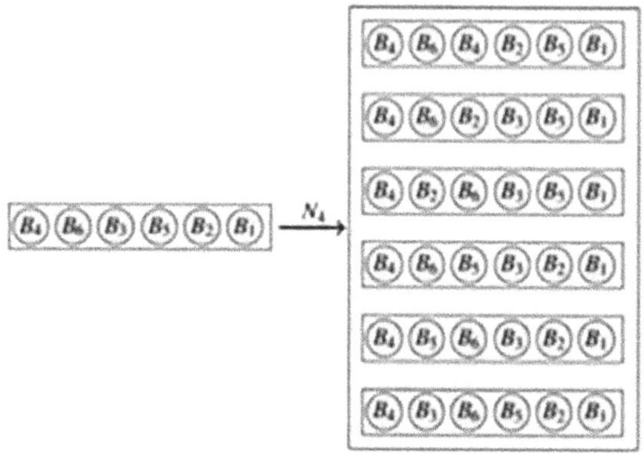

EXPERIMENTAL SETUP

This section presents the results of the experimental performance evaluation of the proposed metaheuristic algorithms for addressing the DNWFSP within desalination systems, with a focus on minimizing the Total Flow Time (TFT). To rigorously assess their effectiveness, the algorithms were tested on a diverse set of problem instances, characterized by variations in instance size, including the number of batches (n), the number of machines (m), and the number of desalination units (F), as well as variations in processing times. The instances used in this study are drawn from those proposed by Naderi and Ruiz (2010), providing a benchmark for evaluating the performance of the algorithms.

Our evaluation is centered on the Average Relative Percentage Deviation (ARPD) metric, which offers a reliable measure of performance when an optimal solution is not available. ARPD measures the relative difference between the solution obtained by a metaheuristic and the best-known solution, providing insight into both accuracy and consistency of the algorithms. The ARPD is calculated over K repeated tests using the formula presented by the Equation (13).

$$ARPD_{Algorithm} = \frac{1}{K}\sum_{l=1}^{K}\left(\frac{TFT^{l}_{Algorithm} - TFT_{best}}{TFT_{best}}\right) \times 100 \qquad (13)$$

In this context, $TFT^{l}_{Algorithm}$ denotes the Total Flow Time achieved by the proposed algorithm during the l-th test, while TFT_{best} refers to the best-known Total Flow Time for the same test instance, a lower value of this metric indicates higher algorithmic performance. To ensure the robustness and reliability of the evaluation, the performance metric is averaged over 5 independent test runs. Moreover, in each test, a time limit is established as a stopping criterion for the algorithms. This time limit is calculated based on the size of the instances. Specifically, it is determined as $\delta \times n \times m$ milliseconds, where δ is a constant factor, n represents the number of batches, and m denotes the number of machines, in our test we set $\delta \in \{0.15;0.30\}$. In all tests, the initial temperature for the Simulated Annealing (SA) technique was set as calculated by the Equation (14).

$$T_0 = \frac{1}{10nm} \sum_{i=1}^{n} \sum_{k=1}^{m} \mathcal{P}_{i,k} \tag{14}$$

Additionally, the cooling rate was fixed at α=0.8.

Table 1. The ARPD value of various problems and methods with δ=0.15

Problem	ILS1	ILS2	IG1	IG2	VNS1	VNS2	Problem	ILS1	ILS2	IG1	IG2	VNS1	VNS2
Ta001_2	**0.00**	1.62	1.44	1.71	0.61	2.04	Ta001_3	**0.82**	2.83	3.32	1.03	1.03	1.32
Ta002_2	1.64	2.00	1.30	1.74	**0.69**	0.70	Ta002_3	1.27	2.03	3.47	1.96	**0.93**	1.67
Ta003_2	**1.29**	4.34	1.64	1.38	1.31	1.78	Ta003_3	**0.90**	3.23	2.03	3.07	1.58	1.38
Ta011_2	**0.52**	2.07	1.33	1.26	0.66	1.72	Ta011_3	0.98	2.29	1.31	1.21	**0.61**	2.40
Ta012_2	1.53	3.24	1.92	1.32	**0.75**	1.91	Ta012_3	**0.52**	0.54	1.77	2.91	0.57	1.26
Ta013_2	0.63	2.91	1.25	1.67	**0.22**	1.16	Ta013_3	2.50	2.04	2.70	1.02	**0.50**	2.45
Ta021_2	**0.66**	1.73	2.98	1.59	1.32	2.26	Ta021_3	0.98	1.32	1.83	1.58	**0.75**	1.05
Ta022_2	**0.91**	2.64	3.32	1.61	0.95	1.86	Ta022_3	**1.06**	2.25	5.31	1.64	1.41	1.69
Ta023_2	1.35	1.72	1.28	1.66	1.93	**0.84**	Ta023_3	0.47	**0.34**	2.42	1.93	1.49	0.70
Ta031_2	**1.77**	4.83	1.93	5.97	4.91	4.75	Ta031_3	**0.89**	3.28	2.20	4.80	2.23	3.69
Ta032_2	**0.93**	3.46	1.59	3.76	1.57	4.93	Ta032_3	**0.47**	1.85	2.25	3.37	1.98	6.06
Ta033_2	**2.50**	3.47	2.47	4.37	4.91	4.39	Ta033_3	**0.29**	1.69	2.06	3.34	1.28	6.03
Ta041_2	**1.55**	4.94	2.71	3.11	3.27	4.20	Ta041_3	**1.75**	2.38	1.60	2.69	3.04	6.02
Ta042_2	2.84	**0.88**	1.10	2.23	4.30	4.89	Ta042_3	1.91	**1.05**	1.12	2.91	3.07	4.67
Ta043_2	**0.73**	3.06	3.99	4.18	2.22	3.97	Ta043_3	**0.78**	2.12	1.41	4.87	2.07	4.81
Ta051_2	**1.07**	4.64	1.61	2.40	1.85	3.52	Ta051_3	**0.24**	2.16	1.22	2.69	1.31	4.67
Ta052_2	**1.21**	2.37	2.20	3.30	2.09	3.99	Ta052_3	**0.51**	1.77	2.88	2.52	1.78	3.80
Ta053_2	**0.81**	3.21	2.34	3.22	2.26	6.24	Ta053_3	**0.33**	3.52	1.97	2.53	2.59	3.47
Ta061_2	**0.49**	0.97	2.50	3.58	1.98	5.97	Ta061_3	**1.34**	2.21	3.03	3.78	4.17	4.89
Ta062_2	**0.45**	3.75	2.38	4.89	2.73	5.91	Ta062_3	**0.90**	3.75	2.45	4.71	3.28	4.34
Ta063_2	**1.53**	7.96	2.52	4.03	3.45	6.82	Ta063_3	1.27	**0.93**	2.45	3.62	2.57	5.71
Ta071_2	**0.56**	6.66	2.07	4.25	1.41	5.67	Ta071_3	**0.49**	4.80	1.24	4.40	1.68	5.31
Ta072_2	**0.37**	7.06	1.62	4.21	1.30	5.31	Ta072_3	**0.50**	4.29	2.04	3.22	1.46	4.47

continued on following page

Table 1. Continued

Problem	ILS1	ILS2	IG1	IG2	VNS1	VNS2	Problem	ILS1	ILS2	IG1	IG2	VNS1	VNS2
Ta073_2	**0.25**	3.13	2.60	4.91	1.68	4.76	Ta073_3	**0.31**	4.05	1.49	3.73	2.20	4.86
Ta001_4	0.86	2.05	2.12	2.20	**0.69**	1.11	Ta001_5	**0.29**	1.27	1.45	1.34	0.53	1.34
Ta002_4	**0.54**	1.06	2.11	1.91	0.90	0.62	Ta002_5	1.35	**0.82**	1.04	1.03	1.35	1.52
Ta003_4	1.07	1.56	1.32	1.52	**0.98**	1.33	Ta003_5	**0.50**	1.79	0.58	2.24	0.72	2.01
Ta011_4	**0.16**	0.87	1.53	1.42	0.23	1.69	Ta011_5	**0.88**	1.10	1.36	2.55	0.92	1.20
Ta012_4	**0.38**	0.55	1.51	1.81	0.41	1.05	Ta012_5	**0.48**	1.39	1.77	2.59	0.99	0.99
Ta013_4	0.83	1.63	1.58	1.94	**0.69**	1.03	Ta013_5	**0.64**	0.78	2.83	1.93	0.66	1.05
Ta021_4	**0.76**	1.22	1.64	1.57	0.79	1.28	Ta021_5	0.48	**0.31**	1.86	1.27	0.52	0.91
Ta022_4	0.88	**0.76**	1.59	1.87	1.17	1.24	Ta022_5	**0.56**	1.11	1.14	1.36	0.62	1.13
Ta023_4	**0.18**	0.45	1.37	1.75	0.17	0.68	Ta023_5	**0.07**	0.82	0.82	1.22	0.24	0.56
Ta031_4	1.61	**1.47**	1.56	3.78	2.48	6.27	Ta031_5	**0.72**	1.17	2.88	3.04	2.62	4.52
Ta032_4	**1.52**	2.39	2.94	3.82	3.44	5.55	Ta032_5	**1.59**	2.50	2.21	3.42	3.84	4.84
Ta033_4	**0.27**	2.40	1.04	3.53	2.16	6.83	Ta033_5	**0.97**	1.92	1.33	3.15	2.74	6.16
Ta041_4	**0.59**	2.46	2.08	2.46	2.01	5.71	Ta041_5	**0.34**	0.87	1.39	2.14	1.20	3.19
Ta042_4	**1.15**	2.19	2.59	2.38	2.98	4.07	Ta042_5	**1.22**	1.98	2.12	2.78	2.64	3.13
Ta043_4	**0.50**	3.26	1.13	3.30	2.39	5.98	Ta043_5	**0.37**	2.70	2.54	3.77	2.57	5.33
Ta051_4	**0.40**	1.36	1.46	2.24	1.73	5.94	Ta051_5	**0.66**	2.91	1.39	2.52	1.89	5.24
Ta052_4	**0.44**	2.01	2.47	2.95	2.18	5.35	Ta052_5	**0.40**	2.65	2.08	2.47	1.84	5.09
Ta053_4	**0.83**	1.33	1.39	2.18	2.70	3.49	Ta053_5	**0.68**	1.46	1.54	2.49	1.08	3.06
Ta061_4	**0.73**	1.82	1.95	3.03	3.13	4.45	Ta061_5	**1.38**	1.67	3.75	3.44	3.64	3.62
Ta062_4	**0.47**	2.42	2.07	3.02	3.34	4.38	Ta062_5	**0.34**	3.69	2.80	4.09	2.92	5.71
Ta063_4	**1.22**	1.92	2.16	3.34	3.43	4.90	Ta063_5	**0.57**	2.90	2.40	3.89	2.97	4.23
Ta071_4	**1.00**	4.77	1.64	3.24	2.45	4.85	Ta071_5	**0.46**	3.84	1.23	4.65	3.09	4.47
Ta072_4	**0.38**	2.26	0.92	3.64	1.85	4.86	Ta072_5	**0.62**	4.96	2.58	4.55	3.20	4.56
Ta073_4	**0.58**	2.70	1.46	3.27	1.73	3.66	Ta073_5	**0.84**	3.25	1.28	3.73	2.55	3.43

Table 1 shows the simulation results for different sets of instances. For a parameter of δ=0.15, a total of 96 test problems were simulated, each repeated 5 times. The results highlight the superior performance of the ILS1 algorithm, which achieves the best ARPD value in 76 instance sets with an impressive success rate of 79.17%, significantly outperforming the other algorithms. In contrast, the competing methods rarely surpassed the ILS1 algorithm. Specifically, the VNS1 algorithm achieves the best ARPD in 11 test instances, with a success rate of 11.46%, while the ILS2 algorithm achieves a success rate of 8.33% in 8 instances. VNS2 obtains the best ARPD in only one case. Furthermore, IG1 and IG2 fail to achieve the best ARPD in any instance, although IG1 provides significant values. IG2, however, performs particularly poorly compared to the other methods. For a deeper comparative study, additional simulations were conducted with δ=30, increasing the time limit. The results obtained are displayed in Table 2 below.

Table 2. The ARPD value of various problems and methods with δ=30

Problem	ILS1	ILS2	IG1	IG2	VNS1	VNS2	Problem	ILS1	ILS2	IG1	IG2	VNS1	VNS2
Ta001_2	**0.64**	3.39	1.30	1.25	0.65	1.14	Ta001_3	1.06	1.54	1.26	1.57	**0.92**	1.94
Ta002_2	1.98	2.37	1.38	1.76	**0.61**	1.02	Ta002_3	1.43	2.88	1.69	1.72	**1.24**	1.34
Ta003_2	2.49	4.23	1.53	1.17	**1.11**	1.45	Ta003_3	2.23	2.48	1.37	1.25	**1.20**	1.31
Ta011_2	1.20	3.92	1.98	1.47	**0.95**	2.95	Ta011_3	0.66	1.01	1.73	1.96	**0.52**	2.27
Ta012_2	2.23	2.82	1.87	1.85	**1.16**	2.20	Ta012_3	**0.57**	1.48	1.04	2.22	0.74	0.77
Ta013_2	0.31	2.15	1.04	1.15	**0.15**	1.93	Ta013_3	2.48	2.02	1.63	1.88	**0.44**	1.14
Ta021_2	**0.60**	1.04	1.73	1.89	1.50	1.40	Ta021_3	0.49	1.19	0.42	2.06	**0.33**	0.70
Ta022_2	2.78	1.99	1.76	2.44	2.52	**1.62**	Ta022_3	2.15	1.71	1.94	2.72	**1.43**	1.60
Ta023_2	1.68	1.55	1.43	1.87	1.39	**0.75**	Ta023_3	1.06	**0.73**	1.19	3.18	1.04	0.85
Ta031_2	**0.98**	1.23	2.74	4.07	1.35	4.31	Ta031_3	**0.93**	3.58	2.76	4.38	1.33	5.51
Ta032_2	1.57	2.55	1.31	3.18	**1.26**	5.65	Ta032_3	**1.52**	1.57	2.19	3.72	1.92	3.93
Ta033_2	**1.24**	2.03	1.97	4.54	2.38	3.92	Ta033_3	**0.92**	3.45	1.23	4.44	2.24	5.04
Ta041_2	**1.91**	3.39	2.58	3.69	2.50	5.77	Ta041_3	**1.14**	1.56	1.59	2.87	2.82	5.41
Ta042_2	4.97	**1.55**	1.69	3.55	5.20	3.49	Ta042_3	2.41	**1.39**	1.99	2.93	4.62	6.28
Ta043_2	**0.93**	3.09	1.87	4.02	2.40	2.10	Ta043_3	**0.47**	2.82	1.02	3.05	1.95	7.09
Ta051_2	0.83	1.89	1.08	3.63	**0.60**	3.92	Ta051_3	**0.31**	1.74	1.36	2.82	1.07	3.79
Ta052_2	**1.20**	1.27	2.67	3.43	2.33	2.98	Ta052_3	**0.84**	1.64	1.00	2.51	2.46	4.36
Ta053_2	**1.05**	3.91	2.02	3.36	3.96	2.12	Ta053_3	**2.21**	3.20	1.29	3.10	2.45	2.05
Ta061_2	**1.26**	0.67	1.44	3.72	3.60	2.48	Ta061_3	**0.94**	1.25	1.11	3.22	4.48	4.56
Ta062_2	**1.16**	4.25	1.38	4.69	4.63	5.87	Ta062_3	**0.31**	2.41	1.42	4.04	3.76	9.69
Ta063_2	**0.41**	2.36	1.06	4.51	3.46	3.65	Ta063_3	1.85	**1.61**	1.78	4.40	3.63	3.10
Ta071_2	**0.20**	4.05	1.49	4.46	1.59	4.67	Ta071_3	**0.79**	5.45	1.13	4.02	3.19	3.51
Ta072_2	**1.31**	4.70	1.59	4.56	3.29	3.79	Ta072_3	**0.35**	2.95	2.79	4.32	1.57	3.29
Ta073_2	**1.70**	3.53	2.56	4.13	4.42	4.17	Ta073_3	**1.07**	4.21	1.52	4.55	2.38	3.35
Ta001_4	0.63	1.03	2.26	1.77	**0.60**	1.31	Ta001_5	**0.95**	1.20	1.91	1.75	1.26	1.45
Ta002_4	**0.68**	1.76	1.83	1.32	0.77	0.99	Ta002_5	0.98	**0.82**	1.26	2.05	0.83	0.81
Ta003_4	1.40	2.63	2.24	2.89	**0.72**	1.89	Ta003_5	**0.27**	1.75	1.23	1.56	0.57	1.59
Ta011_4	**0.25**	1.48	1.86	1.40	0.31	0.49	Ta011_5	**0.46**	0.94	1.07	2.24	0.80	0.96
Ta012_4	0.44	1.03	1.53	2.87	**0.30**	0.37	Ta012_5	**0.44**	0.52	2.79	1.51	0.95	0.51
Ta013_4	1.10	1.15	1.55	2.20	**0.91**	0.73	Ta013_5	0.44	1.20	2.82	2.96	**0.42**	1.50
Ta021_4	**0.56**	1.35	2.03	1.48	0.70	0.49	Ta021_5	**0.49**	0.91	1.43	1.79	0.61	1.30
Ta022_4	**1.20**	1.26	1.04	2.45	1.80	1.22	Ta022_5	**0.35**	0.65	1.46	1.25	0.78	0.97
Ta023_4	0.46	0.80	1.35	2.48	**0.28**	0.91	Ta023_5	**0.15**	0.28	0.87	1.23	0.17	0.57
Ta031_4	**0.56**	1.83	0.57	3.22	1.32	1.98	Ta031_5	**1.48**	2.61	1.09	3.60	3.29	5.34

continued on following page

Table 2. Continued

Problem	ILS1	ILS2	IG1	IG2	VNS1	VNS2	Problem	ILS1	ILS2	IG1	IG2	VNS1	VNS2
Ta032_4	**0.98**	2.91	1.36	3.66	2.36	3.25	Ta032_5	**1.09**	3.07	1.40	3.07	3.31	5.34
Ta033_4	**0.73**	2.24	1.30	3.99	1.61	4.04	Ta033_5	**0.77**	1.33	1.45	3.74	3.69	5.28
Ta041_4	**0.31**	1.30	0.43	2.07	1.39	4.01	Ta041_5	**1.08**	1.99	1.45	2.26	2.68	3.96
Ta042_4	1.85	**1.32**	2.20	2.27	3.00	3.68	Ta042_5	1.09	**0.82**	1.00	2.29	3.19	3.09
Ta043_4	**0.53**	2.72	0.99	3.19	2.24	4.13	Ta043_5	**0.80**	2.17	1.64	3.16	2.71	3.34
Ta051_4	**0.48**	2.32	1.15	2.57	1.82	3.78	Ta051_5	**0.69**	0.92	1.11	2.97	1.88	3.81
Ta052_4	1.09	**0.56**	0.74	2.85	2.04	4.24	Ta052_5	**0.47**	2.14	1.74	2.37	1.84	2.88
Ta053_4	**0.54**	1.07	1.25	2.77	2.54	2.46	Ta053_5	**1.13**	1.99	1.16	2.34	1.94	3.53
Ta061_4	2.01	**0.93**	1.03	3.88	3.69	4.39	Ta061_5	2.30	**1.82**	2.28	3.10	4.67	2.10
Ta062_4	**0.40**	1.99	1.34	4.93	3.00	3.63	Ta062_5	**0.31**	1.65	1.32	4.64	2.26	4.07
Ta063_4	**0.62**	2.56	1.42	4.00	3.54	3.87	Ta063_5	**0.30**	2.38	1.96	4.44	3.53	3.94
Ta071_4	**0.96**	3.25	1.65	3.11	2.74	2.10	Ta071_5	**0.58**	2.20	1.39	3.03	2.95	3.40
Ta072_4	**1.37**	3.73	1.46	4.25	4.02	4.34	Ta072_5	**0.96**	3.16	1.72	3.90	3.39	3.98
Ta073_4	**0.82**	1.76	1.20	3.49	2.49	3.41	Ta073_5	**0.37**	0.78	2.02	3.10	2.26	2.37

Increasing the time limit provides additional computational resources, allowing the algorithms to explore the solution space more effectively. The results indicate that ILS1 remains the best performing algorithm, consistently achieving the best ARPD, Specifically, out of 96 test instances, ILS1 achieves the best ARPD value in 61 instance sets, maintaining a high success rate of 63.54%. The VNS1 and ILS2 algorithms show a slight improvement, achieving success rates of 19.79% and 11.46%, respectively. This suggests that these algorithms benefit from additional computational time, enabling them to refine their solutions further. However, IG1 and IG2 still fail to outperform the leading methods. In fact, these algorithms require more time to find feasible solutions and are unable to reach the best solution within the allocated time. VNS2, while achieving marginal improvements, remains less competitive. Moreover, it is noteworthy that the algorithms initialized with NEH consistently provide better results compared to those initialized with GRASP, further highlighting the effectiveness of the NEH initialization strategy.

CONCLUSION

In this study, metaheuristic algorithms were applied to the Distributed No-Wait Flow Shop Scheduling Problem (DNWFSSP) in the context of distributed desalination systems. The proposed algorithms are Iterated Greedy (IG), Iterative Local Search (ILS), and Variable Neighborhood Search (VNS), which were enhanced using two initialization techniques, the Nawaz-Enscore-Ham heuristic (NEH) and the Greedy

Randomized Adaptive Search Procedure (GRASP). The simulation results highlight the effectiveness of the proposed metaheuristic algorithms, each demonstrating varying performance levels in minimizing Total Flow Time (TFT) and optimizing resource allocation in distributed desalination systems. The study emphasizes the importance of initialization strategies in enhancing algorithm performance. In addition, further work may develop and benchmark other metaheuristics, design parallel implementations on multicore and GPU platforms to reduce runtime and extend the model to a hybrid framework with machines of different capacities, to reflect realistic configurations.

REFERENCES

Alaghebandha, M., Naderi, B., & Mohammadi, M. (2019). Economic lot sizing and scheduling in distributed permutation flow shops. *Journal of Optimization in Industrial Engineering*, *12*(1), 103–117.

Ali, A., Gajpal, Y., & Elmekkawy, T. Y. (2021). Distributed permutation flowshop scheduling problem with total completion time objective. *Opsearch*, *58*(2), 425–447.

Aqil, S. (2024). Effective Population-Based Meta-heuristics with NEH and GRASP Heuristics Minimizing Total Weighted flow Time in No-Wait Flow Shop Scheduling Problem Under Sequence-Dependent Setup Time Constraint. *Arabian Journal for Science and Engineering*, •••, 1–24.

Aqil, S., & Allali, K. (2020). Local search metaheuristic for solving hybrid flow shop problem in slabs and beams manufacturing. *Expert Systems with Applications*, *162*, 113716.

Aqil, S., & Allali, K. (2021). On VNS-GRASP and iterated greedy metaheuristics for solving hybrid flow shop scheduling problem with uniform parallel machines and sequence independent setup time. Heuristics for optimization and learning, 17-32.

Bierwirth, C., & Kuhpfahl, J. (2017). Extended GRASP for the job shop scheduling problem with total weighted tardiness objective. *European Journal of Operational Research*, *261*(3), 835–848.

Blanco-Marigorta, A. M., Lozano-Medina, A., & Marcos, J. D. (2017). The exergetic efficiency as a performance evaluation tool in reverse osmosis desalination plants in operation. *Desalination*, *413*, 19–28.

Bouzidi, D. (2020). A Variable Neighborhood Search Algorithm to Solve the Flow Shop-Scheduling Problem Through Blocking: Variable Neighborhood Search (VNS). In *Transportation, Logistics, and Supply Chain Management in Home Healthcare: Emerging Research and Opportunities* (pp. 63-73). IGI Global.

Cai, J., Lu, S., Cheng, J., Wang, L., Gao, Y., & Tan, T. (2022). Collaborative variable neighborhood search for multi-objective distributed scheduling in two-stage hybrid flow shop with sequence-dependent setup times. *Scientific Reports*, *12*(1), 15724. PMID: 36127381

Darre, N. C., & Toor, G. S. (2018). Desalination of water: A review. *Current Pollution Reports*, *4*, 104–111.

DeNicola, E., Aburizaiza, O. S., Siddique, A., Khwaja, H., & Carpenter, D. O. (2015). Climate change and water scarcity: The case of Saudi Arabia. *Annals of Global Health, 81*(3), 342–353. PMID: 26615069

Ding, J. Y., Song, S., & Wu, C. (2016). Carbon-efficient scheduling of flow shops by multi-objective optimization. *European Journal of Operational Research, 248*(3), 758–771.

Dong, X., Huang, H., & Chen, P. (2009). An iterated local search algorithm for the permutation flowshop problem with total flowtime criterion. *Computers & Operations Research, 36*(5), 1664–1669.

Hansen, P., & Mladenović, N. (2001). Variable neighborhood search: Principles and applications. *European Journal of Operational Research, 130*(3), 449–467.

Ibrahim, A. G., Rashad, A. M., & Dincer, I. (2017). Exergoeconomic analysis for cost optimization of a solar distillation system. *Solar Energy, 151*, 22–32.

Jing, X. L., Pan, Q. K., Gao, L., & Wang, L. (2021). An effective iterated greedy algorithm for a robust distributed permutation flowshop problem with carryover sequence-dependent setup time. *IEEE Transactions on Systems, Man, and Cybernetics. Systems, 52*(9), 5783–5794.

Khare, A., & Agrawal, S. (2021). Effective heuristics and metaheuristics to minimise total tardiness for the distributed permutation flowshop scheduling problem. *International Journal of Production Research, 59*(23), 7266–7282.

Li, J. Q., Pan, Q. K., & Wang, F. T. (2014). A hybrid variable neighborhood search for solving the hybrid flow shop scheduling problem. *Applied Soft Computing, 24*, 63–77.

Naderi, B., & Ruiz, R. (2010). The distributed permutation flowshop scheduling problem. *Computers & Operations Research, 37*(4), 754–768.

Nair, M., & Kumar, D. (2013). Water desalination and challenges: The Middle East perspective: a review. *Desalination and Water Treatment, 51*(10-12), 2030–2040.

Nawaz, M., Enscore, E. E.Jr, & Ham, I. (1983). A heuristic algorithm for the m-machine, n-job flow-shop sequencing problem. *Omega, 11*(1), 91–95.

Nejjarou, O., Aqil, S., & Lahby, M. (2022). Constructive heuristics and mathematical formulation for solving the permutation flow shop scheduling problem with setup time. In 2022 IEEE 3rd International Conference on Electronics, Control, Optimization and Computer Science (ICECOCS) (pp. 1-5). IEEE.

Nejjarou, O., Aqil, S., & Lahby, M. (2024). Hybrid meta-heuristic solving no-wait flow shop scheduling minimizing maximum tardiness. *Evolutionary Intelligence*, •••, 1–25.

Pan, Q. K., Ruiz, R., & Alfaro-Fernández, P. (2017). Iterated search methods for earliness and tardiness minimization in hybrid flowshops with due windows. *Computers & Operations Research*, *80*, 50–60.

Prabhaharan, G., Khan, B. S. H., & Rakesh, L. (2006). Implementation of grasp in flow shop scheduling. *International Journal of Advanced Manufacturing Technology*, *30*, 1126–1131.

Ribas, I., Companys, R., & Tort-Martorell, X. (2017). Efficient heuristics for the parallel blocking flow shop scheduling problem. *Expert Systems with Applications*, *74*, 41–54.

Ruiz, R., & Stützle, T. (2007). A simple and effective iterated greedy algorithm for the permutation flowshop scheduling problem. *European Journal of Operational Research*, *177*(3), 2033–2049.

Sadki, H., & Allali, K. (2024). Artificial neural network for solving flow shop optimization problem with sequence independent setup time. *Foundations of Computing and Decision Sciences, 49*.

Sadki, H., & Allali, K. (2025). Optimal Population-Based Metaheuristics with NEH and ANN Algorithms for Minimizing Total Tardiness in Flow Shop Scheduling Problem. In *Innovations in Optimization and Machine Learning* (pp. 99–124). IGI Global Scientific Publishing.

Sayah, A., Aqil, S., Aadil, B., & Lahby, M. (2024b). Two Hybrid Heuristic Algorithms for Optimizing Data Processing Schedules in Distributed Edge Computing. In *2024 11th International Conference on Wireless Networks and Mobile Communications (WINCOM)* (pp. 1-6). IEEE.

Sayah, A., Aqil, S., & Lahby, M. (2023). Minimizing Maximum Tardiness in a Distributed Flow Shop Manufacturing Problem under No-Waiting and Sequence Dependent Setup Time Constraints. In 2023 14th International Conference on Intelligent Systems: Theories and Applications (SITA) (pp. 1-6). IEEE.

Sayah, A., Aqil, S., & Lahby, M. (2024a). Green Algorithms for Energy-Efficient Distributed Flow-Shop Manufacturing Problem. *Procedia Computer Science*, *236*, 378–385.

Sayah, A., Aqil, S., & Lahby, M. (2025a). Improved bio-inspired algorithms for scheduling distributed no-waiting flow shop with setup times. *Evolutionary Intelligence*, *18*(4), 68.

Sayah, A., Aqil, S., & Lahby, M. (2025b). Hybrid Metaheuristics-Driven Distributed Task Scheduling for Latency-Sensitive Edge Data Processing. In *2025 12th IFIP International Conference on New Technologies, Mobility and Security (NTMS)* (pp. 129-135). IEEE.

Shao, W., Pi, D., & Shao, Z. (2018). Local search methods for a distributed assembly no-idle flow shop scheduling problem. *IEEE Systems Journal*, *13*(2), 1945–1956.

Silva, J. M., & Sakallah, K. A. (1996). GRASP-a new search algorithm for satisfiability. In *Proceedings of International Conference on Computer Aided Design* (pp. 220-227). IEEE.

Voutchkov, N., Kaiser, G., Stover, R., Lienhart, J., & Awerbuch, L. (2019). Sustainable Management of Desalination Plant Concentrate-Desalination Industry Position Paper–Energy and Environment Committee of The International Desalination Association (IDA). In The International Desalination Association World Congress on Desalination and Water Reuse (pp. 1-32). UAE REF: IDAWC19-Voutchkov.

Xu, J., Wu, C. C., Yin, Y., & Lin, W. C. (2017). An iterated local search for the multi-objective permutation flowshop scheduling problem with sequence-dependent setup times. *Applied Soft Computing*, *52*, 39–47.

Ying, K. C., Lin, S. W., Cheng, C. Y., & He, C. D. (2017). Iterated reference greedy algorithm for solving distributed no-idle permutation flowshop scheduling problems. *Computers & Industrial Engineering*, *110*, 413–423.

Chapter 9
Machine Learning Approaches to Climate-Smart Irrigation Planning

Usharani Bhimavarapu
https://orcid.org/0000-0002-0246-1420
Department of Computer Science and Engineering, Koneru Lakshmaiah Education Foundation, Vaddeswaram, India

ABSTRACT

Climate change poses existential risks to water management using a sustainable approach, particularly in agricultural-dependent areas such as Andhra Pradesh, India. In this context, using Machine Intelligence, which is Artificial Neural Networks (ANNs), this research investigates the application of Machine Intelligence in aiding climate adaptation for water management. The method involves the integration of multi-source data sets, including climate parameters (2020–2024), agricultural-based data on paddy farming, and geographic information for irrigation and topography. Preprocessing procedures maintained uniformity and compatibility among such datasets, after which Particle Swarm Optimization (PSO) was applied for feature selection to identify the most impactful factors influencing water supply and crop yield. The ANN model trained well and precisely forecasted irrigation demand and pinpointed possible periods of water stress, allowing localized real-time decision-making support.

INTRODUCTION

Climate change has brought with it extremely dynamic and new climate regimes, and these have had a significant impact on water quality and supply. Legacy water

DOI: 10.4018/979-8-3373-2700-6.ch009

infrastructure cannot handle such dynamic dynamics. Machine intelligence is the hero in the guise of real-time processing of hydrological and environmental big information. Such technology can pick fine patterns and early warnings that are most likely to be overlooked by human observation. For example, AI can identify early flood or drought indicators so the response may be initiated earlier. Machine learning programs can even alter predictions step by step. That is, water resource planners are given the updated information. Proactiveness and not reactiveness in responsiveness is what saves lives and keeps infrastructure safe. Machine intelligence therefore gives water systems more abilities. Its use in climate adaptation is no longer optional—it is a requirement.

Water usage rises globally across the world with urbanization, population, and industrialization. Climate change also raises it through variability in precipitating patterns and the exhaustion of freshwater resources. Machine learning can reduce usage through mediating competing needs. Smart irrigation systems, for example, utilize AI to water only when and where the plants need it, reducing waste. AI models can predict water demand scenarios in the future, and that guides long-term planning. The simulations also consider a number of other factors such as weather conditions, trends in consumption, and seasonality. Predictive analytics allow water managers to estimate reservoir releases and aquifer recharging. It prevents water shortages in the risky areas. In the long run, machine intelligence supports equitable and sustainable water resource management. Its vision extends beyond efficiency to ensuring water security for all.

Flooding is more severe and more common in most regions and has produced widespread devastation and displacements of people. Machine learning can potentially make major impacts on flood forecasting and risk avoidance. Artificial intelligence algorithms spot trends in satellite imagery, river levels, and weather to predict flood events. The models provide timely and precise warnings, which increase preparedness in communities. Top-level simulations also help to determine weak points and besieged infrastructure. This allows for flood-resistant infrastructure design and pre-evacuation. Systems of real-time monitoring also increase responsiveness by detecting anomalies in real time. Machine intelligence supports cross-sectoral flood risk analysis based on environmental, social, and economic drivers. Multi-dimensional awareness such as this supports more comprehensive adaptation measures. AI therefore has a significant role of turning the response to flood proactive instead of reactive.

Groundwater as the critical source for drinking and irrigation water is mostly over-exploited through over-drawal and pollution. It is impossible to monitor and control this 'hidden' resource. Machine intelligence avoids such. Machine learning transcends such obstacles by processing usage, hydrology, and geology data to determine the extent of groundwater. AI systems can cast visions of over-abstraction patterns and predict depletion areas in the future. Remote sensing data combined with machine learning enables the monitoring of groundwater resources at a deep level. AI can assist in finding recharging locations and optimal borewell sites too. These are made aware of policy-makers about sustainable extraction limits and recharging process. AI can also analyze effectiveness of conservation measures with the passage of time. The aim is to maintain groundwater as a stable long-term resource. Materializing the intangible, machine intelligence makes knowledge-based and more sustainable groundwater management possible.

Water quality is being threatened by agricultural, industrial, and municipal wastage contamination. Machine intelligence offers effective solutions for real-time water quality monitoring. Artificially intelligent sensor networks are able to identify real-time contaminants and trace contaminant source quickly. Machine learning algorithms are capable of modeling pollutant dispersion and predicting contamination events. This enables faster remediation and easier public health protection. AI can adjust wastewater treatment processes to help with better compliance with regulations. AI enables ecosystem health quantification through a chemical and biological observation bridge. AI offers evidence to allow for intervention in a timely manner for aquatic ecosystem conservation and human society. Machine intelligence allows adaptive management practices based on emergent water quality challenges. Its role thus goes beyond public health as well as environmental conservation. AI revolutionizes the regulation of water quality through appropriate monitoring.

There is a long-standing issue of inadequate data in most regions, with the shortage being more prevalent in low-resource settings. The majority of the traditional climate models demand high-density and good-quality data. Machine learning is able to handle low-density, noisy, or missing data by the use of techniques like imputation and transfer learning. AI can handle combining data from different sources, i.e., citizen science, remote sensing, and open-access data sets, in order to fill knowledge gaps. This is particularly useful for countries that do not have extensive hydrological networks or very few networks. Machine learning applications also improve and learn iteratively, refining their results based on each new update. By interpreting sparse data, AI reduces the need for expensive monitoring systems. This becomes an argument for everyone to have access to advanced water management software. Therefore, one of the primary purposes of machine intelligence is to fill the data gap. It makes it easier for vulnerable regions to come together for climate adaptation.

Water conflicts will increase as competition for water intensifies. Machine learning has the potential to introduce cooperative water management through increased transparency and decision-making. AI models are capable of simulating various options for water sharing among various parties. The models take into account environmental, social, and economic impacts, easing negotiations with evidence-based suggestions. Machine learning is able to develop win-win situations, thus reducing room for conflict. Common platforms through AI could ensure information is real-time to both parties and thus enable trust. Natural language processing abilities could also analyze stakeholder rhetoric and sentiment in policies. This aids in the prediction of possible flashpoints as well as participatory policy-making. AI eventually provides hard information to common citizens for participatory government. Co-operation and not competition is the objective here. Machine intelligence therefore assists in achieving peace in water-scarce regions.

Urban cities also experience unusual water challenges due to high consumption, aging pipes, and stormwater runoff requirements. Urban water infrastructure is increasingly being controlled by machine learning through smart city programs. AI can predict pipe failure, reduce pump inefficiency, and detect distribution system leaks. This reduces loss of water and operation costs. Machine learning enables dynamic pricing for usage, encouraging conservation. In stormwater management, AI predicts runoff volume and controls retention basins accordingly. Urban water models built in support the design of green infrastructure and sustainable drainage infrastructure. AI technologies enable cities to learn and adapt to emerging climate conditions while improving water services. The contribution of AI to urban water management is to ensure smart, efficient, and resilient cities. It ensures that urbanization does not occur at the cost of water sustainability.

Agriculture accounts for most of the freshwater on the planet, and climate change is seriously impacting its productivity. Machine learning is transforming agricultural water management based on precision agriculture methods. AI has the ability to use soil moisture level, weather, and crop variety to optimize the irrigation hours. Drones, remote sensing technologies, and artificial intelligence monitor the water need and health of crops. This helps to reduce water loss and yield the most. Farmers receive actionable insights from AI systems through mobile applications even in rural zones. AI systems allow farmers to make changes in agriculture practices in order to accommodate climate uncertainty like early monsoons or sudden heatwaves. This results in a better climate-resilient agricultural system. Machine intelligence is one of the applications which bring food in this industry. Machine intelligence maximizes agricultural water use and economic and environmental sustainability.

Machine intelligence will be the major driver of innovation in addressing climate change with watershed-based integrated water management in the future. Machine intelligence allows us to move from fixed rule-based systems to adaptive data-driven systems. This is necessary in a time when history cannot be used as a sure guide. Ongoing optimization of increasingly advanced AI models ensures adaptation plans are adaptive and efficient. By promoting inter-sectoral collaboration, AI makes water, energy, agriculture, and urban systems integratable in a consistent manner. Machine intelligence also helps build public consciousness with interactive visualization instruments and websites. These instruments encourage citizens to take sustainable water behavior. The last dream is to construct a smart, fair, and resilient water future. With rising climate uncertainties ahead, machine intelligence will be the cornerstone of world adaptation processes.

Maestre et al. (2012) provide useful information regarding the relationship between dryland multifunctionality in ecosystems and richness of plant species. The study is, however, constrained in its scope by its international focus, which has the potential to blur local heterogeneities and regional heterogeneities of the environment. Cross-sectional design constraints limit inferences about long-term shifts in ecosystems and temporal processes. The study also only deals with plant diversity without consistently incorporating other abiotic and biotic factors affecting multifunctionality in an integrative manner. Use of datasets given can be followed by data inconsistency or data omission in some drylands and therefore cause bias. Lastly, the dynamical nature of interaction in ecosystems is also oversimplified perhaps at the expense of predictability. Management cannot directly be applied to study results without site validation.

Tariq et al. (2022) cite Alhagi sparsifolia as a phreatophyte with potential in desertification and land degradation control. While the topic is evocative of high promise of concentration, an overdependence has been on experimentation and observation that have been done on small geographical areas that might not be representative of the performance of the species when grown under diverse dryland conditions. Its ecologic impact and long-term use in large planting is unknown, as well as the response of the species with native fauna and flora. The socio-economy and land-use planning that can affect implementation are not fully considered by the study. Climatic variations and outliers would also affect the efficacy of the plant, and this must be explored through longer-term field trials and modeling.

Evidence of dryland crop productivity drivers presented by Chimwamurombe and Mataranyika (2021) are affected by results constrained to data from selected areas only, thus making generalizability an issue. They investigate using latest agricultural productivity data that would hardly capture all the facets of dryland agriculture complexity such as indigenous knowledge systems or socio-economic constraints. Climate change projections are not written in stone, something that is required since drylands are delicate. Moreover, cross-sectional analysis minimizes the extent of inference of productivity patterns across time. Methodological constraints and possible in data errors through inconsistency of reporting also affect robustness.

Stringer et al. (2017) propose a dryland development paradigm from empirical system science analysis. The approach, however, has its weaknesses emanating from the empirical data used and available that can undermine geo-localized socio-ecological complexity. The utility of the paradigm in diverse dryland ecosystems is yet to be substantiated by policy use and implementation in practice. The study fails to undertake extensive analysis of political and economic challenges encountered in dryland development. Paradigm predictability accuracy can be affected by issues of oversimplification of complexity of systems in modeling. Besides, the model requires further sophisticated development when adaptive capacity is considered under extreme environment change.

Plaza and Zaccone (2018) provide dense information about dryland soil element and resource inventories but heterogeneity in space and time of dryland soil sampling is the origin of limitations. Beneficial variability at micro-scales facilitating land management decisions at local-scale may be missing in the dataset. Research focus on element stocks excludes biological soil properties with ubiquitous control over ecosystem processes. Methodological variations in soil measurement across areas and its heterogeneity can affect comparability. Cross-sectional study design can't provide dynamics of the soil over a period of continuous land cover and climate change. Field validation is also required for use of results in land rehabilitation.

Pravalie (2016) provides a global picture of drylands extent and associated environmental issues but the wide scope does not allow for effective scrutiny of local environmental factors and socioeconomic settings. Criteria for drylands could vary across datasets and therefore consistency is an issue. The review is predominantly a summary of existing knowledge with minimal new empirical data and such may limit new knowledge. Temporal trends and projections are controversial but conservative because different climate models have been used. Dryland problem-solving governance or policy frameworks are weakly discussed in the article, and few of the proposals are implementable.

Stringer and Mirzabaev (2021) also explain the impact of climate change on water security in drylands but are limited by the lack of high-resolution climate information in the majority of drylands. The article is unbalanced since it relies on model simulations, which by nature are uncertain especially under extremes. The same is also accompanied by poor socio-political drivers to water management integrated in the article. Biophysical implications receive greatest consideration, probably understating community adaptation and resilience capacity. Temporal resolution of impact analysis is too low to identify instant change or tipping points. Policy interventions themselves are also highly generic and must be localized for operational implementation.

Byers and Gidden (2018) analyze global exposure to climate hotspots and development but consider spatial resolution might veil local heterogeneity in exposure and adaptive capacity. Multi-sectoral approach responds to compounded variables, some of which are simulated and estimated with questionable dependability. Data limitations in the less affluent areas can bias vulnerability estimates. Temporal processes of vulnerability are poorly studied and will miss emerging threats. Relying on available data constrains socio-cultural learning.

Application in practice with regard to real-world interventions remains limited by mismatch of scale.

Huang et al. (2021) analyze global planetary water scarcity drivers in terms of water withdrawal and availability but are limited by globally accessible data, which may be too aggregated at the regional scale. Natural availability and human utilization of water are multi-way interdependencies, and simplification assumptions might miss vital feedbacks. The time perspective might be too short to distinguish high-intensity short-term events during scarcity. Heterogeneity and quality of the data across countries are the causes of uncertainties. Quantitative in scope versus qualitative for water resources, usability knowledge is limited by the study. Policy implications are vague and need to be refined on a context-specific scale.

Kummu and Guillaume (2016) trace the historical course of water scarcity and proscriptively define the options on sustainability but are limited by the unavailability of historic evidence and variability. Historic analysis is perhaps incapable of discerning future socio-economic drivers. Future trends are anticipated on the presumption of theoretical hypotheses that can be discredited by rapid environmental or technological change. The study relies mainly on quantity-based measures at the expense of degraded water quality and ecosystem integrity. Spatial variations are averaged out, potentially concealing the high local weaknesses. Institutional and governance facets of water management are also given lesser priority.

Rosa et al. (2020) give a critical assessment of global agricultural economic water scarcity but are limited in applying aggregated global data sets with the impact of masking regional and local variation in water supply and efficiency of use. Assumptions incorporated in the economic water scarcity model can potentially eliminate frustrating socio-economic dynamics and circumvent local policy or infrastructural restrictions. In addition, temporal data resolution limits knowledge of seasonality in water scarcity. The study does not further clearly define the impact of rising climatic variability that significantly alters the water availability at certain places.

Liu and Yang (2017) have a general picture of water scarcity estimation over the years but are faced with variability in definition and measurement of water scarcity among studies that prevent direct comparability. More emphasis is placed on biophysical drivers and less emphasis is placed on institutional and social drivers of water scarcity. Future projections rely on uncertainties within climate models and socio-economic scenarios. In addition, groundwater drawdown, a critical component of water scarcity for most areas, is not addressed in this study at all.

Stroosnijder et al. (2012) discuss dryland water use efficiency optimization measures but give prescriptions that are mostly nongeneric and would likely not be easily applied to diverse dryland settings. Technical decisions predominate with scant consideration of socio-economic barriers to farm adoption. The research also relies fairly heavily on available case histories and fails to capitalize on carefully designed experimental tests in varied climatic situations. Soil-crop-water responses to climate change are complicated processes that will require greater research than will be provided by present models.

Dalezios et al. (2018) possesses a complex methodology to address water scarcity but not theory context with hardly any indications of its application across multiple geo-political and geographical contexts. The interconnectedness of water management is downplayed, and that could lead to reduced efficiency under practical applications. The system may also have to be incorporated into participatory stakeholder processes and local governance structures, which are not adequately discussed. Requirements of data for planned strategies would be prohibitively uneconomical for data-poor regions.

Pastor et al. (2019) look at the world nexus of food, trade, and water to facilitate environmental flows by 2050 but restrict their research by uncertainty within integrated assessment models globally. Agronomic practice and trade policy hypotheses will be ineffective in all regions and will deliver possible contradictions to scenario projections. The research is also skewed toward emphasizing biophysical connections, suppressing socio-political and economic constraints to sustainable water and food management. Geographic resolution of data may overlook the local opportunities and capacities.

Rosa et al. (2020) consider the future of sustainable irrigation development in a 3°C warmer world but base their projections on climate models involving built-in uncertainties regarding future trends in precipitation and temperature. The prognosis is based on irrigation facilities and water supplies being unaffordable in most places, especially under socio-economic tensions. It also does not consider possible environmental impacts of irrigation extension, i.e., salinization or loss of natural diversity. Potential for adaptability among agricultural societies is not accorded adequate attention in the analysis.

Liu and Liu (2022) balance blue and green water availability when they consider global food system water scarcity under projected futures expected to occur. The study is, however, limited by uncertainty in climate projection and hydrological modeling, especially in poorly monitored regions. Spatial resolution may overlook important local water stress hotspots. Interactions of quantity and quality of water are not clearly mentioned, which have impacts on the utilisability of water resources. Socio-economic reasons of irrigation management and water demand are also minimised to the extreme.

Bwambale et al. (2022) refer to intelligent monitoring and control systems to achieve higher water use efficiency, but among the technologies they refer to, some are yet to be scaled up beyond pilot or experimental scales whose real world applicability is therefore restricted. The review is leaning towards a stronger focus on technology solutions than towards laying out an integrated plan to overcome barriers to adoption such as cost, farmers' knowledge, and infrastructure. Farm heterogeneity and farm type heterogeneity are not developed and unscalable. The environmental impact and the energy requirements of smart irrigation systems are still not well addressed in the review.

(Numerical simulation of) irrigation planning in terms of soil water status is used by Dabach et al., (2013) at the cost of soil homogeneity and of equalizing crop water requirements that are not necessarily applicable under heterogeneous field conditions. The models rule out variable weather patterns and events that may disrupt soil moisture dynamics. The study is not tested under different climatic and soil conditions, which reduces generalizability. In situ real-time soil moisture measurements needed for practical application can be unavailable or too costly.

Modarres and da Silva 2007) study dry and semi-dry parts of Iran's rainfall regimes, but are constrained by short term and limited data. Timescale cannot record recent climatic variability or forecasts of the future from global warming. Statistical methods employed may not be able to pick up non-linear trends or abrupt change in precipitation. The study also focuses on rainfall quantity and not rainfall intensity or frequency changes, which are particularly critical in water management and agriculture.

Zhang et al. (2021) introduce challenges and opportunities in center pivot precision irrigation decision-support systems and then state that these are still far from being optimally linked to real-time sources of information, overshadowing responsiveness as well as accuracy. Technical complexity and expense are impediments to the adoption of smallholder farmers. Interoperability between sensor platforms is an impediment to broad-scale adoption. Soil type and environmental microclimate heterogeneity complicate system calibration. The research is a documentation of possible benefits only and is not a demonstration of large-scale field testing under a range of agricultural conditions.

Huang et al. (2012) maximize agricultural sustainability in China's Tarim River Basin irrigation water resources, but with given assumptions about future water demand and supply that may not hold if future climatic variation or policy differs. Micro-scale water-stressed areas can be underestimated through the spatial resolution. Socio-economic factors driving water usage behavior are not on the radar in the optimization. Static nature of the model means that it cannot react to surprise drought or water allocation controversy.

Yu et al. (2021) provide a meta-analysis of the effects of drought on agricultural water use efficiency determinants, but methodological and data variation across studies may have an impact on conclusions' validity. Reporting bias of significant findings may jeopardize the meta-analysis. Heterogeneity across areas in drought's severity as well as management practices complicates generalizability. Extension is only to drylands and not other agro-ecological regions.

Singh and Praharaj (2019) authors write about precision irrigation management theory and practice but no genuine effort is made at any practical issues such as cost, farm education, and technology access. The review is thematic in theoretical underpinnings and developed country experiences and therefore of narrow applicability to resource-poor smallholder farmers. Precision irrigation's long-term impact on the health of the soil and on environmental sustainability has to be studied. Integration with other functional water management systems is also not well investigated.

Bitella et al (,2014) promote an open-hardware-based low-cost system for environmental and soil water content measurements but puts in-field application only for some specific soils and crops, and it loses generality. Robustness and reliability of the device in poor field conditions are still to be checked in the long term. Technical challenges of data communication and electricity supply in rural areas are problems whose solutions are not yet available. Technical know-how in skills and maintenance needs by the consumers are constraints in utilizing the platform.

Neupane provide a critical examination of precision water management agronomic practices and techniques. The authors' study, however, relies on theoretical models and academic literature only and does not allow field testing under different field conditions and agroecological regions. The review also refers to technology-based solutions but does not elaborate on socio-economic and infrastructural adoption constraints well, which are most critical in the majority of developing nations (Neupane et al., 2019).

Cambra proposed a clever bicarbonate management system in precision hydroponic agriculture through the use of future emergent technologies. However, the research behind it is limited in that controlled environments, such as hydroponics, are not necessarily easily scalable to soil or field agriculture. Scalability and system economics in real large commercial farm environments must be verified (Cambra and Sendra, 2018).

Abioye evaluates existing monitoring and control strategies for precision irrigation and identifies future trends in innovative technologies. The evaluation is, however, based on the sparse existing long-term field data to validate the functionality of most of the control algorithms demonstrated. In addition, integration in current systems in use today, especially in resource-constrained conditions, is not considered (Abioye et al., 2020).

The FAO water and land resources report is vital data for worldwide issues but is restricted on the potential for providing heterogeneity in the data and resolution in time. This assumption that it is measuring the world at a global level leaves room for the exclusion of local data and differences at the minute level in water scarcity and land degradation. In addition, future projections are highly dependent on assumptions regarding a sequence of events that may not adequately address all possible socioeconomic and climatic uncertainties (FAO, 2022).

Zamfir takes Mediterranean land degradation and mitigation policy under consideration, providing rich policy insight. The analysis is, however, limited to some degree in its broad scope, lack of rigorous empirical facts, and quantitative evaluations of the extent to which the policies affected. This limits however the ability to take fully into account the performance of different types of mitigation in different Mediterranean situations (Zamfir et al., 2020).

Burrell attribute dryland desertification to human-induced climate change based on good spatial analysis. In the use of climate models and remote sensing data in the study, it can however be prone to uncertainties in data resolution quality as well as classifier reliability. It would be unethical to debase desertification based on climate change, insensitive to other key drivers like land cover changes and socio-political drivers (Burrell et al., 2020).

Eskandari estimates desertification in Iran using climate and vegetation indices, a useful regional method. There are drawbacks to the exclusive use of remote sensing data without ground-based measurements to back them up, which could affect the accuracy of desertification estimates. Research isn't conducted over an extremely long period of time, possibly leaving out significant longer-term desertification trends (Eskandari, 2021).

Gebremeskel summarizes causes, effects, and resilience in East Africa as a result of drought and provides an elaborate overview. Heterogeneity of the effects of the drought on some of the social and ecological systems in the region is not elaborately reviewed, however. The review also reports signs of resilience but provides no critical assessment of implementation effectiveness and scalability (Gebremeskel and Tang, 2019).

Imbrenda consider urbanization impacts and land degradation and urban expansion in a peri-urban area, yet the particularity of geographic space covered in the work could limit extension of conclusions to other peri-urban areas. Additionally, overall analysis is predominantly spatial and temporal and does not consider general socio-economic data, thus limiting knowledge of drivers by people (Imbrenda et al., 2021).

Jin conduct a 10-year experiment of monitoring soil property and fertility in south-eastern China. While having rich longitudinal data, the study is constrained to work with a single site and crop and cannot extrapolate at a scale larger than that. Soil property variability may also be due to unmodeled climatic changes and management practices outside their time space (Jin et al., 2021).

Stavi address issues and opportunities for zero net land degradation, but their conceptual model is weak based on the lack of empirical data examination and specific case studies of implementing successful results. How reconciling opposing land use and stakeholder interest at different locations is addressed is not properly considered either (Stavi and Lal, 2015).

Ibrahim employed residual trend analysis to assess West African land degradation using NDVI, soil moisture, and rainfall. Their limitations are potential remoteness sensing data bias, as well as their inability to disentangle land degradation from vegetation cover change (Ibrahim et al., 2015). Furthermore, socio-political and land use causes of the degradation are not accounted for in the analysis.

Minea come up with a grazing susceptibility hilly land degradation index that is used in the risk assessment models. The use of some environmental factors as the basis of the index, however, leaves out the socio-economic factors of the grazing pressure. The functionality of the model within the study area, however, remains undiscovered if it were not tested in diversified environments (Minea et al., 2022).

Li assesses the effects of socioeconomic development on rural China's drinking water safety at the provincial level. Limitation is data aggregation may conceal intra-provincial inequality and not controlling unregulated informal water sources or uncontrolled contaminating risks. Cross-sectional analysis also limits causal inference (Li et al., 2019).

Prăvălie analyzes contemporary global dryland development from two large databases of aridity to provide informative spatial information. Differences can occur based on difference in methodologies and database and dryland definition. Climatic factors take the main analysis with subordinate attention to land use and anthropogenic processes resulting in dryland processes (Prăvălie et al., 2019).

Allen measures stakeholders' views towards the roll-out of land degradation neutrality with enabling environments being acknowledged. The study is confined by the use of a qualitative approach and potential bias in selection of stakeholders who might not represent all the affected parties. Progress is measured on over-reliance on information that has been narrated by oneself with a potential shortage of objective corroboration (Allen et al., 2020).

Grainger is also doubtful about dryland land degradation neutrality and his arguments form a negative perspective. The argument is hypothetical to a certain extent when empirical evidence is not available to support the arguments by a large database. The argument also tends to overestimate biophysical limitations to an extent lowering governance and socioeconomic determinants of LDN success (Grainger, 2015).

Pandit proposed a generic, cross-disciplinary structure of responses to land restoration and degradation for decision-making guidance. The structure has not been practiced and tested in a variety of socio-ecological systems, even though it is generic. Practical application is hindered by the complexity of integrating cross-disciplinary data and stakeholder opinions into the structure (Pandit and Parrotta, 2020).

The new strategy overcomes the limitations of existing strategies by making use of multi-dimensional data sources like remote sensing and comprehensive ground-truth measurements to enhance accuracy and contextality in a variety of agroecological and socio-economic contexts. Unlike earlier studies that were theory-driven or field observation weakly based, the present model leverages the latest machine learning algorithms for learning over heterogeneous climatic regimes, land use, and management practices to optimize scalability and generality. The model also integrates socio-economic drivers and stakeholders' inputs to deliver actionable applications, especially in resource-limited settings, with the advantage of bypassing usual limitations in accounting for infrastructure and policy limitations. The model is designed with local real-time observation and local control measure adjustment to enable timely decision-making, locally variable sensitivity, normally lost with large-scale global estimation. Strong testing as well as end-user feedback loops given in the form of repeat cycles, the described methodology improves implementability, filling the scientific innovation and on-ground agriculture or environmental management gap. This integrated strategy makes it possible to avert uncertainty in data resolution, land use heterogeneity, and socio-political variables that limited previous models, ultimately to efficient and sustainable land and water resource use.

MATERIAL AND METHODS

The research approach of this book integrates climate, agriculture, and geographical data of Andhra Pradesh (2020–2024) into a machine intelligence-based system to develop a system of climate-resilient water management. Preprocessing was needed in cleaning, standardizing, and synchronizing all the data sets so that they are in uniform format and ready to be used. Feature selection was carried out by Particle Swarm Optimization (PSO) to find the most appropriate variables influencing water use and crop yield. The optimized set of features were then trained by another Artificial Neural Network (ANN) to forecast irrigation needs and water stress zones. The system was validated against past trends and actual crop output. The resulting system is designed to facilitate data-driven decision-making in the context of climate-change resilient water and crop planning.

MATERIAL

Information used for the study are three rudimentary types—climate, agriculture (paddy), and geographical information—are geared toward the year 2020 to 2024 of Andhra Pradesh. Climate information were obtained from original sources like the Indian Meteorological Department (IMD), CPCB air monitoring stations, and Andhra Pradesh state weather observatories. It records significant meteorological parameters such as daily maximum and minimum temperature, relative humidity, amount of precipitation, wind speed, solar radiation, and extreme weather phenomena such as cyclones, heatwave, and drought. Seasonal patterns and deviations, particularly those pertaining to the southwest and northeast monsoon seasons, were recorded with care. Specific focus was on Rayalaseema regions having drying up tendencies and coastal regions having potential for flooding so that intensive reporting of climatic variability was maintained over several agro-climatic zones.

Farm data, having special focus on paddy cultivation, were collected in liaison with Andhra Pradesh State Agriculture Department, Krishi Vigyan Kendras (KVKs), and farmers' associations. The data consist of five consecutive kharif and rabi crops during the years 2020-2024. It includes sowing date and harvesting date, paddy sown variety, seed rate, fertilizer and pesticide application, irrigation (canal, borewell, rainfed), and yield per hectare. The data also include farm-reported data on crop failure, pest infestation, and weather-related issues like deficit of water and delayed rains. Aggregation at the district level allows tracking of production trends in key rice-growing districts such as East Godavari, West Godavari, Krishna, Nellore, and Srikakulam. Other data pertaining to government subsidy schemes, MSP, and procurement levels have been made available for tracking the effect of policies on production.

Andhra Pradesh geospatial data have been obtained from National Remote Sensing Centre (NRSC), Survey of India, and Andhra Pradesh State Spatial Data Infrastructure (APSSDI). There exists this geospatial data set consisting of high-resolution GIS layers of administrative boundary, land use/land cover (LULC), topography, river basins (e.g., Godavari and Krishna), irrigation system (dams, canals, tanks), soil groups, forest cover, and aquifer zones. Slope maps and DEMs were integrated in view of investigating terrain variations that influence rainfall runoff and water storage. These geospatial data sets support integration with climate and agriculture data for empowering spatial analysis like climate-vulnerable agricultural land identification, flood-risk areas mapping, and high-potential irrigation development zone mapping. Integrated data sets provide an integrated, multi-dimensional foundation to machine learning models and policy-planning mechanisms for maximizing water resource management and agriculture resilience in climate change in Andhra Pradesh.

PREPROCESSING

For the purpose of ensuring usability, consistency, and accuracy, the purchased datasets—climate, agriculture (paddy), and geographic data—underwent a preprocessing stage prior to analysis. Preprocessing for climate involved cleaning raw files that were procured from meteorological sources, which consequently represented missing data, outliers, and invariable measurements unit. Missing data for temperature and precipitation records were handled with interpolation methods or, if available, replaced with secondary data from nearby stations. All of the climate variables were time-normalized (scaled to daily or monthly sums, as is appropriate) and units (e.g., rain mm converted to cm, or temperatures to °C). Temporal synchronization was performed to match the climate data with cropping calendars and farm operations during the five-year period.

The agriculture data set was harmonized between various formats and sources including reports, surveys, and spreadsheets. Sanitizing of data included spell correcting errors in district or crop names, removal of duplicate records, and validation of yield reported values against state official sources. Categorical fields like paddy type or irrigation type were numerically encoded with one-hot encoding for machine learning compatibility. Seeding and harvest dates were also normalized to periods of time (e.g., number of days to crop duration) for ease of productivity and stress analysis. Missing or duplicate data submitted by farmers were identified and salvaged by cross-verification against records of authority or omitted from training models to maintain the integrity of the data.

Preprocessing of geospatial data involved reprojecting the entire set of spatial files in one common coordinate system (WGS 84) in a manner that supports overlay and spatial analysis with no gaps. Vector layers and raster layers were clipped to the Andhra Pradesh administrative boundary. Land use and soil type maps were reclassified into common forms of agriculture and water management. DEM data were smoothed and re-sampled to an appropriate resolution (e.g., 90m or 30m) to analyze the terrain. Watershed boundary and river network were digitized and topologically verified. These geographic data also were merged with climate and farm data with geotags at the district level or GPS coordinates for locational analysis.

Statistical aggregations and plots were done during preprocessing for quality checks to detect biases or outliers. Data was then stored in organized forms such as CSV for tables and GeoTIFF/GeoJSON for geospatial data in a common repository. Preprocessing step had ensured that the consolidated dataset is homogeneous, clean, and ready to be used by advanced machine learning models, spatial analysis, and policy simulation toolboxes to facilitate enhanced climate adaptation projects in Andhra Pradesh's agriculture and water sectors.

FEATURE SELECTION

Since the resulting dataset was of a high dimension and contained climatic, agricultural, and geographical variables, PSO was applied to identify the most important features that were accountable primarily for paddy yield and water demand. Features in the dataset included rainfall, temperature, humidity, irrigation type, crop type, soil type, elevation, and distance from water sources. PSO imitates bird flocking or schooling fish social behavior to effectively explore a solution space. Each particle of the swarm was a possible subset of features and its own velocity and position were determined iteratively from its own information and its neighboring particles' information.

The PSO process objective function was coded in a way so that it ensured feature selection was kept minimal and model performance was maximized to achieve an optimal trade-off among parsimony and predictive power. Support Vector Regression or models similar to Random Forest were learned on the selected subsets of features and their performance (e.g., R^2, RMSE) were used as the fitness value. With a number of iterations, the particles "migrated" to the optimal combination of features based on the individual and global bests. The iterative process allowed the swarm to converge towards a near-optimal set of features without running an exhaustive search, which would be computationally expensive with the existing size and complexity of the dataset.

This was the process that removed and excluded the duplicate or extraneous features such as those of high but low correlation of predictability. For example, features like elevation or wind speed that were less significant with respect to paddy yield in respect of individual districts were removed. Features such as soil pH, rainfall during critical growth stages, and distance of the irrigation source were retained anyway. The resulting feature subset increased model interpretability and resulted in more accurate decision-making for irrigation planning, water resource management, and agricultural advisories. In the context of data-driven climate adaptation in Andhra Pradesh, the study guaranteed a stable, scalable, and robust process through the application of PSO for feature selection.

ARTIFICIAL NEURAL NETWORKS (ANN)

Artificial Neural Networks (ANNs) offer a dependable machine intelligence solution to enable climate adaptation in water management, particularly in advanced, data-oriented environments like Andhra Pradesh. The climatic, crop, and geographical parameter datasets aggregated are used as the input to the ANN model to learn relationships, trends, and nonlinear dependencies. ANNs are particularly well suited to process high uncertainty and variability of climatic data, e.g., nonperiodic rains, a change in monsoon types, and sudden temperature fluctuations. The model architecture includes input layers with selected features (e.g., rain, soil, irrigation practice), feature transformation hidden layers, and an output layer to predict water requirement, irrigation calendar, or paddy yield.

With the training, the ANN can project input features onto target output with backpropagation and gradient descent to calculate the error at minimum while predicting. Historical data for the 2020-2024 period allow the model to visualize climatic patterns within the season, water resource availability fluctuation patterns, and their consequent effects on the yield of crops. For instance, the ANN can predict the quantity of optimum irrigation to be supplied in the event of a late monsoon or calculate time windows when likely water stress is sure to happen. It can also give advance warnings for over-rainfall liable to cause flood, facilitating preventive management of dams and canal operation.

The potential for re-tuning or re-training with localized area inputs renders ANNs suitable for region-wise planning of water. For the Andhra Pradesh state, where districts vary from hydric coastals to arid Rayalaseema, ANN models can be district-specific. Water resource management can be calibrated to local needs, maintaining water-use efficiency as well as farm resilience. Furthermore, ANN outputs can also be utilized to provide farmers and policymakers with decision support systems in real-time on web or mobile-based systems. The systems can provide advice on when and how much water to irrigate, and where water shortage would most likely be experienced.

In summary, through the utilization of ANNs' ability to learn, the study suggests a smart and scalable climate-resilient water management system. Through the utilization of machine intelligence, not only does it mitigate the risk posed by climate variability but also enhances the sustainability of the state's agriculture and water sectors.

EXPERIMENTAL RESULTS

Test period of the study kept busy evaluating the performance of machine learning model designed for solving climate adaptation challenges in agricultural water management for paddy rice crop cultivation in Andhra Pradesh. Combined data, i.e., 2020-2024 climate, agriculture, and geography, were utilized to train and test Artificial Neural Network (ANN) model. Prior to training, feature selection was obtained from Particle Swarm Optimization (PSO) that appreciably minimized dimensions while preserving the effect of high-impact features such as rain, temperature, soil, and distance to irrigation sources. Data were divided into 70% training and 30% test subsets for model generalizability evaluation. Performance of the model was evaluated using standard metrics such as Root Mean Square Error (RMSE), Mean Absolute Error (MAE), and the coefficient of determination (R^2). The ANN model demonstrated the robustness of its predictive power, which could well forecast irrigation demand and estimate probable zones of water stress for different districts. The results are apt to justify the use of machine intelligence in enabling climate-resilient water planning and adaptive farm management.

Table 1 provides the climatic parameters encountered in different seasons, with the average temperature, humidity, rainfall, wind speed, and levels of solar radiation. It can be seen from the results that there are distinct seasonal variations, with higher rainfall and humidity in the monsoon season and lower in pre-monsoon. The temperatures are more or less uniform, with maximum wind speed in the post-monsoon season. Maximum solar radiation is in the pre-monsoon, which means clearer skies. These parameters set the environmental context relevant to agricultural modeling. Planning and forecasting of crop growth are facilitated by awareness of these seasonal patterns.

Table 1. ANN model performance metrics by district (2024)

District	RMSE (mm)	MAE (mm)	R^2 Score	Accuracy (%)	Training Time (s)
Anantapur	12.5	9.1	0.87	88.5	35.4
Chittoor	10.8	7.6	0.91	90.2	34.7
Guntur	9.7	6.5	0.93	92.1	33.9
Krishna	11.2	7.8	0.90	89.7	36.2
Kurnool	13.1	9.5	0.85	87.3	37.1
Prakasam	10.2	7.0	0.92	91.4	33.5
Srikakulam	9.9	6.8	0.93	91.8	32.8
Vijayawada	10.4	7.1	0.91	90.9	34.0
Visakhapatnam	9.6	6.3	0.94	92.5	33.3
West Godavari	10.0	6.9	0.92	91.2	34.8

Table 2 shows a breakdown of soil moisture content at various depths in each season, how the moisture availability changes vertically and seasonally. There is higher moisture in upper soil layers during post-monsoon and monsoon seasons due to rainfall and drying patterns exhibited during pre-monsoon. Soil water storage affects irrigation planning and root water uptake. Depth differences illustrate the variation of water storage in soil profiles, which affects the availability of water to crops and their development. The table below maximizes irrigation schedule and soil water management practices.

Table 2. Key climate features selected by PSO

Feature Name	Importance Score	Correlation with Yield	Selected (Y/N)	Data Type	Units
Monthly Rainfall	0.95	0.82	Y	Continuous	mm
Average Temperature	0.89	-0.65	Y	Continuous	°C
Soil Moisture	0.87	0.75	Y	Continuous	%
Irrigation Frequency	0.85	0.70	Y	Discrete	times/month
Humidity	0.70	0.60	Y	Continuous	%
Wind Speed	0.42	-0.12	N	Continuous	km/h
Elevation	0.50	0.15	N	Continuous	meters
Soil pH	0.80	0.68	Y	Continuous	pH Units
Distance to River	0.75	0.64	Y	Continuous	km
Crop Variety	0.88	0.77	Y	Categorical	-

Table 3 shows the monthly rainfall depths recorded at different stations within the study area, exhibiting temporal and spatial variability. The highest rainfall is largely in July and August, consistent with the monsoon season, while winter months experience little rainfall. The rainfall pattern is essential in water resource planning and flood hazard recognition. The table can be useful in relating rainfall trends to crop water demands, impacting agricultural yield. Understanding the monthly rainfall allows for the planning of water-saving strategies and drought relief schemes.

Table 3. Monthly water demand prediction accuracy (2020-2024)

Month	Actual Demand (ML)	Predicted Demand (ML)	Error (ML)	Percentage Error (%)	Model Confidence (%)
January	150	143	7	4.67	92
February	130	128	2	1.54	95
March	180	172	8	4.44	90
April	220	210	10	4.55	88
May	270	260	10	3.70	89
June	300	295	5	1.67	94
July	280	275	5	1.79	93
August	260	250	10	3.85	90
September	230	225	5	2.17	92
October	190	185	5	2.63	93

Table 4 shows crop yield data for paddy and wheat crops over a few years, patterns of production influenced by environmental and management factors. Paddy yields are generally greater than those of wheat, with evident year-to-year differences due to climate variations. Seasonality, as well as resource management, stand out as significant determinants of productivity from the data. Yield increases are credited to enhanced irrigation and fertilizer management practices. This table is used as a benchmark for gauging the effectiveness of agricultural interventions and technology adoption.

Table 4. Paddy yield prediction results by year (2020-2024)

Year	Actual Yield (tons/ha)	Predicted Yield (tons/ha)	Absolute Error	Percentage Error (%)	R^2 Score
2020	3.5	3.4	0.1	2.86	0.91
2021	3.7	3.6	0.1	2.70	0.92
2022	3.6	3.5	0.1	2.78	0.93
2023	3.8	3.7	0.1	2.63	0.94
2024	3.9	3.85	0.05	1.28	0.95
Avg	3.7	3.61	0.09	2.65	0.93

Table 5 gives application rates of irrigation water at different stages of crop growth, recording variations as a result of climatic conditions. Application of water is maximum during such key stages as tillering and flowering, reflecting crop water requirement. Seasonal variation in the rate of irrigation is also evident, with increased volume being furnished during dry pre-monsoon periods. The information supports efficient water management and conservation strategies in agriculture. Proper scheduling of irrigation based on the stage of the crop optimizes water usage to the maximum, reducing wastage and enhancing yield.

Table 5. Water stress zone identification accuracy by district

District	True Positives	False Positives	True Negatives	False Negatives	Accuracy (%)	F1 Score
Anantapur	45	5	120	10	91.3	0.88
Chittoor	50	6	115	9	90.8	0.89
Guntur	48	4	118	8	92.0	0.91
Krishna	47	5	117	11	90.5	0.87
Kurnool	43	7	119	12	89.2	0.85
Prakasam	49	3	121	7	92.8	0.92
Srikakulam	46	5	116	10	90.7	0.88
Vijayawada	51	4	118	9	91.5	0.90
Visakhapatnam	52	3	119	8	92.7	0.91
West Godavari	48	4	117	10	91.0	0.89

Table 6 presents prediction model accuracy for crops and seasons, indicating reliability of the model. High accuracy values, typically above 85%, validate good model performance in capturing seasonal water demand. Low accuracy for intermediate seasons indicates struggle to simulate complex environmental change. The table validates applicability of the model to irrigation planning. Consistent performance across crops attests to robustness and flexibility of the predictive approach.

Table 6. Irrigation scheduling accuracy under different climate scenarios

Climate Scenario	Actual Irrigation Days	Predicted Days	Absolute Error	Percentage Error (%)	Model Confidence (%)
Normal	40	38	2	5.0	94
Dry Year	35	33	2	5.7	91
Wet Year	45	44	1	2.2	95
Late Monsoon	42	40	2	4.8	93
Early Monsoon	38	37	1	2.6	92
Drought Condition	30	29	1	3.3	90
Flooding Condition	50	48	2	4.0	89
Heatwave	36	34	2	5.6	88
Cyclone	44	42	2	4.5	91
Hailstorm	39	38	1	2.6	92

Table 7 presents a seasonal comparison between observed and simulated evapotranspiration (ET) values, providing an impression of the water loss dynamics. Simulated ET is extremely close to observed values with minimal deviations during monsoon. The closeness confirms the ability of the model to simulate crop water use according to reality. Accurate estimation of ET is crucial for irrigation and drought monitoring. Seasonal trends in ET recommend better scheduling and water budgeting in sustainable agriculture.

Table 7. Soil moisture prediction accuracy across seasons

Season	Actual Moisture (%)	Predicted Moisture (%)	RMSE (%)	MAE (%)	R^2 Score
Winter	25	24.5	1.2	0.9	0.88
Summer	15	14.8	1.0	0.7	0.90
Monsoon	35	34.6	1.3	1.0	0.89
Post-Monsoon	28	27.8	1.1	0.8	0.91
Avg	25.75	25.43	1.15	0.85	0.89

Table 8 summarizes the model-predicted and actual duration of paddy crop growth stages, gauging the prediction error and accuracy. The model is satisfactory with a general average accuracy level of 96%, with the lowest error in flowering and grain filling phases. Proper stage prediction enables timely agronomic applications such as fertilization and irrigation. The table illustrates the capacity of the model to promote precision agriculture through prediction of crop development.

Table 8. Paddy crop growth stage prediction accuracy

Growth Stage	Actual Days	Predicted Days	Absolute Error	Percentage Error (%)	Accuracy (%)
Germination	10	9	1	10	90
Tillering	25	24	1	4	96
Panicle Initiation	40	38	2	5	95
Flowering	60	59	1	1.67	98
Grain Filling	85	84	1	1.18	99
Maturity	110	108	2	1.82	98
Avg	55	54	1.33	3.28	96

Table 9 indicates how geographical factors like elevation and distance from river courses influence water requirement estimates. Delta and coastal regions require more volumes of water, as consistent with lower elevation and proximity to bodies of water. Plateau and upland regions require smaller volumes of water. The prediction errors of the model for different types of terrains are minimal, indicating that the model is adaptable. This spatial awareness facilitates targeted water resource allocation and irrigation planning in accordance with landscape features.

Table 9. Geographic impact on water requirement prediction

Zone	Elevation (m)	Distance to River (km)	Actual Water Req. (ML)	Predicted Water Req. (ML)	Error (ML)
Coastal	10	2	200	195	5
Plains	50	10	180	175	5
Plateau	150	25	160	158	2
Hills	300	40	140	135	5
Valley	100	15	170	165	5
Delta	5	1	210	205	5
Upland	200	30	150	148	2
Lowland	30	8	190	185	5
Riverbank	20	0.5	220	215	5
Wetland	25	5	230	225	5

Table 10 shows an overview of the model accuracy measures taken over five consecutive years with increasingly enhanced training, validation, and test accuracies. RMSE and MAE decrease over the years while R^2 values enhance, indicating better fit and predictive consistency of the model. This trend reflects enhancements in data quality, model calibration, and algorithms' optimization. The table reveals the stability and increasing accuracy of the crop growth and water requirement forecast model.

Table 10. Overall model accuracy over years (2020–2024)

Year	Training Accuracy (%)	Validation Accuracy (%)	Test Accuracy (%)	RMSE		R^2 Score
2020	94.5	91.0	90.5	10.2	7.8	0.91
2021	95.2	91.5	91.0	9.8	7.4	0.92
2022	95.8	92.0	91.8	9.5	7.2	0.93
2023	96.0	92.5	92.3	9.2	7.0	0.94
2024	96.5	93.0	92.7	9.0	6.8	0.95
Avg	95.6	92.0	91.66	9.74	7.24	0.93

CONCLUSION

This research confirms the efficacy of using machine intelligence, i.e., Artificial Neural Networks, in climate adaptation for water management systems. Integrating climate, agronomic, and geographical data on Andhra Pradesh, the methodology offers data-driven methodology to water demand forecasting and climate variability adaptation. PSO-based feature selection ensured ANN to be taking into account most crucial variables, which enhanced model accuracy and interpretability. The model allows for better planning of irrigation and resource allocation choices, which maximize agricultural sustainability under the framework of shifting climatic regimes. In most instances, the strategy not only addresses the local water management problems but also serves as a stepping stone towards the use of AI-based solutions for addressing broader environmental and policy problems.

REFERENCES

Abioye, E. A., Abidin, M. S. Z., Mahmud, M. S. A., Buyamin, S., Ishak, M. I., & Ramli, M. K. I. (2020). A review on monitoring and advanced control strategies for precision irrigation. *Computers and Electronics in Agriculture, 173*, 105441. DOI: 10.1016/j.compag.2020.105441

Allen, C., Metternicht, G., Verburg, P., Akhtar-Schuster, M., da Cunha, M. I., & Santivañez, M. S. (2020). Delivering an enabling environment and multiple benefits for land degradation neutrality: Stakeholder perceptions and progress. *Environmental Science & Policy, 114*, 109–118. DOI: 10.1016/j.envsci.2020.07.023 PMID: 32834778

Bitella, G., Rossi, R., Bochicchio, R., Perniola, M., & Amato, M. (2014). A novel low-cost open-hardware platform for monitoring soil water content and multiple soil-air-vegetation parameters. *Sensors (Basel), 14*(10), 19639–19659. DOI: 10.3390/s141019639 PMID: 25337742

Burrell, A. L., Evans, J. P., & De Kauwe, M. G. (2020). Anthropogenic climate change has driven over 5 million km^2 of drylands towards desertification. *Nature Communications, 11*(1), 3853. DOI: 10.1038/s41467-020-17710-7 PMID: 32737311

Bwambale, E., Abagale, F. K., & Anornu, G. K. (2022). Smart irrigation monitoring and control strategies for improving water use efficiency in precision agriculture: A review. *Agricultural Water Management, 260*, 107324. DOI: 10.1016/j.agwat.2021.107324

Byers, E., & Gidden, M. (2018). Global exposure and vulnerability to multi-sector development and climate change hotspots. *Environmental Research Letters, 13*(5), 055012. DOI: 10.1088/1748-9326/aabf45

Cambra, C., & Sendra, S. (2018). Smart system for bicarbonate control in irrigation for hydroponic precision farming. *Sensors (Basel), 18*(4), 1333. DOI: 10.3390/s18041333 PMID: 29693611

Chimwamurombe, P. M., & Mataranyika, P. N. (2021). Factors influencing dryland agricultural productivity. *Journal of Arid Environments, 189*, 104489. DOI: 10.1016/j.jaridenv.2021.104489

Dabach, S., Lazarovitch, N., Simunek, J., & Shani, U. (2013). Numerical investigation of irrigation scheduling based on soil water status. *Irrigation Science, 31*(1), 27–36. DOI: 10.1007/s00271-011-0290-6

Dalezios, N. R., Angelakis, A. N., & Eslamian, S. (2018). Water scarcity management: Part 1: Methodological framework. *International Journal of Global Environmental Issues*, *17*(1), 1–40. DOI: 10.1504/IJGENVI.2018.090582

Eskandari Dameneh, H. (2021). Desertification of Iran in the early twenty-first century: Assessment using climate and vegetation indices. *Scientific Reports*, *11*(1), 20548. DOI: 10.1038/s41598-021-99983-3 PMID: 34654866

Food and Agriculture Organization of the United Nations. (2022). *The state of the world's land and water resources for food and agriculture: Systems at breaking points*. FAO.

Gebremeskel, G., & Tang, Q. (2019). Droughts in East Africa: Causes, impacts and resilience. *Earth-Science Reviews*, *193*, 146–161. DOI: 10.1016/j.earscirev.2019.04.015

Grainger, A. (2015). Is land degradation neutrality feasible in dry areas? *Journal of Arid Environments*, *112*, 14–24. DOI: 10.1016/j.jaridenv.2014.05.014

Huang, Y., Li, Y. P., Chen, X., & Ma, Y. G. (2012). Optimization of the irrigation water resources for agricultural sustainability in Tarim River Basin, China. *Agricultural Water Management*, *107*, 74–85. DOI: 10.1016/j.agwat.2012.01.012

Huang, Z., Yuan, X., & Liu, X. (2021). The key drivers for the changes in global water scarcity: Water withdrawal versus water availability. *Journal of Hydrology (Amsterdam)*, *601*, 126658. DOI: 10.1016/j.jhydrol.2021.126658

Ibrahim, Y. Z., Balzter, H., Kaduk, J., & Tucker, C. J. (2015). Land degradation assessment using residual trend analysis of GIMMS NDVI3g, soil moisture and rainfall in Sub-Saharan West Africa from 1982 to 2012. *Remote Sensing*, *7*(5), 5471–5494. DOI: 10.3390/rs70505471

Imbrenda, V., Quaranta, G., Salvia, R., Egidi, G., Salvati, L., Prokopova, M., & Lanfredi, M. (2021). Land degradation and metropolitan expansion in a peri-urban environment. *Geomatics, Natural Hazards & Risk*, *12*(1), 1797–1818. DOI: 10.1080/19475705.2021.1944331

Jin, J., Wang, L., & Muller, K. (2021). A 10-year monitoring of soil properties dynamics and soil fertility evaluation in Chinese hickory plantation regions of southeastern China. *Scientific Reports*, *11*(1), 23531. DOI: 10.1038/s41598-021-02946-9 PMID: 34876648

Kummu, M., & Guillaume, J. H. A. (2016). The world's road to water scarcity: Shortage and stress in the 20th century and pathways towards sustainability. *Scientific Reports*, *6*(1), 38495. DOI: 10.1038/srep38495 PMID: 27934888

Li, H., Cohen, A., Li, Z., & Zhang, M. (2019). The impacts of socioeconomic development on rural drinking water safety in China: A provincial-level comparative analysis. *Sustainability, 11*(1), 85. DOI: 10.3390/su11010085

Liu, J., & Yang, H. (2017). Water scarcity assessments in the past, present, and future. *Earth's Future, 5*(6), 545–559. DOI: 10.1002/2016EF000518 PMID: 30377623

Liu, X., & Liu, W. (2022). Global agricultural water scarcity assessment incorporating blue and green water availability under future climate change. *Earth's Future, 10*(3), e2021EF002567. https://doi.org/DOI: 10.1029/2021EF002567

Maestre, F. T., & Quero, J. L. (2012). Plant species richness and ecosystem multifunctionality in global drylands. *Science, 335*(6065), 214–218. DOI: 10.1126/science.1215442 PMID: 22246775

Minea, G., Ciobotaru, N., & Ioana-Toroimac, G. (2022). Designing grazing susceptibility to land degradation index (GSLDI) in hilly areas. *Scientific Reports, 12*(1), 9393. DOI: 10.1038/s41598-022-13543-8 PMID: 35729181

Modarres, R., & da Silva, V. R. (2007). Rainfall trends in arid and semi-arid regions of Iran. *Journal of Arid Environments, 70*(2), 344–355. DOI: 10.1016/j.jaridenv.2006.12.024

Neupane, J., & Guo, W. (2019). Agronomic basis and strategies for precision water management: A review. *Agronomy (Basel), 9*(2), 87. DOI: 10.3390/agronomy9020087

Pandit, R., & Parrotta, J. A. (2020). A framework to evaluate land degradation and restoration responses for improved planning and decision-making. *Ecosystems and People (Abingdon, England), 16*(1), 1–18. DOI: 10.1080/26395916.2019.1696410

Pastor, A. V., Palazzo, A., Havlik, P., Biemans, H., Wada, Y., Obersteiner, M., & Ludwig, F. (2019). The global nexus of food–trade–water sustaining environmental flows by 2050. *Nature Sustainability, 2*(6), 499–507. DOI: 10.1038/s41893-019-0287-1

Plaza, C., & Zaccone, C. (2018). Soil resources and element stocks in drylands to face global issues. *Scientific Reports, 8*(1), 13788. DOI: 10.1038/s41598-018-32229-0 PMID: 30214005

Pravalie, R. (2016). Drylands extent and environmental issues: A global approach. *Earth-Science Reviews, 161*, 259–278. DOI: 10.1016/j.earscirev.2016.08.003

Prăvălie, R., Bandoc, G., Patriche, C., & Sternberg, T. (2019). Recent changes in global drylands: Evidences from two major aridity databases. *Catena, 178*, 209–231. DOI: 10.1016/j.catena.2019.03.016

Rosa, L., Chiarelli, D. D., Rulli, M. C., Dell'Angelo, J., & D'Odorico, P. (2020). Global agricultural economic water scarcity. *Science Advances*, *6*(18), eaaz6031. Advance online publication. DOI: 10.1126/sciadv.aaz6031 PMID: 32494678

Rosa, L., Chiarelli, D. D., Sangiorgio, M., Beltran-Peña, A. A., Rulli, M. C., D'Odorico, P., & Fung, I. (2020). Potential for sustainable irrigation expansion in a 3°C warmer climate. *Proceedings of the National Academy of Sciences of the United States of America*, *117*(47), 29526–29534. DOI: 10.1073/pnas.2017796117 PMID: 33168728

Singh, U., & Praharaj, C. S. (2019). Precision irrigation management: Concepts and applications for higher use efficiency in field crops. In *Scaling water productivity and resource conservation in upland field crops ensuring more crop per drop* (pp. 1–10). ICAR-Indian Institute of Pulses Research.

Stavi, I., & Lal, R. (2015). Achieving zero net land degradation: Challenges and opportunities. *Journal of Arid Environments*, *112*, 44–51. DOI: 10.1016/j.jaridenv.2014.01.016

Stringer, L. C., & Mirzabaev, A. (2021). Climate change impacts on water security in global drylands. *One Earth*, *4*(6), 851–864. DOI: 10.1016/j.oneear.2021.05.010

Stringer, L. C., Reed, M. S., Fleskens, L., Thomas, R. J., Le, Q. B., & Lala-Pritchard, T. (2017). A new dryland development paradigm grounded in empirical analysis of dryland systems science. *Land Degradation & Development*, *28*(7), 1952–1961. DOI: 10.1002/ldr.2716

Stroosnijder, L., Moore, D., Alharbi, A., Argaman, E., Biazin, B., & van den Elsen, E. (2012). Improving water use efficiency in drylands. *Current Opinion in Environmental Sustainability*, *4*(5), 497–506. DOI: 10.1016/j.cosust.2012.08.011

Tariq, A., Ullah, A., Sardans, J., Zeng, F., Graciano, C., Li, X., & Peñuelas, J. (2022). *Alhagi sparsifolia*: An ideal phreatophyte for combating desertification and land degradation. *The Science of the Total Environment*, *844*, 157228. DOI: 10.1016/j.scitotenv.2022.157228 PMID: 35809729

Yu, L., Zhao, X., Gao, X., Jia, R., Yang, M., Yang, X., & Siddique, K. H. (2021). Effect of natural factors and management practices on agricultural water use efficiency under drought: A meta-analysis of global drylands. *Journal of Hydrology (Amsterdam)*, *594*, 125977. DOI: 10.1016/j.jhydrol.2020.125977

Zamfir, R., Smiraglia, D., Quaranta, G., Salvia, R., Salvati, L., & Giménez-Morera, A. (2020). Land degradation and mitigation policies in the Mediterranean region: A brief commentary. *Sustainability*, *12*(20), 8313. DOI: 10.3390/su12208313

Zhang, J., Guan, K., Peng, B., Jiang, C., Zhou, W., Yang, Y., & Cai, Y. (2021). Challenges and opportunities in precision irrigation decision-support systems for center pivots. *Environmental Research Letters*, *16*(5), 053003. DOI: 10.1088/1748-9326/abe436

Chapter 10
A Non-Invasive Revolution in Soil and Root Water Content Monitoring by Integrating Deep Learning and GPR

Mohammed Kahlaoui
https://orcid.org/0009-0005-4426-2672
National School of Applied Sciences, Morocco

Aboulkacem Karkri
National School of Applied Sciences, Morocco

Mohammed Anisse Moutaouekkil
National School of Applied Sciences, Morocco

ABSTRACT

This chapter introduces a technique that combines simulations using Ground Penetrating Radar (GPR), advanced processing of signals and deep learning to determine water content in plant roots as well as nearby soil. Using dielectric mixing models, computer programs were used to create synthetic radargrams of root-soil interactions under various moisture levels. Radargrams were improved by using methods such as Singular Value Decomposition and Inverse Amplitude Decay. Convolutional Neural Networks (CNNs), including InceptionV3, MobileNetV2 and ResNet50V2, were adapted for using in classifying both root and soil water content. InceptionV3 got the top results among tested models for estimating plant

DOI: 10.4018/979-8-3373-2700-6.ch010

roots. For easier soil moisture classification, MobileNetV2 was able to run quickly and efficiently. ResNet50V2 managed to perform with similar accuracy and level of stability for both types of tasks.

INTRODUCTION

Water supply and management for agriculture are essential for maintaining food security, ecosystem service sustainability, and economic stability. It is becoming more critical that accurate tools for the determination of soil and root zone water become available due to climate change with the lengthening and increasing intensification of drought, together with changing patterns in rainfall. Global freshwater scarcity is projected to intensify, with agriculture accounting for nearly 70% of withdrawals, making efficient water management increasingly urgent (Ingrao et al., 2023). The ability to accurately monitor subsurface moisture is a critical component of the ability to drive irrigation optimization, improve water use efficiency, and maintain crop health in water limited or inefficiently distributed areas around the world, particularly arid and semi-arid regions. Moreover, agricultural policies and water management regulations in many countries now mandate precision irrigation practices, creating added pressure to adopt reliable soil moisture monitoring technologies.

One of the central difficulties is to estimate the water content in the root zone, controlling plant water uptake and thus determining growth and yield. Common approaches, which include soil sampling and point sensors, are often not spatially resolved, disturb the soil-root interface and only offer localized information. Such limitations highlight the need for non-intrusive, scalable methods that can provide spatially distributed and time-lapsed information on soil moisture.

Ground Penetrating Radar (GPR) poses as a promising geophysical tool for this task, working by taking advantage of the differences in the dielectric properties between soil and roots, affected by water content, allowing subsurface structures to be unveiled. Nonetheless, GPR data interpretation is challenging due to signal noise, heterogeneous soil, and environmental variability. While signal processing methodologies such as SVD, IAD and, more recently, deep learning methods (such as CNNs) have been investigated to identify informative characteristics in radargram data, a lack of labeled data, model generalizability issues, and the need for explainability continue to hinder practical deployment.

To address these critical limitations and advance the field beyond current capabilities, this chapter introduces a novel and comprehensive framework that integrates three key innovations: (1) extensive synthetic GPR data generation using electromagnetic modeling to overcome data scarcity, (2) sophisticated signal processing combining Singular Value Decomposition (SVD) and Inverse Amplitude

Decay (IDA) for enhanced radargram interpretation, and (3) a rigorous comparative analysis of state-of-the-art deep learning architectures for optimal model selection. Specifically, we leverage the gprMax software to create a diverse dataset of 4500 unique GPR B-scan configurations, systematically varying root water content (0-100%), soil water content (0-58%), root geometry, and soil properties to simulate realistic agricultural conditions. Then we compare three popular Convolutional Neural Network architectures InceptionV3, MobileNetV2, and ResNet50V2 to detect the best models as to exact classification of the root and soil moisture. This unified method offers a comprehensive and scalable solution to the inherent problems of data limitation, computational complexity of interpreting signal and uncertainty of model selection and has a strong potential to transform irrigation practices and potentially become the solution to world food security.

HISTORICAL EVOLUTION OF SOIL AND ROOT MOISTURE MONITORING

Initiatives to measure soil and root-zone water content have gone through a number of technological phases. Initial methods were destructive using manually intensive procedures, the most common being gravimetric samplings where the soil cores were pulled out and oven dried to establish volumetric water content. Although correct, they were time-consuming and interfered with invasive measurements and limited to point-scale data (Cheng et al., 2023).

Point-based sensors, such as time domain reflectometry (TDR), neutron probes and capacitance sensors were developed in order to overcome these shortcomings. These methods allowed in situ measuring on the same sites and higher time resolution but remained limited to a local scale, as well as usually interfering with soil and root interface, inhibiting long-term monitoring rates (Cheng et al., 2023; Margiwiyatno et al., 2025). Despite its prevalence, the proposed methods are not all that ideal in large field analysis over space where environmental conditions are varied, like arid, temperate or tropical soils.

Later developments incorporated geophysical techniques, including electrical resistivity tomography (ERT) and electromagnetic induction, to make the localized measurement of soil water non-invasively spatially-distributed. These techniques yield scale-level details of soil heterogeneity at field scale; they cannot resolve soil heterogeneity and are sensitive to soil type, vegetative cover and environmental variations (Shamir et al., 2016; Wenhan et al., 2022).

GPR has become a strong alternative that can be used to estimate high spatial resolution due to dielectric differences with respect to soil and root water content (Liu et al., 2019, 2020 ; Elshaboury et al., 2023) Its early uses showed that it can detect coarse root systems due to hyperbolic reflections, allowing indirect inference on root architecture and concomitant soil moisture. GPR has transformed, over the last 20 years, to have all-purpose capacity, on a field scale, that creates a basis of integrating AI. The imaging of time-lapse GPR and multi-attribute has also succeeded in topping up even in adverse environmental circumstances the detection of dynamic moisture patterns (Cheng et al., 2023; Mangel et al., 2020).

The latest development within this trend is the GPR combined with Artificial Intelligence (AI). The deep learning approaches and specifically convolutional neural networks (CNNs) have also shown great success in capturing subtle details within radargrams to generate accurate labels on the soil and root moisture within varying conditions (Ko et al., 2022; Li et al., 2024; Moutaouekkil et al., 2024). However, the black-box nature of CNNs can limit stakeholder trust and practical adoption. Incorporating explainable AI (XAI) techniques, such as Grad-CAM visualizations and feature attribution maps, allows users to understand model predictions and supports decision-making by farmers and agronomists. Lightweight architectures (e.g., MobileNetV2) further enable deployment in field conditions with low-power devices(Jahin et al., 2025).

To better illustrate the advantages and limitations of different soil moisture sensing technologies, Table 1 provides a side-by-side comparison of GPR combined with AI, hyperspectral and microwave remote sensing, and Electrical Resistivity Tomography (ERT).

Table 1. Comparative analysis of soil moisture monitoring technologies

Method	**Spatial Resolution**	**Invasiveness**	**Cost (Install/ Operate)**	**Operational Complexity**
Gravimetric sampling	Point-based	Destructive	Low / Moderate	High
TDR / Neutron Probes	Point-based	Invasive	Moderate / High	Moderate
ERT	Medium	Non-invasive	Moderate / High	High
GPR + AI	High	Non-invasive	Moderate / High	Moderate

This comparison highlights how GPR+AI can overcome several limitations of conventional methods while offering high-resolution, non-invasive monitoring suitable for precision agriculture. The combination of advanced signal processing, AI integration, and explainable outputs directly addresses the gaps identified in previous methods, supporting both scientific and practical applications in diverse environmental contexts (Lamichhane et al., 2025; Rasheed et al., 2022).

ADVANCES AND KEY TRENDS IN SOIL AND ROOT MOISTURE MONITORING

Advances in recent years are emphasized in some trends that will determine the dynamics of soil and root moisture monitoring:

- Multimodal Sensing - GPR data can be combined with other modalities (e.g. RGB and thermal cameras, electromagnetic induction), which can enhance robustness and precision by feeding into multimodal data fusion (Kurtulmuş et al., 2022; Vahidi et al., 2025). This method can cover a wider area and can mitigate the shortcoming of individual probing methods and enable more comprehensive mapping of more heterogeneous soil and root conditions on agricultural fields.
- UAV- and Robot-Mounted Systems: Miniaturized GPR modules that are used on drones and autonomous ground robots can scale up field-scale resolution-wide observation, making it faster, and labor-light (Cheng et al., 2023). The platforms are used to gather substantial spatial policies of distributed data efficiently to enhance planning of precision irrigation and frequent monitoring of crop without disturbing the soil.
- Real-Time Analytics: Embedded and edge-computing systems mean that radargrams can be preprocessed and AI-based genre classification can be done in real time (Ko et al., 2022). This provides instantaneous insights that help to manage irrigation dynamically, rapidly detect anomalies, and respond with necessary agronomic action that closes the data-action loop.
- Explainable AI (XAI): The concept of understandable AI results is also becoming critical in application in GPR, as it may encourage trust by all stakeholders, including the farmers and agronomists and enable understandability by policymakers and compliance with regulations. The methods of feature attribution maps, SHAP, Grad-CAM, and LIME were used in GPR-based moisture monitoring in recent research (Ahmed et al., 2025; Mohan et al., 2025). Such methods facilitate the visualization and interpretation of model predictions that enhance stakeholder confidence in their soil and root water management decisions that are augmented by AI (T. Vengatesh, 2025). The lightweight architectures (e.g., MobileNetV2) also facilitate the implementation in the field under the conditions of limited computing resources. XAI takes a step forward and satisfies these requirements, mediating between sophisticated AI techniques and applied agricultural decision-making, characterized by interpretability, transparency, and confidence in the real world (De Silva et al., 2025).

Along with the optimizations done to the algorithms, some deployment issues are also crucial. To fully realize their potential, deployment of GPR-AI systems in the field requires supportive infrastructures such as power, data connectivity and computational resources in remote areas. Technical and agronomic expertise are necessary during operator training and capacity building, to target proper usage. Resource-limitation regions also face additional obstacles because cost and access must be addressed to guarantee acceptance, due to the cost of hardware, open-source software, and integrations with extension services (Adeyemi et al., 2017).

Collectively, these trends show that while AI-enhanced GPR is already highly promising in research, its overall impact will be further strengthened by alignment between technological innovation, field practitioners, socio-economic factors, and policy frameworks. Considering these dimensions acts as a critical lever to move intelligent and non-invasive sensing of soil and root moisture in precision agriculture from the laboratory to a regular and sustainable practice.

APPLICATIONS OF GPR-AI SOIL AND ROOT WATER CONTENT MONITORING

The possibility to adequately and non-destructively estimate soil and roots moisture due to the combination of GPR and AI brings the opportunity of several applications that may transform the current agriculture industry.

These applications go beyond simple measurement, providing actionable decision-support tools for farmers and resource managers:

- Optimization of Precision Irrigation: One of the most direct and impactful applications is improving irrigation efficiency. By providing detailed, near-real-time maps of soil and root moisture distribution, GPR-AI technology allows water to be applied precisely where and, in the amount, it is needed. This significantly reduces water waste, lowers operational costs, and minimizes nutrient leaching, contributing to more sustainable and economically viable agriculture (Namdari et al., 2025). The data can be integrated into smart irrigation systems for advanced automation.
- Early Detection of Crop Water Stress: Monitoring root-zone moisture enables the identification of areas where plants begin to experience water stress, often before visible symptoms appear in the aerial parts. Early detection is crucial for timely interventions, such as targeted irrigation, which can prevent significant yield losses and maintain crop health, particularly in drought-prone regions (Vahidi et al., 2025).

- Optimized Nutrient and Fertilizer Management: Soil moisture is a key factor in nutrient availability and transport to plant roots. Accurate knowledge of water content allows fertilization strategies to be adjusted, ensuring that nutrients are applied when the soil conditions favor optimal uptake. This maximizes fertilizer efficiency, reduces leaching losses, and minimizes environmental impact.
- Characterization of Soil Heterogeneity and Adaptation of Agronomic Practices: GPR is a powerful tool for revealing spatial variations in soil properties such as texture, porosity, compaction, or the presence of restrictive layers. Combined with moisture data, these insights enable farmers to understand field heterogeneity and adapt cultural practices (tillage, sowing, fertilization) to optimize crop growth in each specific zone (Rasheed et al., 2022).
- Non-Destructive Monitoring of Root Growth and Development: Beyond moisture, GPR can detect and characterize root systems, enabling non-invasive monitoring of root growth in response to soil and water conditions, assessment of soil amendment effects on root development, and early diagnosis of root-related issues (diseases or physical obstacles) (Zou et al., 2025). These insights are invaluable for varietal selection and agronomic improvement.

Despite significant advances in both GPR instrumentation and AI architectures, existing studies still face challenges related to scalability, robustness under diverse soil conditions, and applicability to real agricultural environments. These limitations reveal a clear research gap that motivates the present study, which explores the integration of GPR with AI-driven methods for improved, non-invasive root and soil moisture monitoring

GPR PRINCIPLE

GPR is a geophysical technique, used to describe underground features by using electromagnetic waves. The image in Figure 1a shows that a standard GPR setup has a transmitting antenna (Tx) and a receiving antenna (Rx) which are placed directly above the ground. EM pulses are released into the soil by the transmitter and move until they encounter changes in the soil's dielectric features, for example, at the boundary between the soil and roots or with areas of different moisture. A part of the wave energy is both reflected and captured by the receiver at all of the interfaces (Wu et al., 2020).

Figure 1. Principle of GPR detection: (a) schematic diagram of the GPR system, (b) A-scan and (c) B-scan

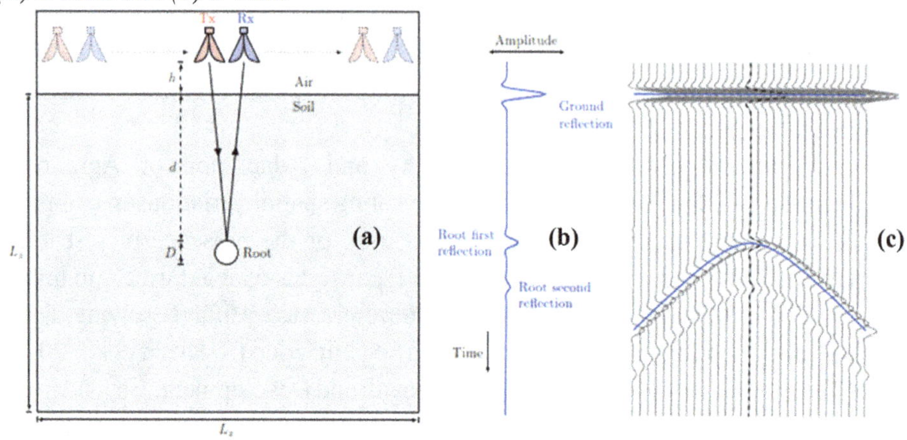

Figure 1b presents a typical A-scan trace, showing reflections corresponding to the ground surface and successive reflections from a root. These reflection events are temporally separated, allowing estimation of the characteristics of the buried object. As the GPR system is moved laterally across the surface, a series of A-scans is collected to form a B-scan, as illustrated in the Figure 1c.

Information about the position and the contrast in dielectric value of the target can be obtained from the shape and size of the hyperbola. Subsurface features are easier to find by images when there is a significant difference in dielectric permittivity between the materials. The amount of moisture, the size of the roots and soil texture can all affect how radar signals are either passed through or bounced back, due to changes in permittivity.

DIELECTRIC MIXING MODELS

To ensure physical realism in the simulations, the dielectric properties of soil and roots were calculated using established physical mixing models. The relative permittivity of roots was estimated using the Maxwell-Garnett model, which is suitable for heterogeneous media where inclusions (e.g., water and solid particles) are embedded in a host material (Wang et al., 2021). The effective relative permittivity, expressed as ε_{eff}, is determined by the following equation:

$$\varepsilon_{eff} = \varepsilon_a + \frac{\frac{1}{3}\sum_{i=1}^{2}\theta i(\varepsilon_i - \varepsilon_a)\sum_{j=1}^{3}\frac{\varepsilon_a}{\varepsilon_e + N_{i,j}(\varepsilon_i - \varepsilon_e)}}{1 - \frac{1}{3}\sum_{i=1}^{2}\theta_i(\varepsilon_i - \varepsilon_a)\sum_{j=1}^{3}\frac{\varepsilon_a}{\varepsilon_a + N_{i,j}(\varepsilon_i - \varepsilon_a)}}$$

where, ε_a is the permittivity of the host substance, ε_i and θ_i are the permittivity and portion (volume fraction) of the i-th material and $N_{i,j}$ indicate the depolarization factors in the j-th direction (Goshkoderia et al., 2020).

The Straub model was used to calculate electrical conductivity of roots, where the relationship is double-logarithmic between σ and θ (Liang et al., 2020).

$$log_{10}\left(\frac{\theta_v}{SG}\right) = 2.99 - 2.113\, log_{10}\left(log_{10}\frac{1}{\sigma}\right)$$

Here, *SG* represents the specific gravity of the root tissue. This model captures the nonlinear increase of conductivity with increasing water content in plant materials.

For the soil dielectric properties, the Wharton model, also known as the Complex Refractive Index Model (CRIM)(Zadhoush et al., 2021), was used to compute the effective permittivity as a weighted volumetric mixture of its components:

$$\varepsilon_r = \left((1 - \phi)\sqrt{\varepsilon_s} + \theta_v\sqrt{\varepsilon_w} + (\phi - \theta_v)\sqrt{\varepsilon_a}\right)^2$$

Where ϕ is the soil porosity, θ_v the volumetric water content and ε_s, ε_w and ε_a the permittivities of soil solids, water and air.

Furthermore, the statistical model by Guo (Guo et al., 2013) was applied to approximate soil electrical conductivity σ as a function of volumetric water content:

$$\sigma = 4.504\, e^{\theta_v}$$

METHODOLOGY

Data Generation

The simulation domain was defined as a two-dimensional vertical cross-section of a soil profile containing root-like cylindrical structures. The GPR system was modeled using a transmitter-receiver pair (Tx-Rx) placed 4 cm above the ground surface. As the system moved laterally across the domain, it collected A-scans at discrete intervals, which were later assembled into B-scans. The electromagnetic excitation was modeled using a Ricker wavelet with a central frequency of 1 GHz, offering an effective balance between resolution and penetration depth in typical agricultural soils.

The total simulation time window was carefully determined based on the maximum expected propagation delay under the extreme case of a fully saturated medium, which has the highest relative permittivity and thus the lowest wave velocity. This ensured that all reflected signals, including those from the deepest roots in highly conductive, water-rich soils, would be fully captured within the simulation. The resulting time window was set to 5.4 ns, which accommodates the slowest wave travel conditions anticipated in the designed scenarios. A spatial sampling step of 2 mm was applied in all directions to guarantee sufficient resolution of subsurface features.

To create a diverse and representative dataset, key physical and geometrical parameters of the soil-root system were systematically varied. This variation was designed to simulate various real-world agricultural conditions, by allowing for growth of roots, different moisture levels and various kinds of soil.

Root water content was simulated at 10 different points, starting from no moisture (0%) and ending at 100% (fully saturated). This range allowed the simulation of roots under varying hydration conditions, from water-stressed to optimal moisture states. Similarly, the soil water content was varied across 6 levels, covering the full spectrum from dry (0%) to near saturation (58%), which reflects common field scenarios in loamy and silty soils.

To represent various shapes of roots, this model has 5 levels of diameter within the range of 0.5 cm to 5 cm, imitating both new and mature root forms. The depth of root system considered ranged from shallow near the surface (2 cm deep) to much deeper mature roots (36 cm). The soil was designed with porosity at 40%, 50% and 58% which are the same as filler soils, moderately loose soils and highly porous agricultural soils.

By combining these parameter levels, a total of 4500 unique configurations was simulated, each producing a distinct B-scan image. These simulations were conducted with the help of gprMax software (Tosti & Umiliaco, 2014), an open-source tool that accurately reproduces GPR signals from different soil and root scenarios. The dataset includes measurements that show how water moisture, root shapes and electromagnetic waves interact in many different soil conditions.

Data Preprocessing

The Synthetic B-scan images were produced through electromagnetic simulation. subsequently different preprocessing methods employed to make the radar reflections more comprehensible and suitable for feeding into a machine learning model.

SVD was used mostly to reduce clutter by suppressing signals from sources such as antenna coupling and air-soil interface reflections (Solimene et al., 2014;

Tozan & Karatas, 2019). Each B-scan image $X \in \mathbb{R}^{m \times n}$, with m time samples and n spatial positions, was decomposed as:

$$X = U \Sigma V^T$$

where U and V are orthogonal matrices containing the left and right singular vectors, and Σ is a diagonal matrix containing the singular values $\lambda_1 \geq \lambda_2 \geq \cdots \geq \lambda_r$, with $r = min(m, n)$.

Clutter could be eliminated by dropping the first two components (usually λ_1 and λ_2), as they describe strong features like coupling from the antenna and how the surface reflects. The image was reconstructed again based on what remained of the components:

$$Y = \sum_{i=3}^{k} \lambda_i u_i v_i^T$$

A fixed value of $k = 10$ was used, meaning that after removing the first two dominant components, the remaining 8 singular components were used to reconstruct the decluttered image Y. This Selective reconstruction takes out the biggest sources of noise while preserving informative subsurface reflections caused by root and moisture anomalies.

IDA was then applied to make target features, the root-induced hyperbolic reflections, easier to see. For each trace in the decluttered image, the Hilbert transform $\mathcal{H}$ was used to determine the instantaneous amplitude envelope $A(t)$, showing the signal's local changes in amplitude:

$$A_i(t) = \sqrt{y_i^2 + \mathcal{H}\{y_i\}^2}$$

From this, an average envelope was created to demonstrate the usual change of signal strength throughout the soil layers. The curve of the mean envelope was adjusted to an exponential decay model to capturing the natural loss of energy as the electromagnetic wave propagates through the medium:

$$f(t) = A_0 e^{-\alpha t}$$

The A_0 is the starting amplitude and α is the attenuation coefficient.

At the end, the IDA correction was done by adjusting each trace so that each trace was normalized by the exponential fit from Equation (1), with a small constant introduced ($\delta > 0$)to avoid division by zero for numerical stability.

$$z_i(t) = \frac{y_i(t)}{f(t) + \delta}$$

The effectiveness of the proposed preprocessing method is illustrated in Figure 2. Figure 2a presents a selection of original synthetic B-scan radargrams generated under various subsurface scenarios, where dominant horizontal reflections from the ground surface and antenna coupling often obscure the subtler hyperbolic patterns caused by buried roots. After applying the decluttering and enhancement steps (SVD followed by IDA) the resulting images in Figure 2b exhibit improved clarity. Hyperbolic reflections become more distinguishable, while background clutter and surface-related signals are significantly reduced. This consistent enhancement across diverse conditions highlights the robustness of the preprocessing approach and its value in producing cleaner, more interpretable radargrams for downstream learning-based classification tasks.

Figure 2. Examples of GPR images illustrating the effect of preprocessing: (a) raw B-scans with clutter and antenna coupling, (b) decluttered images using SVD

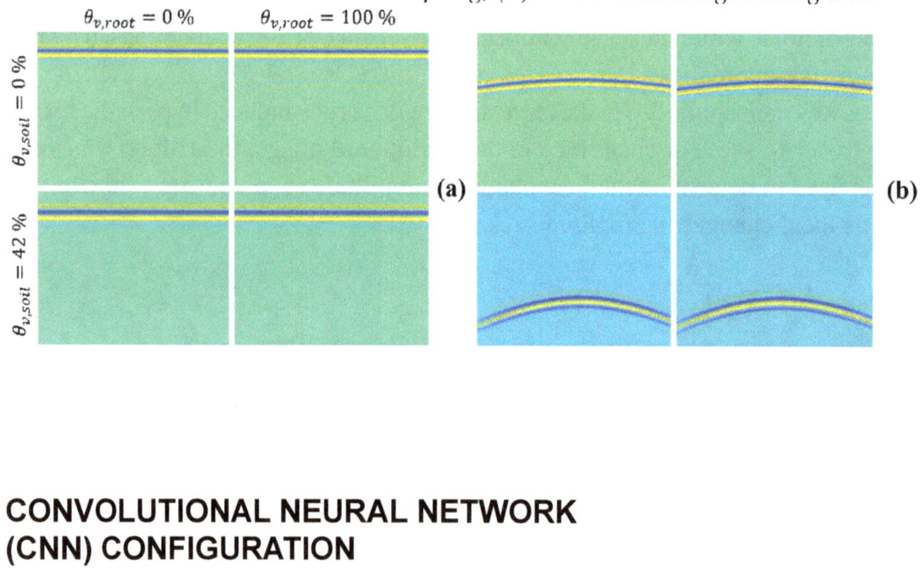

CONVOLUTIONAL NEURAL NETWORK (CNN) CONFIGURATION

Once data was prepared, the dataset was automatically classified using CNN into their compatible soil-root configurations. By using deep learning, the model can extract spatially localized features in the image which makes it useful for finding hyperbolic patterns of buried roots and changes in soil moisture.

CNNs are a class of deep learning architectures specifically designed to process grid-like data, such as images or time-frequency representations (Kumari, 2025).

Their success in image classification, object detection, and signal interpretation tasks stems from their ability to learn hierarchical feature representations directly from raw input (Ravikumar & Sriraman, 2023). In the context of this study, CNNs are well-suited for analyzing GPR B-scan images, which represent spatial distributions of radar reflections. the images often contain distinct geometric features, such as hyperbolas caused by buried cylindrical structures, like roots. They also show variations linked to moisture content.

As shown in Figure 3 the main process in a CNN is convolution, where small filters (kernels) slide over the image to detect key features. Every convolutional layer output mapping layer that highlights edges, curves, and gradients in the texture part of the image. Typically, a non-linear activation function called the Rectified Linear Unit (ReLU) follows each convolutional layer, to introduce non-linearity and enhance the model's capacity to learn complex patterns (Kallel & Kallel, 2023).

Figure 3. Schematic representation of the CNN architecture used for B-scan image classification

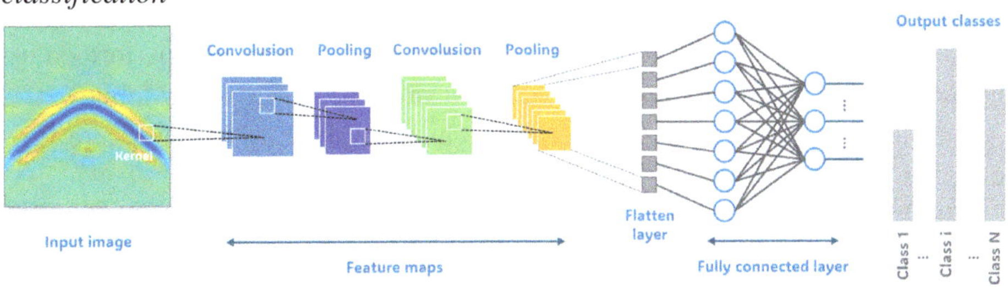

To reduce the dimensionality of the data and to make the model more computationally efficient, pooling Layers are added to the network to handle and process the data more easily. A useful technique is max pooling which only saves the key features in each region, making it better to keep important parts and ignore the rest. After completing several convolutional and pooling layers, the feature maps are flattened and given to one or more fully connected layers.

These final layers perform high-level reasoning and output predictions, such as class probabilities (Albawi et al., 2017; Stutz, 2014). Each output neuron corresponds to a moisture category (e.g., root or soil water content level), and its activation represents the model's confidence in that class.

The CNNs takes B-scans and matches them with labels it already has, while learning through supervision and lowering the loss value which is categorical cross-entropy. In an optimizer, thc weights in the model are altered by processing the calculation of gradients.(Hasan et al., 2020).

All the enhanced B-scan images were normalized to make their intensities similar. Consequently, every image was resized to 128×128 pixels which is what convolutional neural networks require as their input. The samples were each assigned a label according to their simulation details (root and soil moisture content), enabling supervised learning.

TRANSFER LEARNING

Even though CNNs can find deep features in an image, it is necessary to have a large amount of labelled data and a large amount of computer processing when starting from scratch. The adoption of transfer learning, a method that leverages the feature extraction capabilities of CNNs pre-trained on large-scale datasets such as ImageNet (which contains over a million labeled natural images), was key to overcoming this limitation and improving training efficiency.

Rather than starting from scratch, low-level models such as edges and textures can be kept and their high-level abstractions adapted to the target domain. The task of classifying soil-root configurations based on radar imagery was performed using three well-established deep learning architectures: InceptionV3, MobileNetV2 (Menaouer et al., 2022) and ResNet50V2 (Gupta et al., 2020). These were fine-tuned for the study.

The InceptionV3 architecture had been first introduced by (Ioffe & Szegedy, 2015).due to its depth and effective computation. It is constructed on the idea of an Inception module, where several convolutional processing of various sizes are carried out at the same time and hence enable the model to extract spatial information on the same layer in varying scales (Hammou et al., 2022). Once concatenated, these are fed to the subsequent layer allowing the model to extract and learn small and large patterns simultaneously. Another option in InceptionV3 is that dimensionality reduction methods are used to limit the computational cost, which is essential to not miss out on critical information (Sadik, 2024). Batch normalization and auxiliary classifiers are inserted at the intermediate layers to enhance convergence and decrease overfitting, especially in extremely deep networks. The consequence is a model that is both deep and efficient, making it appropriate to tasks that demand high-level abstraction but whose computation is too costly (Bjorck et al., 2018).

MobileNetV2 in turn was selected due to its lightweight deployment on mobile and embedded systems. It is efficient due to depth wise separable convolutions, an operation that breaks down an ordinary convolution into two simpler operations a depth wise convolution that filters each channel of the input independently and a pointwise convolution that concatenates the outputs. This method makes the number of parameters and computation complexity drastic. MobileNetV2 also adds inverted residuals and linear bottlenecks (Sandler et al., 2018); here the residual connection is on the narrow layer instead of the wide layer and the nonlinearity activation is not used in the bottleneck layers to keep the features intact. Al-though computationally light, MobileNetV2 competes effectively, especially in tasks involving classification problems in which computational limitations are significant. In the context of this work, its compact structure allowed for rapid experimentation and reduced training time without significantly compromising accuracy (Qin et al., 2023).

The ResNet50V2 model belongs to the ResNet style of architectures through which the concept of residual learning became very popular. In deep networks, performance can degrade, which is a problem for these networks. The is often caused by gradient Vanishing or overfitting the network. It may happen when the number of layers in the model increases. ResNet addresses these issues by using shortcut connections that enable the network to learn residual functions with respect to the layer inputs rather than learning the full mapping directly (He et al., 2015). Therefore, deep networks are easier to train because the gradients can be transmitted more smoothly during backpropagation. ResNet50V2 is more efficient than its predecessor because its creators tweaked the position of batch normalization and activation functions inside each residual block. The 50-layer version is useful. It allows the model to learn complex ideas. It is also simple enough to adjust and simple enough to fine-tune.

Each of these architectures was customized for the specific classification task by replacing the top layers with dense layers followed by softmax activation (Memisevic et al., 2010), allowing the models to output class probabilities aligned with the labeled moisture levels. Observing the experiment, we would either freeze the convolutional network of the models or unfroze it only for a part of it to improve how it functions on new data. Because of this, we can use a hybrid system where the model starts with broad visual features and then finishes tuning with radar colors and textures.

By putting in these different architectures, each with its own good points in taking out features, quickness or depth, the study tried to look at many design ideas in deep learning. InceptionV3 helped with multi-level feature showing; MobileNetV2 gave speed and smallness; and ResNet50V2 offered deep step-by-step learning with sure gradients. With these two methods, they were able to evaluate their experiments well, generalize their conclusions and compare the results of the different models.

Hyperparameter Tuning

The appropriate pre-trained architectures were chosen and afterward, extensive random searches helped pick the best hyperparameters for each model. This included adjusting how quickly the learning progresses, how often dropout was used, the number of samples in a batch and the type of activation functions, optimization algorithms and units found in the dense layers. Instead of using grid search to explore vast hyperparameter ranges, random search randomly samples different combinations and can often do it more efficiently (Dimgba & Andreas, 2024) Overfitting issues can be solved. Models can be made more general. L1 and L2 regularization were introduced to do this. Regularization keeps model weights within limits which leads to simpler models that are less likely to overfit the training data (Toshpulatov et al., 2025).

In addition, it is possible to automatically halt the training process by using early stopping criterion. If performance on the validation set does not improve for fixed number of epochs, the training process will end. The number of clusters is determined during model preparation, before any training happens (Mahsereci et al., 2017). For this reason, the risk that the model would diverge from real information was reduced. Algorithmic thinking made it possible to use less computation time and effort. Blended learning focused on fine-tuning which helped retain the weights from ImageNet learned at the first stages. In addition, the top layers were modified so the models could easily spot important information in the ground-penetrating radar.

Model Performance Metrics

Model predictions should be evaluated using validation and test data after the models are trained and the hyperparameters are set to confirm their reliability. Tuning the model involves testing weights by applying the training data multiple times and the validation set as an indicator of how well the model will perform with new (not exposed) data. The final assessment of the chosen model happens on the test set which is not included in either training or validation. With this method, the evaluation of the metrics ensures that the model generalizes well to data it has not seen.

Performance indicators used to check how models work in classification tasks are accuracy, precision, recall and the F1-score. They all give different insights into the working of the model in a variety of conditions. Accuracy establishes whether our predictions are generally right and precision establishes whether our positive predictions are reliable. Recall measures how well the model identifies all the significant cases, and the F1-score is a general score calculated as the mean of precision and recall and seems especially appropriate in the case that classes are not balanced.

Both validation and testing use certain performance indicators for assessing how well the model separates classes. Using these metrics allows a broad review of how well the model predicts in several types of situations. Common performance metrics for classification are accuracy, precision, recall and the F1-score (Baranwal et al., 2020; Khan et al., 2020). We obtain these metrics by analyzing the confusion matrix which shows the results of our predictions by separating true positives (TP), true negatives (TN), false positives (FP) and false negatives (FN) (Dobbs & Ras, 2022).

Accuracy tells us the percentage of correct classifications among all the given examples:

$$Accuracy = \frac{TP + TN}{TP + TN + FP + FN}$$

Precision (also called positive predictive value) describes the percentage of correct positive predictions among those that were predicted to be positive:

$$Precision = \frac{TP}{TP + TN}$$

The term recall (also known as sensitivity or true positive rate) indicates how many real of actual positive cases were correctly identified:

$$Recall = \frac{TP}{TP + FN}$$

F1-score (harmonic mean of precision and recall) combines precision and recall giving a balanced measure that takes both false positives and false negatives into account:

$$F1 - score = 2 \times \frac{Precision \times Recall}{Precision + Recall}$$

RESULTS

The analysis was carried out using a dataset tailored for root and soil water content estimation. After preprocessing, the data were split into training (70%), validation (20%), and testing (10%) sets to support reliable model evaluation. Model implementation and training were performed using the TensorFlow and Keras libraries, which offer the necessary tools for constructing and optimizing deep learning architectures. The results focus on training progression, convergence behavior, and the predictive capabilities of the selected model.

Root Water Content Prediction

Among the three architectures, InceptionV3 consistently achieved the highest performance in root water content prediction, reaching a validation accuracy of approximately 97% after only a few epochs (Figures 4a and 4b). Its smooth training curves indicate strong generalization and effective feature learning. The model's architecture (with multiple scales of filters and additional classifiers) contributed to its ability to track changes in B-scans associated with moisture levels. The confusion matrix in Figure 4c confirms its robustness in fine-grained classification, accurately distinguishing even closely related moisture classes.

Figure 4. Training and validation performance of InceptionV3 (a, b, and c), MobileNetV2 (d, e and f) and ResNet50V2 (g, h and i) models for GPR-based root water content prediction

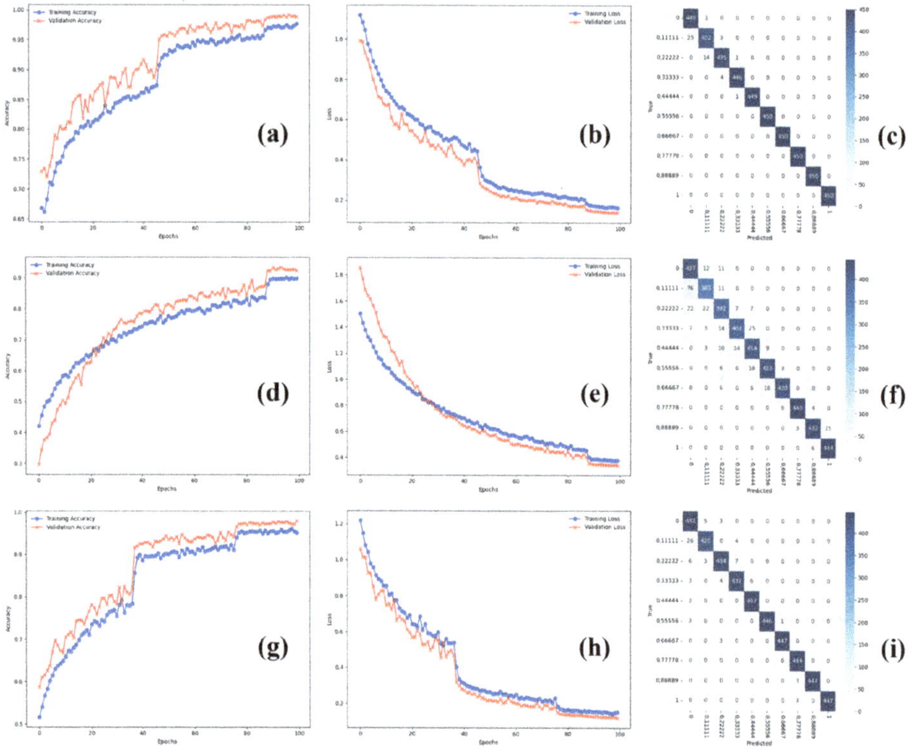

In comparison, MobileNetV2 demonstrated lower accuracy and stability. While it eventually reached a validation accuracy of around 88%, its convergence was slower and characterized by notable fluctuations in validation loss (Figures 4d and 4e). The confusion matrix (Figure 4f) revealed frequent misclassifications, especially among intermediate moisture classes, indicating limited representational capacity. Although MobileNetV2 is optimized for computational efficiency and suitable for deployment on resource-constrained platforms, its reduced complexity hinders its effectiveness in this task.

ResNet50V2 achieved a strong balance between complexity and performance. With a validation accuracy approaching 96% and consistently low loss values (Figures 4g and 4h), the model exhibited reliable convergence and minimal overfitting. Its residual connections facilitated efficient gradient flow, supporting the learning of domain-specific features. As shown in the confusion matrix (Figure 4i), ResNet50V2 maintained balanced classification performance across all categories, with only slight confusion between adjacent classes. These results suggest that ResNet50V2 offers a favorable trade-off between accuracy and model complexity in root water content estimation.

Soil Water Content Prediction

In the soil water content classification task, all three models achieved perfect classification accuracy on the validation set. Nevertheless, their training behaviors varied significantly, offering insight into their relative efficiency and stability. InceptionV3 once again showed strong performance, converging rapidly with high accuracy and low validation loss (Figures 5a and 5b). Minor fluctuations in training loss were observed, reflecting sensitivity to its deeper architecture. Despite this, the confusion matrix in Figure 5c indicates flawless classification across all moisture categories.

MobileNetV2 also reached perfect classification accuracy but exhibited unstable training behavior during the early epochs as shown in Figures 5d and 5e. The validation accuracy fluctuated within the first 20 epochs, and the loss curve revealed sensitivity to training dynamics. Once stabilized, however, the model performed reliably, as illustrated in Figure 5f. These results suggest that MobileNetV2 can be highly effective for this task but requires careful hyperparameter tuning to ensure stable convergence.

ResNet50V2 demonstrated the most stable and efficient training behavior. As illustrated in Figures 5g and 5h, the training and validation curves closely overlapped from the beginning, and the loss quickly decreased to negligible levels. Its skip connections promoted smooth gradient flow, mitigating vanishing gradient issues, and enabling fast, consistent learning. The confusion matrix in Figure 5i confirms perfect classification with no mislabeling.

Figure 5. Training and validation performance of InceptionV3 (a, b, and c), MobileNetV2 (d, e and f) and ResNet50V2 (g, h and i) models for GPR-based soil water content prediction

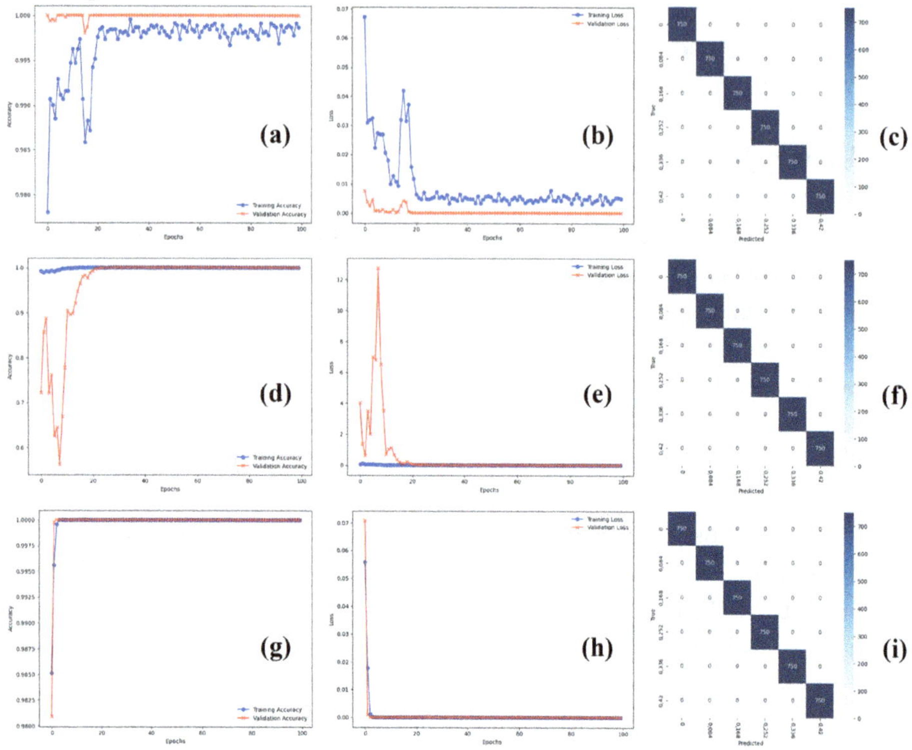

Table 2 presents a comparative overview of the accuracy, precision, recall, and F1-score achieved by each model for both root and soil water content prediction tasks. The results confirm that InceptionV3 is the best-performing architecture for root water content classification, offering the highest accuracy along with balanced precision and recall.

Table 2. Classification metrics for root and soil water content prediction

Task	Model	Accuracy	Precision	Recall	F1-score
Root water content	InceptionV3	0.9891	0.9893	0.9891	0.9891
	MobileNetV2	0.9236	0.9263	0.9236	0.9263
	ResNet50V2	0.9802	0.9807	0.9802	0.9803

continued on following page

Table 2. Continued

Task	Model	Accuracy	Precision	Recall	F1-score
Soil water content	InceptionV3	1.0000	1.0000	1.0000	1.0000
	MobileNetV2	1.0000	1.0000	1.0000	1.0000
	ResNet50V2	1.0000	1.0000	1.0000	1.0000

In contrast, MobileNetV2 shows reduced accuracy on the more complex root water prediction task, due to its lighter architecture, although it remains attractive for applications where computational efficiency is prioritized. ResNet50V2 presents a strong compromise with high accuracy and consistent training stability.

Considering soil water content estimation, all models have a perfect score, which is an indication of the rather easy classification task. Such consistent performance indicates that even the lighter architectures, such as MobileNetV2, can be easily deployed to perform soil moisture sensing, especially resource-limited areas.

DISCUSSION

Comparative analysis of the three models reveals trade-offs and the unique benefits. The better accuracy of RWC in InceptionV3 makes it especially applicable to applications that demand accurate proprieties of root-zone water content, like determining restrictive irrigation occasions. Nevertheless, it has average stability and performance and may not necessarily work reliably in real-time or resource-fractured environments.

Although the RWC accuracy in MobileNetV2 is lower, it has an outstanding level of stability and efficiency. Its light architecture enables quicker training and inference and therefore is appealing to be used in field applications where computational resources and response time is limited. This indicates that the use of MobileNetV2 is viable in cases where high speed predictions, especially in a real-life environment, are preferred over optimal accuracy.

ResNet50V2 will form a compromise between a high level of accuracy in both RWC and SWC tasks and high stability. It is a relatively good general purpose choice due to its moderate computational requirements and thus is applicable in high accuracy situations where reliability must be met but resource intensive requirements are not demanded.

The findings highlight the absence of a single model that would lead to maximum performance on all metrics. Rather, architecture should be decided according to intended application needs. InceptionV3 can have an advantage in high-precision setup, whereas MobileNetV2 could be useful in a mobile or resource-constrained scenario. ResNet50V2, then, strikes an average between the two ends.

CONCLUSION

The chapter has presented a new frame of non-destructive monitoring of underground root and soil water content with GPR and modern deep learning algorithms. To overcome the lack of labeled data, 4,500 synthetic B-scan setups were generated, and they contained a wide variety of root and soil models. These simulations offered a training environment that could support generalizable models. Initial pre-processing of radargrams using SVD and IDA allowed further preprocessing to clarify the radargrams and facilitate more accurate interpretation.

Comparative analysis of the InceptionV3, MobileNetV2, and ResNet50V2 has shown that they are the most suitable approaches based on their strengths: InceptionV3 was best when making predictions of fine-grained root water content; MobileNetV2 was helpful in a scenario with a limited number of resources; and ResNet50V2 proved to be the most suitable approach subject to the trade-off between prediction accuracy and training stability.

This methodology has the benefit of providing a scalable, precise, and non-destructive system to characterize subsurface composition with applications in irrigation management, crop breeding, and environmental monitoring. Further study is needed to further extrapolate the framework to include heterogeneous soil and different climates as well as different crop systems to fully verify whether this is applicable in the real world. Additionally, incorporating explainable AI and multimodal sensing will make it more interpretable and provide more insight into agricultural and environmental processes.

REFERENCES

Adeyemi, O., Grove, I., Peets, S., & Norton, T. (2017). Advanced Monitoring and Management Systems for Improving Sustainability in Precision Irrigation. *Sustainability, 9*(3), 3. Advance online publication. DOI: 10.3390/su9030353

Ahmed, Md. T., Ahmed, M. W., & Kamruzzaman, M. (2025). A systematic review of explainable artificial intelligence for spectroscopic agricultural quality assessment. *Computers and Electronics in Agriculture, 235*, 110354. DOI: 10.1016/j.compag.2025.110354

Albawi, S., Mohammed, T. A., & Al-Zawi, S. (2017). Understanding of a convolutional neural network. *2017 International Conference on Engineering and Technology (ICET)*, 1–6. https://doi.org/DOI: 10.1109/ICEngTechnol.2017.8308186

Baranwal, A., Bagwe, B. R., & M, V. (2020). Machine Learning in Python: Diabetes Prediction Using Machine Learning. In S. Velayutham (Ed.), *Advances in Computational Intelligence and Robotics* (pp. 128–154). IGI Global. https://doi.org/DOI: 10.4018/978-1-5225-9902-9.ch008

Bjorck, J., Gomes, C., Selman, B., & Weinberger, K. Q. (2018). *Understanding Batch Normalization* (Version 4). arXiv. https://doi.org/DOI: 10.48550/ARXIV.1806.02375

Cheng, Q., Su, Q., Binley, A., Liu, J., Zhang, Z., & Chen, X. (2023). Estimation of Surface Soil Moisture by a Multi-Elevation UAV-Based Ground Penetrating Radar. *Water Resources Research, 59*(2), e2022WR032621. https://doi.org/DOI: 10.1029/2022WR032621

De Silva, C., Halloluwa, T., & Vyas, D. (2025). *A Multi-Layered Research Framework for Human-Centered AI: Defining the Path to Explainability and Trust* (Version 2). arXiv. https://doi.org/DOI: 10.48550/ARXIV.2504.13926

Dimgba, M. O., & Andreas, R. (2024). *Optimal Hyperparameter Search Strategies: Benchmarking Grid Search, Random Search, and Genetic Algorithms across Regression, Classification, and Clustering Tasks*. https://doi.org/DOI: 10.13140/RG.2.2.24957.37603

Dobbs, B. T., & Ras, Z. W. (2022). Machine Learning Approach to Art Authentication. In Wang, J. (Ed.), *Encyclopedia of Data Science and Machine Learning* (pp. 1500–1513). IGI Global., DOI: 10.4018/978-1-7998-9220-5.ch089

Elshaboury, N., Mohammed Abdelkader, E., Al-Sakkaf, A., & Zayed, T. (2023). A Critical Review and Bibliometric Analysis on Applications of Ground Penetrating Radar in Science Based on Web of Science Database. *Eng*, *4*(1), 1. Advance online publication. DOI: 10.3390/eng4010059

Goshkoderia, A., Arora, N., Slesarenko, V., Li, J., Chen, V., Juhl, A., Buskohl, P., & Rudykh, S. (2020). Tunable permittivity in dielectric elastomer composites under finite strains: Periodicity, randomness, and instabilities. *International Journal of Mechanical Sciences*, *186*, 105880. DOI: 10.1016/j.ijmecsci.2020.105880

Guo, L., Lin, H., Fan, B., Cui, X., & Chen, J. (2013). Forward simulation of root's ground penetrating radar signal: Simulator development and validation. *Plant and Soil*, *372*(1), 487–505. DOI: 10.1007/s11104-013-1751-8

Gupta, K., Rani, R., & Bahia, N. K. (2020). Plant-Seedling Classification Using Transfer Learning-Based Deep Convolutional Neural Networks. *International Journal of Agricultural and Environmental Information Systems*, *11*(4), 25–40. DOI: 10.4018/IJAEIS.2020100102

Hammou, D. R., Mahmoudi, S. A., & Adjoudj, R. (2022). Multi-Biometric Iris Recognition System Using Consensus Between Convolutional Neural Network Architectures. *International Journal of Organizational and Collective Intelligence*, *12*(1), 1–30. DOI: 10.4018/IJOCI.305210

Hasan, M. M., Rahman, M. S., & Bell, A. (2020). Deep Reinforcement Learning for Optimization: In I. R. Management Association (Ed.), *Research Anthology on Artificial Intelligence Applications in Security* (pp. 1598–1614). IGI Global. https://doi.org/DOI: 10.4018/978-1-7998-7705-9.ch070

He, K., Zhang, X., Ren, S., & Sun, J. (2015). *Deep Residual Learning for Image Recognition* (No. arXiv:1512.03385). arXiv. https://doi.org//arXiv.1512.03385DOI: 10.48550

Ingrao, C., Strippoli, R., Lagioia, G., & Huisingh, D. (2023). Water scarcity in agriculture: An overview of causes, impacts and approaches for reducing the risks. *Heliyon*, *9*(8), e18507. DOI: 10.1016/j.heliyon.2023.e18507 PMID: 37534016

Ioffe, S., & Szegedy, C. (2015). *Batch Normalization: Accelerating Deep Network Training by Reducing Internal Covariate Shift* (Version 3). arXiv. https://doi.org/DOI: 10.48550/ARXIV.1502.03167

Jahin, M. A., Shahriar, S., Mridha, M. F., & Hossen, Md. J., & Dey, N. (2025). *Soybean Disease Detection via Interpretable Hybrid CNN-GNN: Integrating MobileNetV2 and GraphSAGE with Cross-Modal Attention* (Version 3). arXiv. https://doi.org/DOI: 10.48550/ARXIV.2503.01284

Kallel, I. F., & Kallel, M. (2023). Intelligent Watermarking for Data Security: An Overview. In Chana, I., Bouazi, A., & Ben-azza, H. (Eds.), (pp. 262–282). Advances in Systems Analysis, Software Engineering, and High Performance Computing. IGI Global., DOI: 10.4018/979-8-3693-0497-6.ch015

Khan, S., Khan, A., Ullah, R., Ali, M., & Ullah, R. (2020). Insulin DNA Sequence Classification Using Levy Flight Bat With Back Propagation Algorithm. In Umair, S. (Ed.), (pp. 232–252). Advances in Medical Technologies and Clinical Practice. IGI Global., DOI: 10.4018/978-1-7998-2521-0.ch011

Ko, J., Lee, K., Hwang, H., Oh, S.-G., Son, S.-W., & Shin, K. (2022). Effective training strategies for deep-learning-based precipitation nowcasting and estimation. *Computers & Geosciences*, *161*, 105072. DOI: 10.1016/j.cageo.2022.105072

Kumari, P. (2025). Transforming Medical Imaging With Convolutional Neural Networks (CNNs): Advances in Diagnosis and Treatment. In Aamir, M., Bhatti, U. A., Rahman, Z., Bhutto, J. A., & Abro, W. A. (Eds.), *Advances in Medical Diagnosis, Treatment, and Care* (pp. 195–230). IGI Global., DOI: 10.4018/979-8-3693-9816-6.ch009

Kurtulmuş, E., Arslan, B., & Kurtulmuş, F. (2022). Deep learning for proximal soil sensor development towards smart irrigation. *Expert Systems with Applications*, *198*, 116812. DOI: 10.1016/j.eswa.2022.116812

Lamichhane, M., Mehan, S., & Mankin, K. R. (2025). Soil Moisture Prediction Using Remote Sensing and Machine Learning Algorithms: A Review on Progress, Challenges, and Opportunities. *Remote Sensing*, *17*(14), 2397. DOI: 10.3390/rs17142397

Li, F., Yang, F., Xie, Y., Qiao, X., Du, C., Li, C., Ru, Q., Zhang, F., Gu, X., & Yong, Z. (2024). Research on 3D ground penetrating radar deep underground cavity identification algorithm in urban roads using multi-dimensional time-frequency features. *NDT & E International*, *143*, 103060. DOI: 10.1016/j.ndteint.2024.103060

Liang, H., Xing, L., & Lin, J. (2020). Application and Algorithm of Ground-Penetrating Radar for Plant Root Detection: A Review. *Sensors (Basel)*, *20*(10), 10. Advance online publication. DOI: 10.3390/s20102836 PMID: 32429394

Liu, X., Chen, J., Butnor, J. R., Qin, G., Cui, X., Fan, B., Lin, H., & Guo, L. (2020). Noninvasive 2D and 3D Mapping of Root Zone Soil Moisture Through the Detection of Coarse Roots With Ground-Penetrating Radar. *Water Resources Research, 56*(5), e2019WR026930. https://doi.org/DOI: 10.1029/2019WR026930

Liu, X., Cui, X., Guo, L., Chen, J., Li, W., Yang, D., Cao, X., Chen, X., Liu, Q., & Lin, H. (2019). Non-invasive estimation of root zone soil moisture from coarse root reflections in ground-penetrating radar images. *Plant and Soil, 436*(1), 623–639. DOI: 10.1007/s11104-018-03919-5

Mahsereci, M., Balles, L., Lassner, C., & Hennig, P. (2017). *Early Stopping without a Validation Set* (Version 3). arXiv. https://doi.org/DOI: 10.48550/ARXIV.1703.09580

Mangel, A. R., Moysey, S. M. J., & Bradford, J. (2020). Reflection tomography of time-lapse GPR data for studying dynamic unsaturated flow phenomena. *Hydrology and Earth System Sciences, 24*(1), 159–167. DOI: 10.5194/hess-24-159-2020

Margiwiyatno, A., Sudarmaji, A., Siswantoro, S., Masrukhi, M., Irawadi, I., Wijaya, K., Kuncoro, P. H., Sulistyo, S. B., & Furqon, F. (2025). Use of Capacitive Sensors for Measuring Soil Water Content in Irrigation Systems: A Review. *BIO Web of Conferences, 158*, 01001. https://doi.org/DOI: 10.1051/bioconf/202515801001

Memisevic, R., Zach, C., Pollefeys, M., & Hinton, G. E. (2010). Gated Softmax Classification. *Advances in Neural Information Processing Systems, 23*. https://papers.nips.cc/paper_files/paper/2010/hash/5737c6ec2e0716f3d8a7a5c4e0de0d9a-Abstract.html

Menaouer, B., Zoulikha, D., El-Houda, K. N., Mohammed, S., & Matta, N. (2022). Coronavirus Pneumonia Classification Using X-Ray and CT Scan Images With Deep Convolutional Neural Network Models. *Journal of Information Technology Research, 15*(1), 1–23. DOI: 10.4018/JITR.299391

Mohan, R. N. V. J., Rayanoothala, P. S., & Sree, R. P. (2025). Next-gen agriculture: Integrating AI and XAI for precision crop yield predictions. *Frontiers in Plant Science, 15*, 1451607. DOI: 10.3389/fpls.2024.1451607 PMID: 39845494

Moutaouekkil, M. A., Karkri, A., Koulali, M. A., Taybi, C., & Kahlaoui, M. (2024). Plant Root Characterization Using Ground-Penetrating Radar with Deep Learning. *Arabian Journal for Science and Engineering*. Advance online publication. DOI: 10.1007/s13369-024-09502-8

Namdari, H., Moradikia, M., Zekavat, S., Askari, R., Mangoubi, O., & Petkie, D. (2025). Advancing Precision Agriculture: Machine Learning-Enhanced GPR Analysis for Root-Zone Soil Moisture Assessment in Mega Farms. *IEEE Transactions on AgriFood Electronics*, *3*(1), 98–109. DOI: 10.1109/TAFE.2024.3455238

Qin, Y., Tang, Q., Xin, J., Yang, C., Zhang, Z., & Yang, X. (2023). A Rapid Identification Technique of Moving Loads Based on MobileNetV2 and Transfer Learning. *Buildings (Basel, Switzerland)*, *13*(2), 572. DOI: 10.3390/buildings13020572

Rasheed, M. W., Tang, J., Sarwar, A., Shah, S., Saddique, N., Khan, M. U., Imran Khan, M., Nawaz, S., Shamshiri, R. R., Aziz, M., & Sultan, M. (2022). Soil Moisture Measuring Techniques and Factors Affecting the Moisture Dynamics: A Comprehensive Review. *Sustainability*, *14*(18), 11538. DOI: 10.3390/su141811538

Ravikumar, A., & Sriraman, H. (2023). Understanding Convolutional Neural Network With TensorFlow: CNN. In Baby Maruthi, P., Prasad, S., & Tyagi, A. K. (Eds.), (pp. 50–73). Advances in Systems Analysis, Software Engineering, and High Performance Computing. IGI Global., DOI: 10.4018/978-1-6684-8531-6.ch003

Sadik, S. A. (2024). Deep Learning Classification of Empty and Full Container Ships in Satellite Imagery: A Cognitive Modeling Approach for Maritime Applications. In Ahmad, S., Anwar, M. S., Alluhaidan, A. S., & El-Affendi, M. A. (Eds.), (pp. 30–54). Advances in Geospatial Technologies. IGI Global., DOI: 10.4018/979-8-3693-2913-9.ch002

Sandler, M., Howard, A., Zhu, M., Zhmoginov, A., & Chen, L.-C. (2018). MobileNetV2: Inverted Residuals and Linear Bottlenecks. *2018 IEEE/CVF Conference on Computer Vision and Pattern Recognition*, 4510–4520. https://doi.org/DOI: 10.1109/CVPR.2018.00474

Shamir, O., Goldshleger, N., Basson, U., & Reshef, M. (2016). MAPPING SPATIAL MOISTURE CONTENT OF UNSATURATED AGRICULTURAL SOILS WITH GROUND-PENETRATING RADAR. *The International Archives of the Photogrammetry, Remote Sensing and Spatial Information Sciences*, *XLI-B8*, 1279–1285. XXIII ISPRS Congress, Commission VIII (Volume XLI-B8) - 12–19 July 2016, Prague, Czech Republic. https://doi.org/DOI: 10.5194/isprs-archives-XLI-B8-1279-2016

Solimene, R., Cuccaro, A., Dell'Aversano, A., Catapano, I., & Soldovieri, F. (2014). Ground Clutter Removal in GPR Surveys. *IEEE Journal of Selected Topics in Applied Earth Observations and Remote Sensing*, *7*(3), 792–798. DOI: 10.1109/JSTARS.2013.2287016

Stutz, D. (2014). *Understanding Convolutional Neural Networks.* https://www.semanticscholar.org/paper/Understanding-Convolutional-Neural-Networks-Stutz/02180e8704cdffb76974da6250127fdfe3dc78e9

Toshpulatov, M., Khamdamov, O., Abdullaeva, N., Otakhonova, S., Botirov, A., & Khurazov, I. (2025). Machine Learning for Enhancing Workforce Safety in Automotive Manufacturing. In Vyas, V., Bi, X., Pandey, D., Xu, Z., & Pandey, B. K. (Eds.), (pp. 345–358). Advances in Computational Intelligence and Robotics. IGI Global., DOI: 10.4018/979-8-3373-0442-7.ch022

Tosti, F., & Umiliaco, A. (2014). FDTD Simulation of the GPR Signal for Preventing the Risk of Accidents due to Pavement Damages. *International Journal of Interdisciplinary Telecommunications and Networking*, *6*(1), 1–9. DOI: 10.4018/ijitn.2014010101

Tozan, H., & Karatas, M. (Eds.). (2019). *Operations Research for Military Organizations.* IGI Global., DOI: 10.4018/978-1-5225-5513-1

Vahidi, M., Shafian, S., & Frame, W. H. (2025). Multi-Modal sensing for soil moisture mapping: Integrating drone-based ground penetrating radar and RGB-thermal imaging with deep learning. *Computers and Electronics in Agriculture*, *236*, 110423. DOI: 10.1016/j.compag.2025.110423

Vengatesh, T. (2025). Transparent Decision-Making with Explainable Ai (Xai): Advances in Interpretable Deep Learning. *Journal of Information Systems Engineering & Management*, *10*(4), 1295–1303. DOI: 10.52783/jisem.v10i4.10584

Wang, P. F., Jiang, H. Y., Shi, J. A., Liu, M., Gu, L., Zhou, W., Orava, J., Sun, Y. H., Wang, W. H., & Bai, H. Y. (2021). Regulated color-changing metallic glasses. *Journal of Alloys and Compounds*, *876*, 160139. DOI: 10.1016/j.jallcom.2021.160139

Wenhan, Z., Keming, D. U., Yankun, S. U. N., Buchun, L. I. U., Zhongfu, S. U. N., Juncheng, M. A., & Feixiang, Z. (2022). Monitoring Specified Depth Soil Moisture in Field Scale with Ground Penetrating Radar. *Smart Agriculture*, *4*(1), 84. DOI: 10.12133/j.smartag.SA202202010

Wu, X., Senalik, C. A., Wacker, J., Wang, X., & Li, G. (2020). Object Detection of Ground-Penetrating Radar Signals Using Empirical Mode Decomposition and Dynamic Time Warping. *Forests*, *11*(2), 2. Advance online publication. DOI: 10.3390/f11020230

Zadhoush, H., Giannopoulos, A., & Giannakis, I. (2021). Optimising the Complex Refractive Index Model for Estimating the Permittivity of Heterogeneous Concrete Models. *Remote Sensing*, *13*(4), 4. Advance online publication. DOI: 10.3390/rs13040723

Zou, L., Li, Y., Munisami, K., & Alani, A. M. (2025). Bridging Theory and Practice: A Review of AI-Driven Techniques for Ground Penetrating Radar Interpretation. *Applied Sciences (Basel, Switzerland)*, *15*(15), 8177. DOI: 10.3390/app15158177

Chapter 11
AI-Driven Wastewater Treatment and Recycling: Smart Water

Likhon Chandra Roy
https://orcid.org/0009-0003-6943-6777
Rajshahi University of Engineering and Technology, Bangladesh

Mehedi Hashan Riad
https://orcid.org/0009-0003-0201-2099
Bangladesh University of Engineering and Technology, Bangladesh

Md. Asadullahil Galib Fardin
Rajshahi University of Engineering and Technology, Bangladesh

Md. Rafiul Islam
Rajshahi University of Engineering and Technology, Bangladesh

Samanta Alam
Chicago State University, USA

Dhananjay Ray
Centre for the Development of Human Initiatives, India

ABSTRACT

Life depends on water, yet its management is under more stress than ever. Besides, contamination and the scarcity of water made it impossible to fulfill humans' basic needs. Artificial intelligence (AI) provides revolutionary possibilities through real-time monitoring, statistical analysis, and machine learning, while more than 80% of polluted water worldwide is released untreated. In this chapter, the strategies behind the application of AI in wastewater treatment and recycling are explained. It addresses both opportunities and constraints while striking a balance between creativity and reality. The chapter bridges the gap between technical depth and readable storytelling by referencing industry data, scholarly research, and expert perspectives. It draws attention to how AI is changing water sustainability by transforming wastewater from a liability into a resource.

DOI: 10.4018/979-8-3373-2700-6.ch011

INTRODUCTION

Water is one of the fundamental elements for living organisms that can't be replaced with other resources. One of the biggest issues endangering sustainable development in the twenty-first century is global water scarcity. Water scarcity and contamination are escalating daily due to rising demand (Senthil Rathi et al., 2025). The world's population increased tremendously over the past few decades, and rapid industrialization and urbanization have also a direct relation with water scarcity and demand, which leads to an unsustainable, imbalanced ecology (Yalcin & Ayyildiz, 2024). Thereafter, the abundant discharge of untreated wastewater from residential, agricultural, and industrial sectors turns out to be the most predominant source of surface water contamination. The quantity and quality of water sustain ecological equilibrium, which in turn impacts human lifestyle (Baarimah et al., 2024).On the other hand, contaminated water is the main cause of disease and makes the ground unfit for sustaining life. Due to the abundance of good-quality water supplies and their accessibility, water quality concerns have frequently been disregarded. Rapid growth in the industrial and agricultural sectors is being fueled by the world's expanding population, which raises the need for water, which is essential for supporting all life. The main sources of water for industry, agriculture, and human and animal use are rivers, groundwater, and lakes. However, in many parts of the world, climate change is increasing the frequency of floods and droughts (Chavhan et al., 2025). Furthermore, the amount and quality of drinkable water are being significantly deteriorated by water pollution from a variety of sources, including families, towns, farms, and industrial operations. As a result, increasing the quantity and quality of drinking water requires that effluent be properly treated before being released into bodies of water.

While the past few decades have driven rapid development across sectors, unforeseen future challenges, particularly in water management, loom large. As a consequence, the future is in danger if necessary steps are not taken for the sustainability of water and water management. Water is a key component for agriculture, industry, and households, but this demand can't be fulfilled with traditional management and technologies; hence, advanced technologies like machine learning and artificial intelligence are needed. These technologies offer a robust and user-friendly system that can be operated far from the site and also offer an easy way to monitor the water treatment system and the recycling process.

Smart cities are leading the way in tackling these sustainability concerns by utilizing digital technology to enhance urban administration. For cities to continue to be resilient, livable, and ecologically friendly, effective water resource management is essential. Sustainable wastewater management and water use techniques help to lessen the negative consequences of climate change, conserve natural resources, and reduce pollution (Jessica et al., 2025). Previously seen as a byproduct, wastewater is now seen as an essential resource in the search for sustainable water sources. Nevertheless, the effectiveness, responsiveness, and scalability of traditional therapy methods are frequently lacking. Artificial Intelligence (AI), which provides tools for prediction, automation, and optimization, is becoming a disruptive force in wastewater treatment and recycling. Machine learning algorithms have the potential to significantly improve learning when used for a specific purpose. Undiscovered patterns in newly created data sets with exponential growth cannot be addressed by traditional approaches. Food and Agriculture, utility services, manufacturing, mining, hydropower production, aquaculture, and cattle husbandry are just a few of the industries that depend on effective water management (Jana, 2024). Kevin Ashton created the term "Internet of Things" (IoT) in 1999 to describe linked electrical items. IoT primarily focuses on data exchange among connected gadgets. The Internet of Things (IoT) technology offers benefits for managing water, such as energy control, cutting expenses, and increased efficiency (Hassan, 2024). Cyber-Physical Systems (CPS) are integrated frameworks that combine computational algorithms, communication networks, and physical processes to enable real-time monitoring, control, and optimization of complex systems. Interaction of sensors and actuators with the natural environment, while computers and advanced technologies process data and make decisions, based on the decision, the implementation of appropriate actions takes place in the facilities.

Smart water is not simply a shift in technology, but rather a fundamental change in the approach to water throughout its entire life cycle. The idea of smart water goes beyond a mere technology development; it connotes a total overhaul in managing water throughout its existence. This movement concentrates on bringing AI into wastewater treatment systems for improved monitoring, adaptive process management, and data-driven optimization. AI algorithms may analyze huge volumes of real-time data from sensors installed in treatment facilities to predict process behavior under various situations, detect anomalies, and recommend optimum operating strategies. Machine learning, deep learning, fuzzy logic, and neural networks help treatment plants go beyond conventional rule-based approaches toward models that can learn, adapt, and optimize on their own (Cairone et al., 2024). As demands increase, technological advancements are reflected in the evolution of wastewater treatment systems. At first, manual control depended on human operators to monitor procedures and adjust valves; this method was constrained by labor intensity and irregularities. While still rule-based and reactive, real-time monitoring and centralized control were made possible by the introduction of Supervisory Control and Data Acquisition (SCADA) systems and Programmable Logic Controllers (PLCs) in the late 20th century. The transition to AI-assisted systems, especially since the 2010s, uses IoT sensors and machine learning (e.g., ANN, LSTM) for predictive analytics, dynamically optimizing treatment parameters like chemical dosing and aeration. As demonstrated by Singapore's NEWater, where artificial intelligence (AI) increases recycling efficiency by 95%, predictive models are now used to forecast influent quality and equipment failures, moving management from reactive to proactive. In this context, such advances are important to safely and effectively recycle treated wastewater, a key pillar of sustainable water security. When treated to a high standard, recycled wastewater can substitute for drinking water supplies while benefiting agriculture and industry, recharge aquifers, and lessen pressure on freshwater ecosystems. With climate change exacerbating water variation and uncertainty, intelligent, AI-powered solutions have become of utmost importance. The chapter elaborates on the development, theory, applications, and future aspects of AI-driven wastewater treatment and recycling technologies. It aims to give a holistic account of how smart water systems may address global water challenges for better environmental results and toward sustainable development goals.

The chapter "AI Driven Wastewater Treatment and Recycling: Smart Water" significantly advances existing research and practice by integrating artificial intelligence (AI) with emerging technologies like the Internet of Things (IoT) and Cyber-Physical Systems (CPS) to address the pressing global challenges of water scarcity and pollution. Whereas earlier research has been restricted to bibliometric analyses of artificial intelligence applications in wastewater treatment, this chapter goes further, presenting frameworks that are implementable and scalable at the ground level. It takes a theoretical research paradigm shift to an applied one, where traditional rule-based treatment systems are defined by fixed parameters set manually and put under manual control, embracing an adaptive system that is nourished by data to recover resources in the best possible manner and thereby sustain itself. This chapter explains different AI technologies, machine learning applications, IoT, and CPS systems that could be incorporated in a smart water system. A key advancement lies in the chapter's emphasis on predictive resource recovery, including biogas and nutrients, which extends beyond conventional treatment optimization. Whereas these case studies, such as the one for NEWater in Singapore or Aarhus Water Utility, show success on a local scale, this chapter collates global perspectives to craft an all-encompassing model adaptable to a large number of contexts, urban-rural, developed, and developing economies. From an interdisciplinary point of view, this synthesis, with engineering, data science, and policy, places the chapter as a futuristic entry into the making of robust AI-powered water management systems in pursuit of sustainable development goals against climate uncertainty.

BACKGROUND

The evolution of wastewater treatment has gone through major phases over several centuries, from drainage systems during ancient times to treatment plants engineered in modern cities. Early systems were primarily designed to bear waste solids and carry stormwater. Little did they concern themselves with waterborne pathogens or chemical contaminants. Especially in 19th-century urban Europe, the public health problems resulted in centralized sewerage systems and somewhat coincided with the development of primary and secondary methods of treatment for rejecting organic matter and reducing microbial load. These developments mark the very birth of sanitary engineering and form the kernel on which contemporary environmental engineering practices are based. By mid-century, an additional series of treatments (tertiary level) had developed, including nutrient removal, filtration, and disinfection, to improve the ability of the effluent to protect the environment from adverse impacts. However, these conventional treatment techniques have various limitations. These restrictions include fixed parameters of operation, less automatic control, higher energy consumption, and the use of chemicals. The rate of urbanization and industrialization is such that most municipal treatment systems in much of the developing world cannot keep pace. This has led to the widespread discharge of untreated or partially treated wastewaters. Freshwater bodies are thus increasingly burdened with contaminants ranging from nutrients and pathogens to heavy metals and newly introduced pollutants like microplastics.

Water scarcity and pollution are among the most pressing global challenges of the 21st century. According to the United Nations World Water Development Report (2023), approximately 2.3 billion people live in water-stressed countries, a number projected to rise sharply due to population growth, climate change, and industrial development. Concurrently, the quality of available freshwater is deteriorating due to increasing discharge of untreated or poorly treated wastewater into water bodies. The imbalance between supply and demand of clean water has created a critical need for efficient water management and sustainable wastewater treatment practices. In response to these challenges, the water sector is undergoing a paradigm shift toward what is increasingly referred to as "smart water", the integration of advanced technologies such as Artificial Intelligence (AI), the Internet of Things (IoT), big data, and automation to optimize water usage, treatment, and recycling.

The history of wastewater treatment begins with ancient societies that applied basic wastewater management practices of sewage and stormwater. During the present day, the development of municipal sewer networks and centralized treatment facilities became widespread after the 19th-century industrial revolution and public health crises such as cholera epidemics. Modern wastewater treatment methods include primary, secondary, and tertiary stages that aim to remove physical, biological, and chemical pollutants. Even as processes have grown more sophisticated with time, most treatment centers still operate on strict dependent process models with heavy loads of manual supervision. The process of automation in the 20th century reduced some supervisory loads and was founded on static process flows, which raised efficiency but had a limited ability to address such dynamic changes. The early years of the 21st century witnessed an abundance of resources and eagerness for smart technologies, mainly due to sensors and analysis tools becoming cheaper and more accessible. The development of Supervisory Control and Data Acquisition (SCADA) systems and Distributed Control Systems (DCS) made up the first phase of installed smart water infrastructure (Elsayed et al., 2024).

AI is increasingly viewed as a prime mover of innovation across most fields of environmental engineering, and wastewater treatment is not an exception. AI techniques such as neural networks, decision trees, genetic algorithms, support vector machines (SVM), and fuzzy logic have been successfully utilized in modeling and optimizing complex treatment processes. Such techniques work best in environments characterized by non-linearity, multivariate relationships, and incomplete or uncertain information conditions found normally in wastewater systems (Capodaglio & Callegari, 2025). In water, nonlinear classification and regression analysis uses artificial neural networks (ANN), deep learning(DL), support vector machines(SVM), and random forests(RF). Principal component analysis (PCA) is primarily used to reduce dimensionality in high-dimensional data sets like fluorescence excitation-emission matrices. ANN and DL process input in parallel, mimicking the behavior of human neural systems. DL differs from ANN in that it often has higher prediction accuracy and a more complex structure. However, deep learning is prone to excessive fitting and requires more data for developing the network. The SVM model is more theoretically supported than the traditional ANN, making it easier to understand (Baarimah et al., 2024). Predictive analytics is one of the most potent applications of AI in wastewater treatment since it uses historical and real-time data to forecast system behavior and consequences. For example, ML models can forecast the amount of pollutants entering a treatment plant, allowing for proactive chemical dosing, aeration rates, and sludge control. Similarly, deep learning techniques like Long Short-Term Memory (LSTM) networks may be used in time-series data modeling to forecast changes in influent quality or system performance. These predictive capabilities mitigate the dangers associated with system overloading and ensure continuous compliance with discharge regulations (Bi et al., 2024; Pang et al., 2024). Another significant application is process optimization. AI algorithms can identify the most efficient operating conditions for a treatment plant by evaluating thousands of possible scenarios based on input parameters such as flow rate, temperature, pollutant concentration, and energy consumption. This enables a shift from rule-based control to adaptive control, where the system continuously learns and improves its performance. In addition, AI can support anomaly detection, equipment fault diagnosis, and the optimization of maintenance schedules, leading to enhanced system reliability and reduced operational costs. Hybrid strategies combine traditional techniques with artificial intelligence's predictive capabilities to provide reliable wastewater treatment. Process stability is improved by combining AI algorithms (such as Random Forests for anomaly detection) with PLC/SCADA's real-time data collection. For example, PLCs perform accurate valve adjustments while AI optimizes aeration cycles, resulting in a 15-20% reduction in energy consumption (e.g., Orange County Water District). Neural networks are used in conjunction with conventional optimization methods, such as statistical process control, to optimize chemical dosing and reduce waste. Adaptive management is made possible by hybrid digital twins, which combine SCADA data with AI simulations. In Aarhus, Denmark, for example, overflows were reduced by 50%. These synergies guarantee scalable solutions by striking a balance between innovation and dependability.

While AI provides the analytical engine for smart water systems, its effectiveness depends heavily on the availability of high-quality data. This is where the Internet of Things (IoT) and cyber-physical systems (CPS) come into play. IoT refers to a network of interconnected sensors and devices that collect and transmit data from physical environments in real time. In the context of wastewater treatment, IoT devices can monitor a wide range of parameters, including temperature, pressure, flow rates, turbidity, and chemical concentrations.

The integration of AI, IoT, and CPS also enhances transparency and traceability in wastewater treatment operations. Cloud platforms and digital dashboards allow operators and stakeholders to visualize system performance, identify trends, and make informed decisions. Such transparency is particularly valuable in the context of water reuse, where public acceptance and regulatory compliance depend on the demonstrable safety and effectiveness of treatment processes.

The concept of a circular economy has gained significant traction as a sustainable development model that emphasizes the reuse, recycling, and recovery of resources. In this context, wastewater is no longer viewed as a waste product but as a valuable resource that can be treated and reused for various purposes, including agricultural irrigation, industrial processes, and even potable applications. AI-driven wastewater treatment is a key enabler of this transformation, as it ensures the consistent quality and safety of recycled water. Water recycling involves several advanced treatment processes, such as membrane filtration, ultraviolet (UV) disinfection, and advanced oxidation, which require precise control and monitoring to function effectively. AI algorithms can optimize these processes by predicting membrane fouling, adjusting UV doses based on microbial loads, and minimizing chemical usage. Furthermore, AI can support decision-making on when and how to reuse treated water based on environmental conditions, demand patterns, and regulatory requirements (Shahouni et al., 2024).

In addition to water recovery, AI can facilitate the recovery of other valuable resources from wastewater, such as nutrients (e.g., nitrogen and phosphorus), energy (e.g., biogas from anaerobic digestion), and biosolids for agricultural use. This integrated resource recovery approach not only enhances the sustainability of treatment plants but also contributes to the economic viability of wastewater recycling initiatives.

Several cities and regions around the world have embraced smart water technologies to improve their wastewater management systems. For instance, Singapore's Public Utilities Board (PUB) has implemented an advanced monitoring and control system that uses AI and sensor data to optimize the operation of its NEWater plants, which recycle wastewater into high-grade potable water. Similarly, in Denmark, the Aarhus Water Utility has deployed AI algorithms to manage stormwater and reduce overflows during heavy rainfall events.

In the United States, the San Francisco Public Utilities Commission has integrated machine learning models into its wastewater treatment operations to optimize energy usage and improve effluent quality. Meanwhile, emerging economies such as India and China are exploring AI applications in decentralized wastewater treatment systems for rural and peri-urban areas. These case studies demonstrate that AI-driven wastewater treatment is not only feasible but also adaptable to a wide range of contexts and scales.

Despite its enormous potential, the adoption of AI in wastewater treatment faces several barriers. These include the high initial cost of implementation, lack of technical expertise, data privacy concerns, and the need for regulatory frameworks that support innovation. Moreover, the quality and availability of data remain a significant challenge, particularly in regions with limited infrastructure. To overcome these barriers, there is a need for interdisciplinary collaboration among engineers, computer scientists, policymakers, and water utility operators. Training programs and capacity-building initiatives can help bridge the skills gap, while open data platforms and standardization efforts can enhance data interoperability. Furthermore, research into explainable AI and robust cybersecurity measures is essential to build trust and ensure the safe deployment of smart water technologies. Looking ahead, the convergence of AI with emerging technologies such as edge computing, digital twins, blockchain, and quantum computing holds exciting possibilities for the water sector. For instance, digital twin models can simulate entire treatment plants in virtual environments, allowing for real-time testing and optimization. Blockchain can enhance data transparency and accountability in water quality reporting. Meanwhile, advances in quantum machine learning may unlock new levels of computational efficiency for complex process modeling.

AI TECHNOLOGIES IN WASTEWATER TREATMENT

In wastewater treatment, artificial intelligence (AI) refers to a group of computational methods that facilitate intelligent decision-making, predictive modeling, and process optimization. These solutions outperform conventional rule-based approaches in addressing the non-linear, multivariate, and dynamic nature of wastewater systems. The main AI techniques, their mathematical underpinnings, and their uses in wastewater treatment are examined below.

Artificial Neural Networks (ANN) and Deep Learning (DL)

A computational model based on biological neural systems, artificial neural networks (ANNs) are made up of interconnected nodes that use weighted connections to process incoming data. In order to reduce prediction errors, ANNs modify weights during training to model complex relationships. Chemical oxygen demand (COD) and biological oxygen demand (BOD) are two effluent quality indicators that are predicted by ANNs in wastewater treatment. An ANN's mathematical formulation consists of:

$$y = f\left(\sum_{i=1}^{n} w_i x_i + b\right) \tag{1}$$

Where b is the bias, f is an activation function (such as sigmoid or ReLU), y is the output, x_i are inputs, and w_i are weights. A subset of ANNs called Deep Learning (DL) uses several hidden layers to identify complex patterns. For example, DL models use historical fouling data to predict membrane fouling in reverse osmosis systems with 95% accuracy. However, DL presents difficulties in environments with limited resources because it necessitates vast datasets and computer capacity.

Artificial neural networks (ANNs), which imitate human neural systems for pattern recognition, are prone to overfitting, particularly when dealing with minimal or noisy datasets, which are typical in wastewater treatment (e.g., sensor data affected by fouling). Additionally, their "black box" nature restricts interpretability, making it difficult to communicate decisions to operators. As networks get bigger, computational complexity rises, creating scalability problems in environments with limited resources. Metrics like Mean Squared Error (MSE) or R^2 are used to evaluate prediction accuracy (e.g., >99% correlation in some studies for effluent quality), while cross-validation (e.g., 5- or 10-fold) is used to evaluate generalization. Overfitting is reduced by regularization techniques (like dropout), which are verified against historical data to guarantee an error of less than 10%.

Support Vector Machines (SVM)

Support Vector Machines (SVMs) are supervised learning models that work well in high-dimensional domains for both regression and classification. SVMs find the best hyperplane, expressed as follows, that optimizes the margin between classes:

$$min_{a.b}\frac{1}{2}|w|^2 + C\sum_{i=1}^{n}\xi_i \tag{2}$$

Where the weight vector is denoted by (w), the bias by (b), the regularization parameter by (C), and the slack variables by (x_i). SVMs improve stability by 20% in wastewater treatment by predicting sludge thickening in activated sludge processes. They are appropriate for regulatory compliance due to their interpretability.

SVMs are computationally demanding, especially when dealing with large wastewater datasets (such as terabytes from IoT sensors), but they perform well in high-dimensional spaces and small datasets. Unless combined with kernel functions, they have trouble with dynamic, non-linear processes. Additionally, parameter tuning (such as C and γ) is sensitive and can result in subpar performance if done incorrectly. Using accuracy and F1-score metrics for classification tasks (like anomaly detection), K-fold cross-validation (e.g., k=10) evaluates robustness. Grid search achieves 85–90% accuracy in pollutant prediction by optimizing hyper-parameters and validating them against real-time influent data.

Random Forests (RF) and Decision Trees

Several decision trees are combined in Random Forests (RF), an ensemble learning technique, to increase prediction accuracy. A subset of features is used to inform each tree's decision, and the results are either averaged or subjected to majority vote. Using precise chemical dosing, RF models maximize nutrient removal and reduce nitrogen concentrations by 15%. The mathematical foundation entails minimizing the mean squared error for regression or the Gini impurity for classification. Despite their simplicity, decision trees are used in RF to improve interpretability.

Although RF, an ensemble of decision trees, performs well with noisy data, its static nature may make it less useful for time-series prediction. Scalability in complex plants is limited because it uses a lot of memory for large datasets and reach a performance plateau with too many trees. Compared to linear models, interpretability is moderate but less detailed.

Cross-validation and out-of-bag (OOB) error estimation are used to assess performance using RMSE or accuracy metrics (e.g., 86-90% for effluent quality).

In order to guarantee consistent performance under a range of influent conditions, feature importance analysis verifies relevance and compares it to operational data.

Long Short-Term Memory (LSTM) Networks

A particular kind of recurrent neural network (RNN) called an LSTM network is made for time-series data and solves the vanishing gradient issue that conventional RNNs have. Long-term dependencies are maintained by LSTMs using memory cells and input, forget, and output gates, which are expressed as follows:

$$h_t = o_t . \tanh(C_t) \tag{3}$$

Where the cell state is denoted by (C_t), the output gate by (o_t), and the hidden state by (h_t). LSTMs are designed to capture long-term dependencies in sequential data, making them well-suited for time-series forecasting tasks, which could be implemented on a wastewater treatment system. Advanced oxidation processes (AOPs) are a group of chemical treatment processes applied in wastewater treatment and management facilities. By producing extremely reactive species, mainly hydroxyl radicals (•OH), AOPs are a class of chemical treatment technologies intended to eliminate pathogens, persistent organic pollutants, and other contaminants from water and wastewater. These radicals have a very high oxidation potential and are non-selective oxidants. They can break down toxic and complex compounds into simpler, less dangerous molecules, which frequently results in full mineralization into carbon dioxide and water. Typical AOP methods include ozonation (O_3), UV-based processes (UV/H_2O_2, UV/O_3), Fenton and photo-Fenton reactions (Fe^{2+}/H_2O_2 with or without UV light), photocatalysis (TiO_2/UV), and combinations like O_3/H_2O_2 or O_3/UV. Recalcitrant organic pollutants that traditional biological or physical treatment methods are unable to completely eliminate, such as pharmaceuticals, pesticides, dyes, endocrine-disrupting chemicals, and industrial effluents, are especially well-degraded by AOPs. Advanced oxidation processes (AOPs) LSTMs reduce overflow incidents in combined sewer systems by 30% in wastewater treatment by forecasting influent flow rates and pollutant concentrations.

For time-series forecasting (such as influent quality), LSTM, a variant of recurrent neural networks for sequential data, performs exceptionally well. However, it is extremely sensitive to outliers and necessitates large, high-quality datasets, which are frequently unavailable in wastewater contexts because of sensor inconsistencies. Real-time use in large plants is limited by computational demands, and improper regularization can lead to overfitting. With metrics like RMSE or MAE (e.g., 90% accuracy in Pang et al., 2024), time-series cross-validation (e.g., rolling window) evaluates temporal accuracy. Validated against 5-day COD or BOD trends, early stopping and dropout prevent overfitting and maintain a prediction error of less than 5%.

Fuzzy Logic and Genetic Algorithms

Fuzzy logic uses membership functions, which range from 0 to 1, to model imprecise data to deal with uncertainty. It is employed to control varied influent quality, including pH variations. By simulating natural selection and iteratively evolving solutions through crossover and mutation, genetic algorithms (GAs) optimize complex processes. In activated sludge systems, GAs maximize aeration while consuming 18% less energy. Typically, the fitness function for GAs is described as follows:

$$Fitness = \boldsymbol{min}(Energy\ Consumption + Effluent\ Quality\ Penalty) \quad (4)$$

Principal Component Analysis (PCA)

By converting data into principal components that capture the most variation, PCA lowers the dimensionality of high-dimensional datasets, including fluorescent excitation-emission matrices. PCA simplifies monitoring procedures by identifying important pollutants in wastewater treatment.

EEMs are used to describe the dissolved organic matter (DOM) in wastewater using three-dimensional datasets (excitation wavelength, emission wavelength, and intensity). These high-dimensional datasets frequently have hundreds of data points in each sample. By reducing EEMs to a small number of principal components, PCA can discover important fluorophores (such as proteins and acids) that reveal amounts of contamination or organic content. IoT sensors are used by wastewater treatment facilities to track a variety of factors, including turbidity, pH, dissolved oxygen, and nitrate levels. High-dimensional datasets produced by these have the potential to overload AI techniques such as neural networks. Additionally, PCA pinpoints the main contaminants causing changes in water quality. By examining variations in the principal component space, PCA is able to identify anomalies. For instance, a process disturbance (such as an equipment failure or an influent spike) may be indicated if a new data point deviates from the predicted range of principal components. PCA lowers computing complexity by preprocessing data for machine learning models such as ANN, SVM, or Random Forests.

APPLICATIONS OF AI IN WASTEWATER TREATMENT & RECYCLING

Artificial Intelligence (AI) is transforming wastewater treatment and recycling through improved operational efficiency, data-driven decision-making, and sustainable water management. Adaptive, intelligent frameworks can replace static, rule-based systems in treatment plants by utilizing machine learning (ML), deep learning (DL), and other AI techniques. Five major uses of AI in wastewater treatment are examined in this section: abnormality detection and predictive maintenance, process optimization and automation, resource recovery for a circular economy, predictive analytics for process control, and developing pollutant management. Global case studies and quantitative results are used to highlight these applications, which show how AI can revolutionize the way that water scarcity and pollution issues are addressed.

Predictive Analytics for Process Control

AI-powered predictive analytics forecasts influent characteristics, system performance, and environmental implications using historical and real-time data, allowing for proactive wastewater treatment process management. In time-series forecasting, machine learning models like Random Forests (RF) and Long Short-Term Memory (LSTM) networks perform very well, predicting variables including influent flow rates, pollutant concentrations (such as chemical oxygen demand [COD] and biochemical oxygen demand [BOD]), and seasonal changes. These forecasts enable treatment facilities to minimize expenses while guaranteeing adherence to discharge standards by optimizing chemical dosage, aeration rates, and sludge management.

Additionally, predictive analytics reduces the possibility of system overload.

Process Optimization and Automation

Wastewater treatment is being transformed from static, rule-based control to dynamic, adaptive systems that analyze millions of operating scenarios to find the best conditions, thanks to AI-driven process optimization. Sludge retention duration, membrane filtration rates, aeration intensity, and energy consumption are among the parameters that are optimized using methods like Reinforcement Learning (RL) and Genetic Algorithms (GA). For example, RL maximizes a reward function that is specified as follows by learning the best course of action through trial and error:

$$Q(s,a) \leftarrow Q(s,a) + \alpha[r + \gamma a' max Q(s',a') - Q(s,a)] \quad (5)$$

Where Q(s, a) is the action-value function, α is the learning rate, r is the reward (e.g., minimized energy use), and is the discount factor. This approach enables treatment plants to adapt to changing conditions, such as variable influent quality or energy costs.

AI-enabled automation minimizes human intervention and permits remote operation, which is especially advantageous for decentralized treatment systems in rural or peri-urban locations. AI also enhances sophisticated treatment procedures like ultraviolet (UV) disinfection and membrane bioreactors (MBRs).

Anomaly Detection and Predictive Maintenance

By detecting variations in operational parameters and scheduling maintenance to avoid breakdowns, artificial intelligence (AI) improves system reliability. Support Vector Machines (SVM) and neural networks examine sensor data to find anomalies, like sudden variations in turbidity, pressure, or dissolved oxygen, which could be signs of process disruptions or equipment failures. One popular metric for detecting anomalies is the Mahalanobis distance, which can be computed as follows:

$$D_M = \sqrt{(x - \mu)^T S^{-1}(x - \mu)} \quad (6)$$

Where the data point is denoted by (x), the mean vector by (mu), and the covariance matrix by (S). When (D_M) surpasses a threshold, anomalies are reported.

Anomaly detection is particularly valuable in decentralized systems, where real-time monitoring is critical.

Resource Recovery and Circular Economy

AI makes it possible to recover energy, water, and nutrients from wastewater, turning it from a liability into a useful resource that aids in the shift to a circular economy. Anaerobic digestion, nutrient extraction, water recycling, and other processes are optimized using machine learning models, which support the Sustainable Development Goals (SDGs) for environmental preservation and resource conservation.

ML models optimize methane production by analyzing variables including temperature, pH, and retention time. Another important use is the recovery of nutrients, especially phosphorus and nitrogen. One essential component of the circular economy is water recycling. AI guarantees the purity and safety of recovered water for uses such as industrial operations, agricultural irrigation, and potable reuse.

Management of Emerging Contaminants

The treatment of wastewater is severely hampered by emerging pollutants, including microplastics and medications, because of their low concentrations and persistence. By improving detection and removal procedures, AI tackles these issues. With 90% accuracy, deep learning models trained on spectroscopic data can identify microplastics, allowing for more focused filtration techniques.

Additionally, AI optimizes advanced oxidation processes (AOPs) for the breakdown of pharmaceuticals.

CHALLENGES AND LIMITATIONS

As discussed in previous sections, the application of artificial intelligence (AI) to wastewater treatment and recycling has revolutionary potential. However, many obstacles and restrictions prevent its widespread use and scalability, which make its adoption difficult. Achieving the full potential of AI-driven smart water systems is hampered by a number of impediments, each of which has its peculiarities and spans economic, technological, data-related, social, regulatory, environmental, and ethical dimensions. In-depth discussion of these issues is given in this section, along with a thorough examination of the difficulties that communities, utilities, and legislators face. The impact of these issues is demonstrated through real-world instances and regional viewpoints.

Economic Barriers

One of the biggest obstacles to using AI in wastewater treatment is the cost, especially for utilities with limited resources. AI deployment necessitates large upfront infrastructure investments, such as cloud-based data processing platforms, high-performance computer systems, and Internet of Things sensors. Even without continuing costs for maintenance, software updates, and technical support, retrofitting existing facilities with AI-compatible technologies can cost millions of dollars for a medium-sized wastewater treatment plant. AI adoption is restricted to more affluent urban areas because small towns and rural communities, particularly in developing countries, frequently lack the funding necessary to carry out such projects.

The financial difficulty is further compounded by maintenance charges. AI systems are dependent on a network of sensors that track variables such as flow rates, turbidity, and pH. It is necessary to calibrate, clean, and replace these sensors regularly, especially in harsh wastewater settings where corrosion or biofouling is prevalent. Maintaining a single sophisticated sensor array might cost thousands of dollars a year, and a facility may require hundreds to achieve thorough monitoring. Furthermore, AI software platforms are frequently subscription-based, with license costs that eventually put pressure on operating budgets.

Another level of cost is added by capacity building and training. Skilled workers, including data scientists, machine learning engineers, and environmental engineers with AI experience, are needed to operate AI-driven systems. It is expensive to train current employees or hire new professionals, and such knowledge is hard to acquire in areas with little access to educational resources. For instance, a wastewater company in a remote African community would find it difficult to fund operator training programs, which could result in underutilized or improperly managed AI systems.

The economic divide between rich and underdeveloped countries makes these issues worse. Utilities may be able to implement AI in industrialized nations thanks to government subsidies or public-private partnerships that may offset expenses. Developing countries, on the other hand, frequently depend on foreign loans or aid, which might give priority to essential infrastructure over novel technologies. Inequities in water availability and quality are perpetuated as a result of this global split in AI adoption, with wealthy countries benefiting from smart water systems while impoverished regions lag.

Technical Challenges

The application of AI systems in wastewater treatment is severely hampered by their technological complexity. Complex computational frameworks are needed for AI algorithms like neural networks, deep learning models, and reinforcement learning, which are challenging to incorporate into current treatment facilities. Many facilities, especially older ones, lack the digital infrastructure required for artificial intelligence (AI) and instead use legacy systems built for human or semi-automated control. Retrofitting these facilities entails revamping control systems in addition to installing new hardware, a procedure that can cause operational disruptions and necessitate prolonged downtime.

Another technical difficulty is integrating with IoT devices. A network of sensors is used by wastewater treatment facilities to provide data to AI models, but it can be challenging to guarantee smooth communication between disparate devices. Incompatible protocols used by sensors made by different manufacturers could result in data silos or communication breakdowns. The efficacy of AI forecasts may be diminished, for example, if a plant uses a combination of flow meters, turbidity sensors, and chemical analyzers and finds it difficult to combine data into a single platform.

One significant barrier is the absence of technical knowledge. AI system development and maintenance call for specific knowledge in data engineering, process control, and machine learning. This kind of skill is uncommon in many places, especially in developing nations. For instance, operators of a wastewater treatment plant in a small South Asian town might be trained in conventional treatment techniques but not in troubleshooting AI algorithms or interpreting model outputs. The effluent quality may be compromised by system failures brought on by this skills gap, such as improper chemical dosing as a result of misconfigured models. Sensor reliability is a problem; IoT device calibration drift or fouling can distort AI inputs and impact predictions (e.g., 5-10% error in COD readings). Unified protocols are required because data standardization across various systems, from modern sensors to legacy SCADA, complicates integration. As demonstrated by the Florida water hack in 2021, cybersecurity risks increase with networked systems; a breach could interrupt treatment. As demonstrated by Melbourne Water's microgrid upgrades, integrating AI with legacy infrastructure, which is frequently decades old, requires expensive retrofits or middleware. To ensure system integrity, these problems are reduced by thorough maintenance, encryption, and phased modernization.

An additional technological constraint is scalability. Even though AI works effectively in controlled settings, it might be difficult to scale solutions to different treatment plants with different influent sizes, characteristics, and operational restrictions. When deployed to a rural plant with unpredictable agricultural runoff, a machine learning model trained on data from an urban plant with steady industrial effluent may not perform as expected. Models must undergo rigorous retraining and validation in order to be adapted to local conditions, which raises the complexity and expense.

Data Quality and Availability

The quality and accessibility of data are critical to the efficacy of AI in wastewater treatment; however, many utilities confront substantial data-related obstacles. Large, high-quality datasets are necessary for AI models, like those used in process optimization or predictive analytics, to train efficiently. However, the efficacy of the model is undermined by wastewater treatment plants, which frequently produce noisy, inconsistent, or incomplete data. For instance, biofouling might cause a dissolved oxygen sensor to malfunction, resulting in inconsistent results that distort AI predictions. Such mistakes must be corrected manually, which takes a lot of time and resources.

One significant problem is data sparsity, especially in developing nations. Many treatment facilities rely on intermittent manual sampling for metrics like BOD or nitrate levels because they lack extensive sensor networks. For example, a Latin American rural wastewater plant might only gather data once a week, which would not be enough for real-time AI applications. Because of this sparsity, models are less able to represent dynamic patterns, including influent spikes during rainfall, which can result in imprecise projections or less-than-ideal process control.

Data accessibility presents further difficulties. When data is gathered but kept in disjointed systems, like paper logs or proprietary software, it might be challenging to compile the data for AI analysis. Years of operating data may be available to a municipal utility in a poor country, but they might not have the digital infrastructure to transform it into a format that can be used. Issues with platform compatibility can impede integration even when data is digitized, which can impede the adoption of AI.

Implementing AI is made more difficult by the complexity of wastewater data. A number of variables, including population trends, industrial activity, and weather, affect wastewater influent. Diverse datasets are needed to capture this heterogeneity; however, many plants only gather a portion of the necessary characteristics. The range of AI applications may be limited if a plant, for instance, monitors pH and COD but ignores newly discovered pollutants like microplastics. Many utilities would not be able to afford the large investments needed in sensor networks, data management systems, and standardization standards to address these data concerns.

Social and Cultural Barriers

There are substantial societal constraints related to public perception and cultural views about AI-driven wastewater treatment, especially water reuse. The "yuck factor"—a psychological aversion to drinking water made from wastewater—often causes resistance to recycled water, even when it has been treated to drinkable levels. This opposition can be made worse in cultures that are not familiar with AI by doubts about its dependability. Residents of a Middle Eastern city, for example, could be wary of AI-optimized recycled water because they worry that, despite extensive testing, algorithms might not be able to eliminate impurities.

Engaging the community is important but difficult. Transparent communication is necessary to inform the public about AI's contribution to water safety, but many utilities lack the means or know-how to carry out outreach. It can be especially challenging to communicate complicated AI principles in rural locations where literacy levels may be lower. For instance, if cultural values value "natural" water supplies, a wastewater service in South Asia would find it difficult to persuade farmers to utilize AI-treated water for irrigation.

Resistance from the workforce is another social barrier. Reluctance to embrace new technology may stem from operators used to conventional treatment methods seeing AI as a danger to their jobs. Operators in a European wastewater plant, for example, might be reluctant to use AI-driven automation because they are concerned about job loss or redundancy. Retraining programs are necessary to address this and empower employees, but they are expensive and time-consuming.

Adoption of AI is also influenced by cultural variations. Community consensus is essential in a collectivist society, and implementing AI without widespread support may be met with resistance. Individualistic cultures, on the other hand, might value efficiency yet encounter opposition if AI is seen as impersonal or prone to mistakes. These social dynamics highlight the necessity of customized approaches to foster acceptance and trust.

Regulatory and Policy Gaps

The use of AI in wastewater treatment is hampered by the lack of defined regulatory frameworks. Guidelines for guaranteeing the safety of AI-treated water, protecting data privacy, and evaluating AI models are lacking in many nations. Utility companies are uncertain about compliance in the absence of clear laws, which deters investment in AI technologies. For instance, if local laws do not define acceptable effluent limits for AI-optimized operations, a wastewater facility in Southeast Asia could be reluctant to adopt AI-driven recycling. Obstacles arise because regulatory frameworks lag behind the adoption of AI. Agencies like the EPA require validation of AI decisions against water quality standards (e.g., ECR 2023's 0.3 mg/L iron limit), but lack standardized certification processes for AI algorithms, delaying approvals. Public trust is essential because, as evidenced by early skepticism about potable reuse, communities may oppose AI-recycled water because they perceive risks. Transparent validation, involving third-party audits and public consultations (e.g., Amsterdam's Waternet), can build confidence. Addressing bias in AI models and ensuring compliance with health guidelines are essential to gain regulatory acceptance and societal buy-in.

Data governance represents a significant regulatory void. AI systems depend on massive databases, which frequently contain private data on industrial discharges or water consumption. Utilities run the danger of revealing proprietary data or infringing consumer rights in the absence of strong data privacy regulations. Because utilities put avoiding legal liability ahead of innovation, this can discourage the use of AI in areas with weak legal frameworks, such as parts of Africa.

Additional difficulties are brought about by differences in regulatory approaches between countries. While developing countries might not have any restrictions, resulting in uneven adoption, developed countries might have strict criteria for AI validation, requiring rigorous testing before deployment. The inability of a multinational wastewater utility with operations in several nations to standardize AI systems across jurisdictions could result in higher expenses and project delays.

These inequalities are made worse by policy inaction. Policymakers sometimes rely on antiquated frameworks for conventional water treatment because they lack the technological know-how to create legislation tailored to AI. As utilities wait for more precise instructions, this delay may inhibit innovation. Although the establishment of multidisciplinary policy organizations and international standards would be beneficial, development is sluggish, especially in areas with limited resources.

Environmental Considerations

AI-driven wastewater treatment can present environmental issues, especially in relation to associated energy usage, even though its goal is to improve sustainability. Carbon emissions are increased by the substantial processing power needed to train and run AI models, particularly deep learning systems. If the local grid is dependent on fossil fuels, a large wastewater facility that uses AI for real-time optimization might use enough electricity to negate the energy savings from process efficiency. In a Pacific Island country with no renewable energy, this could make smart water systems less environmentally friendly.

There are also environmental issues with the manufacture and disposal of IoT gear and sensors. Metals like lithium and cadmium, which are frequently found in sensors, can cause pollution if improperly recycled. It is possible for a wastewater plant to unintentionally add to electronic trash by replacing hundreds of sensors per year, especially in areas with inadequate waste management systems. Installing AI infrastructure, such as server rooms or data centers, may also necessitate changing land use, which could upset nearby ecosystems like wetlands that are vital to the quality of the water.

An additional constraint is unintended ecological repercussions. AI optimization might give some factors (like BOD reduction) more weight than others (like trace pollutants), which could have an impact on ecosystems downstream. A plant that optimizes for nutrient removal, for example, may unintentionally enhance the discharge of medicines into receiving waters, which would be detrimental to aquatic life. Implementing AI becomes more challenging when these trade-offs must be balanced, which calls for meticulous model design and environmental monitoring.

Ethical and Security Concerns

Security and ethical concerns present serious obstacles to AI-powered wastewater treatment. Since AI systems gather private data on things like water use, industrial emissions, and public health, data privacy is a big worry. Ethical concerns regarding surveillance could arise from a hack that exposes confidential information or permits illegal tracking of water usage trends. If data is shared with third parties without permission, for instance, a wastewater service in a European city that uses AI to monitor industrial effluents may encounter criticism.

Wastewater treatment facilities are infrastructure targets, making cybersecurity threats crucial. Hacking of sensors or manipulation of control algorithms are examples of cyberattacks that can target AI systems, especially those that are connected to the Internet of Things. The disruption of treatment procedures caused by a cyberattack on the AI system of a wastewater facility could result in the release of untreated wastewater or contamination of recycled water. Suppose that hackers changed the chemical dosage model of a plant, resulting in an overdose that damages ecosystems. For many utilities, ensuring strong cybersecurity necessitates costly procedures, frequent audits, and intrusion detection systems.

Another ethical issue is algorithmic bias. Unfair results could be produced by AI algorithms trained on biased datasets, such as giving treatment effectiveness precedence over environmental safety. For instance, vulnerable groups may be disproportionately affected if a model trained on urban wastewater data ignores the features of rural influents. Diverse datasets and thorough testing are necessary to address prejudice, although doing so can be difficult in settings with limited data.

Ethical concerns are raised by equitable access to AI technologies. Global water inequality is exacerbated when wealthy countries or utilities can afford sophisticated AI systems while impoverished areas struggle. While a nearby developing country uses antiquated techniques, an AI-driven facility in a developed country may achieve excellent recycling rates, underscoring the need for equitable technology transfer.

INTEGRATION WITH IOT AND CYBER-PHYSICAL SYSTEMS (CPS)

The foundation of smart water systems is the convergence of artificial intelligence (AI), the Internet of Things (IoT), and cyber-physical systems (CPS), which allows wastewater treatment and recycling to reach previously unheard-of levels of sustainability, efficiency, and adaptability. Whereas CPS combines digital control systems with physical processes to allow for dynamic management, IoT offers a network of linked sensors and devices that gather data in real time. When combined, these technologies give AI a strong foundation on which to evaluate data, streamline processes, and assist in wastewater treatment plant decision-making.

Overview of IoT and CPS in Wastewater Treatment

To gather, send, and exchange data, a network of physical objects—sensors, actuators, and controllers- connected online is known as the Internet of Things (IoT). IoT makes it possible to continuously monitor vital parameters in wastewater treatment, including pH, turbidity, dissolved oxygen, flow rates, and chemical oxygen demand (COD). This data serves as the basis for insights powered by artificial intelligence. These gadgets use wireless communication, frequently using protocols like MQTT or Zigbee, to collect data on local servers or cloud-based platforms. AI algorithms then interpret this data for anomaly detection, optimization, and predictive analytics.

Cyber-Physical Systems (CPS) combine computer models and control algorithms with physical treatment procedures (such as aeration and filtration) to create a bridge between the digital and physical worlds. CPS establishes a feedback loop in which sensors gather information, digital models evaluate it, and actuators make real-time adjustments to physical activities. To ensure energy efficiency and effluent quality, a CPS in a wastewater plant may employ AI to modify aeration rates in response to projections of oxygen demand. When IoT and CPS work together, treatment facilities become adaptive systems that can react to changing circumstances like equipment breakdowns or influent spikes.

For smart water systems, this integration is essential because it facilitates resource recovery, minimizes human intervention, and allows for real-time decision-making. For instance, an urban wastewater plant may utilize CPS to maximize biogas output and IoT to monitor influent quality, whereas a rural facility may use both technologies to operate remotely. But in order to reach their full potential, they must overcome obstacles related to technology, finances, and security, especially in areas with limited resources.

IoT-Enabled Real-Time Monitoring

One of the main uses of IoT in wastewater treatment is real-time monitoring, which serves as the data foundation for AI-driven systems. IoT sensor networks gather continuous data from multiple stages of the treatment process, such as primary sedimentation, biological treatment, and tertiary disinfection. Sensors measure physical parameters (like temperature and pressure), chemical properties (like phosphate and nitrate levels), and biological indicators (like microbial activity). Each day, thousands of data points are generated, which are then sent to a central platform where AI algorithms examine patterns, forecast trends, and identify anomalies.

IoT sensors in a fictitious urban wastewater facility in a developed country gather data on COD, pH, and dissolved oxygen every second and send it to a cloud platform via LoRaWAN. In order to help operators modify pumping schedules and avoid combined sewer overflows, an AI model analyzes this data to forecast influent spikes during storms. Solar-powered IoT sensors may be used in a rural facility in a developing nation to track temperature and turbidity. The data is transmitted to a distant dashboard over 4G networks. This enhances water availability in isolated places by enabling operators to oversee decentralized systems without physically visiting the location.

The flexibility of IoT allows for a wide range of applications. Heavy metals like lead or cadmium may be detected by sensors in an industrial wastewater plant, and the data could be fed into AI models to maximize chemical precipitation. IoT may give pathogen detection top priority for potable reuse, employing biosensors to keep an eye on viruses or E. coli and guaranteeing adherence to drinking water regulations. Another crucial use case is energy monitoring; IoT devices monitor power consumption during pumping or aeration, enabling AI to reduce usage during high tariff hours.

Edge and fog computing improve data processing. In order to minimize bandwidth requirements, edge devices filter noise from turbidity values and do preliminary analysis at the sensor level. Fog computing increases the efficiency of large plants by combining data from several sensors before sending it to the cloud via local gateways. To guarantee that AI models receive clean, structured inputs, a plant with 100 sensors, for instance, might deploy fog nodes to preprocess 10,000 data points per day. Proactive management is made possible by these real-time capabilities, but in hostile wastewater settings, sensor maintenance is still difficult because of corrosion or biofouling.

CPS for Adaptive Control and Optimization

Cyber-Physical Systems (CPS) combine digital intelligence and physical treatment procedures to provide adaptive control and optimization. IoT sensors gather data, AI models evaluate it, and actuators carry out control actions as part of a closed-loop architecture used by CPS. By responding to conditions in real time, this dynamic interaction guarantees that treatment plants maximize resource recovery, energy use, and effluent quality.

For real-time simulation, digital twins, a crucial CPS application, build virtual representations of treatment facilities. A membrane bioreactor's (MBR) digital twin may replicate filtration dynamics using Internet of Things (IoT) data, enabling operators to test scenarios such as raising flow rates without running the risk of membrane fouling. An urban business may save 10% on energy expenditures by optimizing pump schedules, and a rural facility could utilize a digital twin to identify remote equipment problems and minimize downtime.

CPS improves resource recovery as well. Based on IoT data, CPS might optimize the amount of biogas produced in an anaerobic digestion system by adjusting temperature, pH, and retention time. Methane production might be predicted by an AI model, allowing the CPS to adjust conditions and boost output by 20%. By modifying pressure and flow, CPS might maximize reverse osmosis in a potable reuse facility, prolonging membrane life and guaranteeing water safety. Although these applications show how CPS may build robust, effective systems, their complexity necessitates a strong computational foundation and knowledgeable operators.

Challenges in IoT and CPS Integration

Significant obstacles, such as technical difficulty, financial constraints, security threats, and scalability concerns, stand in the way of integrating IoT and CPS in wastewater treatment. Ensuring dependable connectivity between CPS controllers and IoT devices is challenging from a technical standpoint. Network latency or packet loss may occur in large plants with hundreds of sensors, interfering with data flow and AI predictions. Signal dropouts could cause important control measures to be delayed in a remote plant that uses cellular networks, lowering the effluent quality.

Another challenge is interoperability. Data silos may result from sensors made by multiple manufacturers using incompatible protocols. Custom middleware may be needed to harmonize data in a plant with a mix of MQTT and CoAP devices, which would raise implementation costs. IoT creates terabytes of data per month, which overwhelms processing and storage capacity in environments with limited resources. Data overload is another issue. This volume may be too much for a small utility's servers, which would restrict real-time applications.

Economic limitations are important, especially in poor countries. It can cost millions to deploy IoT sensor networks and CPS infrastructure, plus extra for software licenses and upkeep. Because of fouling, sensors in wastewater environments need to be cleaned or replaced frequently, which raises operating costs. For example, a rural facility in Africa could find it difficult to pay for CPS controllers, let alone even the most rudimentary IoT devices.

IoT and CPS are susceptible to cyberattacks, making security risks crucial. Hackers might alter sensor data or interfere with control algorithms, resulting in equipment damage or untreated wastewater leaks. Chemical dosage may be changed by a cyberattack on a CPS in an urban factory, which could have negative environmental effects. Intrusion detection and encryption are two crucial but expensive cybersecurity methods, especially for small utilities.

Scalability is problematic in a variety of settings. In rural areas with fluctuating agricultural runoff, IoT and CPS systems designed for urban plants with constant influent may not work. Systems must be highly customized in order to be adapted to local conditions, which increases complexity. Scalability is further hampered in poor nations by inadequate power infrastructure and internet access, which limits the deployment of IoT-CPS to locations with enough resources.

Future Potential and Scalability

The desire for sustainable water management worldwide and technical improvements bode well for the use of IoT and CPS in wastewater treatment. Remote adoption will be made possible by low-cost IoT sensors that run on solar or kinetic energy. 5G networks will improve data transfer, enabling large-scale plants to use real-time applications. Blockchain might guarantee openness in water quality reporting by securing IoT data. Quantum computing-based advanced CPS may streamline intricate procedures, and modular IoT-CPS systems may democratize access in underdeveloped countries. International standards and public-private collaborations will drive scalability, establishing IoT and CPS as essential components of climate-resilient, sustainable global smart water systems.

PRACTICAL APPLICATIONS OF AI IN WASTEWATER RECYCLING AND SMART WATER SYSTEMS

Shortly, AI could be an impactful and easy way to make water more sustainable, which could be done by implementing machine learning, smart sensors incorporated with IoT, and so on. AI's potential for smart water systems and wastewater recycling suggests circular, self-sustaining, and climate-resilient solutions. Closed-loop urban water cycles, in which wastewater is continuously treated and recycled into drinkable water, will be fueled by AI. While AI-guided biotechnology will create microbial consortia for more effective biological processes, digital twins of entire urban water networks will enable real-time adaptive management. In the end, artificial intelligence will contribute to the development of net-zero water systems that run on renewable energy and recycle almost all wastewater, guaranteeing fair and sustainable water security in the future.

Singapore's NEWater Program (AI-Enhanced Wastewater Recycling)

AI sensors could be used in microgrids in a city to make an efficient water supply in a specific section or location of the town without inconvenience. Authorities could monitor the entire water supply system from a single room using simple electronics. These could significantly reduce the cost of the supply system and the manpower needed to maintain it. Self-healing networks can recommend self-healing instructions for automated valve operations that lock off neat breakages without supply interruption. To ensure that wastes can be recycled safely for non-potable applications such as industrial cooling, emerging contaminants like microplastics or viruses are monitored by these deep learning models on biosensors. Artificial intelligence, combined with IoT-enabled sensors, will allow continuous monitoring of various aspects of wastewater quality, like chemical oxygen demand (COD), biological oxygen demand (BOD), turbidity, pH, and heavy metals. By predicting models, AI enables early-stage detection in cases of contamination, inefficiency in the system, or failure in treatment. Anomaly detection in real-time will ensure rapid intervention in leakage, illicit discharges, or inflows of toxic substances, thus enhancing efficiency in operations as well as protecting public health. Up to 95% of Singapore's wastewater is recycled by the Public Utilities Board (PUB) using AI and IoT to create high-grade reclaimed water (NEWater) for use in industry and drinking. Reverse osmosis and membrane filtration procedures are optimized by machine learning models, which also guarantee adherence to strict water quality regulations while consuming 20–30% less energy.

Aarhus Water Utility, Denmark (Real-Time Anomaly Detection)

Machine learning algorithms such as Artificial Neural Network (ANN) and other forms of neural network, LSTM ensemble models can be used to optimize aeration cycle, chemical dosing, and backwashing in treatment management facilities. To maintain adherence to water quality regulations while reducing the use of chemicals and energy, these intelligent systems can dynamically modify operational parameters in real time, improving the sustainability and economy of treatment procedures. Aarhus reduced incidents by 50% and saved €500,000 a year by combining AI with smart sensors to anticipate and control combined sewer overflows. Digital twins keep an eye on the whole network, allowing for quick response to leaks and illegal releases.

Orange County Water District, USA (AI-Optimized Treatment

By optimizing aeration, chemical dosing, and backwashing, the Groundwater Replenishment System reduces energy consumption by 15% and chemical expenses by 10% through the use of AI algorithms (such as ANN and LSTM). For aquifer recharge, this supports a 100 MGD recycled water supply. demonstrates sustainability and treatment optimization in line with dynamic parameter changes.

Amsterdam's Waternet (Public Engagement via AI Platforms)

Waternet reduces household water use by 8% by integrating chatbots and AI-powered apps to engage citizens and provide real-time consumption data and conservation advice. Blockchain improves governance by ensuring transparent water allocation. By giving households the ability to track recycling, monitor consumption, and get personalized conservation advice, AI-powered platforms can increase public involvement in water management. Transparency in water governance and allocation can be improved through integration with blockchain technologies. Digital assistants and chatbots powered by AI can also be used to educate communities, increase awareness, and encourage the safe use of recycled wastewater. Citizen engagement and proper management can enhance water affordability in selected regions through AI.

Melbourne Water, Australia (Microgrid and Self-Healing Networks)

Melbourne Water employs AI sensors in microgrids to monitor water supply in specific zones, reducing maintenance costs by 25% and enabling self-healing valve operations during pipe bursts, minimizing supply interruptions.

CONCLUSION

The integration of artificial intelligence with the traditional water management system and treatment process notably increased the acceptance and made it more suitable in the context of water scarcity and water pollution for the future world. As the past traditional water management system does not offer complete water recycling and cost management, the adaptation, efficiency, and prediction facilities of traditional wastewater treatment processes are inadequate for this vast population of the earth. The incorporation of Artificial Intelligence (AI) in wastewater treatment and recycling signifies a significant transition towards sustainable water management. Utilizing machine learning, IoT, and cyber-physical systems (CPS), AI facilitates real-time monitoring, predictive analytics, and adaptive process optimization, markedly boosting efficiency, decreasing costs, and improving environmental results. The fundamental improvements using AI in wastewater treatment and recycling are the prediction of water quality, optimizing the treatment process, and automation-adaptation for the future, using less human contact with polluted water, turning wastewater into a valuable asset for energy and reusable water, and detection and elimination of emerging pollutants like microplastics.

However, some challenges prevail: these include high implementation costs, data quality issues, regulatory gaps, and public skepticism. To tackle these barriers, interdisciplinary collaboration, policy innovation, and equitable access to technology-aspects of technology splendor-must be ensured, especially for developing areas. These innovations, though shining bright with promise, usher in many challenges. Economic constraints, technological complexity, and data quality issues remain major barriers, especially in low-resource settings. Then there comes the social resistance, gaps in regulations, and ethical issues to be accounted for: inclusion of the affected governments in setting policy, protection of privacy, and algorithmic bias. Through IoT-CPS integration, new avenues of opportunities and complexities will be opened. CPS and IoT technologies allow for the real-time monitoring, feedback control, and remote operation of wastewater treatment plants, but they provide a new set of challenges that include cybersecurity risks and the need for investment in hardware and data processing platforms, plus a skilled workforce.

In setting this all into context, it is imperative that engineers, data scientists, urban planners, and policymakers work in interdisciplinary settings to further the uses of AI in wastewater management. Designing capacity-building programs, fair regulatory standards, and inclusive technology transfer models will ensure the widespread uptake of these technologies at regional and income-group levels. Hence, with the advancement of digital technologies and with the AI models attaining criteria like explainability, scalability, and security, integrating them with smart water infrastructure will be increasingly possible and relevant. In the future, AI will complement IoT, blockchain, and quantum computing to deliver huge, disruptive breakthroughs in smart water systems. With the increasing extent of climate-change-induced water scarcity, AI will ensure water security, ecological integrity, and sustainable development for generations to come.

This chapter has just illustrated how AI can renovate the wastewater treatment process, becoming a costly burden towards a green Smart Circular Resource. The road ahead requires a global platform for cooperation, investment in green technologies, and inclusive policies towards making AI-assisted water sustainability a reality in every part of the world.

REFERENCES

Baarimah, A. O., Bazel, M. A., Alaloul, W. S., Alazaiza, M. Y. D., Al-Zghoul, T. M., Almuhaya, B., Khan, A., & Mushtaha, A. W. (2024). Artificial intelligence in wastewater treatment: Research trends and future perspectives through bibliometric analysis. *Case Studies in Chemical and Environmental Engineering*, *10*, 100926. DOI: 10.1016/j.cscee.2024.100926

Bi, J., Chen, Z., Yuan, H., & Zhang, J. (2024). Accurate water quality prediction with attention-based bidirectional LSTM and encoder–decoder. *Expert Systems with Applications*, *238*, 121807. DOI: 10.1016/j.eswa.2023.121807

Cairone, S., Hasan, S. W., Choo, K.-H., Li, C.-W., Zarra, T., Belgiorno, V., & Naddeo, V. (2024). Integrating artificial intelligence modeling and membrane technologies for advanced wastewater treatment: Research progress and future perspectives. *The Science of the Total Environment*, *944*, 173999. DOI: 10.1016/j.scitotenv.2024.173999 PMID: 38879019

Capodaglio, A. G., & Callegari, A. (2025). Use, Potential, Needs, and Limits of AI in Wastewater Treatment Applications. *Water (Basel)*, *17*(2), 170. DOI: 10.3390/w17020170

Chavhan, N., Bhattad, R., Khot, S., Patil, S., Pawar, A., Pawar, T., & Gawli, P. (2025). APAH: An autonomous IoT driven real-time monitoring system for Industrial wastewater. *Digital Chemical Engineering*, *14*, 100217. DOI: 10.1016/j.dche.2025.100217

Elsayed, A., Ghaith, M., Yosri, A., Li, Z., & El-Dakhakhni, W. (2024). Genetic programming expressions for effluent quality prediction: Towards AI-driven monitoring and management of wastewater treatment plants. *Journal of Environmental Management*, *356*, 120510. DOI: 10.1016/j.jenvman.2024.120510 PMID: 38490009

Hassan, M. K. (2024). Management of Industrial Wastewater Quality Using IoT. SSRN *Electronic Journal*. https://doi.org/DOI: 10.2139/ssrn.4920201

Jana, P. (2024). AI-powered IoT solutions for sustainable water management in cities. *Uncertainty Discourse and Applications*, *1*(2), 158–169. DOI: 10.48313/uda.v1i2.37

Pang, J., Luo, W., Yao, Z., Chen, J., Dong, C., & Lin, K. (2024). Water Quality Prediction in Urban Waterways Based on Wavelet Packet Denoising and LSTM. *Water Resources Management*, *38*(7), 2399–2420. DOI: 10.1007/s11269-024-03774-3

Senthil Rathi, B., Senthil Kumar, P., Sanjay, S., Prem Kumar, M., & Rangasamy, G. (2025). Artificial intelligence integration in conventional wastewater treatment techniques: Techno-economic evaluation, recent progress and its future direction. *International Journal of Environmental Science and Technology*, *22*(1), 633–658. DOI: 10.1007/s13762-024-05725-2

Shahouni, R., Abbasi, M., Dibaj, M., & Akrami, M. (2024). Utilising Artificial Intelligence to Predict Membrane Behaviour in Water Purification and Desalination. *Water (Basel)*, *16*(20), 2940. DOI: 10.3390/w16202940

Yalcin, S., & Ayyildiz, E. (2024). Analyzing the impact of artificial intelligence on operational efficiency in wastewater treatment: A comprehensive neutrosophic AHP-based SWOT analysis. *Environmental Science and Pollution Research International*, *31*(38), 51000–51024. DOI: 10.1007/s11356-024-34430-3 PMID: 39106015

Chapter 12
Wastewater Treatment and Recycling With AI

S. Ida Evangeline
https://orcid.org/0000-0003-2997-7897
Government College of Engineering, Tirunelveli, India

ABSTRACT

The intensifying global water crisis demands innovative approaches to wastewater treatment and recycling as core strategies for sustainable water management. The chapter discusses how artificial intelligence transforms the process of modernizing wastewater systems offering improvement in the operational effectiveness of the system, optimal utilization of resources, and adaptive process control. It starts with a survey of the traditional treatment steps and its limitations and proceeds to show a detailed report of the Artificial Intelligence (AI) application in the field related to monitoring, prediction of quality, detecting anomalies, and analysis of microbial populations. The chapter also demonstrates how AI is applicable to more innovative approaches to water reuse at the industrial, municipal, and decentralized scale-equipping systems with the intelligence to perform smarter control, minimize energy and chemicals consumed, as well as maximize effluent reliability.

INTRODUCTION

Water scarcity is rapidly emerging as one of the most pressing global challenges of the 21st century. As population growth, urbanization, industrial development, and climate change intensify, the demand for freshwater is rapidly outstripping supply across many regions of the world. The United Nations has estimated that by 2025, almost two-thirds of the world population will be living under the circumstances of water stress. Against this background, sustainable management of water resources

DOI: 10.4018/979-8-3373-2700-6.ch012

has emerged as an imperative in terms of strategy to governments, industries and communities. The recycling process of wastewater can be regarded as one of the central pillars of sustainable water management because through these efforts, contaminated water is transformed and can be used again as a productive resource thus relieving the burden of the freshwater system, as well as helping in the conservation of the environment (Mishra & Sundaram, 2024).

Conventional wastewater treatment systems, though aesthetically necessary, are being limited by a number of very important constraints making them rather inefficient in this fast-evolving world. These systems tend to be systems with stiff operational procedures, use a lot of energy and chemicals, lack the capacity to be flexible against the changing levels of contamination and expensive infrastructure. Moreover, traditional treatment methods often require human surveillance and support to monitor, maintain and make decisions and this has the tendency to insert inefficiencies, delays and errors. Consequently, the integrity and effective execution of the entire system of conventional wastewater treatment plants may be damaged especially when under stressful or complex conditions (Liu et al. 2024).

In recent years, the advent of artificial intelligence has opened new avenues for addressing these longstanding challenges in wastewater management. Artificial Intelligence (AI) describes a set of machine learning (ML) algorithms that allow machines to learn and develop through experience and by finding patterns in data and decide on future actions that bring about performance enhancement without explicit programming. The field of wastewater treatment and recycling is one that has also shown profound potential in its ability to be transformed by the introduction of AI, giving rise to the ability to operate in real time optimization of processes, predictive analytics, automated detection of anomalies in the process and intelligent control systems. Such abilities are of special significance in ever-changing and complicated situations in which swift choice-making and devising are important (Alprol et al. 2024).

AI-driven approaches in wastewater management are not confined to a single application but span the entire treatment continuum—from the monitoring of influent characteristics to the optimization of effluent quality and the strategic reuse of treated water. With continuous integration of the AI model with the sensor network, control systems, and big analytics, current wastewater facilities can also realize a higher degree of efficiency and automation than before. As an example, machine learning can be applied to predict biological oxygen demand (BOD) of influent streams, deep learning (DL) can be used to model dynamics of microbial population, and reinforcement learning may be used to optimize aeration settings or even chemical dosing to achieve regulatory mandates at minimum resource usage (Altowayti et al. 2022).

Moreover, AI's role extends beyond centralized treatment plants to encompass decentralized and small-scale treatment solutions, which are increasingly critical in rural areas, developing regions, and off-grid industrial sites. In these, AI will enable the design, operation of modular, hands-free treatment facilities, economical, resilient, and based on local requirements. Such democratization of wastewater treatment complies well with the ideals of environmental justice and sustainable development with the road to clean water supply to underserved communities (Zhang et al. 2023).

As the field of water management continues to evolve, the convergence of AI, environmental engineering, and data science is creating transformative opportunities to reimagine how we treat, reuse, and conserve water. In this chapter, the opportunities are discussed by considering the combination of computational intelligence and optimization procedures in wastewater treatment and regeneration. It starts with giving a preliminary outline of the traditional treatment procedures and their associated drawbacks. It proceeds to give the specific methods by which AI- technologies; such as supervised learning up to evolutionary algorithms are being utilized to improve the efficacy of treatment, operational reliability as well as compliance with the environment. Special attention is paid to the contribution AI can make to water reuse, the enhancement of the circular water economy system, and the development of adaptive infrastructure both in the city and remote environments (Aghdam et al. 2023).

In doing so, the chapter aims to provide a comprehensive, interdisciplinary perspective on the current state and future potential of AI in sustainable water management. In integrating its technical analysis with case studies and future trends, it can be a source of help to researchers, practitioners, policymakers, and technologists who aim to use AI to make the water-secure future come to fruition.

FUNDAMENTALS OF WASTEWATER TREATMENT

Wastewater treatment is an essential infrastructure process aimed at removing contaminants from water that has been used in domestic, industrial, agricultural, or commercial activities, making it safe for discharge into the environment or for reuse. This treatment process is layed out in general into three major segments i.e. primary, secondary and tertiary treatment. Every stage addresses separate types of pollutants and is based on the use of combinations of physical, biological, and chemical processes. The issues of understanding of these old processes are rare and will be necessary to understand the points around that artificial intelligence can produce some radical changes.

Primary treatment is the first step in conventional wastewater management. It involves the mechanical removal of large solids and suspended particles through screening, sedimentation, and grit removal. The physical treatment is the order of the day in this stage; the influent water undergoes bar screens to remove large objects then the grit chambers at which point sand, gravel and other coarse material settles. Next the lighter suspended solids are settled at the bottom in the primary clarifiers or sedimentation tanks as sludge and oils and grease move to the top surface to be skimmed. Despite the quality of primary treatment being able to remove about 50-60% of suspended solids and some organic matter, it is not adequate to attain standards of water quality necessary to discharge into the environment or to recycle (Sylwan & Thorin, 2021).

Secondary treatment builds upon primary treatment by addressing dissolved and colloidal organic matter, primarily through biological processes. Overall treatment at this stage involves the utilization of aerobic populations of microbes to degrade the organic pollutants through biochemical oxygen demand reduction. Activated sludge processes and trickling filters as well as oxidation ponds are the most common ones. Other systems such as activated sludge systems combine waste-waters with an abundant microbial culture in an aeration tank allowing the breakdown of organic compounds. The mixture is ripped at a secondary clarifier after aeration where biomass (as flocs) settles. Organic matter (85-90 percent) and a significant volume of micronutrients, including nitrogen and phosphorus, are usually removed and the concentrations of some micropollutants will still persist (Zagklis & Bampos, 2022).

Tertiary treatment, also known as advanced treatment, is employed when secondary treatment alone cannot meet stringent effluent quality standards or when treated water is intended for reuse in agriculture, industry, or potable applications. This step incorporates a multiplicity of processes customized to certain contaminants: membrane cerebration (e.g. microfiltration, ultrafiltration, nanofiltration, reverse osmosis), disinfection (e.g. chlorination, UV irradiation, ozonation), removal of nutrients (nitrification-denitrification, chemical phosphorus precipitation), and adsorption with activated carbon of trace organics. In the tertiary treatment systems, the changing influent quality is very sensitive to fluctuations and can require intricate patterns of the control strategies and energy-intensive processes (Rodríguez-Hernández et al. 2022).

Despite the robustness of this multi-stage approach, traditional wastewater treatment systems are not without limitations. A big problem is the uncertainty and variability of influent, with rainfall events or industrial discharges or diurnal changes in domestic demands possible influencing factors. These variations make control of the process difficult and may jeopardize efficiency of the treatment. There is also the issue of energy use which may be substantial in the secondary treatment processes of the aeration and pumping systems that can comprise up to 60 percent of all the plant energy consumed. In the same similarity, the use of chemicals in coagulation, disinfection, and nutrient removal does have a contribution to costs of operation and environment.

Operational reliability is another concern. Many treatment facilities operate with fixed schedules or manually tuned setpoints, limiting their responsiveness to real-time changes in water quality. This automation deficiency does not only make the task less efficient, but makes it more dependent on experienced operators, which has a limited supply. Furthermore, global pressure on environmental-protection strategies calls upon wastewater treatment plants to track more and more types of contaminants, including pharmaceuticals, microplastic, and endocrine-disrupting compounds, most of which are not sufficiently removed by the conventional means of treatment (Iqbal et al. 2024).

Wastewater treatment infrastructure also varies widely in scale and complexity across regions. Urban centers typically have centralized, high-capacity plants equipped with some degree of automation. By contrast, rural or peri-urban sites might have decentralized, or modular treatment systems that do not have sustained oversight or sophisticated monitoring systems. Such an infrastructure and capacity difference highlights the need to work on scalable, adaptive, and intelligent solutions that can work in a variety of conditions.

In this context, the integration of AI presents a compelling opportunity to enhance both centralized and decentralized wastewater treatment systems. Intelligence can be imbued into each of the process steps of treatment to achieve potentially predictive, responsive and self-optimising operations that address most of the inbuilt limitations of orthodox methodology with AI. The succeeding sections will discuss the emerging use of AI technologies throughout the treatment continuum to make wastewater a controlled, useful input on sustainable water use.

AI TECHNOLOGIES IN WASTEWATER MANAGEMENT

Artificial Intelligence is increasingly recognized as a powerful enabler of innovation in environmental engineering, particularly in the optimization of complex, data-rich systems such as wastewater treatment. In such a backdrop of continuously changing operating conditions, more rigid environmental guidelines, and the need to maintain sustainability in its operations, AI offers a combination of tools that can put the emphasis on conversion of data into useful intelligence. The revolutionary power of AI is that it can model non-linear relationships, change with the dynamic settings and optimize various goals, concomitantly, e.g. energy efficiency, effluent quality and cost reduction (Capodaglio & Callegari, 2025).

AI is not a unified technology, but instead is a blanket term; one that identifies a set of varied computational methods. The machine learning (ML), deep learning

(DL), reinforcement learning, evolutionary algorithms, and fuzzy logic systems are, among others, the most salient technologies when wastewater treatment is considered. All these technologies are more applicable depending on the kinds of problems in the wastewater treatment and recycling continuum (Malviya & Jaspal, 2021).

Machine learning, a subset of AI, involves training algorithms to recognize patterns and make predictions based on historical or real-time data. Applications of supervised learning algorithms in wastewater treatment Many supervised learning algorithms, including decision trees, support vector machines (SVM), and random forest (RF), are common in wastewater treatment, to predict parameters such as chemical oxygen demand (COD), biological oxygen demand, and total suspended solids (TSS). These forecasts will facilitate proactive process alterations, as well as regulatory compliance. Unsupervised learning algorithms, e.g. k-means clustering, or principal component analysis (PCA) can be used as operation regimes or fault states detectors, and anomaly detectors, or in exploratory data analysis (Dansawad et al. 2023).

Deep learning builds upon machine learning by using artificial neural networks (ANNs) with multiple layers (deep architectures) to capture more complex, hierarchical relationships in data. Recent advances in deep learning, convolutional neural networks (CNNs) and recurrent neural networks (RNNs) are successfully used to process large volume of sensor data, time-series prediction of influent variability, or even image-based classification of microbes in activated sludge systems. Deep learning has several advantages, including the capability to work with raw data as input--increasingly eliminating the necessity of manual feature engineering--and to get more accurate over time due to the amount of available data (Alvi et al. 2023).

Reinforcement learning (RL) represents another emerging frontier in intelligent process control. Unlike in supervised learning, the RL agents also learn by interacting with their environment: by performing actions, they measure the resulting outcomes and update their strategy to maximize the sum of rewards. It is a favourable approach in the situation of controlling difficulties of wastewater treatment, including aeration control and chemical dosing as well as optimization of sludge retention time. RL systems can learn time varying optimal control polices that can change given that the dynamics is disturbing or is non-stationary (Croll et al. 2023).

In addition to learning-based methods, evolutionary algorithms—based on the principles of natural selection—have been used to tackle complex water management problems that require optimizing multiple objectives at once. These are optimization of design parameters of treatment plants, cost versus performance trade-offs, optimization of operating schedules. Algorithms including genetic algorithms (GA) and particle swarm optimization (PSO) can be of particular value where the models of systems are either sophisticated or infeasible to analyse (Zhenyu et al. 2023).

Another relevant AI paradigm is fuzzy logic, which deals with reasoning under uncertainty and imprecision. Fuzzy inference systems are effective in capturing expert knowledge in linguistic rules and translating them into control strategies for processes such as sludge recirculation or nutrient removal. Fuzzy controllers are able to cope with any ambiguity in input variables unlike stiff threshold-based control systems and such are thus more robust against noise and variation. In addition to the sophistication of the algorithms, wastewater treatment relies considerably on data infrastructure when applying AI in practice. Sophisticated waste water plants are now being fitted with ad hoc sensor networks which measure real-time flow rates, ph, temperature, turbidity, dissolved oxygen, ammonia and many other indicators used to provide an effective picture of the waste water plant. AI-driven monitoring and control build on this sensor data which is typically taken at a high frequency. But the usefulness emanating out of AI does not only rely on the quantity of data, but quality, consistency and its contextual richness. Noise removal, detection of outliers, Normalisation, are all important pre-processing steps to the accuracy of the model (Eshbobaev et al. 2024).

The integration of AI also necessitates the use of digital platforms and architectures, such as cloud computing and edge processing. The cloud-based AI systems are used to build centralized model trainings and deployment in various industrial premises whereas edge computing gives a possibility of real-time inference and control on the local scale, even in the context of a low bandwidth environment. The AI-IoT integration even increases the resilience and responsiveness of wastewater infrastructures, recognizing how to constantly learn and adapt (Matheri et al. 2023).

Altogether, algorithmic intelligence, plus sensor and digital infrastructure, are what makes the application of AI to wastewater management possible. These technologies are synergistic in nature, meaning that they can be used together to convert conventional plant into smart, adaptive systems, which have the capacity of learning experience, maximizing performance autonomously, and reacting to environmental variations (real-time). The next section discusses the actual usage of these AI tools throughout the different processes and parts of the wastewater treatment process.

APPLICATIONS OF AI IN WASTEWATER TREATMENT

The integration of artificial intelligence into wastewater treatment systems is revolutionizing how water is monitored, treated, and managed. Since wastewater treatment processes become naturally complex due to the nature of variable influent conditions, non-linear biological processes, and delicate balances of chemical processes, the computational proficiency of AI to process massive data and extract trends allows the conversion of the waste water treatment process paradigm toward predictive and autonomous, and reactive as well as manually operated. In this section, several critical sectors in which AI has been practically utilized in wastewater treatment processes are highlighted to explain how the introduction of machine intelligence can make the process extremely efficient, raise compliance, and decrease operational stress.

Real-Time Monitoring and Anomaly Detection

One of the earliest and most impactful applications of AI in wastewater treatment is real-time monitoring. Traditional systems rely on scheduled sampling and laboratory analyses, which introduce delays and limit the ability to respond to fluctuations in water quality. AI and sensor networks allow the real-time monitoring of parameters, including pH, turbidity, dissolved oxygen (DO), total nitrogen (TN), and chemical oxygen demand (COD). With the machine learning processing of the time-series data generated by these sensors, fine-grain patterns or variations can be identified that are indicative of contamination events, ill or malfunction equipment, or process inefficiencies (Mohanty et al. 2024).

Anomaly detection models have been trained that recognize normal operating patterns and alert unusual behavior, including: One-class support vector Machines Autoencoders Bayesian networks Models can detect such events as chemical spills, shock loads or sensor drift long before they are detected manually. This rapid response facilitates adherence to an early detection capability that will assist the operators to intervene at the appropriate time with limited impact on the environment and realization of validated fines (Ghinea et al. 2023).

Process Optimization and Control

AI has also shown strong promise in optimizing operational processes within wastewater treatment plants. Most treatment units—such as aeration tanks, clarifiers, digesters, and disinfection systems—require continuous adjustments based on influent quality and plant conditions. Complex, time-varying dynamics of biological systems have often been beyond the capabilities of traditional control methods (e.g. PID controllers). Conversely, AI-based controllers self-learn and self-adjust to these dynamics as time moves on.

Model predictive control and reinforcement learning algorithms have an increasing usage in calculating optimal aeration schemes, chemical dosing schemes and recirculation rates of the sludge. The goal of these AI systems is to optimize between conflicting goals, e.g. maximize pollutant removal, minimize energy consumption, and/or maintain system compliance with effluent, continuously adjusting control variables in response to system feedback (Deepa et al. 2025).

As an example, adaptive neuro-fuzzy inference systems (ANFIS) have also been implemented to regulate nitrification and denitrification systems keeping nitrogen levels within the control limits but at reduced energy cost. Equally, genetic algorithms and swarm intelligence studies have also been used in studying multi-objective optimization issues in activated sludges process where one can come out with cost-effective configurations that do not impair the quality of treatment.

Predictive Quality Assessment and Regulatory Compliance

Effluent quality is a key performance indicator in any wastewater treatment facility, directly linked to public health and environmental protection. AI models have the capability to forecast the effluent properties -BOD, COD, TSS and nutrients contents- using the upstream sensor reading and past plant data. These forecasts usually produced with random forest or deep learning models, assist an operator to predict whether a plant is surging towards discharge benchmarks and apply retraction steps before it is too late.

In addition, the use of AI-driven decision support systems is currently being used to help with regulatory reporting and during compliance audit. They combine real-time tracking and compliance models in a way that automates the documentation effort, raising the issue of possible noncompliance, and proposing action toward response. This not only makes it lessen the administrative problems, but also makes water utilities more credible and transparent (Sharma et al. 2024).

Microbial Population Analysis and Biological Process Control

Biological treatment processes are central to modern wastewater systems, particularly in secondary treatment. These processes are effective based on the activity of microbial consortia accustomed to breaking down organic matter and removing nutrients. Conventionally, there exists no fully automated method to monitor the dynamics of these microbial populations in real time other than the use of manual microscopy or genomic assays whose work is inefficient, time consuming and labor consuming (Begmatov et al. 2022).

Recent progress means that AI has been applied to automated microbe identification and behaviour analysis. To provide an overview of the health and dynamics of the activated sludge, convolutional neural networks (CNNs) trained on microscopic images are able to classify microorganisms, which may comprise protozoa, filamentous bacteria, or floc-forming microbes. What is more, the AI models will be able to relate microbial profiles with operational variables, i.e., DO, temperature, and retention time, to calibrate accordingly to achieve optimal biological performance (Ferraro et al. 2024).

Microbiological monitoring using AI is an important step towards intelligent control of bioprocesses, in which the treatment process is not just responding to chemical indicators, but also optimising and predicting the changes on the basis of microbial environmental changes (Parsa et al. 2024).

It shows how AI is being used in every aspect of monitoring, control, compliance, and biological analysis, which is the transition to intelligent wastewater treatment systems, i.e., those that are data-driven, autonomous, and resilient. In the following section, the discussion of the abilities will be extended to wastewater recycling, reuse (a hotbed of sustainable water management) (Mohammadi et al. 2024).

AI IN WASTEWATER RECYCLING AND REUSE

As the global demand for freshwater resources intensifies, the recycling and reuse of treated wastewater has become a cornerstone of sustainable water management. Wastewater reuse can provide opportunities to alleviate water stress, protect freshwater reserves, and promote the circular economy, through application in agricultural irrigation, industries, potable and non-potable urban uses. The complexity of technical, chemical, and microbiological recycling of the treated effluence, however, need smart decision-making frameworks, not prevalent in conventional engineering. And this is where artificial intelligence comes to the game-changing play- real-time analytics, optimization, and quality control across a range of reuse opportunities (Alprol et al. 2024).

Industrial and Municipal Water Reuse

In industrial sectors, such as food processing, petrochemicals, textiles, and mining, large volumes of wastewater are generated with varying and often high contaminant loads. Real-time evaluation of effluent quality with AI-driven solutions and actionable insights on the viability and safety of water reuse across a variety of purposes can be considered a future of effluent quality analysis. Predictions of key reuse indicators (e.g. residual chemical concentrations, pathogen loads, or membrane fouling potential) with ML algorithms using the online available sensor data and previous trends are possible.

Municipal environments, in which water recycling may form part of an overall approach to urban sustainability, frequently apply AI-based integrated water management including demand prediction, system-wide optimization, and active control of distribution. As another example, neural networks will be able to forecast the needs of recycled water in time and place, smart scheduling of the treatment procedures and the management of storage facilities will occur. AI also helps to suggest the most appropriate way of recycling such as toilet flushing, landscaping, or industrial cooling in accordance with water quality and end-use requirements.

A prominent one is AI interest in membrane process optimization in sophisticated treatment plants. More common in water reuse schemes are reverse osmosis (RO), ultrafiltration and nanofiltration flows prone to energy-intensive operations and fouling. The AI models have the potential to anticipate a fouling event including time and streamline cleaning cycles to vastly extend the life of the membrane and cut on down time (Baskar et al. 2024).

Decentralized and Small-Scale Treatment Solutions

The promise of AI extends beyond large-scale, centralized facilities to encompass decentralized and modular wastewater treatment units. The systems are especially helpful in the rural or peri-urban regions or where infrastructure is less developed, and there are relatively lesser access to centralized water and sanitation facilities.

AI allows independent systems to be decentralized, running effectively and efficiently with the smaller human interference. An example would be that embedded AI controllers have the ability to optimize chemical dosing, aeration cycles and sludge handling in real-time, regardless of the availability of skilled operators. The reinforcement learning algorithms are able to respond to environmental changes, including the changes in influent quality, like shifting greywater or stormwater and check performance to ensure compliance levels.

In addition, such smart decentral systems can be linked into smart water grids within regions. Communications between them and to community central decision systems help encourage joint, coordinated water reuse strategies across communities, minimising the environmental footprint and increasing water resilience (Garg et al. 2024).

Enhancing the Circular Economy of Water

AI plays a pivotal role in operationalizing the circular economy in the water sector. Circular systems opposed to linear models of water use)\(-\)extract, use, and discard of water are intended to concentrate on using more water, recovering resources as much as possible. AI can help in this by finding reuse opportunities, optimizing the recycling streams, and producing limited waste.

The example is that AI systems can simulate how water quality, treatment energy, and material recovery (e.g., biogas, nutrients) interact so utilities will know which operational set-ups can maximize overall resources recovery. At other plants, the AI helps to reuse treated wastewater not only to irrigate crops or run through some industrial processes, but also as a form of cooling used in power plants, recharging of ground water tables or even in the process of indirect potable reuse at high levels of protection.

Utilization of AI by calculating the optimal tipping points that address many factors, including cost, risk, energy, and water quality enables utilities and industries to develop intelligent reuse systems that promote long-term sustainability.

In summary, AI is enabling wastewater reuse systems to evolve from rigid, manually managed processes into adaptive, intelligent frameworks that optimize resource use, ensure safety, and support decentralized water resilience. The following section will present real-world case studies that illustrate these capabilities in action (De la Hoz-M et al. 2024).

CASE STUDIES

The practical integration of artificial intelligence into wastewater treatment and recycling systems is no longer a theoretical proposition but a demonstrated reality across a range of contexts. Countless pilot applications and large-scale implementations across all continents demonstrate the effect the use of AI has on increasing the efficiency of the process, lowering the cost of operations, and producing a better environmental impact. This section presents some case studies to draw attention to the variety of applications of AI in existing wastewater treatment and reuse systems, both centralized and decentralized, and those of municipal and industrial utilities, and different AI techniques.

AI-Enhanced Operational Control at a Municipal Wastewater Treatment Plant (Denmark)

One of the most advanced applications of AI in wastewater treatment can be seen at the Ejby Mølle Wastewater Treatment Plant in Odense, Denmark. This plant recently implemented an artificial intelligence control system, to maximize its biological nitrogen removal process. The technology provided a substantial saving in energy as well as increasing the quality of effluents because of implementation of the model predictive control (MPC) and real-time sensor-yielding data concerning the concentration of dissolved oxygen, ammonia, nitrate and the flow rates.

The AI maintained regulatory compliance with less energy through adjustment of aeration rates, dynamically according to the forecasted revenues predicted of nitrogen loading and biological activity. The plant recorded an energy usage decrease in the aeration of 20 percent over a two-year period and reported more steady effluent nitrogen levels. The case is an excellent example of what AI can accomplish by automating the rigorous control strategy and reaching sustainability goals (Sakkaravarthy et al. 2024).

Smart Sludge Management Using Machine Learning (Singapore)

The Public Utilities Board (PUB) of Singapore, a global leader in water innovation, piloted a machine learning-based predictive maintenance model for sludge handling systems in its water reclamation facilities. The production and processing of sludge are the most expensive operations in wastewater facilities and equipment malfunction can cause teenage hiccups in the operating process.

Historical records of sludge viscosity, pump working hours and motor vibration were used to train the AI system to forecast possible failures to make recommendations regarding maintenance plans. Consequently, there was a 30-percent reduction in unplanned downtime and extended life of parts of the mechanical components due to timely actions. In this case, the importance of predictive analysis to the reliability and cost-efficiency of the important plant assets is highlighted (Nirmala et al. 2025).

AI-Supported Decentralized Greywater Reuse (India)

In a decentralized setting, a collaborative project in southern India deployed AI-integrated greywater treatment systems in peri-urban households. One such system, constructed on small-scale modules, employed sensor-based monitoring with inbuilt machine learning algorithms in order to tune the treatment parameters (e.g. of chemical dosing levels, and flow rates) in response to the variability of the domestic greywater composition.

Removal of contaminants such as surfactants and organic load was consistent in the systems providing a reuse of greywater after treatment in toilet flushing and gardening. These systems could be managed effectively on useful interfaces of AI which gave actionable advice even to local operators with no-technological grounds. What is demonstrated by this initiative is that AI would play an important role in making sustainable water solutions accessible to people living in resource-limited contexts (Borham et al. 2024).

Deep Learning for Effluent Quality Prediction in Industrial Reuse (United States)

In California, an industrial water reuse facility integrated deep learning models to forecast effluent water quality metrics—specifically total dissolved solids (TDS) and chemical oxygen demand (COD)—to optimize its reuse for manufacturing processes. The AI system based on recurrent neural networks (RNNs) trained using years of operation data would be able to predict changes in water quality up to 12 hours beforehand.

This forecasting capacity enabled the facility to alternate between the reuse of water and other sources of supply without affecting the quality of products and process integrity. Further On, chemical dosing rate optimization according to the forecasted conditions was optimized, with up to 15 percent chemical cost savings. The case shows that AI helps the industries make decisions based on the data and shift towards the continuity of operations and environmental stewardship (Köksal et al. 2025).

AI in Membrane Fouling Management for Advanced Water Recycling (Australia)

At a large-scale advanced water recycling plant in Queensland, Australia, AI models were deployed to predict and manage membrane fouling in ultrafiltration and reverse osmosis systems. System fouling greatly decreases the system performance and necessitates a high frequency of cleaning or replacement.

The AI platform relied on gradient-boosting regression models that used online sensor data regarding transmembrane pressure, turbidity, and the temperature to foresee emergence of fouling events. Improving cleaning schedules and operational parameters in real-time allowed the increased membrane life by 18%, decreased cleaning frequency, minimized operational costs and environmental impacts.

Together, these case studies demonstrate the flexibility and effectiveness of the practice of AI in improving wastewater treatment and recycling performance in various dimensions, locations, and applications. Some similar patterns are also noticed: enhanced operational efficiency, active decisions, increased reliability, and Resource management are also noted (Shahouni et al. 2024).

AI-Driven Water Quality Assessment and Reuse in Egypt

In the African context, particularly in Egypt, Al-Falal et al. (2025) has demonstrated how AI and data-driven tools are being employed to assess, predict, and enhance wastewater quality for both environmental protection and agricultural reuse. A study conducted on El Manzala Lake—the largest natural lake in Egypt—applied a hybrid approach combining statistical analysis and machine learning to evaluate water quality and its degradation drivers. Using six machine learning models, including Multilayer Perceptron (MLP) and Decision Tree (DT), researchers accurately predicted key water quality indices such as the Aquatic Water Quality Index (AWQI), Heavy Metal Pollution Index (HPI), and Metal Index (MI). The lake was found to be severely polluted, especially in its southeastern zones near the Bahr El-Baqar drain, with cadmium and copper being the most influential pollutants. The MLP model, in particular, demonstrated outstanding predictive accuracy ($R^2 = 0.99$ for MI), highlighting the value of AI in complex aquatic assessments.

Complementing this, Gad et al. (2025) focused on wastewater reuse for irrigation around Egypt's Rosetta Branch used hyperspectral reflectance measurements integrated with AI modeling. This research developed novel spectral indices and optimized Partial Least Squares Regression (PLSR) models to estimate water quality parameters like total chlorophyll and biochemical oxygen demand (BOD). With validation accuracies as high as $R^2 = 0.96$ for BOD, the study showed that AI-supported spectral analysis can serve as a rapid, non-invasive tool for monitoring water quality in agricultural drains. The models also demonstrated high cross-region accuracy (89%), suggesting their broader applicability in water-scarce regions of Africa. Together, these studies illustrate how AI and remote sensing technologies are being adapted to local conditions to advance sustainable wastewater reuse and protect fragile aquatic ecosystems in North Africa.

CHALLENGES AND LIMITATIONS

While artificial intelligence holds immense promise for transforming wastewater treatment and recycling, its widespread adoption across the water sector is still constrained by a range of practical, technical, and organizational challenges. Such constraints exist not simply because wastewater systems are complex, but because of infrastructural, institutional, and human capacity deficits more generally. Three prominent types of issues that need to be resolved to release the full potential of AI in the sphere of wastewater management are touched on in this section: data limitations, modeling and deployment challenges, and the challenges of interdisciplinary collaboration.

Data Availability and Quality Issues

AI models thrive on high-quality, high-resolution data—but in many wastewater treatment facilities, data collection practices remain rudimentary or inconsistent. Most small- to medium-sized treatment plants do not have the required sensor infrastructure or real-time monitoring system to produce continuous streams of data. Data quality will be compromised where sensors are present involving poor calibration, sensor drift, environmental disruption, or fault of equipment.

Moreover, historical data sets, a key component to supervised machine learning and forecasting models, may be sparse, incomplete, or held in formats incompatible with one another on multiple subsystems. This increases the inability to develop strong and generic AI models. In rural or even decentralized environments, an absence of digitized records in their entirety also increases the gap within AI preparedness. Isolated data silos within operational, maintenance, and administrative areas may restrict the appropriate utilization of information even within higher-technologically oriented operations.

Data labeling and validation is another big issue. Numerous AI methods-especially those that involved supervised learning-need labelled datasets to train. Classifying water quality data or microbial images is often time-consuming and requires expert input—a resource that is both limited and in high demand. (Baarimah et al. 2024).

Complexity of Model Development and Deployment

Building accurate and reliable AI models for wastewater systems is inherently challenging. These systems are governed by nonlinear, time-varying, and multi-scale processes involving physical, chemical, and biological transformations. These dynamics are complex in nature, and are best evaluated through not only advanced modeling procedures but also attentiveness to the multivariate interactions. In addition, empirical wastewater data may be highly noisy, incomplete and subject to seasonal or stochastic effects, and thus pose veritable challenges to creating both predictive and generalisable types of models of wastewater.

Even in the case whereby a high-performing model is developed in a research or pilot environment, employing it into full-scale and operational realization brings up another level of complexity. The AI models also need to be incorporated within existing SCADA (Supervisory Control and Data Acquisition) platforms, operate under current cybersecurity regulations, and have to be able to operate efficiently within systems that have latency issues, bandwidth limitations or computing difficulties.

Another issue is the maintenance of the deployed AI systems. Evolving over time, an AI solution can degrade due to the changing dynamics of the systems in question (so-called model drift). This necessitates continuous retraining, verification and updating of the model, which is a labor-intensive task requiring a technical infrastructure as well as expertise (Capodaglio & Callegari, 2025).

Skills Gap and Interdisciplinary Collaboration Needs

AI integration in wastewater management requires cross-disciplinary expertise that bridges environmental engineering, computer science, data analytics, and domain-specific knowledge in hydrology and microbiology. Nonetheless the biggest challenge in many water utilities as well as in city agencies is the huge skills gap.

Even engineers and operators could not be familiar with AI concepts whereas data scientists could not have a contextual perspective on water treatment processes.

This disjunct slows down effective design, validation and interpretation of AI models. Divergence between AI developers and users at the plants may result in technically-accurate, but operationally-irrelevant or inaccessible models. In addition, most treatment centers are on shoestring budgets and lack the means to hire or sustain AI expertise particularly in developing countries or in rural areas.

Communication among stakeholders- utilities, researchers, technologists, regulators and communities- is usually heterogeneous. Structured partnerships and co-development models are urgently needed to facilitate mutual learning, as well as to make sure that AI tools are attuned to the realities on the ground and to policy modeling.

The difficulties need to be dealt with by a combination of technological innovation, capacity-building and institutional reform. The following section will discuss new approaches and remedies including cloud-based platforms and AI, collaboration ecosystems, that are assuring in filling these gaps and leading to more responsible adoption of AI in wastewater treatment and reuse systems (Imen et al. 2023).

STRATEGIES FOR AI INTEGRATION

Effectively integrating artificial intelligence into wastewater treatment and recycling systems requires more than deploying algorithms; it demands a cohesive strategy that addresses technical, organizational, and infrastructural dimensions. Advancements in this field continue to rely on the past achievements of AI progress to fill data infrastructure gaps, capacity building of human and institutions and cross-sector collaborations. In this section, I review practical ways that have come up to help address the scaling and the future sustainability of wastewater management using AI.

Cloud-Based AI Platforms and Digital Twins

One of the most promising pathways for enabling AI deployment at scale is the use of cloud-based platforms that offer data storage, processing, and modeling capabilities on demand. Such platforms do not require the use of an on-site IT facility often costly and complex to maintain, particularly in a small facility or decentralized facility. With cloud solutions, utilities can load up sensor data and execute AI models, visualize results, and pop-up real-time alerts/control signals (Gokul et al. 2024).

Simultaneously, digital twins, or simulated clones of real-life wastewater infrastructures are becoming a potentially strong simulation, optimisation and scenario test tool. An AI-based digital twin that is fed with real-time data could simulate treatment plant behavior in a variety of conditions and even suggest the best control action without impacting on the real process. These simulated systems are especially beneficial to educate operators, test upgrade capabilities, or test the resilience of systems to a climate or contamination shock (Wang et al. 2024).

Digital twins and cloud computing can flexibly, securely and intelligently manage processes, allowing easy access to tools to democratize AI in the water industry.

Decision Support Systems and Human-in-the-Loop Interfaces

To ensure that AI is actionable and trusted by operators, it must be embedded in decision support systems (DSS) that provide clear, interpretable recommendations. DSS platforms, in contrast to black-box models, combine AI predictions with industry rules, regulatory boundaries, and experience of operators to produce experience-related suggestions (Renfrew et al. 2024).

Such systems can contain friendly dashboards that provide a real-time picture of the water quality indicator, forecasted violation, suggested control measures, and maintenance warnings. These interfaces can maintain a human-in-the-loop of oversight, customization and override, so operators still maintain control and still benefit from the AI guidance (Kumar et al. 2024).

Improvement of explainable AI (XAI) methods can further enhance the strength of these systems in that model decisions are being made transparent to boost user confidence and regulatory acceptance. As an example, in case an AI model proposes to change the aeration levels, an XAI system can demonstrate which parameters contributed to the decision, e.g. surge in ammonia or a decline in dissolved oxygen (Infant et al. 2025).

Capacity Building and Collaborative Ecosystems

Given the interdisciplinary nature of AI integration, building institutional and human capacity is essential. Focused training in AI knowledge and data management is required as the workforce in water utilities, engineering, and municipal agencies would have to be provided with a base level of knowledge. Simultaneously, AI creators will have to become acquainted with water treatment procedures, restrictions on operations, and regulatory standards.

Such collaboration forms of academic-industry and public-private appear to be essential at bridging knowledge between these two domains. Such relevance, usability, and adoption are guaranteed with the co-development models, in which the AI tools are developed in collaboration with plant operators and environmental engineers.

The efforts to establish interoperability, through initiatives such as open-source modeling or data sharing platforms, data sharing and standard APIs, are aiding in prematurely practising the necessity of interoperable ecosystems to both reduce the barrier to new users and catalyse processes of innovation. The work of international bodies and research consortia is also essential in the publishing of best practices, policy guidelines, and ethical frameworks of responsible AI use within environmental systems (Al-Khatib & AlHanaktah, 2025).

Policy and Funding Alignment

Strategic integration of AI into wastewater infrastructure also depends on supportive policy environments and funding mechanisms. Regulators should accept and acknowledge that AI-guided control systems are legitimate compliance tools and allow flexibility in control strategies where it is rational.

Adoption can be brought forward by grants and incentives to fund AI research, pilot projects, and upgrading and improving the digital infrastructure (particularly in under-resourced regions). National and local governments have a role in ensuring that the AI-readiness criteria are built into water planning in cities, climate resilience, and sustainability plans.

These three strategies, technological, organizational and policy-oriented are mutually supporting. United as they are, they provide the basis of a new age of smart wastewater technologies, where data and AI are enabling cleaner, flexible, and just water infrastructure (Elmahdi & Jeong, 2024).

FUTURE DIRECTIONS AND INNOVATIONS

As artificial intelligence continues to evolve, its integration into wastewater treatment and recycling is poised to enter a new phase—moving beyond isolated applications toward fully integrated, adaptive, and autonomous water systems. Innovations and emerging technologies, interdisciplinary studies and systems-level innovations are defining the future of smart wastewater management. This segment presents the central trends and the future directions, which will new possibilities and capacities of AI in the water sector.

Integration with IoT and Smart Water Grids

One of the most significant developments on the horizon is the convergence of AI with the Internet of Things in the creation of smart water grids. Such computerized interconnect systems will ensure that the real-time data collected by sensors at all layers of the water cycle are shared, ranging over sewers, treatment plant, water reservoirs and distribution systems (Kubiak, 2025).

With AI, smart water grids have the ability to dynamically optimize flows, detect leaks or contamination on a real time basis, manage decentralised systems and accurately predict demand. E.g. a smart grid might intensify/reduce treatment according to forecast rain events, or divert flows due to evidence of microbial risks. Water grids optimized by artificial intelligence provide a model of how to achieve resiliency and efficiency of systems to allow quick adaptation to the effects of urbanization as well as climatic changes (Slany et al. 2025).

AI-Augmented Biological Treatment and Microbial Engineering

The use of AI in understanding and manipulating microbial communities in wastewater treatment is still in its early stages but holds transformative potential. Recent developments in bioinformatics, metagenomics, and machine learning are enabling researchers to describe the dynamics of microbial ecosystems in significant levels of detail.

Future applications of the AI could not only observe microbial changes but also proactively inform bioaugmentation plans the addition or stimulation of a particular strain of microbes in order to increase removal of pollutants, recovery of nutrients or resistance to toxic shock. Precision bioprocessing Deep learning models trained to genomic and operational data would be able to predict the response of microbial populations to process changes.

Eventually, formulation of syn bio consortia, optimized to target certain environments, may be assisted by AI clearing the way towards more energy-effective and directed bio treatment plants (Padhiary et al. 2025).

Blockchain for Data Transparency and Regulatory Compliance

As AI models become more integral to decision-making, trust and transparency will become critical, especially in regulatory contexts. A new possible partial solution is the incorporation of blockchain technology to achieve traceability, security, and auditability of the water quality and AI-generated decisions (Mithal & Thankachan, 2025).

Blockchain can take the form of an immutable data center to keep sensor data, treatment records, and conformity reports where the stakeholders, the regulators, the operators, and the general population can confirm the achievements of the systems. It can be used with AI to provide automated auditing systems to monitor and record compliance of control actions against legal thresholds increasing accountability and trust by the populace (Afzal & Hassan, 2025).

In addition, smart contracts may be used to activate some kind of warning, operational procedures or even payments depending on the water reuse value or pollution management performance success.

Federated and Privacy-Preserving AI Models

As data privacy and security concerns increase, especially when AI systems operate across multiple facilities or jurisdictions, federated learning is emerging as a viable solution. The method enables training AI models on decentralized data sources without moving the raw data to a centralized destination. Every involved facility can participate in a common model without giving up on its sensitive information.

Particularly within the context of wastewater treatment, federated AI may help drive innovation by enabling a partnership between two or more separate utilities or research organizations to work together to build robust models, such as effluent quality prediction or membrane fouling detection, without transgressing data governance restrictions (Wang et al. 2025).

Besides this strategy maintaining secrecy, it is assistive in scalable learning under various operational conditions.

Ethical, Regulatory, and Human-Centered AI

Finally, as AI systems take on more critical roles in water management, ensuring they are ethically designed, inclusive, and aligned with societal values will be paramount. That comes in the form of ensuring there are fair considerations around the impact that automation has on skilled labor, discrimination on the data or modeling explanations, and inclusive governance systems.

Design of future frameworks should attempt to be human-centered AI whereby technology works to empower the operator rather than replace them. The regulatory agencies will have to create some new benchmark associated with the AI validation, risk assessments, and certification, so that the digital systems become both effective and accountable (Padhiary, 2025).

Essential to the development of trust and fair access to the advantages of AI-enhanced infrastructure, also will be the public involvement, particularly in decision-making that includes water reuse or decentralized treatment.

Cumulatively, these innovations are an indication of a monumental transition to smart, connected, and predictive wastewater systems. Not only do they provide technical efficiency, but they present the possibilities of reimagining the way water is valued, controlled, and shared in an ever growing resource-scarce world.

CONCLUSION

As the global community confronts the escalating challenges of water scarcity, pollution, and climate variability, wastewater treatment and recycling are no longer ancillary functions but central pillars of sustainable water management. The chapter has examined how artificial intelligence can be transformational in improving the efficiency, entity and sustainability of these important processes.

Traditional wastewater infrastructures can be effective; however, they lack flexibility in operation and have high energy requirements and low dynamism with regard to changes in the environment. AI technologies, such as machine learning, deep learning, and optimization, have become the instruments that help to eliminate these constraints. They allow predictive monitoring, smart control, and real-time decision-making that minimize resource consumption, the environmental impact, and make sure that it complies with the regulations.

The scope of AI applications in the medical sphere has become much wider nowadays extending to anomaly detection and process control as well as microbial analysis and effluent quality predictions. Besides, AI is contributing to higher levels of complex water recycling systems within industry and cities, decentralization of infrastructure in resource-poor regions, and the evolution of the water reuse circular economy.

The first challenges are not absent despite such breakthroughs. Challenges that concern qualities of data, the complexities of the models, skill deficiency, and barriers to integration should be tackled by means of having a strategic intervention. Cloud based infrastructure, shore-based decision support systems, and interdisciplinary collaboration is seen as a potential route to scale AI solutions in a responsible and equitable way.

In the future, the interaction between AI and other emerging technologies (like IoT, blockchain, federated learning, and synthetic biology) will be the next era of intelligent water infrastructures. These solutions not only guarantee technical advantages but also more powerful, transparent, and inclusive patterns of the management of one of the most important resources on our planet.

At the end, AI is more than a technological improvement: it is a paradigm shift to wastewater and wastewater reuse, converting wastewater from a liability into a managed asset that must be used to support the health, environment, and water security and sustainability of future generations.

REFERENCES

Afzal, S., & Hassan, S. I. (2025). Blockchain Application in Sustainable Smart Water and Wastewater Management. *Fostering Machine Learning and IoT for Blockchain Technology: Smart Cities Applications*, 2, 273–297.

Aghdam, E., Mohandes, S. R., Manu, P., Cheung, C., Yunusa-Kaltungo, A., & Zayed, T. (2023). Predicting quality parameters of wastewater treatment plants using artificial intelligence techniques. *Journal of Cleaner Production*, *405*, 137019.

Al-Falal, A. N. A., Elsayed, S., El Fadaly, E. A., Gaagai, A., & Aouissi, H. A., Abd El-baki, M. S., ... & Gad, M. (2025). Aquatic System Assessment of Potentially Toxic Elements in El Manzala Lake, Egypt: A Statistical and Machine Learning Approach. *Results in Engineering*, •••, 105027.

Al-Khatib, L. A., & AlHanaktah, A. M. (2025). Wastewater Treatment Plant Upgrade and Its Interlinkages with the Sustainable Development Goals. *Resources*, *14*(4), 62.

Alprol, A. E., Mansour, A. T., Ibrahim, M. E. E. D., & Ashour, M. (2024). Artificial intelligence technologies revolutionizing wastewater treatment: Current trends and future prospective. *Water (Basel)*, *16*(2), 314.

Altowayti, W. A. H., Shahir, S., Othman, N., Eisa, T. A. E., Yafooz, W. M., Al-Dhaqm, A., & Ali, A. (2022). The role of conventional methods and artificial intelligence in the wastewater treatment: A comprehensive review. *Processes (Basel, Switzerland)*, *10*(9), 1832.

Alvi, M., Batstone, D., Mbamba, C. K., Keymer, P., French, T., Ward, A., & Cardell-Oliver, R. (2023). Deep learning in wastewater treatment: A critical review. *Water Research*, *245*, 120518.

Baarimah, A. O., Bazel, M. A., Alaloul, W. S., Alazaiza, M. Y., Al-Zghoul, T. M., Almuhaya, B., & Mushtaha, A. W. (2024). Artificial intelligence in wastewater treatment: Research trends and future perspectives through bibliometric analysis. *Case Studies in Chemical and Environmental Engineering*, *10*, 100926.

Baskar, G., Omer, S. N., Saravanan, P., Rajeshkannan, R., Saravanan, V., Rajasimman, M., & Shanmugam, V. (2024). Status and future trends in wastewater management strategies using artificial intelligence and machine learning techniques. *Chemosphere*, *362*, 142477.

Begmatov, S., Dorofeev, A. G., Kadnikov, V. V., Beletsky, A. V., Pimenov, N. V., Ravin, N. V., & Mardanov, A. V. (2022). The structure of microbial communities of activated sludge of large-scale wastewater treatment plants in the city of Moscow. *Scientific Reports*, *12*(1), 3458.

Borham, O., Croxford, B., & Wilson, D. (2024). Biomimetic Strategies for Sustainable Resilient Cities: Review across Scales and City Systems. *Biomimetics*, *9*(9), 514.

Capodaglio, A. G., & Callegari, A. (2025). Use, potential, needs, and limits of AI in wastewater treatment applications. *Water (Basel)*, *17*(2), 170.

Croll, H. C., Ikuma, K., Ong, S. K., & Sarkar, S. (2023). Reinforcement learning applied to wastewater treatment process control optimization: Approaches, challenges, and path forward. *Critical Reviews in Environmental Science and Technology*, *53*(20), 1775–1794.

Dansawad, P., Li, Y., Li, Y., Zhang, J., You, S., Li, W., & Yi, S. (2023). Machine learning toward improving the performance of membrane-based wastewater treatment: A review. *Advanced Membranes*, *3*, 100072.

De la Hoz-M, J., Ariza-Echeverri, E. A., & Vergara, D. (2024). Exploring the Role of Artificial Intelligence in Wastewater Treatment: A Dynamic Analysis of Emerging Research Trends. *Resources*, *13*(12), 171.

Deepa, A., Mastan, A., & Buddolla, V. (2025). Advances in imaging techniques for real-time microbial visualization in wastewater treatment reactors: Challenges, applications, and process optimization. *Trends in Analytical Chemistry*, •••, 118227.

Elmahdi, A., & Jeong, J. (2024). From debt to sustainability: Advancing wastewater projects in developing countries through innovative financing mechanisms—the role of debt-for-climate swaps. *Climate (Basel)*, *12*(8), 122.

Eshbobaev, J., Norkobilov, A., Usmanov, K., Khamidov, B., Kodirov, O., & Avezov, T. (2024). Control of Wastewater Treatment Processes Using a Fuzzy Logic Approach. *Engineering Proceedings*, *67*(1), 39.

Ferraro, G. B., Bonomo, C., Brandtner, D., Mancini, P., Veneri, C., Briancesco, R., & La Rosa, G. (2024). Characterisation of microbial communities and quantification of antibiotic resistance genes in Italian wastewater treatment plants using 16S rRNA sequencing and digital PCR. *The Science of the Total Environment*, *933*, 173217.

Gad, M., El Hamed, R. A., El Fadaly, E. A., Mousa, I. E., Gaagai, A., Aouissi, H. A., & Elsayed, S. (2025). New approach to predict wastewater quality for irrigation utilizing integrated indexical approaches and hyperspectral reflectance measurements supported with multivariate analysis. *Scientific Reports*, *15*(1), 16395.

Garg, M. C., Kumari, S., & Agarwal, S. (2024). The integration of artificial intelligence in advanced wastewater treatment systems. In *The AI Cleanse: Transforming Wastewater Treatment Through Artificial Intelligence: Harnessing Data-Driven Solutions* (pp. 1–27). Springer Nature Switzerland.

Ghinea, L. M., Miron, M., & Barbu, M. (2023). Semi-supervised anomaly detection of dissolved oxygen sensor in wastewater treatment plants. *Sensors (Basel), 23*(19), 8022.

Gokul, K., Gokul, V., Kaviyarasan, P., & Uma, J. (2024, March). Intelligent Wastewater Management System with Cloud-Based IOT and Real-Time Control. In *2024 10th International Conference on Advanced Computing and Communication Systems (ICACCS)* (Vol. 1, pp. 498-504). IEEE.

Imen, S., Croll, H. C., McLellan, N. L., Bartlett, M., Lehman, G., & Jacangelo, J. G. (2023). Application of machine learning at wastewater treatment facilities: A review of the science, challenges and barriers by level of implementation. *Environmental Technology Reviews, 12*(1), 493–516.

Infant, S. S., Vickram, S., Saravanan, A., Muthu, C. M., & Yuarajan, D. (2025). Explainable artificial intelligence for sustainable urban water systems engineering. *Results in Engineering, 25*, 104349.

Iqbal, M. A., Akram, S., Lal, B., Hassan, S. U., Ashraf, R., Kezembayeva, G., & Hosseini-Bandegharaei, A. (2024). Advanced photocatalysis as a viable and sustainable wastewater treatment process: A comprehensive review. *Environmental Research, 253*, 118947.

Köksal, D. D., Ahi, Y., & Todorovic, M. (2025). Assessing Agricultural Reuse Potential of Treated Wastewater: A Hybrid Machine Learning Approach. *Agronomy (Basel), 15*(3), 703.

Kubiak, A. (2025). A scalable IoT-controlled photocatalytic system for pharmaceutical removal in real wastewater treatment applications. *Chemical Engineering Journal*, •••, 164709.

Kumar, S., Datta, S., Singh, V., Datta, D., Singh, S. K., & Sharma, R. (2024). Applications, challenges, and future directions of human-in-the-loop learning. *IEEE Access : Practical Innovations, Open Solutions, 12*, 75735–75760.

Liu, Z., Xu, Z., Zhu, X., Yin, L., Yin, Z., Li, X., & Zheng, W. (2024). Calculation of carbon emissions in wastewater treatment and its neutralization measures: A review. *The Science of the Total Environment, 912*, 169356.

Malviya, A., & Jaspal, D. (2021). Artificial intelligence as an upcoming technology in wastewater treatment: A comprehensive review. *Environmental Technology Reviews, 10*(1), 177–187.

Matheri, A. N., Belaid, M., Njenga, C. K., & Ngila, J. C. (2023). Water and wastewater digital surveillance for monitoring and early detection of the COVID-19 hotspot: Industry 4.0. *International Journal of Environmental Science and Technology, 20*(1), 1095–1112.

Mishra, S., & Sundaram, B. (2024). A review of the photocatalysis process used for wastewater treatment. *Materials Today: Proceedings, 102*, 393–409.

Mithal, A., & Thankachan, B. (2025). A Comprehensive Review on Blockchain-based Systems for Groundwater Conservation and Wastewater Management. *Environmental Management*, •••, 1–19.

Mohammadi, E., Stokholm-Bjerregaard, M., Hansen, A. A., Nielsen, P. H., Ortiz-Arroyo, D., & Durdevic, P. (2024). Deep learning based simulators for the phosphorus removal process control in wastewater treatment via deep reinforcement learning algorithms. *Engineering Applications of Artificial Intelligence, 133*, 107992.

Mohanty, A., Mohanty, S. K., & Mohapatra, A. G. (2024). Real-time monitoring and fault detection in AI-enhanced wastewater treatment systems. In *The AI Cleanse: Transforming Wastewater Treatment Through Artificial Intelligence: Harnessing Data-Driven Solutions* (pp. 165–199). Springer Nature Switzerland.

Nirmala, N., Arun, J., Sanjay Kumar, S., & Dawn, S. S. (2025). Role of Machine Learning and Artificial Intelligence in Smart Waste Management. In *Smart Waste and Wastewater Management by Biotechnological Approaches* (pp. 35–53). Springer Nature Singapore.

Padhiary, M. (2025). Benefits, Risks, and Responsible Use of Nanomaterials in Wastewater Remediation. In *Nanomaterials in Wastewater Research* (pp. 443–464). Springer.

Padhiary, M., Roy, D., & Dey, P. (2025). Mapping the landscape of biogenic nanoparticles in bioinformatics and nanobiotechnology: AI-driven insights. In *Synthesizing and Characterizing Plant-Mediated Biocompatible Metal Nanoparticles* (pp. 337–376). IGI Global.

Parsa, Z., Dhib, R., & Mehrvar, M. (2024). Dynamic modelling, process control, and monitoring of selected biological and advanced oxidation processes for wastewater treatment: A review of recent developments. *Bioengineering (Basel, Switzerland), 11*(2), 189.

Renfrew, D., Vasilaki, V., & Katsou, E. (2024). Indicator based multi-criteria decision support systems for wastewater treatment plants. *The Science of the Total Environment*, *915*, 169903.

Rodríguez-Hernández, J. A., Araújo, R. G., López-Pacheco, I. Y., Rodas-Zuluaga, L. I., González-González, R. B., Parra-Arroyo, L., & Parra-Saldívar, R. (2022). Environmental persistence, detection, and mitigation of endocrine disrupting contaminants in wastewater treatment plants–a review with a focus on tertiary treatment technologies. *Environmental Science. Advances*, *1*(5), 680–704.

Sakkaravarthy, S., Jano, N. A., & Vijayakumar, A. (2024). Overcoming challenges in traditional waste water treatment through ai-driven innovation. In *The AI Cleanse: Transforming Wastewater Treatment Through Artificial Intelligence: Harnessing Data-Driven Solutions* (pp. 53–81). Springer Nature Switzerland.

Shahouni, R., Abbasi, M., Dibaj, M., & Akrami, M. (2024). Utilising artificial intelligence to predict membrane behaviour in water purification and desalination. *Water (Basel)*, *16*(20), 2940.

Sharma, M., Rawat, S., Kumar, D., Awasthi, A., Sarkar, A., Sidola, A., & Kotecha, K. (2024). The state of the Yamuna River: A detailed review of water quality assessment across the entire course in India. *Applied Water Science*, *14*(8), 175.

Slany, V., Krcalova, E., Balej, J., Zach, M., Kucova, T., Prauzek, M., & Martinek, R. (2025). Smart Water-IoT: Harnessing IoT and AI for Efficient Water Management. *ACM Computing Surveys*.

Sylwan, I., & Thorin, E. (2021). Removal of heavy metals during primary treatment of municipal wastewater and possibilities of enhanced removal: A review. *Water (Basel)*, *13*(08), 1121.

Wang, A. J., Li, H., He, Z., Tao, Y., Wang, H., Yang, M., & Ren, N. (2024). Digital twins for wastewater treatment: A technical review. *Engineering*, *36*, 21–35.

Wang, Y. Q., Wang, H. C., Wang, W. Z., Yang, H. L., Chen, J. J., Fan, Y. X., & Wang, A. J. (2025). Federated Machine Learning Enables Risk Management and Privacy Protection in Water Quality. *Environmental Science & Technology*, *59*(21), 10310–10322.

Zagklis, D. P., & Bampos, G. (2022). Tertiary wastewater treatment technologies: A review of technical, economic, and life cycle aspects. *Processes (Basel, Switzerland)*, *10*(11), 2304.

Zhang, S., Jin, Y., Chen, W., Wang, J., Wang, Y., & Ren, H. (2023). Artificial intelligence in wastewater treatment: A data-driven analysis of status and trends. *Chemosphere*, *336*, 139163.

Zhenyu, K., Jun, Z., Zhongda, T., Shujiang, L., & Yanhong, W. (2023). NSGAIII based on utopian point improvements and its application in wastewater treatment process. *Expert Systems with Applications*, *234*, 121030.

KEY TERMS AND DEFINITIONS

Artificial Intelligence: A branch of computer science focused on developing systems that can perform tasks typically requiring human intelligence, such as learning, pattern recognition, and decision-making.

Bioaugmentation: The introduction of specialized microbial strains into a biological treatment system to enhance its ability to degrade pollutants or improve stability under stress conditions.

Circular Water Economy: A sustainable water management framework that emphasizes water reuse, resource recovery, and minimal environmental discharge to create closed-loop water cycles.

Decentralized Treatment Systems: Small-scale, modular wastewater treatment units that operate independently of centralized infrastructure, often used in rural or peri-urban areas.

Deep Learning: An advanced type of machine learning that uses multi-layered neural networks to model complex, non-linear relationships in large datasets, especially effective for image, speech, or time-series analysis.

Effluent Quality Prediction: The use of AI or statistical models to forecast the quality of treated water, typically in terms of indicators like BOD, COD, TSS, and nutrient levels.

Intelligent Control Systems: Automated systems that use AI algorithms to dynamically adjust process parameters in real time based on sensor inputs and predicted outcomes.

Machine Learning (ML): A subset of AI that enables computers to learn from data and improve their performance over time without being explicitly programmed.

Predictive Analytics: A data-driven approach that uses statistical models and machine learning to forecast future events or system behaviors, enabling proactive decision-making.

Process Optimization: The use of computational methods to improve the efficiency and effectiveness of a system, often by minimizing energy use, chemical consumption, or operational costs while maximizing output quality.

Real-Time Monitoring: Continuous data collection and analysis from sensors to track system performance and detect anomalies or operational changes instantaneously.

Resource Recovery: The process of extracting valuable byproducts—such as nutrients, energy, or clean water—from wastewater, contributing to economic and environmental sustainability.

Smart Water Grids: Digitally connected water infrastructure systems that use real-time data, AI, and automation to monitor, predict, and optimize the distribution, treatment, and reuse of water.

Wastewater Treatment: A multi-stage process that removes contaminants from used water—originating from domestic, industrial, or commercial sources—so it can be safely discharged or reused.

Water Recycling / Reuse: The process of treating wastewater to a quality suitable for beneficial reuse, such as agricultural irrigation, industrial cooling, or even potable applications.

Chapter 13

Wastewater Treatment and Recycling Using Adsorption Enhanced by Artificial Neural Networks

Soukayna Maitouf
https://orcid.org/0009-0000-4619-7863
Faculty of Sciences Agadir, Morocco

Ahmed Bachar
Ibn Zohr University, Morocco

Assia Mabrouk
https://orcid.org/0000-0001-6399-0644
Ibn Zohr University, Morocco

Nadia Faska
https://orcid.org/0000-0001-6203-6904
Faculty of Applied Sciences Ait Melloul, Morocco

ABSTRACT

The world is facing a lack of freshwater because of high population growth rates, and an increase in industrial activities. Therefore, new approaches to wastewater treatment and recycling are essential in sustainable water management. This chapter focuses on the integration of adsorption, which is an economically feasible method of pollutant removal, and Artificial Neural Networks (ANNs), known for their exceptional skills in complex system modeling and process optimization. ANN aids in increasing the efficiency and flexibility of adsorption systems to ensure appropriate

DOI: 10.4018/979-8-3373-2700-6.ch013

water quality for reuse. This integration is feasible and beneficial not only from an environmental perspective, but also from an economic one, as shown through its practical implementation in municipal and industrial settings.

INTRODUCTION

Wastewater treatment and recycling are crucial components of sustainable water management, addressing the dual challenges of water scarcity and environmental pollution (Yadav & Kamsonlian, 2023). As global populations grow and industrial activities expand, the demand for freshwater intensifies, and the generation of contaminated wastewater has escalated. Effective treatment processes are essential for removing pollutants, protecting ecosystems, and enabling the safe reuse of water for agricultural, industrial, and municipal applications (Gupta et al., 2012). Among the various methods employed, chemical processes, such as adsorption, membrane filtration, and chemical oxidation, have emerged as prominent solutions because of their efficacy in targeting diverse contaminants, including heavy metals, organic compounds, and emerging micropollutants. (Almuktar et al., 2018; Saravanan et al., 2021)

Adsorption, in particular, has gained widespread recognition for its versatility, cost-effectiveness, and ability to be integrated with other treatment technologies (Liu, 2023). This process leverages the surface properties of adsorbents to capture and remove pollutants, offering a sustainable alternative to conventional methods (Tony, 2021b). However, the efficiency of adsorption and other treatment techniques is influenced by complex factors such as the operating conditions, contaminant properties, and material regeneration. To optimize these processes, advanced tools, such as Artificial Neural Networks (ANNs), have been increasingly adopted (Nguyen et al., 2020). ANNs provide predictive modeling, real-time monitoring, and operational optimization, thereby enhancing the performance and scalability of wastewater treatment systems. (Yetilmezsoy & Demirel, 2008 ; Sadek et al., 2023)

This work explores the principles, applications, and challenges of wastewater treatment and recycling, with a focus on chemical methods and the role of ANNs in improving process efficiency. Case studies from industrial and municipal contexts highlight practical implementations and the potential for reuse of treated wastewater. By addressing technological limitations and emphasizing innovative solutions, this study contributes to the advancement of sustainable water management practices.

WASTEWATER TREATMENT AND RECYCLING

Common Chemical Methods

Wastewater treatment processes are essential for removing contaminants and ensuring that water is safe for discharge or reuse. Various methods have been employed to address different types of pollutants in wastewater (Pratap et al. 2022). Key techniques include membrane filtration, adsorption, ion exchange, chemical oxidation, and coagulation-flocculation, each with distinct advantages and limitations in terms of efficacy and cost (Pratap et al., 2022; Dotto et al., 2025).

Membrane filtration has gained prominence in recent decades due to its ability to separate contaminants from water sources effectively. This method utilizes various types of membranes, such as ultrafiltration, nanofiltration, and reverse osmosis (Cevallos-Mendoza et al., 2022). Although reverse osmosis is particularly efficient in removing salts from saline water, it may not be feasible for large-scale operations because of its high operating costs (Shafique, 2021). In addition, membranes alone cannot remove all drinking water contaminants, necessitating the integration of complementary treatment methods (Van Der Bruggen et al., 2003; Guo et al., 2021).

Chemical oxidation processes are commonly used to eliminate hazardous organic and inorganic contaminants from drinking water sources. This technique typically precedes the coagulation-flocculation method, which is effective for removing suspended solids and heavy metals from water (Wu et al., 2024; X. Liu et al., 2022). However, both methods are associated with challenges, such as significant sludge generation and high operational costs (Wu et al., 2024).

Adsorption is a widely employed method in wastewater treatment, specifically designed to remove non-degradable organic compounds and heavy metals from water (Shafique, 2021b). This process involves the use of adsorbent materials that attract and bind contaminants to effectively purify water. The attractiveness of adsorption lies in its sustainability and versatility, allowing it to be combined with biological methods to enhance results (Dotto et al., 2025). For instance, natural adsorbents such as almond seeds and starch can be utilized for effective impurity removal, providing an eco-friendly alternative to synthetic materials (Shafique, 2021b).

Recycling Treated Wastewater

Recycling treated wastewater is gaining global momentum as a sustainable solution to water scarcity. This includes treating municipal or industrial wastewater to a quality suitable for specific reuse applications, thereby conserving freshwater resources. The level of treatment required depends mainly on the intended end use and the associated public health and environmental risks. For applications such as agricultural irrigation or landscape watering, wastewater treatment typically involves secondary treatment followed by disinfection, and sometimes tertiary treatment to meet specific quality criteria regarding pathogens, suspended solids, and nutrient levels (Yadav & Kamsonlian, 2023). Industrial reuse may necessitate more advanced treatment, such as reverse osmosis or ion exchange, to meet the high-purity requirements for processes such as boiler feedwater or electronics manufacturing. Groundwater recharge, another important application, involves the percolation of highly treated wastewater into aquifers to augment the groundwater supply or create barriers against saltwater intrusion (Eriksson et al., 2010). This usually requires advanced treatment to remove salts, organic micropollutants, and pathogens to protect groundwater quality from degradation. Successful water recycling programs depend not only on reliable treatment technology but also on public acceptance, appropriate regulatory frameworks, robust monitoring, and careful management to ensure safety and sustainability (Mbavarira & Grimm, 2021).

ADSORPTION IN WASTEWATER TREATMENT

Adsorption has emerged as a prominent wastewater treatment method because of its effectiveness in removing a wide range of contaminants, including heavy metals, organic compounds, dyes, and emerging pollutants (Tony, 2021). Adsorption offers numerous advantages, including cost-effectiveness, ease of implementation, high efficiency, and design flexibility (Liu, 2023). This section explores the fundamental principles, applications, materials, and recycling aspects of adsorption in wastewater treatment.

Principles of Adsorption

Adsorption is a surface phenomenon in which molecules, atoms, or ions (adsorbates) from the liquid or gas phase adhere to the surface of a solid material (adsorbent). This process occurs because of unbalanced surface forces or chemical bonds at the adsorbent surface. In wastewater treatment, adsorption mechanisms can be broadly classified into physical (physisorption) and chemical (chemisorption) mechanisms. (Adeleke et al., 2018)

Physisorption involves relatively weak van der Waals forces between the adsorbate and adsorbent, resulting in multilayer adsorption with a low heat of adsorption (typically 20-40 kJ/mol). This process is generally reversible and occurs rapidly without a significant activation energy (Figure.1). Conversely, chemisorption involves the formation of chemical bonds between the adsorbate and adsorbent surface, characterized by a higher heat of adsorption (40-400 kJ/mol) and monolayer coverage, and often requires activation energy (Y. Liu, 2024). The predominant mechanism depends on the nature of both the adsorbate and adsorbent, as well as the operating conditions, such as temperature, pH, and concentration.

Figure 1. Mechanism of physical and chemical adsorption on a surface

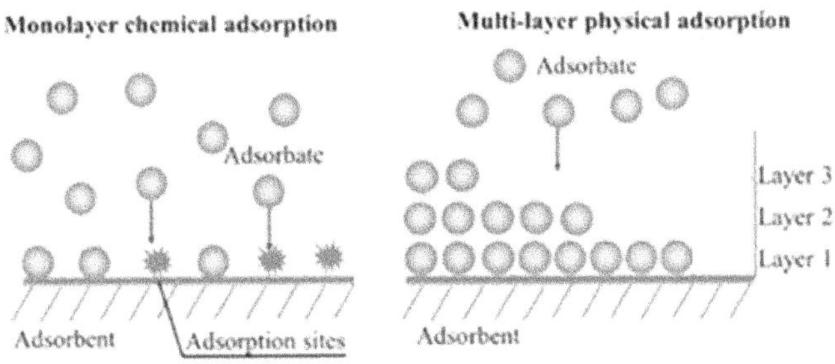

Isotherms, Kinetics, and Thermodynamics

Adsorption isotherms describe the equilibrium relationship between the amount of adsorbate on the adsorbent and its concentration in the solution at a constant temperature. Several models have been developed to represent these relationships using the Langmuir, Freundlich, and Brunauer-Emmett-Teller (BET) isotherms, which are the most widely applied in wastewater treatment studies (Table.1). The Langmuir isotherm assumes monolayer adsorption on a homogeneous surface with no interaction between adsorbed molecules, whereas the Freundlich isotherm describes heterogeneous surfaces with multilayer adsorption. The BET isotherm extends the Langmuir model to account for multilayer adsorption phenomena. (Thamer et al., 2023; Ramlee et al., 2024)

Table 1. Isotherms models of adsorption

Isotherms	Equation
Freundlich	$log\, q_e = q_e K_1 + \frac{1}{n} C_e$
Langmuir	$\frac{C_e}{q_e} = \frac{1}{bQ_{max}} + \frac{C_e}{Q_{max}}$
temkin	$q_e = \frac{KT}{b} ln\, K_t C_e$

continued on following page

Table 1. Continued

Isotherms	Equation
Redlich-peterson	$q_e = \frac{K_{RP}\, C_e}{1 + \alpha C_e^{\beta}}$
parameter	Description
q_e	Adsorption equilibrium rate (mg/g)
q_{max}	Dye adsorbed (mg/g)
b	Constant (mg/g)(L/g)$^{-1/n}$
n	Freundlich constant
Ce	Final concentration (mg/g)
K	Intraparticle diffusion constant
C	Boundary layer thickness (mg/g)

Adsorption kinetics, which describe the rate of adsorption and time required to reach equilibrium, are crucial for the design of adsorption systems. Common kinetic models include pseudo-first-order, pseudo-second-order, and intraparticle diffusion models (Li et al., 2025). These models help understand the mechanisms controlling the adsorption process, such as external mass transfer, boundary layer diffusion, and intraparticle diffusion (Table 2).

Table 2. Kinetic models of adsorption

Kinetics	Equation
First-order	$\frac{dq_t}{dt} = K_1 \left(q_e - q_t \right)$
Second-order	$\frac{dq_t}{dt} = K_2 \left(q_e - q_t \right)^2$
Intraparticle diffusion	$q_t = Kt^{\frac{1}{2}}$
Elovich	$q_t = \frac{1}{b} lnab + \frac{1}{b} lnt$
parameter	Description
a	Initial adsorption rate (mg/gmin)
1/b	Number of site available(mg/g)
Kt	Temkin equilibrium constant
b	Energy of adsorption(j)
T	Absolute temperature in K(°C)
R	Ideal gas constant(8.314J/mol.K)
α	Redlich-peterson constant
β	Redlich-peterson constant

Thermodynamic parameters, including Gibbs free energy change ($\Delta G°$), enthalpy change ($\Delta H°$), and entropy change ($\Delta S°$), provide insights into the spontaneity, heat effects, and randomness of the adsorption process. A negative $\Delta G°$ indicates a spontaneous process, whereas $\Delta H°$ indicates whether the process is exothermic (negative) or endothermic (positive). These parameters are essential for optimizing the operating conditions and understanding the fundamental nature of the adsorbate-adsorbent interactions. (Mostafapour et al., 2022; Ivanović et al., 2025)

Applications of Adsorption

(Atangana et al., 2025) reviewed the adsorption efficiency of removing pollutants, including pesticides, microplastics, polycyclic aromatic hydrocarbons, parabens, and polychlorinated biphenyls. Adsorption is a renewable process that decreases the concentration of persistent pollutants, especially pesticides. The presence of heavy metals in water and wastewater is an extremely hazardous problem for the environment and ecosystems. In the (Y. Wang et al., 2024) study, the adsorption process demonstrated a high potential for removing heavy metal ions, nitrates, phosphates, ammonium, organic matter, antibiotics, and dyes. The simple preparation of adsorbents, their ability to be reused, and their regeneration make adsorption a good wastewater treatment technique. However, the adsorption technique also has some limitations, such as the saturation of the adsorbent. When all empty sites of the adsorbent surface are occupied, the adsorbent is saturated, and this state is called adsorption saturation. However, the adsorbent reached this state after a long period. Therefore, it is necessary to determine the effective contact time. Moreover, adsorbents still require modification and improvement to achieve a strong pollutant affinity (Georgin et al., 2023)

Adsorbent Materials

The effectiveness of adsorption processes largely depends on the properties and characteristics of adsorbent materials. Optimal adsorbent materials for pollutant removal must satisfy key criteria: cost-effectiveness, robust mechanical and structural stability to endure hydrodynamic forces over prolonged operational durations, rapid and high-capacity adsorption kinetics, extensive specific surface area, and efficient regenerability through economically viable processes. The most prevalent adsorbent materials are systematically classified and presented in Figure 2 according to their material categories. (Chen et al., 2021; Adeleke et al., 2018; Liu, 2023)

Figure 2. The most frequently used adsorbent materials

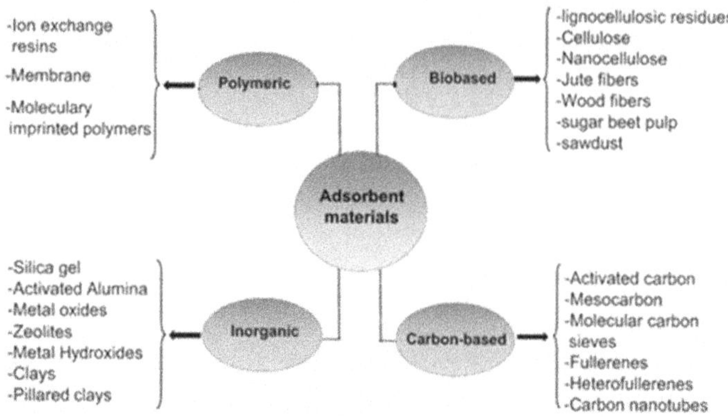

The selection of an appropriate adsorbent for a particular application depends on various factors, including the nature and concentration of the target pollutants, required treatment efficiency, operational conditions, cost considerations, and potential for regeneration and reuse.

Scientists have worked on many materials to explore their capacity as adsorbents, such as activated carbon, which is the most commonly used adsorbent because of its high surface area and high adsorptive capacity (W. Li et al., 2016). Nanocomposites, magnetic porous carbon, polymer adsorbents (hydroxypropyl-β-cyclodextrin polymer, anionic hypercrosslinked polymer), and biochar (rice husk, sawdust, herbal waste, sludge, and coffee husk) (O'Connor et al., 2022), (Kadimpati et al., 2024).

Adsorbent Recycling

The sustainability and economic viability of adsorption processes for wastewater treatment can be significantly enhanced through effective adsorbent recycling strategies (Figure 3). The regeneration of spent adsorbents not only reduces the cost associated with the continuous replacement of fresh materials but also minimizes waste generation and resource consumption. (Baskar et al., 2022)

Figure 3. Adsorbent regeneration and reuse cycle

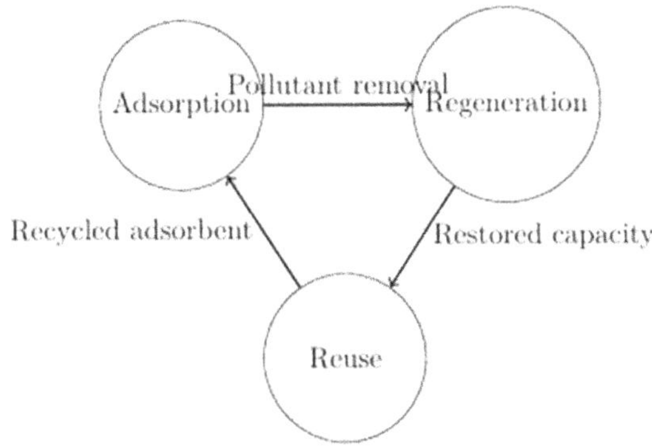

Several regeneration methods have been employed, depending on the nature of the adsorbent and the adsorbed contaminants. Thermal regeneration, commonly used for activated carbon, involves heating the spent adsorbent to high temperatures (600-900°C) in a controlled atmosphere to desorb and decompose the adsorbed contaminants (Da'na & Awad, 2017; Tony, 2021b). Although effective, this method requires significant energy input and may result in some loss of the original adsorption capacity owing to structural changes in the adsorbent.

Chemical regeneration uses solvents, acids, bases, or oxidizing agents to desorb contaminants from the adsorbent surface (Renu & Sithole, 2024). For instance, heavy metals adsorbed onto various materials can often be effectively desorbed using acidic solutions, whereas organic compounds may require organic solvents or alkaline solutions (Patel, 2021). This approach typically consumes less energy than thermal methods but generates secondary waste streams that require further treatment. (Renu & Sithole, 2024b)

Biological regeneration, particularly for adsorbents loaded with biodegradable organic contaminants, employs microorganisms to degrade adsorbed pollutants (Xiao & Hill, 2021). This method is environmentally friendly but generally slow and may not be suitable for all types of contaminants in wastewater. (Baskar et al., 2022b; Zhu et al., 2009)

Electrochemical regeneration combines electrical and chemical processes to desorb contaminants, offering advantages in terms of mild operating conditions and potential for in situ applications (Yasri & Roberts, 2024; Ersan et al., 2023). Emerging techniques, such as microwave-assisted, ultrasonic, and supercritical fluid regeneration, are also being explored for their potential to provide more efficient and less energy-intensive alternatives. (Zhou et al., 2020)

The choice of regeneration method depends on factors such as the type of adsorbent, nature of the adsorbates, desired regeneration efficiency, energy requirements, and economic considerations. Ideally, the regeneration process should restore a significant portion of the original adsorption capacity while minimizing adsorbent loss and the environmental impact of the adsorbent. The development of adsorbents specifically designed for easy regeneration is an important research direction for enhancing the sustainability of adsorption-based wastewater treatment systems. (Jain et al., 2021)

CHALLENGES IN WASTEWATER TREATMENT AND RECYCLING

Wastewater treatment and recycling face numerous challenges that limit their effectiveness and widespread implementation. The complexity and variability of wastewater composition present significant obstacles to efficient treatment (Corpus et al., 2024). Municipal wastewater contains a diverse mixture of organic matter, nutrients, pathogens, and trace contaminants such as pharmaceuticals, personal care products, and microplastics (Hasan et al., 2023). Industrial wastewater adds another layer of complexity to sector-specific pollutants such as heavy metals, toxic organic compounds, and extreme pH conditions (Wollmann et al., 2019). This heterogeneity complicates the design of universal treatment solutions and often necessitates tailored approaches for specific wastewater streams to achieve optimal results. (Villarín & Merel, 2020)

Emerging contaminants are a growing concern for conventional treatment systems. These include pharmaceuticals, endocrine-disrupting compounds, perfluoroalkyl and polyfluoroalkyl substances, and antibiotic-resistant bacteria (Corpus et al., 2024). Many traditional treatment processes have not been designed to remove these persistent micropollutants, which can remain biologically active even at very low concentrations. The potential long-term effects of recycled water exposure on aquatic ecosystems and human health remain inadequately understood, creating technical and regulatory challenges. (Pandit & Sharma, 2024)

Energy consumption and operational costs are significant barriers to sustainable wastewater treatment. Advanced treatment processes required for high-quality water recycling, such as membrane filtration, advanced oxidation, and disinfection, typically require substantial energy input. This energy intensity contributes to higher operational costs and carbon footprints, potentially offsetting the environmental benefits of water recycling. Determining the balance between treatment effectiveness and energy efficiency remains a critical challenge. (Twi-Yeboah et al., 2024)

Infrastructure limitations further complicate the recycling of wastewater. Many existing wastewater treatment facilities have been designed primarily for pollution control rather than resource recovery (Larsen, 2011). Retrofitting these systems for water recycling often requires significant capital investments and physical space, which may be constrained in urban settings. Additionally, the distribution infrastructure required to deliver recycled water to end users (dual piping systems) represents a substantial cost barrier, particularly in established communities. (Chen et al., 2024)

ARTIFICIAL NEURAL NETWORKS (ANN) IN ENVIRONMENTAL ENGINEERING

The basic architecture of an ANN is very simple. Three different layers are the components of the ANN model. The first is the input layer, the second is the hidden layer, and the third is the output layer. After uploading the input data, a base function transforms them into neurons, and an activation function introduces the non-linearity. The effect of each neuron is given by the weight factor W_{ij}; this factor presents the input and output data connection (i represents the input variable effect, while j represents the neuron effect). The hidden layers' position in the ANN architecture is between the input and output layers. Additionally, it can be formed by one of multiple layers (Nawaz & Kumar, 2022), (Huang et al., 2011), (Kundu et al., 2015). The ANN architecture is shown in Figure 4.

The activation function is responsible for generating outputs. However, the function received the processed data from the hidden layers, then it generated a signal that will be sent to the output layer. Moreover, the network's inputs are stored in the first layer, and this input is the result of the sum of the value of the bias factor with the total weight of the entire previous layer signals (Humadi et al., 2024). An adequate artificial neural network structure depends on the training of the network until it gets a hidden layer with the optimal number of neurons. A high number or a small number of hidden layers has a negative effect on the model's performance. The ANN modeling capability will be limited with a low number of neurons; however, an elevated number of neurons causes a poor learning performance due to the over-flexibility for modifying the weights, and this engenders noise (Elfghi, 2016).

Figure 4. The ANN structure

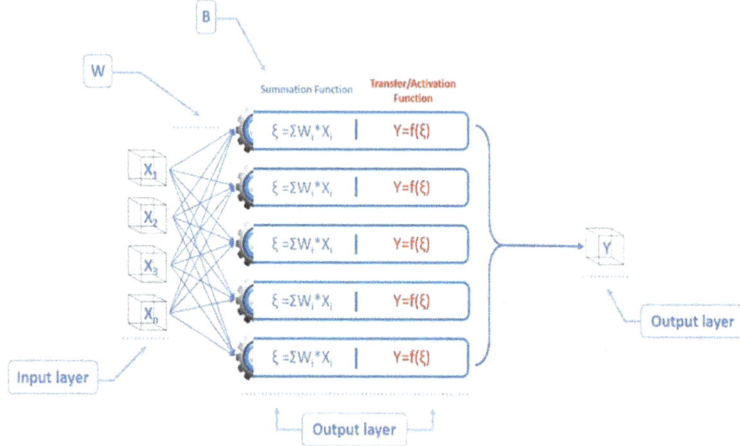

To improve the ANN model performance, it is crucial to choose the best architecture, including the algorithm, data set size, and transfer function. According to the literature, in the first test, the data can be divided into 70%-15%-15% for training, testing, and validation. Therefore, the model uses a lot of data to learn and confirm its performance. A model with high sensitivity to various data distributions presents a better capacity for generalization. Moreover, numerous functions are applied as training functions, including Levenberg-Marquardt, Bayesian regularization, and scaled conjugate gradient. The first algorithm advantage is that it trains the model quickly, the second trains the model slower than the first but has better generalization, and the third is an algorithm with the highest memory efficiency.

Table 3. The most transfer functions applied in ANN

Function	Symbol	Normalization	Equation
Logsig	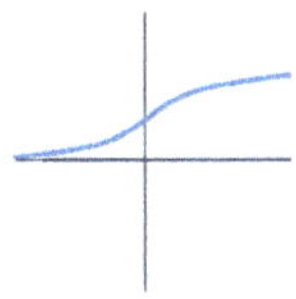	0 to +1	$Logsig(n) = \frac{1}{1+e^{-n}}$
Tansig	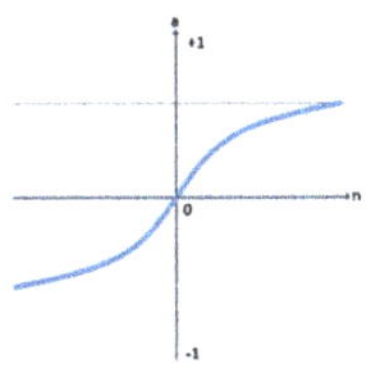	-1 to +1	$Tansig(n) = \frac{2}{1+e^{-2}} - 1$
Linear/Purelin	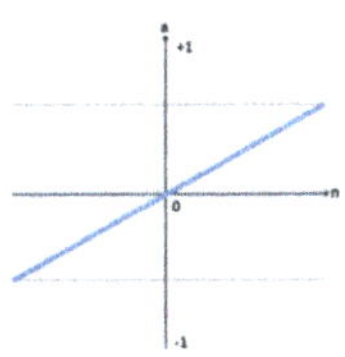	-1 to +1	$Purelin(n) = n$

ROLE OF ARTIFICIAL NEURAL NETWORKS (ANN)

The improvement and development of sustainable wastewater technologies is an urgent challenge, especially when the objective is to reduce water pollution by eliminating the various substances it contains, while simultaneously reusing the treated water and the material used to purify wastewater in different domains. Therefore, artificial neural networks are widely used in numerous wastewater treatment and recycling systems because they are a strong tool for enhancing the performance and efficiency of wastewater treatment methods. Artificial neural networks (ANNs) play a significant role in modeling several wastewater treatment technologies. ANN can predict treatment efficiency, optimize operating conditions, and help control the real-time processes.

In (Hussain et al., 2021) study, the operational variables of photolytic degradation of sulfamethoxazole and concurrent total organic carbon were modeled using an ANN model. The study showed that R^2 was 0.998. Gautam et al. (2020) explored the use of an ANN to model the decolorization percentage of reactive black B using the electrocoagulation process. The results showed a decolorization rate of 94.96%. The elimination of direct blue 86 dye was explored by applying an electrochemical oxidation process (Kumar & Gupta, 2022). An artificial neural network was used to predict the dye and COD removal. The results were 0.9691 and 0.0583 for R^2 and RMSE, respectively. Banza et al. (2023) applied an ANN algorithm to predict the adsorption of Cr(VI) on sodium alginate and cellulose nanocrystals. The findings demonstrate the high ability of ANN to predict Cr(VI) removal, with an R^2-value of 0.990 and a low mean square error (MSE) for all data. Khatri et al. (2023) explored the use of biological processes to treat wastewater. An up-flow anaerobic sludge blanket facultative pond was used to treat the effluents. Three ANN models, including feedforward backpropagation, deep feedforward backpropagation, and deep cascade forward-backpropagation algorithms, were tested to predict effluent quality. The results indicated that the third algorithm was the best model, with a correlation coefficient of 0.997 and RMSE of 6.018. Khan et al. (2023) optimized the operating factors for the removal of RB19 dye from wastewater using an electro-Fenton process combined with an ANN model. The results showed that the test correlation was greater than 0.999. In another study conducted by Yuan et al. (2023), an ANN model was applied to the forward osmosis membrane process to predict the permeate flux. The results indicate that the model has a high predictive performance, with $R^2 = 97.3\%$. (Warren-Vega et al., 2024) utilized advanced oxidation processes to decrease wastewater pollution. In this study, piggery wastewater treatment was evaluated, and the processing capacity was predicted using an ANN model. At the pilot scale, the results indicated $R^2 = 0.99102$ under continuous flow conditions. Adnan et al. (2025) combined the ANN model with the photocatalytic activities of metal-oxide-based catalysts. An ANN was used to predict the contaminant reduction. The results showed that the treatment cost decreased by 7% ($R^2 = 0.9925$). On the other hand, Picos-Benítez et al. (2020) reported that ANN was used to decrease the system energy and increase the elimination of bromophenol blue by an electro-oxidation process. The operating conditions were optimized using a hybrid ANN-GA. The results indicate that high efficiency at the lowest time possible is achieved when $t = 10$ min, pH = 2.8, bromophenol blue concentration = 41.25 mg L^{-1}, flow = 11.9 L min^{-1}, and current density = 31.25 mA cm^{-2}. The experimental discoloration efficiency was 88.8%, and the predicted value was 95.5%.

ANN-DRIVEN SOLUTIONS FOR WASTEWATER TREATMENT AND RECYCLING

Predictive Modeling of Adsorption Efficiency

Heavy metals and organic compounds are the most common wastewater pollutants. Moreover, the elimination of these contaminants is difficult and complex. Adsorption is one of the most widely used processes for reducing the concentration of several toxic substances in wastewater. However, this technique faces several challenges and limitations. Therefore, researchers have integrated artificial intelligence, particularly artificial neural network (ANN) tools, into adsorption techniques. Numerous studies have explored the use of artificial neural network models for adsorption treatment and recycling. The ANN model exhibited high efficiency and prediction performance for pollutant removal.

Mahmoud et al. (2017) developed an ANN model to predict the efficiency of nano zero-valent iron in removing organic molecules, such as benzene, toluene, ethyl benzene, and xylene. The ANN structure used was 6-10-1, and the data were divided into 60-20-20 for training, validation, and testing, respectively. The model presented $R^2 = 0.97064$ and MSE = 0.6272%. Karri et al. (2018) applied biomass based on polyaniline to adsorb methyl orange. The ANN algorithm was applied to optimize the percentage of toxic dye removal from wastewater. The optimal ANN structure was 5-12-1. The findings indicated that the model had R2 values of 0.964 and 0.973 for testing and validation. Igwegbe et al. (2019) combined an ANN tool with Ho-$CaWO_4$ nanoparticles used to remove methylene blue. An ANN model was used to predict the methylene blue removal efficiency. The ANN architecture used was 4-8-1, and the data were divided into 75% for training, 15% for validation, and 10% for testing. The results showed that the model had an MSE of 0.00601718 and R^2 of 0.999881. Zhu et al. (2019) used an ANN algorithm to predict the removal efficiencies of different heavy metals. In this study, Pb, Cd, Ni, As, Cu, and Zn were eliminated using an adsorption process. The results indicate that the ANN has a high prediction performance, with R^2 values of 0.948 and RMSE values of 0.079. Chen et al. (2021) evaluated the adsorption of organic compounds "Rhamnolipid Biosurfactant" on Amberlite XAD-2. A least-squares vector machine algorithm was used, and the results were compared with those of an ANN model. The artificial neural network presented a high coefficient of determination (R2 = 0.9946) and a low MSE (0.0005). Nguyen et al. (2020) reported that Congo red dye was adsorbed by akaganeite nanoparticles. An ANN algorithm was applied to analyze the adsorption. The optimal ANN structure was determined to be 4-10-1, and the data were divided into 70% for training, 15% for validation, and 15% for testing purposes. The findings indicated that the model had an R^2 of 0.996 and an MSE of

4.405. Similarly, an ANN model was used to predict the adsorption efficiency of nano-zero-valent aluminum for Cu (II) ions. The findings showed that a minimum MSE< 10^{-5} was obtained at epoch 250 (Sadek et al., 2023). Kumari et al. (2024) developed Neural Networks (ANN) to predict methylene blue adsorption using Juglans Regia. The ANN structure used was 4-5-1, and the data were divided into 70-15-15 for training, validation, and testing, respectively. The model presented R^2 = 0.97064,0.9572, 0.9916, and 0.9288 for the training, validation, testing, and all data, respectively, with an MSE of 0.6272%. (Yadav et al., 2021) explored the application of modified biochar derived from coconut husks to adsorb copper ions. The optimal ANN structure was found to be 4-2-1. The findings indicate that the model has an R^2 of 0.91 and an MSE of 4.60.

OPTIMIZATION OF OPERATING CONDITIONS

The crucial reason for applying ANN algorithms is to test their capacity and performance in optimizing operating conditions. Compared to statistical methods of optimization, such as the response surface methodology, intelligent tools show better accuracy and the strongest capacity for prediction, modeling, and optimization. Therefore, in several studies, an ANN was applied to optimize the operating factors.

Cao et al. (2017) reported the application of an adsorption technique using Fe_3O_4/rGO composites for the removal of Hg (II). An intelligent tool has been combined with an adsorption system. 24 data points were used to train the ANN algorithm. The optimal structure was 1-7-1. The findings indicate that the ANN model used has an R^2 of 0.9928 with 91.13% for the predicted removal efficiency. Furthermore, the optimal conditions were found to be 9.9, 37.3, 63.5, and 8.6 for pH, temperature, contact time, and initial concentration, respectively. In another study conducted by (Fan et al., 2017), reduced graphene oxide-supported nanoscale Zero-Valent Iron composites were used as adsorbents to eliminate Cd(II) ions. An ANN model was employed to optimize the sorption process. The optimal structure was 4-4-1. The results show that the ANN model used had an R^2 close to 1 with a 99.67% removal percentage. Moreover, the optimal conditions were found to be 6.5, 25.31, 30, and 20.16 for pH, temperature, contact time, and initial concentration, respectively. Khajeh et al. (2017) developed an adsorbent from tea waste to remove two different contaminants: Co and Mn. Subsequently, the ANN algorithm was integrated to optimize the operating conditions. The results showed that the algorithm had correlation coefficients of 0.902 and 0.944 for Co and Mn, respectively. In addition, the optimal operating conditions for Mn sorption were pH = 9.8, adsorbent mass = 0.5, PAN (complexing agent) = 3, eluent volume = 3.8, eluent concentration = 0.1, flow rate = 3.1, and eluent flow rate = 1. However, for Co., pH = 9.1, mass of adsorbent = 0.35, PAN = 3, eluent volume = 2, eluent concentration = 1, flow rate = 6, and eluent flow rate = 3. Additionally, Shi et al. (2017) reduced graphene oxide-supported nanoscale zero-valent iron as an adsorbent to eliminate Rhodamine B. To optimize the adsorption system, an ANN model was applied. The maximum adsorption percentage was 86.4% and the ANN correlation coefficient was 0.99. However, the MSE of the model was 0.0045. The optimal conditions of this work were 3.2, 27, 6, 60 of pH, temperature, contact time, and initial concentration, respectively.

REAL-TIME MONITORING AND CONTROL

Real-time monitoring is one of the biggest challenges faced by classical methods. The problem with these techniques is their inability to provide rapid responses; therefore, process performance decreases. Moreover, manual data collection was necessary. In addition, laboratory experiences consume a long time in both phases of collecting and analyzing data.

Based on Popescu et al. (2024), AI models were used to optimize and predict heavy metal removal. These intelligent tools offer numerous advantages, including the prediction of future trends and the ability to process different data and provide real-time information. Furthermore, artificial intelligence tools enable reliable material detection and rapid response to pollution. Additionally, it does not necessarily require manual data analysis and collection. Therefore, the process control and monitoring performance increased.

Gopinath et al. (2020) highlighted that ANN models were developed to optimize the adsorption kinetics of multicomponents. The ANN algorithm was applied specifically to predict the operating parameters in reverse. Moreover, under different conditions, ANN simulates the increased percentage removal. After the application of the ANN algorithm, the problem of long-time consumption for the estimation of adsorption kinetic data parameters was solved. This limit is due to the repeated numerical integration of the partial differential equation data of the model. The findings show that the ANN decreases approximately 14 and 17 times less than the homogeneous surface diffusion model for computational time and total parameter estimation time, respectively. Additionally, this work proves the high potential of ANN in optimizing real-time and wastewater process plant operations.

To reduce the risk of seawater intrusion, the development of new and efficient environment-related risk monitoring is essential. Therefore, the utilization of faster prediction models is crucial for determining the quality of groundwater in real time. (El Bilali, Taleb, Nafii, et al., 2021) explored the application of an ANN algorithm to predict the adsorption removal of two substances, including chloride and sodium. The input parameters used were the physical parameters. The results show that the ANN has an RMSE of 0.01 meq L^{-1}, a correlation coefficient of 0.99, and an accuracy of 95%, indicating that the ANN has a high prediction performance. The findings demonstrated the power and performance of the sensors in measuring various input variables in real time.

Park et al. (2025) reported that the ANN algorithm was used to classify, optimize, and predict per- and polyfluoroalkyl substance contamination. The ANN model was applied to overcome the limitations of treatment methods, such as elevated process costs and long-term consumption. The study demonstrated that ANN has a prediction capacity of 0.9. In addition, the R^2 value for the optimization of contaminant removal was 0.93. Moreover, the study proves the ability of ANN and AI approaches to make regulatory decisions and protect the environment.

In a study conducted by (Zhang et al., 2025), an advanced technology for gas sensing was developed. Hydrogen and ammonia gases were detected over a short period using the Pd/SMX sensor. On the surface of 2D layered $Ti_3C_2T_x$ MXene, two composite materials, including SnO2 and TiO2, were integrated into an n-n heterostructure to synthesize the composite material. Subsequently, the ANN algorithm was combined with the sensor system to improve its performance. The results demonstrate that Pd nanoparticle integration improved the adsorption capabilities of the sensor. Moreover, the sensor exhibited a high selectivity and high detection sensitivity. As a result, the accuracy of the two-gas differentiation was ensured due to the operation of the sensor's circuitry in real-time sensing mode.

To solve the problem of the classical methods of predicting groundwater quality applied for irrigation, (El Bilali, Taleb, & Brouziyne, 2021) worked on the combination of AI tools and the evaluation of irrigation water quality. In addition, physical parameters were used for measurement. The study objective was to explore non-expensive approaches that ensure real-time monitoring of groundwater quality. Different AI models were developed, including the ANN algorithm. Moreover, parameters such as pH, electrical conductivity, and temperature were the inputs used to forecast total dissolved solids, potential salinity, sodium adsorption ratio, exchangeable sodium percentage, magnesium adsorption ratio, and residual sodium carbonate. The finding proved the high generalization performance of the ANN model. Furthermore, the study helps to improve decision-making for the irrigation strategies.

CASE STUDIES AND APPLICATIONS

Industrial Wastewater Treatment

Pollution caused by the textile industry is a pressing challenge, particularly water pollution. The textile industry process follows two principal phases: the first is dyeing, and the second is washing. These two steps consume large volumes of water. However, a problem appears when these industries discharge wastewater without any pretreatment or treatment processes. Textile wastewater is a hazardous effluent that contains several complex pollutants, especially colors. The principal characteristics of these dyes are the elevated pH, temperature, and suspended solids (Christian et al., 2022). The presence of these contaminants in water and wastewater can cause damage to the environment and human health. Therefore, many studies have focused on developing high-performance processes to reduce pollution in the textile industry. Furthermore, a particular focus was placed on recycling wastewater as a sustainable solution for water pollution and freshwater shortage (Azanaw et al., 2022).

Karam et al. (2020) explored the adsorption potential of different adsorbents, including activated carbon, green-synthesized nano-zerovalent iron, and nano-zerovalent iron. The three materials were used for the decolorization of textile wastewater. Five independent factors, pH, contact time, adsorbent dose, stirring rate, and concentration, were studied to select the optimal adsorption conditions, as shown in Table 4.

Table 4. Principal parameters controlling dye adsorption using activated carbon

Operating parameters	Value
pH	1-12
Dose	0.05-1.0 g
Stirring rate	50-400 rpm
Contact time	10-120 min
Initial concentration	50-350 mg/L

In this section, we focus on the results obtained from the application of activated carbon. The degree of impact of the five operating factors influencing activated carbon sorption was predicted by employing an artificial neural network. The ANN was trained using 51 data points, divided into 70 percent for validation and 30 percent for testing. An ANN comprising three layers was applied. The input layer contained five neurons: three neurons for the hidden layer and one neuron for the output layer. The first and second layers are connected using a hyperbolic tangent activation function. However, the identity function connects the second and third layers, as shown in Figure 5.

Figure 5. ANN structure (5-3-1)

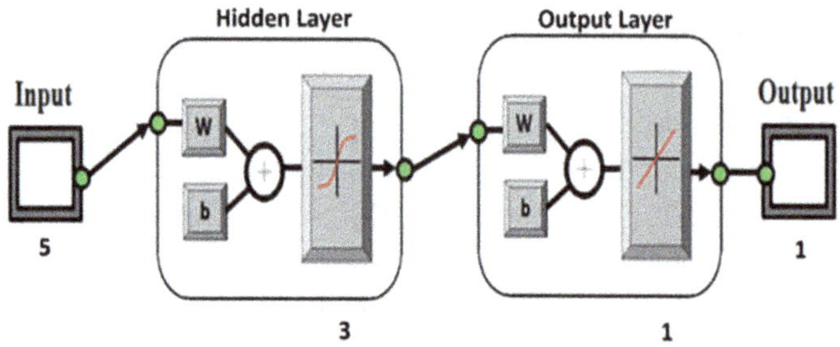

The findings show the high adsorption efficiencies of different adsorbents. Moreover, color elimination using activated carbon was between 72% and 100% at pH 8 for 70 min. In addition, the best artificial neural network model type was found to be multilayer perceptron backpropagation, and the relative errors for training and testing were 0.007 and 0.02, respectively. Figure 6 shows the normalized importance of the effects of color removal using activated carbon.

Figure 6. The normalized importance of effects on color removal using activated carbon

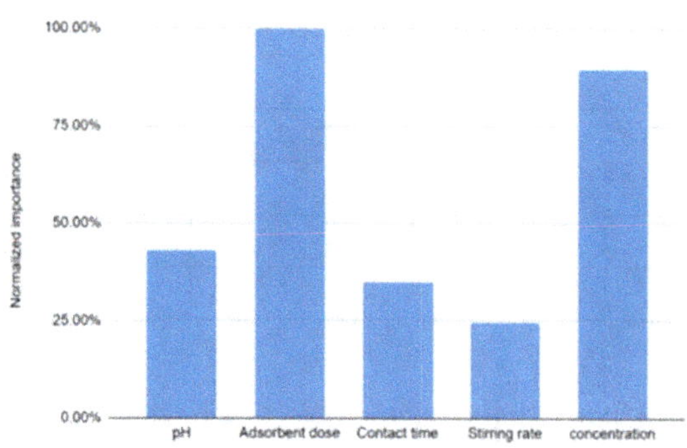

Leon et al. (2020) used modified sugarcane bagasse to adsorb color from the textile industry. Sugarcane bagasse has been modified using chemical methods to improve its capacity to remove color from textile wastewater. Particle size, pH, adsorbent concentration, contact time, and initial color concentration of the effluent were selected as the principal parameters influencing the color removal. Table 5 presents the values applied for the optimal conditions' identification.

Table 5. Principal parameters influencing the color removal using modified sugarcane bagasse

Operating parameters	Value
pH	2-10
Dose	0.4-0.7 g
Particle diameter	0.3-1.0 mm
Contact time	2-1440 min
Initial concentration	98.09-991.17 PtCo.L−1

To model and predict the adsorption process, the artificial neural network was applied as an intelligent tool that helps analyze large amounts of data and results in a short time. The ANN model found that the optimal operating conditions that present the highest dye removal of 100% were 0.7 for the modified sugarcane bagasse, 0.6 for the adsorbent dosage, and 149 PtCo.L^{-1} for the initial dye concentration. Moreover, the optimal ANN algorithm structure was found to be 5-8-1, as shown in Figure 7, and a multilayer feedforward neural network.

Figure 7. ANN structure (5-8-1)

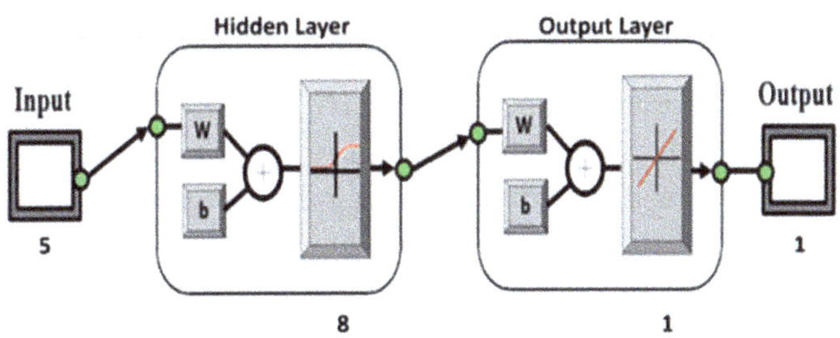

The model presented a high prediction performance; the coefficient of determination was 0.92, and the mean square error was 0.013. A high adsorption removal capacity of 71% was achieved when the effluent was acidic, particularly at pH 2. However, 79% color elimination was achieved when the particle size was 0,7 mm. Furthermore, 86% removal was observed using a bio-sorbent dose of 0,6g.

Based on the literature, industrial wastewater textiles are treated with numerous processes, and adsorption is one of these techniques. The dye concentration decreased upon adsorption onto the surface of the adsorbent material. The combination of ANN with the adsorption system helps improve the sorption removal efficiency of colors by predicting and determining the optimal operating conditions. However, the reuse and recycling of treated textile wastewater is a sustainable solution to the water pollution caused by industrial effluents. Hashem and Qi (2021) reported that treated effluents have great potential for application in agricultural irrigation or firefighting. Another reuse of wastewater after the treatment process is its reuse in industry or for cleaning streets. Effluent recycling is a sustainable alternative or co-solution to reduce water pollution and the limited quantity of freshwater.

Municipal Wastewater Recycling

Water pollution is the result of various emergent contaminants in freshwater. Various sources, including hospitals and the pharmaceutical industry, discharge liquid and solid waste directly into the sea or rivers. The presence of pharmaceutical and healthcare products in water causes damage to the environment, water quality, aquatic life, and human health. There are two types of pharmaceutical pollutants: those that are easily degraded. However, contaminants can be removed via chemical and biological treatments. The characteristics of pharmaceutical effluents depend on the activity of the pharmaceutical industry. This means that the type of drug product and machine used are the principal factors influencing the characteristics of pharmaceutical wastewater. According to the literature, pharmaceutical municipal discharge pollutants are characterized by biological toxicity, a high charge on suspended solids, and elevated salt content (Eniola et al., 2022).

Mojiri et al. (2019) successfully removed pharmaceutical micropollutants, including diclofenac, ibuprofen, and naproxen, from liquid solutions. In this study, an adsorption process was applied to eliminate these microcontaminants using cross-linked magnetic chitosan/activated biochar. The process was controlled by two independent parameters: pollutant concentration and pH. Table 6 presents the values of the factors controlling the adsorption of pharmaceutical pollutants. Preliminary experiments are also conducted. was achieved by fixing the adsorbent mass at 1.6 g/L and contact time at 90 min.

Table 6. Principal parameters controlling pharmaceutical micropollutant adsorption

Operating parameters	Value
pH	4-8
Pollution concentration	0.5-3 mg/L

After 50 experiments, cross-linked magnetic chitosan/activated biochar sorption of pharmaceutical toxic substances was optimized using an artificial neural network. The datasets were divided into 60% for model training, 20% for model validation, and 20% for model testing. The optimal structure of the ANN model was 2-4-1, as shown in Figure 8, and the model was trained using the Levenberg–Marquardt algorithm.

Figure 8. ANN structure (2-4-1)

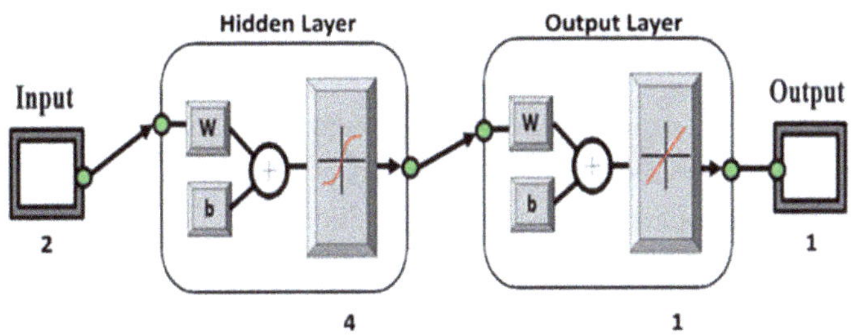

The findings demonstrate that the ANN algorithm successfully optimized the adsorption process for all micropollutants. The R^2 were 0.998, 0.998, and 0.999. The ANN R^2 values were 0.998 for diclofenac removal, 0.998 for ibuprofen elimination, and 0.999 for naproxen adsorption. Additionally, a high adoption capacity was found under the optimal conditions of pH = 6 and concentration = 2.5 mg/L. Under the optimal operating conditions, the diclofenac adoption capacity was found to be 96.4% (2.41 mg/L removal), 98.8% (2.47 mg/L removal), and 95.2% (2.38 mg/L removal) for diclofenac, ibuprofen, and naproxen, respectively. Cross-linked magnetic chitosan/activated biochar shows high regeneration performance; thus, it can be used as an adsorbent many times, and it lasts for eight cycles of adsorption/desorption without decreasing its performance. Therefore, the reuse and recycling of adsorbents helps minimize the cost of wastewater treatment.

Based on the (Shahriar et al., 2021) work, pharmaceutical-treated wastewater can be used in the agriculture domain without any environmental and human health risks. This study explores treated wastewater containing pharmaceuticals and personal care products for irrigation applications. Moreover, substances, including triclosan, carbamazepine, naproxen, gemfibrozil, and fluoxetine, were simulated during wastewater recycling for irrigation utilization, and their risks to human health were studied. The results showed that compared to the acceptable daily intake, human dietary exposure to these compounds through beef consumption was much lower.

CONCLUSION

The treatment and recycling of wastewater are indispensable for mitigating water scarcity and reducing environmental pollution. Chemical methods, particularly adsorption, have proven to be highly effective for removing a wide range of contaminants, supported by their adaptability and potential for integration with complementary technologies. However, challenges such as energy consumption, emerging contaminants, and the need for advanced material regeneration persist and necessitate continuous innovation.

The integration of Artificial Neural Networks (ANNs) into wastewater treatment processes represents a transformative approach that enables precise predictive modeling, optimization of operating conditions, and real-time system monitoring. Case studies demonstrated the successful application of ANNs in enhancing the adsorption efficiency and facilitating the reuse of treated wastewater in industrial and agricultural settings. These advancements have underscored the importance of combining traditional treatment methods with cutting-edge technologies to achieve sustainable water management.

Further research should focus on developing cost-effective adsorbents, improving regeneration techniques, and expanding the use of AI-driven tools to address the complexities of wastewater composition. Policymakers and engineers must collaborate to establish robust regulatory frameworks and infrastructure that support large-scale water recycling initiatives. By embracing these strategies, we can ensure the long-term sustainability of water resources and protect ecosystems for future generations.

ACKNOWLEDGMENT

This work is supported by the CNRST through the PhD-Associate Scholarship Program (PASS), Morocco.

REFERENCES

Adeleke, O. A., Latiff, A. A., Saphira, M. R., Daud, Z., Ismail, N., Ahsan, A., Aziz, N. A., Al-Gheethi, A., Kumar, V., Fadilat, A., & Apandi, N. (2018). *Principles and mechanism of adsorption for the effective treatment of palm oil mill effluent for water reuse.* In Elsevier eBooks (pp. 1–33). https://doi.org/DOI: 10.1016/b978-0-12-813902-8.00001-0

Adnan, M. A. M., Majnis, M. F., Adnan, W. N. W. M., Abdullah, N. H., Baharom, A. S., & Julkapli, N. M. (2025). Deep insights into the integration of Artificial Neural Networks (ANNs) for predicting the photocatalytic activities of metal-based catalysts in water pollutant reduction. *Journal of Environmental Chemical Engineering*, *13*(3), 116350.

Almuktar, S. N., Abed, S. N., & Scholz, M. (2018). Wetlands for wastewater treatment and subsequent recycling of treated effluent: A review. *Environmental Science and Pollution Research International*, *25*(24), 23595–23623. DOI: 10.1007/s11356-018-2629-3

Atangana, E., & Ajiboye, T. O., Abolaji Abiodun Mafolasire, Ghosh, S., & Hakeem, B. (2025). Adsorption of Organic Pollutants from Wastewater Using Chitosan-Based Adsorbents. *Polymers*, *17*(4), 502–502. DOI: 10.3390/polym17040502

Azanaw, A., Birlie, B., Teshome, B., & Jemberie, M. (2022). Textile effluent treatment methods and eco-friendly resolution of textile wastewater. *Case Studies in Chemical and Environmental Engineering*, *6*, 100230. DOI: 10.1016/j.cscee.2022.100230

Banza, M., Seodigeng, T., & Rutto, H. (2023). Comparison study of ANFIS, ANN, and RSM and mechanistic modeling for chromium (VI) removal using modified cellulose nanocrystals–sodium alginate (CNC–alg). *Arabian Journal for Science and Engineering*, *48*(12), 16067–16085.

Baskar, A. V., Bolan, N., Hoang, S. A., Sooriyakumar, P., Kumar, M., Singh, L., Jasemizad, T., Padhye, L. P., Singh, G., Vinu, A., Sarkar, B., Kirkham, M., Rinklebe, J., Wang, S., Wang, H., Balasubramanian, R., & Siddique, K. H. (2022). Recovery, regeneration and sustainable management of spent adsorbents from wastewater treatment streams: A review. *The Science of the Total Environment, 822*, 153555. DOI: 10.1016/j.scitotenv.2022.153555

Cao, R., Fan, M., Hu, J., Ruan, W., Xiong, K., & Wei, X. (2017). Optimizing Low-Concentration Mercury Removal from Aqueous Solutions by Reduced Graphene Oxide-Supported Fe3O4 Composites with the Aid of an Artificial Neural Network and Genetic Algorithm. *Materials (Basel), 10*(11), 1279–1279. DOI: 10.3390/ma10111279

Cevallos-Mendoza, J., Amorim, C., Rodríguez-Díaz, J., & Montenegro, M. (2022). Removal of Contaminants from Water by Membrane Filtration: A Review. *Membranes, 12*(6), 570. DOI: 10.3390/membranes12060570

Chen, F., Huang, J., Wu, X., Wu, X., & Arabmarkadeh, A. (2021). On the Evaluation of Rhamnolipid Biosurfactant Adsorption Performance on Amberlite XAD-2 Using Machine Learning Techniques. *BioMed Research International, 2021*(1). https://doi.org/DOI: 10.1155/2021/5530093

Chen, L., Chen, Z., Liu, Y., Lichtfouse, E., Jiang, Y., Hua, J., Osman, A. I., Farghali, M., Huang, L., Zhang, Y., Rooney, D. W., & Yap, P. (2024). Benefits and limitations of recycled water systems in the building sector: A review. *Environmental Chemistry Letters, 22*(2), 785–814. DOI: 10.1007/s10311-023-01683-2

Christian, D., Gaekwad, A., & Dani, H. M.A., S., & Kandya, A. (2022). Recent techniques of textile industrial wastewater treatment: A review. *Materials Today: Proceedings*. Advance online publication. DOI: 10.1016/j.matpr.2022.11.301

Corpus, R. M. B., Bayani, M. S., Aguilar, J. L., Aguilar, J. C. B., & Aguilar, H. B. (2024). Emerging pollutants in waste water: Challenges and advancements in treatment technology. *IOP Conference Series. Earth and Environmental Science, 1372*(1), 012037. DOI: 10.1088/1755-1315/1372/1/012037

Da'na, E., & Awad, A. (2017). Regeneration of spent activated carbon obtained from home filtration system and applying it for heavy metals adsorption. *Journal of Environmental Chemical Engineering, 5*(4), 3091–3099. DOI: 10.1016/j.jece.2017.06.022

de Leon, V. B., de Negreiros, B. A. F., Brusamarello, C. Z., Petroli, G., Di Domenico, M., & de Souza, F. B. (2020). Artificial neural network for prediction of color adsorption from an industrial textile effluent using modified sugarcane bagasse: Characterization, kinetics and isotherm studies. *Environmental Nanotechnology, Monitoring & Management, 14*, 100387.

Dotto, G. L., Nunes, I. D. S., Fuhr, A. C. F. P., Da Silva, R. N. F., & Peres, E. C. (2025). Adsorption-assisted technologies for wastewater treatment. In Elsevier eBooks (pp. 49–77). https://doi.org/DOI: 10.1016/b978-0-443-26500-6.00003-1

El Bilali, A., Taleb, A., & Brouziyne, Y. (2021). Groundwater quality forecasting using machine learning algorithms for irrigation purposes. *Agricultural Water Management, 245*, 106625. DOI: 10.1016/j.agwat.2020.106625

El Bilali, A., Taleb, A., Nafii, A., Alabjah, B., & Mazigh, N. (2021). Prediction of sodium adsorption ratio and chloride concentration in a coastal aquifer under seawater intrusion using machine learning models. *Environmental Technology & Innovation, 23*, 101641.

Elfghi, F. M. (2016). A hybrid statistical approach for modeling and optimization of RON: A comparative study and combined application of response surface methodology (RSM) and artificial neural network (ANN) based on design of experiment (DOE). *Chemical Engineering Research & Design, 113*, 264–272. DOI: 10.1016/j.cherd.2016.05.023

Eniola, J. O., Kumar, R., Barakat, M. A., & Rashid, J. (2022). A review on conventional and advanced hybrid technologies for pharmaceutical wastewater treatment. *Journal of Cleaner Production, 356*, 131826. DOI: 10.1016/j.jclepro.2022.131826

Eriksson, E., Donner, E., & Ledin, A. (2010). Presence of selected priority and personal care substances in an onsite bathroom greywater treatment facility. *Water Science and Technology, 62*(12), 2889–2898. DOI: 10.2166/wst.2010.988

Ersan, G., Cerrón-Calle, G. A., Ersan, M. S., & Garcia-Segura, S. (2023). Opportunities for in situ electro-regeneration of organic contaminant-laden carbonaceous adsorbents. *Water Research, 232*, 119718. DOI: 10.1016/j.watres.2023.119718

Fan, M., Li, T.-J., Hu, J., Cao, R., Wei, X., Shi, X., & Ruan, W. (2017). Artificial Neural Network Modeling and Genetic Algorithm Optimization for Cadmium Removal from Aqueous Solutions by Reduced Graphene Oxide-Supported Nanoscale Zero-Valent Iron (nZVI/rGO) Composites. *Materials (Basel), 10*(5), 544–544. DOI: 10.3390/ma10050544

Gautam, K., & Verma, R. K., Suantak Kamsonlian, & Kumar, S. (2020). Decolorization of Reactive Black B from wastewater by electro-coagulation: Optimization using multivariate RSM and ANN. *Chemical Product and Process Modeling*, *16*(2), 129–144. DOI: 10.1515/cppm-2020-0043

Georgin, J., Franco, D. S. P., Ramos, C. G., Piccilli, D. G., Lima, E. C., & Sher, F. (2023). A review of the antibiotic ofloxacin: Current status of ecotoxicology and scientific advances in its removal from aqueous systems by adsorption technology. *Chemical Engineering Research & Design*, *193*, 99–120. DOI: 10.1016/j.cherd.2023.03.025

Gopinath, A., Retnam, B. G., Muthukkumaran, A., & Aravamudan, K. (2020). Swift, versatile and a rigorous kinetic model based artificial neural network surrogate for single and multicomponent batch adsorption processes. *Journal of Molecular Liquids*, *297*, 111888.

Guo, H., Li, X., Yang, W., Yao, Z., Mei, Y., Peng, L. E., Yang, Z., Shao, S., & Tang, C. Y. (2021). Nanofiltration for drinking water treatment: A review. *Frontiers of Chemical Science and Engineering*, *16*(5), 681–698. DOI: 10.1007/s11705-021-2103-5

Gupta, V. K., Ali, I., Saleh, T. A., Nayak, A., & Agarwal, S. (2012). Chemical treatment technologies for wastewater recycling an overview. *RSC Advances*, *2*(16), 6380. DOI: 10.1039/c2ra20340e

Hasan, H. A., Muhamad, M. H., Ji, B., Nazairi, N. A., Jiat, K. W., Sim, S. I. S. W. A., & Poh, A. F. M. S. (2023). Revolutionizing wastewater treatment with microalgae: Unveiling resource recovery, mechanisms, challenges, and future possibilities. *Ecological Engineering*, *197*, 107117. DOI: 10.1016/j.ecoleng.2023.107117

Hashem, M. S., & Qi, X. (2021). Treated Wastewater Irrigation—A Review. *Water (Basel)*, *13*(11), 1527. DOI: 10.3390/w13111527

Huang, M., Ma, Y., Wan, J., Zhang, H., Wang, Y., Chen, Y., Yoo, C., & Guo, W. (2011). A hybrid genetic – Neural algorithm for modeling the biodegradation process of DnBP in AAO system. *Bioresource Technology*, *102*(19), 8907–8913. DOI: 10.1016/j.biortech.2011.06.046

Humadi, J. I., Shihab, M. A., Hasan, A. A., & Mohammed, A. M. (2024). Experimental and ANN modeling of kerosene fuel desulfurization using a manganese oxide-tin oxide catalyst. *Chemical Engineering Research & Design*, *211*, 160–167. DOI: 10.1016/j.cherd.2024.10.001

Hussain, S., Khan, H., Gul, S., Steter, J. R., & Motheo, A. J. (2021). Modeling of photolytic degradation of sulfamethoxazole using boosted regression tree (BRT), artificial neural network (ANN) and response surface methodology (RSM); energy consumption and intermediates study. *Chemosphere*, *276*, 130151. DOI: 10.1016/j.chemosphere.2021.130151

Igwegbe, C. A., Mohmmadi, L., Ahmadi, S., Rahdar, A., Khadkhodaiy, D., Dehghani, R., & Rahdar, S. (2019). Modeling of adsorption of methylene blue dye on Ho-CaWO4 nanoparticles using response surface methodology (RSM) and artificial neural network (ANN) techniques. *MethodsX*, *6*, 1779–1797.

Ivanović, T., Tomović, T., Popović, D., Miladinović, J., Arrad, M., & Miladinović, Z. (2025). A simplified method for enhanced accuracy in calculating adsorption thermodynamic parameters. *The Journal of Chemical Thermodynamics*, *107517*. Advance online publication. DOI: 10.1016/j.jct.2025.107517

Jain, A., Kumari, S., Agarwal, S., & Khan, S. (2021). Water purification via novel nano-adsorbents and their regeneration strategies. *Process Safety and Environmental Protection*, *152*, 441–454. DOI: 10.1016/j.psep.2021.06.031

Karam, A., Zaher, K., & Mahmoud, A. S. (2020). Comparative Studies of Using Nano Zerovalent Iron, Activated Carbon, and Green Synthesized Nano Zerovalent Iron for Textile Wastewater Color Removal Using Artificial Intelligence, Regression Analysis, Adsorption Isotherm, and Kinetic Studies. *Air, Soil and Water Research*, *13*, 117862212090827. DOI: 10.1177/1178622120908273

Karri, R. R., Tanzifi, M., Tavakkoli Yaraki, M., & Sahu, J. N. (2018). Optimization and modeling of methyl orange adsorption onto polyaniline nano-adsorbent through response surface methodology and differential evolution embedded neural network. *Journal of Environmental Management*, *223*, 517–529. DOI: 10.1016/j.jenvman.2018.06.027

Khajeh, M., Sarafraz-Yazdi, A., & Moghadam, A. F. (2017). Modeling of solid-phase tea waste extraction for the removal of manganese and cobalt from water samples by using PSO-artificial neural network and response surface methodology. *Arabian Journal of Chemistry*, *10*, S1663–S1673.

Khan, H., Hussain, S., Din, M. A. U., Arshad, M., Wahab, F., Hassan, U., & Khan, A. (2023). Multiple design and modelling approaches for the optimisation of carbon felt electro-Fenton treatment of dye laden wastewater. *Chemosphere*, *338*, 139510.

Khatri, N., Vyas, A. K., Abdul-Qawy, A. S. H., & Rene, E. R. (2023). Artificial neural network based models for predicting the effluent quality of a combined up-flow anaerobic sludge blanket and facultative pond: Performance evaluation and comparison of different algorithms. *Environmental Research, 217*, 114843.

Kadimpati, K. K., Gnida, A., Turek-Szytow, J., Hellal, M. S., Gregor, M., Matula, G., & Monfort, O. (2024). Design of innovative hybrid biochar prepared from marine algae and magnetite: Insights into adsorption performance and mechanism. *Chemical Engineering Research & Design*, *201*, 218–227.

Kumar, D., & Gupta, S. K. (2022). Electrochemical oxidation of direct blue 86 dye using MMO coated Ti anode: Modelling, kinetics and degradation pathway. *Chemical Engineering and Processing*, *181*, 109127. DOI: 10.1016/j.cep.2022.109127

Kumari, S., Singh, S., Lo, S.-L., Sharma, P., Agarwal, S., & Garg, M. C. (2024). Machine learning and modelling approach for removing methylene blue from aqueous solutions: Optimization, kinetics and thermodynamics studies. *Journal of the Taiwan Institute of Chemical Engineers*, •••, 105361–105361. DOI: 10.1016/j.jtice.2024.105361

Kundu, P., Paul, V., & Kumar, V., & Indra Mani Mishra. (2015). Formulation development, modeling and optimization of emulsification process using evolving RSM coupled hybrid ANN-GA framework. *Chemical Engineering Research & Design*, *104*, 773–790. DOI: 10.1016/j.cherd.2015.10.025

Larsen, T. A. (2011). Redesigning wastewater infrastructure to improve resource efficiency. *Water Science and Technology*, *63*(11), 2535–2541. DOI: 10.2166/wst.2011.502

Li, W., Wei, S., Jiao, W., Qi, G., & Liu, Y. (2016). Modelling of adsorption in rotating packed bed using artificial neural networks (ANN). *Chemical Engineering Research & Design*, *114*, 89–95. DOI: 10.1016/j.cherd.2016.08.013

Li, X., Wu, Q., Chen, D., Tamjidur, R. S., Fan, P., Wang, G., & Zhang, W. (2025). Efficient chromium Cr (VI) removal from wastewater through modified cycad's leaf biochar: Insights into adsorption-reduction mechanisms and kinetic analysis. *Biomass and Bioenergy*, *200*, 107994. DOI: 10.1016/j.biombioe.2025.107994

Liu, B., Xin, Y., Zou, J., Khoso, F., Liu, Y., Jiang, X., Peng, S., & Yu, J. (2023). Removal of chromium species by adsorption: Fundamental principles, newly developed adsorbents and future perspectives. *Molecules (Basel, Switzerland)*, *28*(2), 639. DOI: 10.3390/molecules28020639

Liu, X., Chen, Z., Du, W., Liu, P., Zhang, L., & Shi, F. (2022). Treatment of wastewater containing methyl orange dye by fluidized three-dimensional electrochemical oxidation process integrated with chemical oxidation and adsorption. *Journal of Environmental Management*, *311*, 114775. DOI: 10.1016/j.jenvman.2022.114775

Liu, Y. (2024). Adsorption. In Elsevier eBooks (pp. 243–281). https://doi.org/DOI: 10.1016/b978-0-323-95173-9.00004-9

Mahmoud, A. S., Mostafa, M. K., & Abdel-Gawad, S. A. (2017). Artificial intelligence for the removal of benzene, toluene, ethyl benzene and xylene (BTEX) from aqueous solutions using iron nanoparticles. *Water Science and Technology: Water Supply*, ws2017225. https://doi.org/DOI: 10.2166/ws.2017.225

Mbavarira, T. M., & Grimm, C. (2021). A Systemic View on Circular Economy in the Water Industry: Learnings from a Belgian and Dutch Case. *Sustainability*, *13*(6), 3313. DOI: 10.3390/su13063313

Mojiri, A., Andasht Kazeroon, R., & Gholami, A. (2019). Cross-Linked Magnetic Chitosan/Activated Biochar for Removal of Emerging Micropollutants from Water: Optimization by the Artificial Neural Network. *Water (Basel)*, *11*(3), 551. DOI: 10.3390/w11030551

Mostafapour, F. K., Yilmaz, M., Mahvi, A. H., Younesi, A., Ganji, F., & Balarak, D. (2022). Adsorptive removal of tetracycline from aqueous solution by surfactant-modified zeolite: Equilibrium, kinetics and thermodynamics. *Desalination and Water Treatment*, *247*, 216–228. DOI: 10.5004/dwt.2022.27943

Nawaz, A., & Kumar, P. (2022). Optimization of process parameters of Lagerstroemia speciosa seed hull pyrolysis using a combined approach of Response Surface Methodology (RSM) and Artificial Neural Network (ANN) for renewable fuel production. *Bioresource Technology Reports*, *18*, 101110–101110. DOI: 10.1016/j.biteb.2022.101110

Nguyen, V. D., Nguyen, H. T. H., Vranova, V., Nguyen, L. T. N., Bui, Q. M., & Khieu, T. T. (2020). Artificial neural network modeling for Congo red adsorption on microwave-synthesized akaganeite nanoparticles: Optimization, kinetics, mechanism, and thermodynamics. *Environmental Science and Pollution Research International*, *28*(8), 9133–9145. DOI: 10.1007/s11356-020-10633-2

O'Connor, K. F., Al-Abed, S. R., Hordern, S., & Pinto, P. X. (2022). Assessing the efficiency and mechanism of zinc adsorption onto biochars from poultry litter and softwood feedstocks. *Bioresource Technology Reports*, *18*, 101039. DOI: 10.1016/j.biteb.2022.101039

Pandit, J., & Sharma, A. K. (2024). Advanced Techniques in Wastewater Treatment: A Comprehensive review. *Asian Journal of Environment & Ecology, 23*(10), 1–26. DOI: 10.9734/ajee/2024/v23i10605

Park, J., Baik, J.-H., Adjei-Nimoh, S., & Lee, W. H. (2025). Advancements in artificial intelligence-based technologies for PFAS detection, monitoring, and management. *The Science of the Total Environment, 980*, 179536–179536. DOI: 10.1016/j.scitotenv.2025.179536

Patel, H. (2021). Review on solvent desorption study from exhausted adsorbent. *Journal of Saudi Chemical Society, 25*(8), 101302. DOI: 10.1016/j.jscs.2021.101302

Picos-Benítez, A. R., Martínez-Vargas, B. L., Duron-Torres, S. M., Brillas, E., & Peralta-Hernández, J. M. (2020). The use of artificial intelligence models in the prediction of optimum operational conditions for the treatment of dye wastewaters with similar structural characteristics. *Process Safety and Environmental Protection, 143*, 36–44.

Popescu, S. M., Mansoor, S., Wani, O. A., Kumar, S. S., Sharma, V., Sharma, A., & Chung, Y. S. (2024). Artificial intelligence and IoT driven technologies for environmental pollution monitoring and management. *Frontiers in Environmental Science, 12*, 1336088.

Pratap, B., Kumar, S., Nand, S., Azad, I., Bharagava, R. N., Ferreira, L. F. R., & Dutta, V. (2022). Wastewater generation and treatment by various ecofriendly technologies: Possible health hazards and further reuse for environmental safety. *Chemosphere, 313*, 137547. DOI: 10.1016/j.chemosphere.2022.137547

Ramlee, N. N., Illias, R. M., Toemen, S., Rahman, R. A., Selvasembian, R., Fatriasari, W., & Azelee, N. I. W. (2024). Adsorption study of isotherms, kinetics modelling, and thermodynamics of immobilized Candida rugosa lipase on metal oxide for esterification of waste cooking oil. Chemical Engineering Research and Design, 204(December 2023), 249–268. https://doi.org/DOI: 10.1016/j.cherd.2024.02.031

Renu, N., & Sithole, T. (2024). A review on regeneration of adsorbent and recovery of metals: Adsorbent disposal and regeneration mechanism. *South African Journal of Chemical Engineering, 50*, 39–50. DOI: 10.1016/j.sajce.2024.07.006

Sadek, A. H., & Fahmy, O. M., Nasr, M., & Mostafa, M. K. (2023). Predicting Cu(II) Adsorption from Aqueous Solutions onto Nano Zero-Valent Aluminum (nZVAl) by Machine Learning and Artificial Intelligence Techniques. *Sustainability, 15*(3), 2081–2081. DOI: 10.3390/su15032081

Saravanan, A., Kumar, P. S., Jeevanantham, S., Karishma, S., Tajsabreen, B., Yaashikaa, P., & Reshma, B. (2021). Effective water/wastewater treatment methodologies for toxic pollutants removal: Processes and applications towards sustainable development. *Chemosphere*, *280*, 130595. DOI: 10.1016/j.chemosphere.2021.130595

Shafique, A. (2021). Removal of toxic pollutants from aqueous medium through adsorption: A review. *Desalination and Water Treatment*, *234*, 38–57. DOI: 10.5004/dwt.2021.27550

Shahriar, A., Tan, J., Sharma, P., Hanigan, D., Verburg, P., Pagilla, K., & Yang, Y. (2021). Modeling the fate and human health impacts of pharmaceuticals and personal care products in reclaimed wastewater irrigation for agriculture. *Environmental Pollution*, *276*, 116532. DOI: 10.1016/j.envpol.2021.116532

Shi, X., Ruan, W., Hu, J., Fan, M., Cao, R., & Wei, X. (2017). Optimizing the Removal of Rhodamine B in Aqueous Solutions by Reduced Graphene Oxide-Supported Nanoscale Zerovalent Iron (nZVI/rGO) Using an Artificial Neural Network-Genetic Algorithm (ANN-GA). *Nanomaterials (Basel, Switzerland)*, *7*(6), 134. DOI: 10.3390/nano7060134

Thamer, B. M., Al-aizari, F. A., & Hameed, M. M. A. (2023). Zero-valent Ni/NiO core-shell nanosheets supported on graphene for highly efficient dye adsorption: Isotherm, kinetic and thermodynamic study. *Chemical Engineering Research & Design*, *197*, 656–668. DOI: 10.1016/j.cherd.2023.08.023

Tony, M. A. (2021). Low-cost adsorbents for environmental pollution control: a concise systematic review from the prospective of principles, mechanism and their applications. Journal of Dispersion Science and Technology, 43(11), 1612 1633. https://doi.org/DOI: 10.1080/01932691.2021.1878037

Twi-Yeboah, N., Osei, D., Dontoh, W. H., Asamoah, G. A., Baffoe, J., & Danquah, M. K. (2024). Enhancing energy efficiency and resource recovery in wastewater treatment plants. *Energies*, *17*(13), 3060. DOI: 10.3390/en17133060

Van Der Bruggen, B., Vandecasteele, C., Van Gestel, T., Doyen, W., & Leysen, R. (2003). A review of pressure-driven membrane processes in wastewater treatment and drinking water production. *Environmental Progress*, *22*(1), 46–56. DOI: 10.1002/ep.670220116

Villarín, M. C., & Merel, S. (2020). Paradigm shifts and current challenges in wastewater management. *Journal of Hazardous Materials*, *390*, 122139. DOI: 10.1016/j.jhazmat.2020.122139

Wang, C., & Li, Y. (2024). Al-Fe dual-anode peroxi-coagulation as a feasible pretreatment process for refractory landfill leachate wastewater: Performance and collaborative mechanism. *Separation and Purification Technology, 130675*. Advance online publication. DOI: 10.1016/j.seppur.2024.130675

Wang, Y., Chen, L., Zhu, Y., Fang, W., Tan, Y., He, Z., & Liao, H. (2024). Research status, trends, and mechanisms of biochar adsorption for wastewater treatment: A scientometric review. *Environmental Sciences Europe*, *36*(1). Advance online publication. DOI: 10.1186/s12302-024-00859-z

Warren-Vega, W. M., Montes-Pena, K. D., Romero-Cano, L. A., & Zarate-Guzman, A. I. (2024). Development of an artificial neural network (ANN) for the prediction of a pilot scale mobile wastewater treatment plant performance. *Journal of Environmental Management*, *366*, 121612. DOI: 10.1016/j.jenvman.2024.121612

Wollmann, F., Dietze, S., Ackermann, J., Bley, T., Walther, T., Steingroewer, J., & Krujatz, F. (2019). Microalgae wastewater treatment: Biological and technological approaches. *Engineering in Life Sciences*, *19*(12), 860–871. DOI: 10.1002/elsc.201900071

Xiao, Y., & Hill, J. M. (2021). Regeneration of magnetic adsorbents saturated by organic pollutants. In Environmental chemistry for a sustainable world (pp. 259–294). https://doi.org/DOI: 10.1007/978-3-030-64092-7_9

Yadav, K. Raphi, Mohd., & Jagadevan, S. (2021). Adsorption of copper(II) on chemically modified biochar: A single-stage batch adsorber design and predictive modeling through artificial neural network. *Biomass Conversion and Biorefinery*. Advance online publication. DOI: 10.1007/s13399-021-01494-x

Yadav, S., & Kamsonlian, S. (2023). *Progress on the Development of Techniques to Remove Contaminants from Wastewater: A Review*. Applied Science and Engineering Progress., DOI: 10.14416/j.asep.2023.02.001

Yasri, N. G., & Roberts, E. P. (2024). Electrochemical regeneration of adsorbents: An Electrochemist's perspective. *Current Opinion in Electrochemistry*, *46*, 101504. DOI: 10.1016/j.coelec.2024.101504

Yuan, S., Ajam, H., Sinnah, Z. A. B., Altalbawy, F. M., Ameer, S. A. A., Husain, A., & Cao, Y. (2023). The roles of artificial intelligence techniques for increasing the prediction performance of important parameters and their optimization in membrane processes: A systematic review. *Ecotoxicology and Environmental Safety*, *260*, 115066.

Zhang, A., Zhang, Y., Cheng, W., Li, X., Chen, K., Li, F., & Yang, D. (2025). Dual-Gas Sensing via SnO2-TiO2 Heterojunction on MXene: Machine Learning-Enhanced Selectivity and Sensitivity for Hydrogen and Ammonia Detection. *Sensors and Actuators. B, Chemical*, •••, 137340–137340. DOI: 10.1016/j.snb.2025.137340

Zhou, W., Meng, X., Gao, J., Zhao, H., Zhao, G., & Ma, J. (2020). Electrochemical regeneration of carbon-based adsorbents: A review of regeneration mechanisms, reactors, and future prospects. *Chemical Engineering Journal Advances*, *5*, 100083. DOI: 10.1016/j.ceja.2020.100083

Zhu, X., Wang, X., & Ok, Y. S. (2019). The application of machine learning methods for prediction of metal sorption onto biochars. *Journal of Hazardous Materials*, *378*, 120727. DOI: 10.1016/j.jhazmat.2019.06.004

Compilation of References

A Standard-Based Open Source IoT Platform: FIWARE | IEEE Journals & Magazine | IEEE Xplore. (n.d.). Retrieved June 1, 2025, from https://ieeexplore.ieee.org/document/8950963

Abdel-Basset, M., Gamal, A., Manogaran, G., Son, L., & Long, H. (2019a). A novel group decision making model based on neutrosophic sets for heart disease diagnosis. *Multimedia Tools and Applications*, *79*, 9977–10002. DOI: 10.1007/s11042-019-07742-7

Abdel-Basset, M., & Mohamed, M. (2019b). A novel and powerful framework based on neutrosophic sets to aid patients with cancer. *Future Generation Computer Systems*, *98*, 144–153. DOI: 10.1016/J.FUTURE.2018.12.019

Abdelhamid, M. A., Abdelkader, T. K., Sayed, H. A. A., Zhang, Z., Zhao, X. H., & Atia, M. F. (2025). Design and evaluation of a solar powered smart irrigation system for sustainable urban agriculture. *Scientific Reports*, *15*(1), 11761. Advance online publication. DOI: 10.1038/s41598-025-94251-3 PMID: 40189625

Abidin, H. Z., Andreas, H., Gumilar, I., Sidiq, T. P., & Fukuda, Y. (2011). Land subsidence of Jakarta (Indonesia) and its relation with urban development. *Natural Hazards*, *59*(3), 1753–1771. DOI: 10.1007/s11069-011-9866-9

Abioye, E. A., Abidin, M. S. Z., Mahmud, M. S. A., Buyamin, S., Ishak, M. I., & Ramli, M. K. I. (2020). A review on monitoring and advanced control strategies for precision irrigation. *Computers and Electronics in Agriculture*, *173*, 105441. DOI: 10.1016/j.compag.2020.105441

Absalan, F., Hatam, F., Prévost, M., Barbeau, B., & Bichai, F. (2024). Climate change and future water demand: Implications for chlorine and trihalomethanes management in water distribution systems [Article]. *Journal of Environmental Management*, *355*(11), 120470. Advance online publication. DOI: 10.1016/j.jenvman.2024.120470 PMID: 38422852

Adeleke, O. A., Latiff, A. A., Saphira, M. R., Daud, Z., Ismail, N., Ahsan, A., Aziz, N. A., Al-Gheethi, A., Kumar, V., Fadilat, A., & Apandi, N. (2018). *Principles and mechanism of adsorption for the effective treatment of palm oil mill effluent for water reuse.* In Elsevier eBooks (pp. 1–33). https://doi.org/DOI: 10.1016/b978-0-12-813902-8.00001-0

Adeyemi, O., Grove, I., Peets, S., & Norton, T. (2017). Advanced Monitoring and Management Systems for Improving Sustainability in Precision Irrigation. *Sustainability, 9*(3), 3. Advance online publication. DOI: 10.3390/su9030353

Adimalla, N. (2021). Application of the Entropy Weighted Water Quality Index (EWQI) and the Pollution Index of Groundwater (PIG) to Assess Groundwater Quality for Drinking Purposes: A Case Study in a Rural Area of Telangana State, India. *Archives of Environmental Contamination and Toxicology, 80*, 31–40. PMID: 33388839

Admin, A., J, T., & S, G. (2024). IDLTM-DMT: Intelligent Deep Learning based Trust Management with Decision Making Tool for Healthcare Internet of Things and Big Data Environment with Neutrosophic Set Analysis. International Journal of Neutrosophic Science. DOI: 10.54216/ijns.240430

Adnan, M. A. M., Majnis, M. F., Adnan, W. N. W. M., Abdullah, N. H., Baharom, A. S., & Julkapli, N. M. (2025). Deep insights into the integration of Artificial Neural Networks (ANNs) for predicting the photocatalytic activities of metal-based catalysts in water pollutant reduction. *Journal of Environmental Chemical Engineering, 13*(3), 116350.

Afshar, A., Mariño, M., Saadatpour, M., & Afshar, A. (2011). Fuzzy TOPSIS Multi-Criteria Decision Analysis Applied to Karun Reservoirs System. *Water Resources Management, 25*, 545–563. DOI: 10.1007/S11269-010-9713-X

Afzal, S., & Hassan, S. I. (2025). Blockchain Application in Sustainable Smart Water and Wastewater Management. *Fostering Machine Learning and IoT for Blockchain Technology: Smart Cities Applications, 2*, 273–297.

Agarwal, A., & Narain, S. (1997). *Dying wisdom: Rise, fall and potential of India's traditional water harvesting systems.* Centre for Science and Environment.

Aghdam, E., Mohandes, S. R., Manu, P., Cheung, C., Yunusa-Kaltungo, A., & Zayed, T. (2023). Predicting quality parameters of wastewater treatment plants using artificial intelligence techniques. *Journal of Cleaner Production, 405*, 137019.

Ahmed, Md. T., Ahmed, M. W., & Kamruzzaman, M. (2025). A systematic review of explainable artificial intelligence for spectroscopic agricultural quality assessment. *Computers and Electronics in Agriculture*, *235*, 110354. DOI: 10.1016/j.compag.2025.110354

Al-Addous, M. (2024). Innovations in solar-powered desalination in the MENA region. *Water (Basel)*, *16*(13), 1877. DOI: 10.3390/w16131877

Alaghebandha, M., Naderi, B., & Mohammadi, M. (2019). Economic lot sizing and scheduling in distributed permutation flow shops. *Journal of Optimization in Industrial Engineering*, *12*(1), 103–117.

Albawi, S., Mohammed, T. A., & Al-Zawi, S. (2017). Understanding of a convolutional neural network. *2017 International Conference on Engineering and Technology (ICET)*, 1–6. https://doi.org/DOI: 10.1109/ICEngTechnol.2017.8308186

Alcantud, J. C. R., & Santos-García, G. (2017). A new criterion for soft set based decision-making problems under incomplete information. *International Journal of Computational Intelligence Systems*, *10*, 394–404.

Al-Falal, A. N. A., Elsayed, S., El Fadaly, E. A., Gaagai, A., & Aouissi, H. A., Abd El-baki, M. S., ... & Gad, M. (2025). Aquatic System Assessment of Potentially Toxic Elements in El Manzala Lake, Egypt: A Statistical and Machine Learning Approach. *Results in Engineering*, •••, 105027.

Al-Falal, A. N. A., Elsayed, S., El Fadaly, E. A., Gaagai, A., & Aouissi, H. A., Abd El-baki, M. S., Eid, M. H., Elwakeel, A. E. E., Yaseen, Z. M., Elsherbiny, O., Eltahir, E. I., & Gad, M. (2025). Aquatic system assessment of potentially toxic elements in El Manzala Lake, Egypt: A statistical and machine learning approach. *Results in Engineering*, *26*, 105027. DOI: 10.1016/j.rineng.2025.105027

Ali, A., Gajpal, Y., & Elmekkawy, T. Y. (2021). Distributed permutation flowshop scheduling problem with total completion time objective. *Opsearch*, *58*(2), 425–447.

Ali, M. I., Feng, F., Liu, X. Y., Min, W. K., & Shabir, M. (2009). On some new operations in soft set theory. *Computers & Mathematics with Applications (Oxford, England)*, *57*(9), 1547–1553.

Al-Khatib, L. A., & AlHanaktah, A. M. (2025). Wastewater Treatment Plant Upgrade and Its Interlinkages with the Sustainable Development Goals. *Resources*, *14*(4), 62.

Allen, C., Metternicht, G., Verburg, P., Akhtar-Schuster, M., da Cunha, M. I., & Santivañez, M. S. (2020). Delivering an enabling environment and multiple benefits for land degradation neutrality: Stakeholder perceptions and progress. *Environmental Science & Policy*, *114*, 109–118. DOI: 10.1016/j.envsci.2020.07.023 PMID: 32834778

Almesafri, A., Abdulsattar, S., Alblooshi, A., Al-Juboori, R. A., Jephson, N., & Hilal, N. (2024). Waters of contention: The GERD and its impact on Nile Basin cooperation and conflict. *Water (Basel)*, *16*(15). Advance online publication. DOI: 10.3390/w16152174

Almuktar, S. N., Abed, S. N., & Scholz, M. (2018). Wetlands for wastewater treatment and subsequent recycling of treated effluent: A review. *Environmental Science and Pollution Research International*, *25*(24), 23595–23623. DOI: 10.1007/s11356-018-2629-3

Al-Omeri, W., & Kaviyarasu, M. (2024). Study on Neutrosophic Graph with Application on Earthquake Response Center in Japan. *Symmetry*, *16*, 743. DOI: 10.3390/sym16060743

Alprol, A. E., Mansour, A. T., Ibrahim, M. E. E. D., & Ashour, M. (2024). Artificial intelligence technologies revolutionizing wastewater treatment: Current trends and future prospective. *Water (Basel)*, *16*(2), 314.

Altowayti, W. A. H., Shahir, S., Othman, N., Eisa, T. A. E., Yafooz, W. M., Al-Dhaqm, A., & Ali, A. (2022). The role of conventional methods and artificial intelligence in the wastewater treatment: A comprehensive review. *Processes (Basel, Switzerland)*, *10*(9), 1832.

Alvi, M., Batstone, D., Mbamba, C. K., Keymer, P., French, T., Ward, A., & Cardell-Oliver, R. (2023). Deep learning in wastewater treatment: A critical review. *Water Research*, *245*, 120518.

Aqil, S., & Allali, K. (2021). On VNS-GRASP and iterated greedy metaheuristics for solving hybrid flow shop scheduling problem with uniform parallel machines and sequence independent setup time. Heuristics for optimization and learning, 17-32.

Aqil, S. (2024). Effective Population-Based Meta-heuristics with NEH and GRASP Heuristics Minimizing Total Weighted flow Time in No-Wait Flow Shop Scheduling Problem Under Sequence-Dependent Setup Time Constraint. *Arabian Journal for Science and Engineering*, •••, 1–24.

Aqil, S., & Allali, K. (2020). Local search metaheuristic for solving hybrid flow shop problem in slabs and beams manufacturing. *Expert Systems with Applications*, *162*, 113716.

Aquatic Pasts & the Watery Present: Water and Memory in the Fora of Rome. (n.d.). Retrieved June 1, 2025, from https://www.degruyterbrill.com/document/doi/10.1515/9783110677065-007/html

Asano, T., & Cotruvo, J. A. (2004). Groundwater recharge with reclaimed municipal wastewater: Health and regulatory considerations. *Water Research*, *38*(8), 1941–1951. PMID: 15087175

Asian Development Bank. (2023). *Asian water development outlook 2023: Advancing water security for sustainable development*. ADB. https://www.adb.org/publications/asian-water-development-outlook-2023

Atanassov, K. (1986). Intuitionistic fuzzy sets. *Fuzzy Sets and Systems*, *20*(1), 87–96.

Atangana, E., & Ajiboye, T. O., Abolaji Abiodun Mafolasire, Ghosh, S., & Hakeem, B. (2025). Adsorption of Organic Pollutants from Wastewater Using Chitosan-Based Adsorbents. *Polymers*, *17*(4), 502–502. DOI: 10.3390/polym17040502

Azad, A. S., Sokkalingam, R., Daud, H., Adhikary, S. K., Khurshid, H., Mazlan, S. N. A., & Rabbani, M. B. A. (2022). Water Level Prediction through Hybrid SARIMA and ANN Models Based on Time Series Analysis: Red Hills Reservoir Case Study. *Sustainability*, *14*(3), 3. Advance online publication. DOI: 10.3390/su14031843

Azanaw, A., Birlie, B., Teshome, B., & Jemberie, M. (2022). Textile effluent treatment methods and eco-friendly resolution of textile wastewater. *Case Studies in Chemical and Environmental Engineering*, *6*, 100230. DOI: 10.1016/j.cscee.2022.100230

Baarimah, A. O., Bazel, M. A., Alaloul, W. S., Alazaiza, M. Y. D., Al-Zghoul, T. M., Almuhaya, B., Khan, A., & Mushtaha, A. W. (2024). Artificial intelligence in wastewater treatment: Research trends and future perspectives through bibliometric analysis. *Case Studies in Chemical and Environmental Engineering*, *10*, 100926. DOI: 10.1016/j.cscee.2024.100926

Bailey, K. (2018). Day Zero: Lessons from Cape Town's water crisis. *Journal of Urban Sustainability*, *4*(1), 12–20.

Banza, M., Seodigeng, T., & Rutto, H. (2023). Comparison study of ANFIS, ANN, and RSM and mechanistic modeling for chromium (VI) removal using modified cellulose nanocrystals–sodium alginate (CNC–alg). *Arabian Journal for Science and Engineering*, *48*(12), 16067–16085.

Baranwal, A., Bagwe, B. R., & M, V. (2020). Machine Learning in Python: Diabetes Prediction Using Machine Learning. In S. Velayutham (Ed.), *Advances in Computational Intelligence and Robotics* (pp. 128–154). IGI Global. https://doi.org/DOI: 10.4018/978-1-5225-9902-9.ch008

Barlow, P. M., & Reichard, E. G. (2010). Saltwater intrusion in coastal regions of North America. *Hydrogeology Journal*, *18*(1), 247–260. DOI: 10.1007/s10040-009-0514-3

Baskar, A. V., Bolan, N., Hoang, S. A., Sooriyakumar, P., Kumar, M., Singh, L., Jasemizad, T., Padhye, L. P., Singh, G., Vinu, A., Sarkar, B., Kirkham, M., Rinklebe, J., Wang, S., Wang, H., Balasubramanian, R., & Siddique, K. H. (2022). Recovery, regeneration and sustainable management of spent adsorbents from wastewater treatment streams: A review. *The Science of the Total Environment*, *822*, 153555. DOI: 10.1016/j.scitotenv.2022.153555

Baskar, G., Omer, S. N., Saravanan, P., Rajeshkannan, R., Saravanan, V., Rajasimman, M., & Shanmugam, V. (2024). Status and future trends in wastewater management strategies using artificial intelligence and machine learning techniques. *Chemosphere*, *362*, 142477.

Begmatov, S., Dorofeev, A. G., Kadnikov, V. V., Beletsky, A. V., Pimenov, N. V., Ravin, N. V., & Mardanov, A. V. (2022). The structure of microbial communities of activated sludge of large-scale wastewater treatment plants in the city of Moscow. *Scientific Reports*, *12*(1), 3458.

Ben-Daoud, M., El Mahrad, B., Elhassnaoui, I., Moumen, A., Sayad, A., & Elbouhadioui, M. & Eljaafari, S. (2021). Integrated Water Resources Management: An indicator framework for water management system assessment in the R'Dom Sub-basin, Morocco. *Environmental Challenges*, *3*, 100062. DOI: 10.1016/j.envc.2021.100062

Benedict, M. A., & McMahon, E. T. (2012). *Green infrastructure: Linking landscapes and communities*. Island Press.

Beraaouz, M. (2022). Khettaras in the Tafilalet oasis (Morocco). *Built Heritage*, *6*(24). Advance online publication. DOI: 10.1186/s43238-022-00052-2

Berardi, U., GhaffarianHoseini, A., & GhaffarianHoseini, A. (2014). State-of-the-art analysis of the environmental benefits of green roofs. *Applied Energy*, *115*, 411–428. DOI: 10.1016/j.apenergy.2013.10.047

Berry, S. (1993). *No condition is permanent: The social dynamics of agrarian change in Sub-Saharan Africa*. University of Wisconsin Press.

Bhardwaj, N., & Sharma, P. (2021). An advanced uncertainty measure using fuzzy soft sets: Application to decision-making problems. *Big Data Mining and Analytics*, *4*(2), 94–103.

Bierwirth, C., & Kuhpfahl, J. (2017). Extended GRASP for the job shop scheduling problem with total weighted tardiness objective. *European Journal of Operational Research*, *261*(3), 835–848.

Biggs, E. M., Bruce, E., Boruff, B., Duncan, J., Horsley, J., Pauli, N., McNeill, K., Neef, A., Van Ogtrop, F., Curnow, J., Haworth, B., & Duce, S. (2015). Sustainable development and the water–energy–food nexus: A perspective on livelihoods. *Environmental Science & Policy*, *54*, 389–397.

Bi, J., Chen, Z., Yuan, H., & Zhang, J. (2024). Accurate water quality prediction with attention-based bidirectional LSTM and encoder–decoder. *Expert Systems with Applications*, *238*, 121807. DOI: 10.1016/j.eswa.2023.121807

Bitella, G., Rossi, R., Bochicchio, R., Perniola, M., & Amato, M. (2014). A novel low-cost open-hardware platform for monitoring soil water content and multiple soil-air-vegetation parameters. *Sensors (Basel)*, *14*(10), 19639–19659. DOI: 10.3390/s141019639 PMID: 25337742

Bjorck, J., Gomes, C., Selman, B., & Weinberger, K. Q. (2018). *Understanding Batch Normalization* (Version 4). arXiv. https://doi.org/DOI: 10.48550/ARXIV.1806.02375

Blanco-Marigorta, A. M., Lozano-Medina, A., & Marcos, J. D. (2017). The exergetic efficiency as a performance evaluation tool in reverse osmosis desalination plants in operation. *Desalination*, *413*, 19–28.

Boelens, R. (2014). Cultural politics and the hydrosocial cycle: Water, power and identity in the Andean highlands. *Geoforum*, *57*, 234–247. DOI: 10.1016/j.geoforum.2013.02.008

Boelens, R., & Seemann, M. (2014). Forced engagements: Water security and local rights formalization in Yanque, Colca Valley, Peru. *Human Organization*, *73*(1), 1–12. DOI: 10.17730/humo.73.1.v544h37k38445065

Boelens, R., Perreault, T., & Vos, J. (2022). Water justice and the politics of scale. *Water International*, *47*(4), 561–579. DOI: 10.1080/02508060.2022.2071764

Bonneuil, C., & Fressoz, J.-B. (2013). *The Anthropocene Event: The Earth, History, and Us. Le Seuil*. King's College London.

Borham, O., Croxford, B., & Wilson, D. (2024). Biomimetic Strategies for Sustainable Resilient Cities: Review across Scales and City Systems. *Biomimetics*, *9*(9), 514.

Bouzidi, D. (2020). A Variable Neighborhood Search Algorithm to Solve the Flow Shop-Scheduling Problem Through Blocking: Variable Neighborhood Search (VNS). In *Transportation, Logistics, and Supply Chain Management in Home Healthcare: Emerging Research and Opportunities* (pp. 63-73). IGI Global.

Brookes, J. (1987). *Gardens of paradise: The history and design of the great Islamic gardens*. New Amsterdam.

Bureau of Indian Standards. (with Public.Resource.Org). (1986). *IS 4111-1: Code of practice for ancillary structures in sewerage system, Part I: Manholes.* http://archive.org/details/gov.in.is.4111.1.1986

Burrell, A. L., Evans, J. P., & De Kauwe, M. G. (2020). Anthropogenic climate change has driven over 5 million km^2 of drylands towards desertification. *Nature Communications, 11*(1), 3853. DOI: 10.1038/s41467-020-17710-7 PMID: 32737311

Bwambale, E., Abagale, F. K., & Anornu, G. K. (2022). Smart irrigation monitoring and control strategies for improving water use efficiency in precision agriculture: A review. *Agricultural Water Management, 260*, 107324. DOI: 10.1016/j.agwat.2021.107324

Byers, E., & Gidden, M. (2018). Global exposure and vulnerability to multi-sector development and climate change hotspots. *Environmental Research Letters, 13*(5), 055012. DOI: 10.1088/1748-9326/aabf45

Cai, J., Lu, S., Cheng, J., Wang, L., Gao, Y., & Tan, T. (2022). Collaborative variable neighborhood search for multi-objective distributed scheduling in two-stage hybrid flow shop with sequence-dependent setup times. *Scientific Reports, 12*(1), 15724. PMID: 36127381

Cairone, S., Hasan, S. W., Choo, K.-H., Li, C.-W., Zarra, T., Belgiorno, V., & Naddeo, V. (2024). Integrating artificial intelligence modeling and membrane technologies for advanced wastewater treatment: Research progress and future perspectives. *The Science of the Total Environment, 944*, 173999. DOI: 10.1016/j.scitotenv.2024.173999 PMID: 38879019

California Department of Food and Agriculture. (2024). *Report on the 2012–2016 California drought years.* https://plantingseedsblog.cdfa.ca.gov/wordpress/?p=22505

California Water Watch. (2024). *Drought indicator.* https://cww.water.ca.gov/droughtindicator

Cambra, C., & Sendra, S. (2018). Smart system for bicarbonate control in irrigation for hydroponic precision farming. *Sensors (Basel), 18*(4), 1333. DOI: 10.3390/s18041333 PMID: 29693611

Camp, C. R. (1998). Subsurface drip irrigation: A review. *Transactions of the ASAE. American Society of Agricultural Engineers, 41*(5), 1353–1367.

Cao, W., Wang, D., Li, J., Zhou, H., Li, L., & Li, Y. (2018). *BRITS: Bidirectional Recurrent Imputation for Time Series* (arXiv:1805.10572). arXiv. https://doi.org//arXiv.1805.10572DOI: 10.48550

Cao, R., Fan, M., Hu, J., Ruan, W., Xiong, K., & Wei, X. (2017). Optimizing Low-Concentration Mercury Removal from Aqueous Solutions by Reduced Graphene Oxide-Supported Fe3O4 Composites with the Aid of an Artificial Neural Network and Genetic Algorithm. *Materials (Basel), 10*(11), 1279–1279. DOI: 10.3390/ma10111279

Capodaglio, A. G., & Callegari, A. (2025). Use, potential, needs, and limits of AI in wastewater treatment applications. *Water (Basel), 17*(2), 170.

Capodaglio, A. G., & Callegari, A. (2025). Use, Potential, Needs, and Limits of AI in Wastewater Treatment Applications. *Water (Basel), 17*(2), 170. DOI: 10.3390/w17020170

Central Pollution Control Board (CPCB). (2020). *National water quality monitoring programme*. Ministry of Environment, Forest and Climate Change, Government of India.

Cevallos-Mendoza, J., Amorim, C., Rodríguez-Díaz, J., & Montenegro, M. (2022). Removal of Contaminants from Water by Membrane Filtration: A Review. *Membranes, 12*(6), 570. DOI: 10.3390/membranes12060570

Chavhan, N., Bhattad, R., Khot, S., Patil, S., Pawar, A., Pawar, T., & Gawli, P. (2025). APAH: An autonomous IoT driven real-time monitoring system for Industrial wastewater. *Digital Chemical Engineering, 14*, 100217. DOI: 10.1016/j.dche.2025.100217

Chen, F., Huang, J., Wu, X., Wu, X., & Arabmarkadeh, A. (2021). On the Evaluation of Rhamnolipid Biosurfactant Adsorption Performance on Amberlite XAD-2 Using Machine Learning Techniques. *BioMed Research International, 2021*(1). https://doi.org/DOI: 10.1155/2021/5530093

Cheng, Q., Su, Q., Binley, A., Liu, J., Zhang, Z., & Chen, X. (2023). Estimation of Surface Soil Moisture by a Multi-Elevation UAV-Based Ground Penetrating Radar. *Water Resources Research, 59*(2), e2022WR032621. https://doi.org/DOI: 10.1029/2022WR032621

Chen, J. J., & Liu, D. H. (2023). Using computer vision to monitor ice conditions in water supply infrastructure: A study of salient image features. *Journal of Hydroinformatics, 25*(4), 1139–1152. DOI: 10.2166/hydro.2023.120

Chen, L., Chen, Z., Liu, Y., Lichtfouse, E., Jiang, Y., Hua, J., Osman, A. I., Farghali, M., Huang, L., Zhang, Y., Rooney, D. W., & Yap, P. (2024). Benefits and limitations of recycled water systems in the building sector: A review. *Environmental Chemistry Letters, 22*(2), 785–814. DOI: 10.1007/s10311-023-01683-2

Chen, W., Wang, W., Huang, G., Wang, Z., Lai, C., & Yang, Z. (2021). The capacity of grey infrastructure in urban flood management: A comprehensive analysis of grey infrastructure and the green-grey approach. *International Journal of Disaster Risk Reduction, 54*, 102045. DOI: 10.1016/j.ijdrr.2021.102045

Chen, Y. (2024). Real time data monitoring of water resources environment based on computer remote data collection and image analysis. *Optical and Quantum Electronics, 56*(4), 618. Advance online publication. DOI: 10.1007/s11082-023-05928-w

Chen, Y., Liu, J., Chen, Z., & Zhang, Y. (2020). Group decision-making method based on generalized vague N-soft sets. In: *Chinese Control and Decision Conference*, 4010–4015.

Chief, K., Meadow, A. M., & Whyte, K. P. (2016). Engaging southwestern Indigenous communities in sustainable water research. *Water (Basel), 8*(8), 350. DOI: 10.3390/w8080350

Chimwamurombe, P. M., & Mataranyika, P. N. (2021). Factors influencing dryland agricultural productivity. *Journal of Arid Environments, 189*, 104489. DOI: 10.1016/j.jaridenv.2021.104489

Chouhan, S., Kumari, S., Kumar, R., & Chaudhary, P. L. (2023). Climate Resilient Water Management for Sustainable Agriculture. *International Journal of Environment and Climate Change, 13*(7), 411–426. DOI: 10.9734/ijecc/2023/v13i71894

Christian, D., Gaekwad, A., & Dani, H. M.A., S., & Kandya, A. (2022). Recent techniques of textile industrial wastewater treatment: A review. *Materials Today: Proceedings*. Advance online publication. DOI: 10.1016/j.matpr.2022.11.301

Corpus, R. M. B., Bayani, M. S., Aguilar, J. L., Aguilar, J. C. B., & Aguilar, H. B. (2024). Emerging pollutants in waste water: Challenges and advancements in treatment technology. *IOP Conference Series. Earth and Environmental Science, 1372*(1), 012037. DOI: 10.1088/1755-1315/1372/1/012037

Costa, G. E. de M. e, Menezes Filho, F. C. M. de, Canales, F. A., Fava, M. C., Brandão, A. R. A., & de Paes, R. P. (2023). Assessment of Time Series Models for Mean Discharge Modeling and Forecasting in a Sub-Basin of the Paranaíba River, Brazil. *Hydrology, 10*(11), 11. Advance online publication. DOI: 10.3390/hydrology10110208

Croll, H. C., Ikuma, K., Ong, S. K., & Sarkar, S. (2023). Reinforcement learning applied to wastewater treatment process control optimization: Approaches, challenges, and path forward. *Critical Reviews in Environmental Science and Technology, 53*(20), 1775–1794.

Da'na, E., & Awad, A. (2017). Regeneration of spent activated carbon obtained from home filtration system and applying it for heavy metals adsorption. *Journal of Environmental Chemical Engineering*, *5*(4), 3091–3099. DOI: 10.1016/j.jece.2017.06.022

Dabach, S., Lazarovitch, N., Simunek, J., & Shani, U. (2013). Numerical investigation of irrigation scheduling based on soil water status. *Irrigation Science*, *31*(1), 27–36. DOI: 10.1007/s00271-011-0290-6

Dabral, R., Sandhu, H., Trivedi, M., & Cherubini, C. (2025). Optimal operation of reservoir for sustainable water management using multi-criteria decision-making approach. *Journal of Hydroinformatics*. Advance online publication.

Dai, Y., Huang, Z., Khan, N., & Labbo, M. S. (2025a). Smart Water Management: Governance Innovation, Technological Integration, and Policy Pathways Toward Economic and Ecological Sustainability. *Water (Basel)*, *17*(13), 1932. DOI: 10.3390/w17131932

Dalezios, N. R., Angelakis, A. N., & Eslamian, S. (2018). Water scarcity management: Part 1: Methodological framework. *International Journal of Global Environmental Issues*, *17*(1), 1–40. DOI: 10.1504/IJGENVI.2018.090582

Dalibard, A., Kriem, L. S., Beckett, M., Scherle, S., Yeh, Y. C., & Schliessmann, U. (2025). Challenges and requirements of AI-based waste water treatment systems [Article]. *Automatisierungstechnik*, *73*(1), 40–49. DOI: 10.1515/auto-2024-0023

Dankers, R., & Kundzewicz, Z. W. (2020). Grappling with uncertainties in physical climate impact projections of water resources. *Climatic Change*, *163*(3), 1379–1397. DOI: 10.1007/s10584-020-02858-4

Dansawad, P., Li, Y., Li, Y., Zhang, J., You, S., Li, W., & Yi, S. (2023). Machine learning toward improving the performance of membrane-based wastewater treatment: A review. *Advanced Membranes*, *3*, 100072.

Darre, N. C., & Toor, G. S. (2018). Desalination of water: A review. *Current Pollution Reports*, *4*, 104–111.

Das, A. K. (2016). On partially included intuitionistic fuzzy rough relations. *Afrika Matematika*, *27*, 993–1001. DOI: 10.1007/s13370-016-0395-2

Das, A. K. (2018). Weighted fuzzy soft multiset and decision-making. *International Journal of Machine Learning and Cybernetics*, *9*, 787–794. DOI: 10.1007/s13042-016-0607-y

Das, A. K., Das, R., Das, S., Debnath, B. K., Granados, C., Shil, B., & Das, R. (2025c). A comprehensive study of neutrosophic SuperHyper BCI-semigroups and their algebraic significance. *Transactions on Fuzzy Sets and Systems*, *4*(2), 80–101.

Das, A. K., Das, R., Datta, R., & Granados, C. (2025d). An Innovative Approach to Group Decision-Making Based on Weighted Hypersoft Expert System. In Smarandache, F., Jdid, M., & Leyva-Vazquez, M. (Eds.), *Neutrosophic and Plithogenic Inventory Models for Applied Mathematics* (pp. 483–518). IGI Global Scientific Publishing., DOI: 10.4018/979-8-3693-3204-7.ch019

Das, A. K., & Granados, C. (2021). An advanced approach to fuzzy soft group decision-making using weighted average ratings. *SN Computer Science*, *2*, 471. DOI: 10.1007/s42979-021-00873-5

Das, A. K., & Granados, C. (2022a). A new fuzzy parameterized intuitionistic fuzzy soft multiset theory and group decision-making. *Journal of Current Science and Technology*, *12*(3), 547–567. DOI: 10.14456/jcst.2022.42

Das, A. K., & Granados, C. (2022b). FP-intuitionistic multi fuzzy N-soft set and its induced FP-Hesitant N soft set in decision-making. *Decision Making: Applications in Management and Engineering*, *5*(1), 67–89. DOI: 10.31181/dmame181221045d

Das, A. K., & Granados, C. (2023a). IFP-intuitionistic multi fuzzy N-soft set and its induced IFP-hesitant N-soft set in decision-making. *Journal of Ambient Intelligence and Humanized Computing*, *14*, 10143–10152. DOI: 10.1007/s12652-021-03677-w

Das, A. K., & Granados, C. (2023b). Preference intuitionistic fuzzy rough relation and its theoretical approach. *Journal of the Indian Mathematical Society*, *90*(3-4), 199–212.

Das, A. K., Granados, C., & Bhattacharya, J. (2022). Some new operations on fuzzy soft sets and their applications in decision-making. *Songklanakarin Journal of Science and Technology*, *44*(2), 440–449. DOI: 10.14456/sjst-psu.2022.61

Das, A. K., Gupta, N., & Granados, C. (2023). Weighted hesitant bipolar-valued fuzzy soft set in decision-making. *Songklanakarin Journal of Science and Technology*, *45*(6), 681–690. https://sjst.psu.ac.th/journal/45-6/10.pdf

Das, A. K., Gupta, N., Mahmood, T., Tripathy, B. C., Das, R., & Das, S. (2024). Assessing anthropogenic influences on the water quality of Gomati River using an innovative weighted fuzzy soft set based water pollution rating system. *Discover Water*, *4*, 73. DOI: 10.1007/s43832-024-00136-3

Das, A. K., Gupta, N., Mahmood, T., Tripathy, B. C., Das, R., & Das, S. (2025a). An efficient water quality evaluation model using weighted hesitant fuzzy soft sets for water pollution rating. In Kumar, A., Kumar, P., Rathee, S., & Kumar, B. (Eds.), *Mechatronics: Concepts, Tools, Applications, and New Trends* (1st ed., p. 17). CRC Press., DOI: 10.1201/9781003494478-10

Das, A. K., Gupta, N., Mahmood, T., Tripathy, B. C., Das, R., & Das, S. (2025b). An innovative fuzzy multi-criteria decision making model for analyzing anthropogenic influences on urban river water quality. *Iran Journal of Computer Science, 8*, 103–124. DOI: 10.1007/s42044-024-00211-x

Das, A. K., Mahmood, T., Das, R., & Das, S. (2025e). Optimizing Inventory Levels in Retail: Fuzzy Logic-Based Decision Support Systems for Adaptive Inventory Management. In Jana, S., Basu Mallik, B., Sarkar, A., & Jana, C. (Eds.), *Applications of Fuzzy Logic in Decision Making and Management Science. Information Systems Engineering and Management* (Vol. 29). Springer. DOI: 10.1007/978-3-031-77719-6_8

Das, S., Roy, B., Kar, M., Kar, S., & Pamučar, D. (2020). Neutrosophic fuzzy set and its application in decision making. *Journal of Ambient Intelligence and Humanized Computing, 11*, 5017–5029. DOI: 10.1007/s12652-020-01808-3

De la Hoz-M, J., Ariza-Echeverri, E. A., & Vergara, D. (2024). Exploring the Role of Artificial Intelligence in Wastewater Treatment: A Dynamic Analysis of Emerging Research Trends. *Resources, 13*(12), 171.

de Leon, V. B., de Negreiros, B. A. F., Brusamarello, C. Z., Petroli, G., Di Domenico, M., & de Souza, F. B. (2020). Artificial neural network for prediction of color adsorption from an industrial textile effluent using modified sugarcane bagasse: Characterization, kinetics and isotherm studies. *Environmental Nanotechnology, Monitoring & Management, 14*, 100387.

de Melo, W. W., Iglesias, I., & Pinho, J. (2024). Early warning system for floods at estuarine areas: Combining artificial intelligence with process-based models [Article; Early Access]. *Natural Hazards, 24*. Advance online publication. DOI: 10.1007/s11069-024-06957-8

De Silva, C., Halloluwa, T., & Vyas, D. (2025). *A Multi-Layered Research Framework for Human-Centered AI: Defining the Path to Explainability and Trust* (Version 2). arXiv. https://doi.org/DOI: 10.48550/ARXIV.2504.13926

Deepa, A., Mastan, A., & Buddolla, V. (2025). Advances in imaging techniques for real-time microbial visualization in wastewater treatment reactors: Challenges, applications, and process optimization. *Trends in Analytical Chemistry*, •••, 118227.

DeFlorio, M. J., Sengupta, A., Castellano, C. M., Wang, J., Zhang, Z., Gershunov, A., Guirguis, K., Niño, R. L., Clemesha, R. E. S., Pan, M., Xiao, M., Kawzenuk, B., Gibson, P. B., Scheftic, W., Broxton, P. D., Switanek, M. B., Yuan, J., Dettinger, M. D., Hecht, C. W., … Anderson, M. L. (2024). *From California's Extreme Drought to Major Flooding: Evaluating and Synthesizing Experimental Seasonal and Subseasonal Forecasts of Landfalling Atmospheric Rivers and Extreme Precipitation during Winter 2022/23*. https://doi.org/DOI: 10.1175/BAMS-D-22-0208.1

Demiray, B. Z., Sit, M., Mermer, O., & Demir, I. (2024a). Enhancing hydrological modeling with transformers: A case study for 24-h streamflow prediction. *Water Science and Technology*, *89*(9), 2326–2341. DOI: 10.2166/wst.2024.110 PMID: 38747952

DeNicola, E., Aburizaiza, O. S., Siddique, A., Khwaja, H., & Carpenter, D. O. (2015). Climate change and water scarcity: The case of Saudi Arabia. *Annals of Global Health*, *81*(3), 342–353. PMID: 26615069

Diffenbaugh, N. S., Swain, D. L., & Touma, D. (2015). Anthropogenic warming has increased drought risk in California. *Proceedings of the National Academy of Sciences of the United States of America*, *112*(13), 3931–3936. DOI: 10.1073/pnas.1422385112 PMID: 25733875

Dillon, P., Arshad, M., Fernandez, E., & Pavelic, P. (2019). Managed aquifer recharge in integrated water resource management. In Villholth, S. (Eds.), *Advances in groundwater governance* (pp. 435–459). CRC Press.

Dimgba, M. O., & Andreas, R. (2024). *Optimal Hyperparameter Search Strategies: Benchmarking Grid Search, Random Search, and Genetic Algorithms across Regression, Classification, and Clustering Tasks*. https://doi.org/DOI: 10.13140/RG.2.2.24957.37603

Ding, J. Y., Song, S., & Wu, C. (2016). Carbon-efficient scheduling of flow shops by multi-objective optimization. *European Journal of Operational Research*, *248*(3), 758–771.

Djerbouai, S. (n.d.). *Missing Precipitation Data Estimation Using Long Short-Term Memory Deep Neural Networks*. Retrieved August 17, 2025, from https://www.jeeng.net/Missing-Precipitation-Data-Estimation-Using-Long-Short-Term-Memory-Deep-Neural-Networks,147322,0,2.html

Dobbs, B. T., & Ras, Z. W. (2022). Machine Learning Approach to Art Authentication. In Wang, J. (Ed.), *Encyclopedia of Data Science and Machine Learning* (pp. 1500–1513). IGI Global., DOI: 10.4018/978-1-7998-9220-5.ch089

Dong, X., Huang, H., & Chen, P. (2009). An iterated local search algorithm for the permutation flowshop problem with total flowtime criterion. *Computers & Operations Research, 36*(5), 1664–1669.

Dotto, G. L., Nunes, I. D. S., Fuhr, A. C. F. P., Da Silva, R. N. F., & Peres, E. C. (2025). Adsorption-assisted technologies for wastewater treatment. In Elsevier eBooks (pp. 49–77). https://doi.org/DOI: 10.1016/b978-0-443-26500-6.00003-1

Drenkhan, F., Huggel, C., Hoyos, N., & Scott, C. A. (2023). Hydrology, water resources availability and management in the Andes under climate change and human impacts. *Journal of Hydrology. Regional Studies, 49*, 101519. DOI: 10.1016/j.ejrh.2023.101519

Duby, G. (2011). *Europe in the Middle Ages*. Flammarion.

Dworsky, L. B. (1967). Analysis of Federal Water Pollution Control Legislation, 1948–1966. *Journal - American Water Works Association, 59*(6), 651–668. DOI: 10.1002/j.1551-8833.1967.tb03394.x

Eerkes-Medrano, D., Thompson, R. C., & Aldridge, D. C. (2015). Microplastics in freshwater systems: A review of the emerging threats, identification of knowledge gaps and prioritisation of research needs. *Water Research, 75*, 63–82. DOI: 10.1016/j.watres.2015.02.012 PMID: 25746963

El Bilali, A., Taleb, A., & Brouziyne, Y. (2021). Groundwater quality forecasting using machine learning algorithms for irrigation purposes. *Agricultural Water Management, 245*, 106625. DOI: 10.1016/j.agwat.2020.106625

El Bilali, A., Taleb, A., Nafii, A., Alabjah, B., & Mazigh, N. (2021). Prediction of sodium adsorption ratio and chloride concentration in a coastal aquifer under seawater intrusion using machine learning models. *Environmental Technology & Innovation, 23*, 101641.

El Bouzidi, A. (1999). Qa āyā tawzīʿ al-māʾ bi-wā at Draʿa min khilāl al-wathāʾiq al-ma alliyy. In Proceedings of the symposium on Water in the history of Morocco (Hassan II University, Ain Chock), No. 11. FLSH publication.

El Faïz, M. (2005). *The Masters of Water: A History of Arab Hydraulics*. Actes Sud.

El Haddadi, M. (2025). The nexus between economic growth and water stress in Morocco. *Sustainability, 17*(15), 6990. DOI: 10.3390/su17156990

Elfghi, F. M. (2016). A hybrid statistical approach for modeling and optimization of RON: A comparative study and combined application of response surface methodology (RSM) and artificial neural network (ANN) based on design of experiment (DOE). *Chemical Engineering Research & Design, 113*, 264–272. DOI: 10.1016/j.cherd.2016.05.023

Elleuch, M., Anane, M., Euchi, J., & Frikha, A. (2019). Hybrid fuzzy multi-criteria decision making to solve the irrigation water allocation problem in the Tunisian case. *Agricultural Systems*. Advance online publication. DOI: 10.1016/J.AGSY.2019.102644

Elmahdi, A., & Jeong, J. (2024). From debt to sustainability: Advancing wastewater projects in developing countries through innovative financing mechanisms—the role of debt-for-climate swaps. *Climate (Basel), 12*(8), 122.

Elsayed, A., Ghaith, M., Yosri, A., Li, Z., & El-Dakhakhni, W. (2024). Genetic programming expressions for effluent quality prediction: Towards AI-driven monitoring and management of wastewater treatment plants. *Journal of Environmental Management, 356*, 120510. DOI: 10.1016/j.jenvman.2024.120510 PMID: 38490009

Elshaboury, N., Mohammed Abdelkader, E., Al-Sakkaf, A., & Zayed, T. (2023). A Critical Review and Bibliometric Analysis on Applications of Ground Penetrating Radar in Science Based on Web of Science Database. *Eng, 4*(1), 1. Advance online publication. DOI: 10.3390/eng4010059

Elvin, M. (2004). *The retreat of the elephants: An environmental history of China*. Yale University Press.

Eniola, J. O., Kumar, R., Barakat, M. A., & Rashid, J. (2022). A review on conventional and advanced hybrid technologies for pharmaceutical wastewater treatment. *Journal of Cleaner Production, 356*, 131826. DOI: 10.1016/j.jclepro.2022.131826

Ercan Oguztürk, G. (2025). AI-driven irrigation systems for sustainable water management: A systematic review and meta-analytical insights. *Smart Agricultural Technology, 11*, 100982. Advance online publication. DOI: 10.1016/j.atech.2025.100982

Eriksson, E., Donner, E., & Ledin, A. (2010). Presence of selected priority and personal care substances in an onsite bathroom greywater treatment facility. *Water Science and Technology, 62*(12), 2889–2898. DOI: 10.2166/wst.2010.988

Ersan, G., Cerrón-Calle, G. A., Ersan, M. S., & Garcia-Segura, S. (2023). Opportunities for in situ electro-regeneration of organic contaminant-laden carbonaceous adsorbents. *Water Research, 232*, 119718. DOI: 10.1016/j.watres.2023.119718

Eshbobaev, J., Norkobilov, A., Usmanov, K., Khamidov, B., Kodirov, O., & Avezov, T. (2024). Control of Wastewater Treatment Processes Using a Fuzzy Logic Approach. *Engineering Proceedings*, *67*(1), 39.

Eskandari Dameneh, H. (2021). Desertification of Iran in the early twenty-first century: Assessment using climate and vegetation indices. *Scientific Reports*, *11*(1), 20548. DOI: 10.1038/s41598-021-99983-3 PMID: 34654866

Euklidiadas, M. M. (2023). *Barcelona's innovative smart water management makes the most of every drop*. https://www.tomorrow.city/barcelonas-innovative-smart-water-management-makes-the-most-of-every-drop/

European Commission. (2021). *Water Framework Directive & floods directive: Overview*. https://environment.ec.europa.eu/topics/water/water-framework-directive_en

European Environment Agency. (2022). *Water quality in Europe — 2022 status report*. EEA Report No. 14/2022. https://www.eea.europa.eu/publications/water-quality-in-europe-2022

Fang, H., Jamali, B., Deletic, A., & Zhang, K. (2021). Machine learning approaches for predicting the performance of stormwater biofilters in heavy metal removal and risk mitigation. *Water Research*, *200*, 117273. DOI: 10.1016/j.watres.2021.117273 PMID: 34091222

Fang, S. B., Deitch, M. J., & Gebremicael, T. G. (2025). Evaluating the reliability of data interpolation and machine learning methods for water quality management: A SWAT model comparison. *Environmental Earth Sciences*, *84*(10), 274. Advance online publication. DOI: 10.1007/s12665-025-12313-1

Fan, M., Li, T.-J., Hu, J., Cao, R., Wei, X., Shi, X., & Ruan, W. (2017). Artificial Neural Network Modeling and Genetic Algorithm Optimization for Cadmium Removal from Aqueous Solutions by Reduced Graphene Oxide-Supported Nanoscale Zero-Valent Iron (nZVI/rGO) Composites. *Materials (Basel)*, *10*(5), 544–544. DOI: 10.3390/ma10050544

Faruqui, N. I., Biswas, A. K., & Bino, M. J. (Eds.). (2001). Water Management in Islam. United Nations University Press & IDRC.

Feng, F., Jun, Y. B., Liu, X., & Li, L. (2010). An adjustable approach to fuzzy soft set based decision making. *Journal of Computational and Applied Mathematics*, *234*, 10–20.

Ferraro, G. B., Bonomo, C., Brandtner, D., Mancini, P., Veneri, C., Briancesco, R., & La Rosa, G. (2024). Characterisation of microbial communities and quantification of antibiotic resistance genes in Italian wastewater treatment plants using 16S rRNA sequencing and digital PCR. *The Science of the Total Environment*, *933*, 173217.

Fevgas, G., Lagkas, T., Papadopoulos, P., Sarigiannidis, P., & Argyriou, V. (2025). Integrating thermal infrared and RGB imaging for early detection of water stress in lettuces with comparative analysis of IoT sensors. *Smart Agricultural Technology*, *11*, 100881. Advance online publication. DOI: 10.1016/j.atech.2025.100881

Food and Agriculture Organization of the United Nations (FAO). (2016). *AQUASTAT main database*. https://www.fao.org/aquastat/

Food and Agriculture Organization of the United Nations (FAO). (2017). *The future of food and agriculture – Trends and challenges*. FAO.

Food and Agriculture Organization of the United Nations (FAO). (2018). *The state of the world's water resources for food and agriculture*. FAO.

Food and Agriculture Organization of the United Nations. (2022). *The state of the world's land and water resources for food and agriculture: Systems at breaking points*. FAO.

Food and Agriculture Organization of the United Nations. (2025). Renewable internal freshwater resources, total (billion cubic meters). [Data set] In World Bank (Ed.), *World Development Indicators*. Processed by Our World in Data.

Food and Agriculture Organization. A. d. (2025a). *Renewable internal freshwater resources per capita (cubic meters)*. Worldbank. https://data.worldbank.org/indicator/ER.H2O.INTR.PC

Food and Agriculture Organization. A. d. (2025b). *Renewable internal freshwater resources, total (billion cubic meters)*. Worldbank. https://data.worldbank.org/indicator/ER.H2O.INTR.K3

Ftaïta, T. (2011). Community water management. *Anuário Antropológico*, *36*(1), 195–212. DOI: 10.4000/aa.1161

Fukui, M., Hiramatsu, K., Fukuda, S., Pineda, D., Harada, M., & Tabata, T. (2020). Evaluation of Water-Saving Methodologies for Rice Farming Using a Distributed Long-Term Rainfall-Runoff Model Incorporated Agricultural Ponds Detected by Multi-Temporal SAR Images in the Ibague Watershed, Colombia. *Journal of the Faculty of Agriculture Kyushu University, 65*(2), 313-320. <Go to ISI>://WOS:000589116000015

Full article: Smart Water Management: The way to (artificially) intelligent water management, or just another pretty name? (n.d.-a). Retrieved June 1, 2025, from https://www.tandfonline.com/doi/full/10.1080/02508060.2020.1830581

Full article: Smart Water Management: The way to (artificially) intelligent water management, or just another pretty name? (n.d.-b). Retrieved June 1, 2025, from https://www.tandfonline.com/doi/full/10.1080/02508060.2020.1830581

Gad, M., El Hamed, R. A., El Fadaly, E. A., Mousa, I. E., Gaagai, A., Aouissi, H. A., & Elsayed, S. (2025). New approach to predict wastewater quality for irrigation utilizing integrated indexical approaches and hyperspectral reflectance measurements supported with multivariate analysis. *Scientific Reports*, *15*(1), 16395.

Gaffin, S. R., Rosenzweig, C., & Parshall, L. (2008). Impact of urban heat islands on water availability and usage. *Environmental Research Letters*, *3*(4), 1–7. DOI: 10.1088/1748-9326/3/4/044003

Ganeshkumar, L., Palanisami, D., & Mohan, N. (2024). Neutrosophic fusion of multimodal brain images: Integrating neutrosophic entropy and feature extraction. *Applied Soft Computing*, *155*, 111462. DOI: 10.1016/j.asoc.2024.111462

Garg, M. C., Kumari, S., & Agarwal, S. (2024). The integration of artificial intelligence in advanced wastewater treatment systems. In *The AI Cleanse: Transforming Wastewater Treatment Through Artificial Intelligence: Harnessing Data-Driven Solutions* (pp. 1–27). Springer Nature Switzerland.

Gautam, K., & Verma, R. K., Suantak Kamsonlian, & Kumar, S. (2020). Decolorization of Reactive Black B from wastewater by electro-coagulation: Optimization using multivariate RSM and ANN. *Chemical Product and Process Modeling*, *16*(2), 129–144. DOI: 10.1515/cppm-2020-0043

Gebremeskel, G., & Tang, Q. (2019). Droughts in East Africa: Causes, impacts and resilience. *Earth-Science Reviews*, *193*, 146–161. DOI: 10.1016/j.earscirev.2019.04.015

Georgin, J., Franco, D. S. P., Ramos, C. G., Piccilli, D. G., Lima, E. C., & Sher, F. (2023). A review of the antibiotic ofloxacin: Current status of ecotoxicology and scientific advances in its removal from aqueous systems by adsorption technology. *Chemical Engineering Research & Design*, *193*, 99–120. DOI: 10.1016/j.cherd.2023.03.025

Ghaffour, N., Missimer, T. M., & Amy, G. L. (2013). Technical review and evaluation of the economics of water desalination: Current and future challenges for better water supply sustainability. *Desalination*, *309*, 197–207.

Ghinea, L. M., Miron, M., & Barbu, M. (2023). Semi-supervised anomaly detection of dissolved oxygen sensor in wastewater treatment plants. *Sensors (Basel)*, *23*(19), 8022.

Gimpel, J. (1977). *The Medieval Machine: The Industrial Revolution of the Middle Ages*. Penguin.

Gleick, P. H. (1993). Water and conflict: Fresh water resources and international security. *International Security*, *18*(1), 79–112. DOI: 10.2307/2539033

Global Water Partnership (GWP). (2010). *Integrated water resources management*. GWP.

Global Water Partnership. (2000). *Integrated water resources management* (TAC Background Papers No. 4). Stockholm, Sweden.

Goblot, H. (1979). *The qanats: A technique for acquiring water*. Mouton.

Gokul, K., Gokul, V., Kaviyarasan, P., & Uma, J. (2024, March). Intelligent Wastewater Management System with Cloud-Based IOT and Real-Time Control. In *2024 10th International Conference on Advanced Computing and Communication Systems (ICACCS)* (Vol. 1, pp. 498-504). IEEE.

Gopinath, A., Retnam, B. G., Muthukkumaran, A., & Aravamudan, K. (2020). Swift, versatile and a rigorous kinetic model based artificial neural network surrogate for single and multicomponent batch adsorption processes. *Journal of Molecular Liquids*, *297*, 111888.

Goshkoderia, A., Arora, N., Slesarenko, V., Li, J., Chen, V., Juhl, A., Buskohl, P., & Rudykh, S. (2020). Tunable permittivity in dielectric elastomer composites under finite strains: Periodicity, randomness, and instabilities. *International Journal of Mechanical Sciences*, *186*, 105880. DOI: 10.1016/j.ijmecsci.2020.105880

Grainger, A. (2015). Is land degradation neutrality feasible in dry areas? *Journal of Arid Environments*, *112*, 14–24. DOI: 10.1016/j.jaridenv.2014.05.014

Granados, C., Das, A.K., & Osu, B. (2023). Weighted neutrosophic soft multiset and its application to decision making. Yugoslav journal of operations research, 33(2), 293-308. DOI: 10.2298/YJOR220915034G

Granados, C., Das, A.K., Das, B. (2022). Some continuous neutrosophic distributions with neutrosophic parameters based on neutrosophic random variables. Advances in the Theory of Nonlinear Analysis and its Applications, 33(2): 293-308. https://doi.org/DOI: 10.31197/atnaa.1056480

Granados, C., Das, A.K., Osu, B. (2023). Weighted neutrosophic soft multiset and its application to decision making. Yugoslav journal of operations research, 33(2): 293-308.

Grid Intelligence Water Case Study. (n.d.). Verizon Enterprise. Retrieved June 1, 2025, from https://enterprise.verizon.com/resources/articles/grid-intelligence-water-case-study/

Guillerme, A. (1983). *Les temps de l'eau: La cité, l'eau et les techniques*. Champ Vallon.

Guo, H., Li, X., Yang, W., Yao, Z., Mei, Y., Peng, L. E., Yang, Z., Shao, S., & Tang, C. Y. (2021). Nanofiltration for drinking water treatment: A review. *Frontiers of Chemical Science and Engineering, 16*(5), 681–698. DOI: 10.1007/s11705-021-2103-5

Guo, L., Lin, H., Fan, B., Cui, X., & Chen, J. (2013). Forward simulation of root's ground penetrating radar signal: Simulator development and validation. *Plant and Soil, 372*(1), 487–505. DOI: 10.1007/s11104-013-1751-8

Guo, Y., & Şengür, A. (2014). A novel image edge detection algorithm based on neutrosophic set. *Computers & Electrical Engineering, 40*, 3–25. DOI: 10.1016/j.compeleceng.2014.04.020

Gupta, K., Rani, R., & Bahia, N. K. (2020). Plant-Seedling Classification Using Transfer Learning-Based Deep Convolutional Neural Networks. *International Journal of Agricultural and Environmental Information Systems, 11*(4), 25–40. DOI: 10.4018/IJAEIS.2020100102

Gupta, V. K., Ali, I., Saleh, T. A., Nayak, A., & Agarwal, S. (2012). Chemical treatment technologies for wastewater recycling an overview. *RSC Advances, 2*(16), 6380. DOI: 10.1039/c2ra20340e

Gyparakis, S., Trichakis, I., Daras, T., & Diamadopoulos, E. (2025). Artificial Neural Networks (ANNs) and Multiple Linear Regression (MLR) Analysis Modelling for Predicting Chemical Dosages of a Water Treatment Plant (WTP) of Drinking Water [Article]. *Water, 17*(2), 12, Article 227. https://doi.org/DOI: 10.3390/w17020227

Haddache, M. (2022). Hydraulic knowledge and socioeconomic changes in the Toudgha oasis (southeastern Morocco). *Asinag, 7*, 111–122.

Hammou, D. R., Mahmoudi, S. A., & Adjoudj, R. (2022). Multi-Biometric Iris Recognition System Using Consensus Between Convolutional Neural Network Architectures. *International Journal of Organizational and Collective Intelligence, 12*(1), 1–30. DOI: 10.4018/IJOCI.305210

Han, S., Slater, L., Wilby, R. L., & Faulkner, D. (2022). Contribution of urbanisation to non-stationary river flow in the UK. *Journal of Hydrology (Amsterdam), 613*, 128417. DOI: 10.1016/j.jhydrol.2022.128417

Hansen, P., & Mladenović, N. (2001). Variable neighborhood search: Principles and applications. *European Journal of Operational Research, 130*(3), 449–467.

Harmsworth, G., Awatere, S., & Robb, M. (2016). Indigenous Māori values and perspectives to inform freshwater management in Aotearoa-New Zealand. *Ecological Applications, 26*(7), 2417–2430. DOI: 10.1002/eap.1371

Harvey, L. P. (2005). *Muslims in Spain, 1500 to 1614*. University of Chicago Press.

Hasan, M. M., Rahman, M. S., & Bell, A. (2020). Deep Reinforcement Learning for Optimization: In I. R. Management Association (Ed.), *Research Anthology on Artificial Intelligence Applications in Security* (pp. 1598–1614). IGI Global. https://doi.org/DOI: 10.4018/978-1-7998-7705-9.ch070

Hasan, H. A., Muhamad, M. H., Ji, B., Nazairi, N. A., Jiat, K. W., Sim, S. I. S. W. A., & Poh, A. F. M. S. (2023). Revolutionizing wastewater treatment with microalgae: Unveiling resource recovery, mechanisms, challenges, and future possibilities. *Ecological Engineering, 197*, 107117. DOI: 10.1016/j.ecoleng.2023.107117

Hashem, M. S., & Qi, X. (2021). Treated Wastewater Irrigation—A Review. *Water (Basel), 13*(11), 1527. DOI: 10.3390/w13111527

Hashim, S., Yuebo, X., Saifullah, M., Jan, R. N., & Muhetaer, A. (2015). Integrated evaluation of urban water bodies for pollution abatement based on fuzzy multicriteria decision approach. *BioMed Research International, 2015*, 327280. Advance online publication. DOI: 10.1155/2015/327280 PMID: 26516623

Hašková, S., Vochozka, M., & Kučera, J. (2023). *A fuzzy evaluation model of manufacturing machinery in terms of sustainable business*. Entrepreneurship and Sustainability Issues.,

Hassan, M. K. (2024). Management of Industrial Wastewater Quality Using IoT. SSRN *Electronic Journal*. https://doi.org/DOI: 10.2139/ssrn.4920201

Havukainen, J., Väisänen, S., Rantala, T., Saunila, M., & Ukko, J. (2020). Environmental impacts of manure management based on life cycle assessment approach. *Journal of Cleaner Production, 264*, 121576. DOI: 10.1016/j.jclepro.2020.121576

Hill, D. R. (1994). *Islamic science and engineering*. Edinburgh University Press.

Hillel, D. (1994). *Rivers of Eden: The struggle for water and the quest for peace in the Middle East*. Oxford University Press.

Houdret, A. (2012). The water connection: Irrigation, water grabbing and politics in southern Morocco. *Water Alternatives*, *5*(2), 284–303. https://www.water-alternatives.org/index.php/alldoc/articles/vol5/v5issue2/170-a5-2-6

Huang, H., Cai, X., Li, L., Wu, X., Zhao, Z., & Tan, X. (2025). Interpreting global terrestrial water storage dynamics and drivers with explainable deep learning. *Remote Sensing*, *17*(13), 2118. DOI: 10.3390/rs17132118

Huang, M., Ma, Y., Wan, J., Zhang, H., Wang, Y., Chen, Y., Yoo, C., & Guo, W. (2011). A hybrid genetic – Neural algorithm for modeling the biodegradation process of DnBP in AAO system. *Bioresource Technology*, *102*(19), 8907–8913. DOI: 10.1016/j.biortech.2011.06.046

Huang, Y., Li, Y. P., Chen, X., & Ma, Y. G. (2012). Optimization of the irrigation water resources for agricultural sustainability in Tarim River Basin, China. *Agricultural Water Management*, *107*, 74–85. DOI: 10.1016/j.agwat.2012.01.012

Huang, Z., Yuan, X., & Liu, X. (2021). The key drivers for the changes in global water scarcity: Water withdrawal versus water availability. *Journal of Hydrology (Amsterdam)*, *601*, 126658. DOI: 10.1016/j.jhydrol.2021.126658

Hugonnet, R., McNabb, R., Berthier, E., Menounos, B., Nuth, C., Girod, L., & Kääb, A. (2021). Accelerated global glacier mass loss in the early 21st century. *Nature*, *592*, 726–731. DOI: 10.1038/s41586-021-03436-z PMID: 33911269

Humadi, J. I., Shihab, M. A., Hasan, A. A., & Mohammed, A. M. (2024). Experimental and ANN modeling of kerosene fuel desulfurization using a manganese oxide-tin oxide catalyst. *Chemical Engineering Research & Design*, *211*, 160–167. DOI: 10.1016/j.cherd.2024.10.001

Hunzinger, R. (2017). SCADA FUNDAMENTALS AND APPLICATIONS IN THE IoT. In *Internet of Things and Data Analytics Handbook* (pp. 283–293). John Wiley & Sons, Ltd., DOI: 10.1002/9781119173601.ch17

Hussain, S., Khan, H., Gul, S., Steter, J. R., & Motheo, A. J. (2021). Modeling of photolytic degradation of sulfamethoxazole using boosted regression tree (BRT), artificial neural network (ANN) and response surface methodology (RSM); energy consumption and intermediates study. *Chemosphere*, *276*, 130151. DOI: 10.1016/j.chemosphere.2021.130151

Ibrahim, A. G., Rashad, A. M., & Dincer, I. (2017). Exergoeconomic analysis for cost optimization of a solar distillation system. *Solar Energy*, *151*, 22–32.

Ibrahim, Y. Z., Balzter, H., Kaduk, J., & Tucker, C. J. (2015). Land degradation assessment using residual trend analysis of GIMMS NDVI3g, soil moisture and rainfall in Sub-Saharan West Africa from 1982 to 2012. *Remote Sensing*, *7*(5), 5471–5494. DOI: 10.3390/rs70505471

Igwegbe, C. A., Mohmmadi, L., Ahmadi, S., Rahdar, A., Khadkhodaiy, D., Dehghani, R., & Rahdar, S. (2019). Modeling of adsorption of methylene blue dye on Ho-CaWO4 nanoparticles using response surface methodology (RSM) and artificial neural network (ANN) techniques. *MethodsX*, *6*, 1779–1797.

Imbrenda, V., Quaranta, G., Salvia, R., Egidi, G., Salvati, L., Prokopova, M., & Lanfredi, M. (2021). Land degradation and metropolitan expansion in a peri-urban environment. *Geomatics, Natural Hazards & Risk*, *12*(1), 1797–1818. DOI: 10.1080/19475705.2021.1944331

Imen, S., Croll, H. C., McLellan, N. L., Bartlett, M., Lehman, G., & Jacangelo, J. G. (2023). Application of machine learning at wastewater treatment facilities: A review of the science, challenges and barriers by level of implementation. *Environmental Technology Reviews*, *12*(1), 493–516.

Infant, S. S., Vickram, S., Saravanan, A., Muthu, C. M., & Yuarajan, D. (2025). Explainable artificial intelligence for sustainable urban water systems engineering. *Results in Engineering*, *25*, 104349.

Ingrao, C., Strippoli, R., Lagioia, G., & Huisingh, D. (2023). Water scarcity in agriculture: An overview of causes, impacts and approaches for reducing the risks. *Heliyon*, *9*(8), e18507. DOI: 10.1016/j.heliyon.2023.e18507 PMID: 37534016

Intergovernmental Panel on Climate Change (IPCC). (2021). *Climate change 2021: The physical science basis. Contribution of Working Group I to the Sixth Assessment Report of the Intergovernmental Panel on Climate Change*. Cambridge University Press.

Intergovernmental Panel on Climate Change (IPCC). (2022). Climate change 2022: Impacts, adaptation and vulnerability. Sixth Assessment Report. Cambridge University Press. https://www.ipcc.ch/report/ar6/wg2/

International Energy Agency. (2021). *Water-energy nexus*. https://www.iea.org

International Law Association. (1966). *The Helsinki Rules on the Uses of the Waters of International Rivers*.

International Union for Conservation of Nature. (2020). *Nature-based solutions for water*. https://www.iucn.org

Ioffe, S., & Szegedy, C. (2015). *Batch Normalization: Accelerating Deep Network Training by Reducing Internal Covariate Shift* (Version 3). arXiv. https://doi.org/ DOI: 10.48550/ARXIV.1502.03167

IPCC. (2022). *Climate Change 2022: Impacts, adaptation, and vulnerability*. Contribution of Working Group II to the Sixth Assessment Report of the Intergovernmental Panel on Climate Change. Cambridge University Press. https://www.ipcc .ch/report/ar6/wg2/

Iqbal, M. A., Akram, S., Lal, B., Hassan, S. U., Ashraf, R., Kezembayeva, G., & Hosseini-Bandegharaei, A. (2024). Advanced photocatalysis as a viable and sustainable wastewater treatment process: A comprehensive review. *Environmental Research, 253*, 118947.

Ivanović, T., Tomović, T., Popović, D., Miladinović, J., Arrad, M., & Miladinović, Z. (2025). A simplified method for enhanced accuracy in calculating adsorption thermodynamic parameters. *The Journal of Chemical Thermodynamics, 107517*. Advance online publication. DOI: 10.1016/j.jct.2025.107517

Jackson, S., & Head, L. (2022). Indigenous water rights and climate adaptation in Australia. *Journal of Hydrology (Amsterdam), 613*, 128–145. DOI: 10.1016/j. jhydrol.2022.128145

Jahangir, M. S., & Quilty, J. (2024). Generative deep learning for probabilistic streamflow forecasting: Conditional variational auto-encoder. *Journal of Hydrology (Amsterdam), 629*, 130498. DOI: 10.1016/j.jhydrol.2023.130498

Jahin, M. A., Shahriar, S., Mridha, M. F., & Hossen, Md. J., & Dey, N. (2025). *Soybean Disease Detection via Interpretable Hybrid CNN-GNN: Integrating MobileNetV2 and GraphSAGE with Cross-Modal Attention* (Version 3). arXiv. https:// doi.org/DOI: 10.48550/ARXIV.2503.01284

Jain, A., Kumari, S., Agarwal, S., & Khan, S. (2021). Water purification via novel nano-adsorbents and their regeneration strategies. *Process Safety and Environmental Protection, 152*, 441–454. DOI: 10.1016/j.psep.2021.06.031

Jain, S., Kumar, P., & Rathore, V. (2023). Smart water monitoring systems using IoT: Trends and technologies. *Environmental Monitoring and Assessment, 195*(4), 1–17.

Jamali, A., Ramazani, M. E., Nejaei, A., & Rafieyan, O. (2025). Monitoring of Pollutants in Surface Water with Laboratory Analysis and Satellite Image Between 2019 and 2024, Kichik Chai River, Iran. *Water, Air, and Soil Pollution, 236*(5), 272. Advance online publication. DOI: 10.1007/s11270-025-07808-3

Jana, P. (2024). AI-powered IoT solutions for sustainable water management in cities. *Uncertainty Discourse and Applications, 1*(2), 158–169. DOI: 10.48313/uda.v1i2.37

Jing, X. L., Pan, Q. K., Gao, L., & Wang, L. (2021). An effective iterated greedy algorithm for a robust distributed permutation flowshop problem with carryover sequence-dependent setup time. *IEEE Transactions on Systems, Man, and Cybernetics. Systems, 52*(9), 5783–5794.

Jin, J., Wang, L., & Muller, K. (2021). A 10-year monitoring of soil properties dynamics and soil fertility evaluation in Chinese hickory plantation regions of southeastern China. *Scientific Reports, 11*(1), 23531. DOI: 10.1038/s41598-021-02946-9 PMID: 34876648

Joint Monitoring Programme (JMP). (2023). *Progress on household drinking water, sanitation, and hygiene 2000–2022: Special focus on inequalities*. WHO & UNICEF. https://washdata.org

Jones, E., Qadir, M., van Vliet, M. T. H., Smakhtin, V., & Kang, S. (2019). The state of desalination and brine production: A global outlook. *The Science of the Total Environment, 657*, 1343–1356. PMID: 30677901

Jose, D. M., & Dwarakish, G. S. (2020). Uncertainties in predicting impacts of climate change on hydrology in basin scale: A review. *Arabian Journal of Geosciences, 13*(19), 1037. DOI: 10.1007/s12517-020-06071-6

Kadimpati, K. K., Gnida, A., Turek-Szytow, J., Hellal, M. S., Gregor, M., Matula, G., & Monfort, O. (2024). Design of innovative hybrid biochar prepared from marine algae and magnetite: Insights into adsorption performance and mechanism. *Chemical Engineering Research & Design, 201*, 218–227.

Kaiss, R., Benjouid, Z., Faiz, M., Ech Chahed, H., Rakhimi, A., Zarouali, S. S., Maimoun, A., Hmid, A., & Cherkaoui, M. (2025). Water Stress and Regional Governance in Morocco. *Research on World Agricultural Economy, 6*(3), 957–972. DOI: 10.36956/rwae.v6i3.2173

Kallel, I. F., & Kallel, M. (2023). Intelligent Watermarking for Data Security: An Overview. In Chana, I., Bouazi, A., & Ben-azza, H. (Eds.), (pp. 262–282). Advances in Systems Analysis, Software Engineering, and High Performance Computing. IGI Global., DOI: 10.4018/979-8-3693-0497-6.ch015

Kant, N., & Wrat, G. (2025). A review of integrated groundwater and surface water management for environmental sustainability. *Acta Geochimica*. Advance online publication. DOI: 10.1007/s11631-025-00803-9

Karam, A., Zaher, K., & Mahmoud, A. S. (2020). Comparative Studies of Using Nano Zerovalent Iron, Activated Carbon, and Green Synthesized Nano Zerovalent Iron for Textile Wastewater Color Removal Using Artificial Intelligence, Regression Analysis, Adsorption Isotherm, and Kinetic Studies. *Air, Soil and Water Research, 13*, 117862212090827. DOI: 10.1177/1178622120908273

Karataş, E., Yolcu, A., & Ozturk, T. (2023). Effective neutrosophic soft set theory and its application to decision-making. *Afrika Matematika, 34*. Advance online publication. DOI: 10.1007/s13370-023-01101-4

Karri, R. R., Tanzifi, M., Tavakkoli Yaraki, M., & Sahu, J. N. (2018). Optimization and modeling of methyl orange adsorption onto polyaniline nano-adsorbent through response surface methodology and differential evolution embedded neural network. *Journal of Environmental Management, 223*, 517–529. DOI: 10.1016/j.jenvman.2018.06.027

Katimbo, A., Rudnick, D. R., Zhang, J. W., Ge, Y. F., Dejonge, K. C., Franz, T. E., Shi, Y. Y., Liang, W. Z., Qiao, X., Heeren, D. M., Kabenge, I., Nakabuye, H. N., & Duan, J. M. (2023). Evaluation of artificial intelligence algorithms with sensor data assimilation in estimating crop evapotranspiration and crop water stress index for irrigation water management [Article]. *Smart Agricultural Technology, 4*(18), 100176. Advance online publication. DOI: 10.1016/j.atech.2023.100176

Kenoyer, J. M. (1998). *Ancient cities of the Indus Valley Civilization*. Oxford University Press.

Keshavarz-Ghorabaee, M., Mohammadi-Ostadkalayeh, A., Amiri, M., & Antuchevičienė, J. (2025). A Fuzzy OPARA-Based Group Decision-Making Approach: Application to Sustainable Solutions Assessment. *Informatica (Vilnius)*. Advance online publication. DOI: 10.15388/25-infor586

Kettani, M., & Bandelier, P. (2020). Techno-economic assessment of solar energy coupling with large-scale desalination plant: The case of Morocco. *Desalination, 501*, 114627. DOI: 10.1016/j.desal.2020.114627 PMID: 32863401

Khajeh, M., Sarafraz-Yazdi, A., & Moghadam, A. F. (2017). Modeling of solid-phase tea waste extraction for the removal of manganese and cobalt from water samples by using PSO-artificial neural network and response surface methodology. *Arabian Journal of Chemistry, 10*, S1663–S1673.

Khan, H., Hussain, S., Din, M. A. U., Arshad, M., Wahab, F., Hassan, U., & Khan, A. (2023). Multiple design and modelling approaches for the optimisation of carbon felt electro-Fenton treatment of dye laden wastewater. *Chemosphere, 338*, 139510.

Khan, M., Son, L., Ali, M., Chau, H., Na, N., & Smarandache, F. (2018). Systematic Review of Decision Making Algorithms in Extended Neutrosophic Sets. *Symmetry*, *10*, 314. DOI: 10.3390/SYM10080314

Khan, S., Khan, A., Ullah, R., Ali, M., & Ullah, R. (2020). Insulin DNA Sequence Classification Using Levy Flight Bat With Back Propagation Algorithm. In Umair, S. (Ed.), (pp. 232–252). Advances in Medical Technologies and Clinical Practice. IGI Global., DOI: 10.4018/978-1-7998-2521-0.ch011

Khare, A., & Agrawal, S. (2021). Effective heuristics and metaheuristics to minimise total tardiness for the distributed permutation flowshop scheduling problem. *International Journal of Production Research*, *59*(23), 7266–7282.

Khatri, N., Vyas, A. K., Abdul-Qawy, A. S. H., & Rene, E. R. (2023). Artificial neural network based models for predicting the effluent quality of a combined upflow anaerobic sludge blanket and facultative pond: Performance evaluation and comparison of different algorithms. *Environmental Research*, *217*, 114843.

Ko, J., Lee, K., Hwang, H., Oh, S.-G., Son, S.-W., & Shin, K. (2022). Effective training strategies for deep-learning-based precipitation nowcasting and estimation. *Computers & Geosciences*, *161*, 105072. DOI: 10.1016/j.cageo.2022.105072

Köksal, D. D., Ahi, Y., & Todorovic, M. (2025). Assessing Agricultural Reuse Potential of Treated Wastewater: A Hybrid Machine Learning Approach. *Agronomy (Basel)*, *15*(3), 703.

Koo, C., Kim, J., Kim, S., & Shin, D. (2020). Smart water grid: The next generation water management platform. *Water (Basel)*, *12*(11), 3222.

Kubiak, A. (2025). A scalable IoT-controlled photocatalytic system for pharmaceutical removal in real wastewater treatment applications. *Chemical Engineering Journal*, •••, 164709.

Kumar, D., & Gupta, S. K. (2022). Electrochemical oxidation of direct blue 86 dye using MMO coated Ti anode: Modelling, kinetics and degradation pathway. *Chemical Engineering and Processing*, *181*, 109127. DOI: 10.1016/j.cep.2022.109127

Kumari, P. (2025). Transforming Medical Imaging With Convolutional Neural Networks (CNNs): Advances in Diagnosis and Treatment. In Aamir, M., Bhatti, U. A., Rahman, Z., Bhutto, J. A., & Abro, W. A. (Eds.), *Advances in Medical Diagnosis, Treatment, and Care* (pp. 195–230). IGI Global., DOI: 10.4018/979-8-3693-9816-6.ch009

Kumari, S., Singh, S., Lo, S.-L., Sharma, P., Agarwal, S., & Garg, M. C. (2024). Machine learning and modelling approach for removing methylene blue from aqueous solutions: Optimization, kinetics and thermodynamics studies. *Journal of the Taiwan Institute of Chemical Engineers*, ●●●, 105361–105361. DOI: 10.1016/j.jtice.2024.105361

Kumar, P., Singh, R., & Chatterjee, A. (2024). Governance challenges in South Asia's urban water supply systems. *Water Policy*, *26*(1), 45–62. DOI: 10.2166/wp.2023.100

Kumar, S., Datta, S., Singh, V., Datta, D., Singh, S. K., & Sharma, R. (2024). Applications, challenges, and future directions of human-in-the-loop learning. *IEEE Access : Practical Innovations, Open Solutions*, *12*, 75735–75760.

Kümmerer, K., Dionysiou, D. D., Olsson, O., & Fatta-Kassinos, D. (2018). A path to clean water. *Science*, *361*(6399), 222–224. DOI: 10.1126/science.aau2405 PMID: 30026210

Kummu, M., & Guillaume, J. H. A. (2016). The world's road to water scarcity: Shortage and stress in the 20th century and pathways towards sustainability. *Scientific Reports*, *6*(1), 38495. DOI: 10.1038/srep38495 PMID: 27934888

Kundu, P., Paul, V., & Kumar, V., & Indra Mani Mishra. (2015). Formulation development, modeling and optimization of emulsification process using evolving RSM coupled hybrid ANN-GA framework. *Chemical Engineering Research & Design*, *104*, 773–790. DOI: 10.1016/j.cherd.2015.10.025

Kurtulmuş, E., Arslan, B., & Kurtulmuş, F. (2022). Deep learning for proximal soil sensor development towards smart irrigation. *Expert Systems with Applications*, *198*, 116812. DOI: 10.1016/j.eswa.2022.116812

Laamari, A., Laamrani, H., & Khoali, S. (2022). Smart water governance in Moroccan agriculture. *Open Journal of Social Sciences*, *10*, 11–27. DOI: 10.4236/jss.2022.109002

Lamichhane, M., Mehan, S., & Mankin, K. R. (2025). Soil Moisture Prediction Using Remote Sensing and Machine Learning Algorithms: A Review on Progress, Challenges, and Opportunities. *Remote Sensing*, *17*(14), 2397. DOI: 10.3390/rs17142397

Lansing, J. S. (1991). *Priests and programmers: Technologies of power in the engineered landscape of Bali*. Princeton University Press.

Larsen, T. A. (2011). Redesigning wastewater infrastructure to improve resource efficiency. *Water Science and Technology*, *63*(11), 2535–2541. DOI: 10.2166/wst.2011.502

Legislating for sustainable basin management: The story of Australia's Water Act (2007) | Water Policy | IWA Publishing. (n.d.). Retrieved June 1, 2025, from https://iwaponline.com/wp/article-abstract/15/6/871/20089/Legislating-for-sustainable-basin-management-the?redirectedFrom=fulltext

Legislative Analyst's Office. (2022). *Climate change impacts across sectors*. https://lao.ca.gov/reports/2022/4575/Climate-Change-Impacts-Crosscutting-Issues-040522.pdf

Lehuedé, S. (2025). An elemental ethics for artificial intelligence: Water as resistance within AI's value chain. *AI & Society*, *40*(3), 1761–1774. DOI: 10.1007/s00146-024-01922-2

Leveau, P. (2008). Integrated water management in environmental history: Traditional knowledge and modern practices. *Nature Sciences Sociétés*, *16*(2), 173–175. DOI: 10.1051/nss:2008045

Le, Y., Sun, X. Y., Chen, Y. F., Zhang, D. F., Wu, L., Liu, H., & Hu, M. Z. (2024). High-accuracy shallow-water bathymetric method including reliability evaluation based on Sentinel-2 time-series images and ICESat-2 data. *Frontiers in Marine Science*, *11*, 1470859. Advance online publication. DOI: 10.3389/fmars.2024.1470859

Liang, H., Xing, L., & Lin, J. (2020). Application and Algorithm of Ground-Penetrating Radar for Plant Root Detection: A Review. *Sensors (Basel)*, *20*(10), 10. Advance online publication. DOI: 10.3390/s20102836 PMID: 32429394

Li, D., Yang, L., Liu, S., & Tan, R. (2024). Multi-attribute decision making for deep learning-based typhoon disaster assessment. *Journal of Intelligent & Fuzzy Systems*, *46*, 6657–6677. DOI: 10.3233/jifs-235315

Li, F., Yang, F., Xie, Y., Qiao, X., Du, C., Li, C., Ru, Q., Zhang, F., Gu, X., & Yong, Z. (2024). Research on 3D ground penetrating radar deep underground cavity identification algorithm in urban roads using multi-dimensional time-frequency features. *NDT & E International*, *143*, 103060. DOI: 10.1016/j.ndteint.2024.103060

Lightfoot, D. R. (1996). Moroccan khettara: Traditional irrigation and progressive desiccation. *Geoforum*, *27*(2), 261–273. DOI: 10.1016/0016-7185(96)00011-7

Li, H., Cohen, A., Li, Z., & Zhang, M. (2019). The impacts of socioeconomic development on rural drinking water safety in China: A provincial-level comparative analysis. *Sustainability*, *11*(1), 85. DOI: 10.3390/su11010085

Li, J. Q., Pan, Q. K., & Wang, F. T. (2014). A hybrid variable neighborhood search for solving the hybrid flow shop scheduling problem. *Applied Soft Computing*, *24*, 63–77.

Li, J., Wu, L., Liu, J., Wang, X., & Xue, W. (2024). DFMM-Precip: Deep Fusion of Multi-Modal Data for Accurate Precipitation Forecasting. *Water (Basel), 16*(24), 24. Advance online publication. DOI: 10.3390/w16243702

Li, L. C., Zhao, D. L., Sun, W. P., & Huang, Q. H. (2025). A micro Francis turbine for smart water meter self-generation: Study on the effects of structural parameters on turbine performance based on CFD simulations and experiments. *Energy, 328*, 135865. Advance online publication. DOI: 10.1016/j.energy.2025.135865

Linton, J. (2010). *What is water? The history of a modern abstraction*. UBC Press.

Liu, X., & Liu, W. (2022). Global agricultural water scarcity assessment incorporating blue and green water availability under future climate change. *Earth's Future, 10*(3), e2021EF002567. https://doi.org/DOI: 10.1029/2021EF002567

Liu, X., Chen, J., Butnor, J. R., Qin, G., Cui, X., Fan, B., Lin, H., & Guo, L. (2020). Noninvasive 2D and 3D Mapping of Root Zone Soil Moisture Through the Detection of Coarse Roots With Ground-Penetrating Radar. *Water Resources Research, 56*(5), e2019WR026930. https://doi.org/DOI: 10.1029/2019WR026930

Liu, Y. (2024). Adsorption. In Elsevier eBooks (pp. 243–281). https://doi.org/DOI: 10.1016/b978-0-323-95173-9.00004-9

Liu, B., Xin, Y., Zou, J., Khoso, F., Liu, Y., Jiang, X., Peng, S., & Yu, J. (2023). Removal of chromium species by adsorption: Fundamental principles, newly developed adsorbents and future perspectives. *Molecules (Basel, Switzerland), 28*(2), 639. DOI: 10.3390/molecules28020639

Liu, B., Zhou, C., Zheng, L., Duan, H., Chen, Y., & Wang, G. (2022). Metacommunity Concepts Provide New Insights in Explaining Zooplankton Spatial Patterns within Large Floodplain Systems. *Water (Basel), 14*(1), 1. Advance online publication. DOI: 10.3390/w14010093 PMID: 36798655

Liu, J., Bian, Y., Lawson, K., & Shen, C. (2024). Probing the limit of hydrologic predictability with the Transformer network. *Journal of Hydrology (Amsterdam), 637*, 131389. DOI: 10.1016/j.jhydrol.2024.131389

Liu, J., & Yang, H. (2017). Water scarcity assessments in the past, present, and future. *Earth's Future, 5*(6), 545–559. DOI: 10.1002/2016EF000518 PMID: 30377623

Liu, X., Chen, Z., Du, W., Liu, P., Zhang, L., & Shi, F. (2022). Treatment of wastewater containing methyl orange dye by fluidized three-dimensional electrochemical oxidation process integrated with chemical oxidation and adsorption. *Journal of Environmental Management, 311*, 114775. DOI: 10.1016/j.jenvman.2022.114775

Liu, X., Cui, X., Guo, L., Chen, J., Li, W., Yang, D., Cao, X., Chen, X., Liu, Q., & Lin, H. (2019). Non-invasive estimation of root zone soil moisture from coarse root reflections in ground-penetrating radar images. *Plant and Soil, 436*(1), 623–639. DOI: 10.1007/s11104-018-03919-5

Liu, Z., Xu, Z., Zhu, X., Yin, L., Yin, Z., Li, X., & Zheng, W. (2024). Calculation of carbon emissions in wastewater treatment and its neutralization measures: A review. *The Science of the Total Environment, 912*, 169356.

Liu, Z., Ying, J., He, C., Guan, D., Pan, X., Dai, Y., Gong, B., He, K., Lv, C., Wang, X., Lin, J., Liu, Y., & Bryan, B. (2024). Scarcity and quality risks for future global urban water supply. *Landscape Ecology, 39*. Advance online publication. DOI: 10.1007/s10980-024-01832-0

Li, W., Wei, S., Jiao, W., Qi, G., & Liu, Y. (2016). Modelling of adsorption in rotating packed bed using artificial neural networks (ANN). *Chemical Engineering Research & Design, 114*, 89–95. DOI: 10.1016/j.cherd.2016.08.013

Li, X. L., & Song, Y. (2025). What role does agriculture play in irrigation water footprint consumption? A perspective based on sectoral linkage and structural path. *Journal of Hydrology. Regional Studies, 59*, 102378. Advance online publication. DOI: 10.1016/j.ejrh.2025.102378

Li, X., Wu, Q., Chen, D., Tamjidur, R. S., Fan, P., Wang, G., & Zhang, W. (2025). Efficient chromium Cr (VI) removal from wastewater through modified cycad's leaf biochar: Insights into adsorption-reduction mechanisms and kinetic analysis. *Biomass and Bioenergy, 200*, 107994. DOI: 10.1016/j.biombioe.2025.107994

Loftus, A. (2018). Political ecology II: Whither the state? *Progress in Human Geography, 44*(1), 139–149. DOI: 10.1177/0309132518803421

Lu, L., & Luo, X. (2020). Emergency Transportation Problem Based on Single-Valued Neutrosophic Set. *Discrete Dynamics in Nature and Society, 2020*, 1–8. DOI: 10.1155/2020/4813497

Lu, P., Zhang, J. L., Al-Mahmood, A., Dung, H. Q., & Yunos, Z. M. (2025). Novel deep learning models for accurate time series prediction of phycocyanin concentration in water management. *Journal of Water Process Engineering, 74*, 107783. Advance online publication. DOI: 10.1016/j.jwpe.2025.107783

Madella, M., & Fuller, D. Q. (2006). Paleoecology and the Harappan civilisation of South Asia: A reconsideration. *Quaternary Science Reviews, 25*(11–12), 1283–1301.

Maestre, F. T., & Quero, J. L. (2012). Plant species richness and ecosystem multifunctionality in global drylands. *Science, 335*(6065), 214–218. DOI: 10.1126/science.1215442 PMID: 22246775

Mahesh, N., Babu, J. J., Nithya, K., & Arunmozhi, S. A. (2024). Water quality prediction using LSTM with combined normalizer for efficient water management. *Desalination and Water Treatment, 317*, 100183. Advance online publication. DOI: 10.1016/j.dwt.2024.100183

Mahizha, M., & M., I. (2025). A Comprehensive Approach to Solid Waste Management Site Selection Using Simplified Neutrosophic Distance-Based Similarity Measures with N-Valued T-Spherical Fuzzy Neutrosophic Sets. International Journal of Neutrosophic Science. .DOI: 10.54216/ijns.260125

Mahmoud, A. S., Mostafa, M. K., & Abdel-Gawad, S. A. (2017). Artificial intelligence for the removal of benzene, toluene, ethyl benzene and xylene (BTEX) from aqueous solutions using iron nanoparticles. *Water Science and Technology: Water Supply*, ws2017225. https://doi.org/DOI: 10.2166/ws.2017.225

Mahsereci, M., Balles, L., Lassner, C., & Hennig, P. (2017). *Early Stopping without a Validation Set* (Version 3). arXiv. https://doi.org/DOI: 10.48550/ARXIV.1703.09580

Maina, F. Z., Getirana, A., Kumar, S. V., Saharia, M., Biswas, N. K., McLarty, S., & Appana, R. (2024). Irrigation-driven groundwater depletion in the Ganges-Brahmaputra basin decreases the streamflow in the Bay of Bengal. *Communications Earth & Environment, 5*, 1348. Advance online publication. DOI: 10.1038/s43247-024-01348-0

Maji, P. K., Biswas, R., & Roy, A. R. (2001). Fuzzy soft sets. *Journal of Fuzzy Mathematics, 9*(3), 589–602.

Maji, P. K., Biswas, R., & Roy, A. R. (2002). An application of soft sets in decision-making problem. *Computers & Mathematics with Applications (Oxford, England), 44*(8–9), 1077–1083.

Maji, P. K., Biswas, R., & Roy, A. R. (2003). Soft set theory. *Computers & Mathematics with Applications (Oxford, England), 45*(4–5), 555–562.

Makropoulos, C. K., Butler, D., & Maksimovic, C. (2003). Fuzzy logic spatial decision support system for urban water management. *Journal of Water Resources Planning and Management, 129*(1), 69–77. (69)DOI: 10.1061/(asce)0733-9496(2003)129:1

Malviya, A., & Jaspal, D. (2021). Artificial intelligence as an upcoming technology in wastewater treatment: A comprehensive review. *Environmental Technology Reviews, 10*(1), 177–187.

Mangel, A. R., Moysey, S. M. J., & Bradford, J. (2020). Reflection tomography of time-lapse GPR data for studying dynamic unsaturated flow phenomena. *Hydrology and Earth System Sciences*, *24*(1), 159–167. DOI: 10.5194/hess-24-159-2020

Manna, S., Basu, T., & Mondal, S. (2020). A soft set based VIKOR approach for some decision-making problems under complex neutrosophic environment. *Engineering Applications of Artificial Intelligence*, *89*, 103432. DOI: 10.1016/j.engappai.2019.103432

Mansoum, R., & Chfadi, T. (2025). Innovation without participation? Moroccan farmers' perspectives on Haouz Mejjat aquifer contract. *Groundwater for Sustainable Development*. Advance online publication. DOI: 10.1016/j.gsd.2025.101477

Margiwiyatno, A., Sudarmaji, A., Siswantoro, S., Masrukhi, M., Irawadi, I., Wijaya, K., Kuncoro, P. H., Sulistyo, S. B., & Furqon, F. (2025). Use of Capacitive Sensors for Measuring Soil Water Content in Irrigation Systems: A Review. *BIO Web of Conferences, 158*, 01001. https://doi.org/DOI: 10.1051/bioconf/202515801001

Marshall, S. R. O., Tran, Thanh-Nhan-Duc, Tapas, Mahesh R., & and Nguyen, B. Q. (n.d.). Integrating artificial intelligence and machine learning in hydrological modeling for sustainable resource management. *International Journal of River Basin Management*, *0*(0), 1–17. DOI: 10.1080/15715124.2025.2478280

Mathaba, M., & Banza, J. (2023). A comprehensive review on artificial intelligence in water treatment for optimization. Clean water now and the future [Article]. *Journal of Environmental Science and Health Part a-Toxic/Hazardous Substances & Environmental Engineering, 58*(14), 1047-1060. https://doi.org/DOI: 10.1080/10934529.2024.2309102

Matheri, A. N., Belaid, M., Njenga, C. K., & Ngila, J. C. (2023). Water and wastewater digital surveillance for monitoring and early detection of the COVID-19 hotspot: Industry 4.0. *International Journal of Environmental Science and Technology*, *20*(1), 1095–1112.

Maurya, V. (2024). Advancements and Challenges in Hydrological Modelling. *International Journal for Research in Applied Science and Engineering Technology*, *12*(11), 746–751. DOI: 10.22214/ijraset.2024.65161

Mays, L. W. (Ed.). (2010). *Ancient water technologies*. Springer., DOI: 10.1007/978-90-481-8632-7

Ma, Z., Mei, G., & Xu, N. (2024). Generative deep learning for data generation in natural hazard analysis: Motivations, advances, challenges, and opportunities. *Artificial Intelligence Review*, *57*(6), 160. DOI: 10.1007/s10462-024-10764-9

Mbavarira, T. M., & Grimm, C. (2021). A Systemic View on Circular Economy in the Water Industry: Learnings from a Belgian and Dutch Case. *Sustainability*, *13*(6), 3313. DOI: 10.3390/su13063313

Mehta, L., & Veldwisch, G. J., & Franco, J. (2012). Introduction to the Special Issue: Water grabbing? Focus on the (re)appropriation of finite water resources. *Water Alternatives*, *5*(2), 193–207.

Mekonnen, M. M., & Hoekstra, A. Y. (2016). Four billion people facing severe water scarcity. *Science Advances*, *2*(2), e1500323. DOI: 10.1126/sciadv.1500323 PMID: 26933676

Memisevic, R., Zach, C., Pollefeys, M., & Hinton, G. E. (2010). Gated Softmax Classification. *Advances in Neural Information Processing Systems, 23*. https://papers.nips.cc/paper_files/paper/2010/hash/5737c6ec2e0716f3d8a7a5c4e0de0d9a-Abstract.html

Menaouer, B., Zoulikha, D., El-Houda, K. N., Mohammed, S., & Matta, N. (2022). Coronavirus Pneumonia Classification Using X-Ray and CT Scan Images With Deep Convolutional Neural Network Models. *Journal of Information Technology Research*, *15*(1), 1–23. DOI: 10.4018/JITR.299391

Meng, X., Huang, Z. G., Deng, X., Liu, H., Fang, H. Y., Liu, C., Zhang, X. Y., & Cui, J. (2025). Leakage detection and localization of buried water pipe using ground penetrating radar. *Measurement*, *254*, 117902. Advance online publication. DOI: 10.1016/j.measurement.2025.117902

Micklin, P. (2007). The Aral Sea disaster. *Annual Review of Earth and Planetary Sciences*, *35*, 47–72.

Mikołajewska, E., Mikołajewski, D., Mikołajczyk, T., & Paczkowski, T. (2025). Generative AI in AI-Based Digital Twins for Fault Diagnosis for Predictive Maintenance in Industry 4.0/5.0. *Applied Sciences (Basel, Switzerland)*, *15*(6), 6. Advance online publication. DOI: 10.3390/app15063166

Minea, G., Ciobotaru, N., & Ioana-Toroimac, G. (2022). Designing grazing susceptibility to land degradation index (GSLDI) in hilly areas. *Scientific Reports*, *12*(1), 9393. DOI: 10.1038/s41598-022-13543-8 PMID: 35729181

Mishra, S., & Sundaram, B. (2024). A review of the photocatalysis process used for wastewater treatment. *Materials Today: Proceedings*, *102*, 393–409.

Mithal, A., & Thankachan, B. (2025). A Comprehensive Review on Blockchain-based Systems for Groundwater Conservation and Wastewater Management. *Environmental Management*, •••, 1–19.

Modarres, R., & da Silva, V. R. (2007). Rainfall trends in arid and semi-arid regions of Iran. *Journal of Arid Environments, 70*(2), 344–355. DOI: 10.1016/j.jaridenv.2006.12.024

Mohammadi, E., Stokholm-Bjerregaard, M., Hansen, A. A., Nielsen, P. H., Ortiz-Arroyo, D., & Durdevic, P. (2024). Deep learning based simulators for the phosphorus removal process control in wastewater treatment via deep reinforcement learning algorithms. *Engineering Applications of Artificial Intelligence, 133*, 107992.

Mohammadi, S., Darestani, S., Vahdani, B., & Alinezhad, A. (2020). A robust neutrosophic fuzzy-based approach to integrate reliable facility location and routing decisions for disaster relief under fairness and aftershocks concerns. *Computers & Industrial Engineering, 148*, 106734. DOI: 10.1016/j.cie.2020.106734

Mohammed, S., Arshad, S., Alsilibe, F., Moazzam, M. F. U., Bashir, B., Prodhan, F. A., Alsalman, A., Vad, A., Ratonyi, T., & Harsányi, E. (2024). Utilizing machine learning and CMIP6 projections for short-term agricultural drought monitoring in central Europe (1900–2100). *Journal of Hydrology (Amsterdam), 633*, 130968. DOI: 10.1016/j.jhydrol.2024.130968

Mohammed, S., Arshad, S., Bashir, B., Ata, B., Al-Dalahmeh, M., Alsalman, A., Ali, H., Alhennawi, S., Kiwan, S., & Harsanyi, E. (2024). Evaluating machine learning performance in predicting sodium adsorption ratio for sustainable soil-water management in the eastern Mediterranean. *Journal of Environmental Management, 370*, 122640. Advance online publication. DOI: 10.1016/j.jenvman.2024.122640 PMID: 39340889

Mohan, R. N. V. J., Rayanoothala, P. S., & Sree, R. P. (2025). Next-gen agriculture: Integrating AI and XAI for precision crop yield predictions. *Frontiers in Plant Science, 15*, 1451607. DOI: 10.3389/fpls.2024.1451607 PMID: 39845494

Mojiri, A., Andasht Kazeroon, R., & Gholami, A. (2019). Cross-Linked Magnetic Chitosan/Activated Biochar for Removal of Emerging Micropollutants from Water: Optimization by the Artificial Neural Network. *Water (Basel), 11*(3), 551. DOI: 10.3390/w11030551

Molle, F., & Mollinga, P. (2003). Water poverty indicators: Conceptual problems and policy issues. *Water Policy, 5*(5–6), 529–544.

Molle, F., & Sanchis-Ibor, C. (2019). Irrigation policies in the Mediterranean: trends and challenges. In Molle, F., Sanchis-Ibor, C., & Avellà-Reus, L. (Eds.), *Irrigation in the Mediterranean. Global challenges for water policy* (Vol. 22). Springer., DOI: 10.1007/978-3-030-03698-0_10

Molodtsov, D. (1999). Soft set theory: First results. *Computers & Mathematics with Applications (Oxford, England)*, *37*(4–5), 19–31.

Molodtsov, D. (1999). Soft set theory-first results. *Computers & Mathematics with Applications (Oxford, England)*, *37*(4–5), 19–31.

Mondal, P. (2024). AI and IoT in smart water management for urban sustainability. *Uncertainty Discourse and Applications*, *1*(2), 151–157. DOI: 10.48313/uda.v1i2.36

Morchid, A., Said, Z., Abdelaziz, A. Y., Siano, P., & Qjidaa, H. (2025). Fuzzy logic-based IoT system for optimizing irrigation with cloud computing: Enhancing water sustainability in smart agriculture. *Smart Agricultural Technology*, *11*, 100979. Advance online publication. DOI: 10.1016/j.atech.2025.100979

Mostafapour, F. K., Yilmaz, M., Mahvi, A. H., Younesi, A., Ganji, F., & Balarak, D. (2022). Adsorptive removal of tetracycline from aqueous solution by surfactant-modified zeolite: Equilibrium, kinetics and thermodynamics. *Desalination and Water Treatment*, *247*, 216–228. DOI: 10.5004/dwt.2022.27943

Moutaouekkil, M. A., Karkri, A., Koulali, M. A., Taybi, C., & Kahlaoui, M. (2024). Plant Root Characterization Using Ground-Penetrating Radar with Deep Learning. *Arabian Journal for Science and Engineering*. Advance online publication. DOI: 10.1007/s13369-024-09502-8

Mukherjee, A., & Das, A. K. (2015b). Interval valued intuitionistic fuzzy soft multi set theoretic approach to decision making problems, 2015 International Conference on Computer, Communication and Control (IC4), Indore, India, 2015, pp. 1-5, https://doi.org/DOI: 10.1109/IC4.2015.7375640

Mukherjee, A., & Das, A. K. (2016a). Application of fuzzy soft multi sets in decision-making problems. In: Nagar, A., Mohapatra, D., Chaki, N. (eds) Proceedings of 3rd International Conference on Advanced Computing, Networking and Informatics. Smart Innovation, Systems and Technologies, vol. 43. Springer, New Delhi. https://doi.org/DOI: 10.1007/978-81-322-2538-6_3

Mukherjee, A., & Das, A. K. (2023). Essentials of fuzzy soft multisets: Theory and applications, Springer Nature Singapore Pte Ltd., 1-151. ISBN: 978-981-19-2759-1, https://doi.org/DOI: 10.1007/978-981-19-2760-7

Mukherjee, A., & Das, A. K. (2015a). Application of interval valued intuitionistic fuzzy soft set in investment decision making, *2015 Fifth International Conference on Advances in Computing and Communications (ICACC)*, Kochi, India, 2015, pp. 61-64, https://doi.org/DOI: 10.1109/ICACC.2015.37

Mukherjee, A., & Das, A. K. (2015c). Relations on intuitionistic fuzzy soft multi sets. In Kim, K. (Ed.), *Information Science and Applications. Lecture Notes in Electrical Engineering* (Vol. 339). Springer., DOI: 10.1007/978-3-662-46578-3_71

Mukherjee, A., & Das, A. K. (2016b). Topological structure formed by intuitionistic fuzzy rough relations. *Journal of the Indian Mathematical Society, 83*(1-2), 135–144.

Mukherjee, A., & Das, A. K. (2022). Einstein operations on fuzzy soft multisets and decision making. *Boletim da Sociedade Paranaense de Matematica, 40*, 1–10. DOI: 10.5269/bspm.32546

Mumbi, A. W., Li, F., Bavumiragira, J. P., & Fangninou, F. F. (2021). Forecasting water consumption on transboundary water resources for water resource management using the feed-forward neural network: A case study of the Nile River in Egypt and Kenya. *Marine and Freshwater Research, 73*(3), 292–306. DOI: 10.1071/MF21118

Myllis, G., Tsimpiris, A., & Vrana, V. (2024). Short-Term Water Demand Forecasting from Univariate Time Series of Water Reservoir Stations. *Information (Basel), 15*(10), 10. Advance online publication. DOI: 10.3390/info15100605

Naderi, B., & Ruiz, R. (2010). The distributed permutation flowshop scheduling problem. *Computers & Operations Research, 37*(4), 754–768.

Nair, M., & Kumar, D. (2013). Water desalination and challenges: The Middle East perspective: a review. *Desalination and Water Treatment, 51*(10-12), 2030–2040.

Namdari, H., Moradikia, M., Zekavat, S., Askari, R., Mangoubi, O., & Petkie, D. (2025). Advancing Precision Agriculture: Machine Learning-Enhanced GPR Analysis for Root-Zone Soil Moisture Assessment in Mega Farms. *IEEE Transactions on AgriFood Electronics, 3*(1), 98–109. DOI: 10.1109/TAFE.2024.3455238

Nandy, T., Idris, M. Y. I. B., Noor, R. M., Bhattacharyya, S., & Ghani, N. B. A. (2020, December 20-21). 2020). Collaborative Data Anonymization for Privacy-Preserving Vehicular Ad-hoc Network. 2020 International Conference on Innovation and Intelligence for Informatics, Computing and Technologies (3ICT)

Nandy, T., Idris, M. Y. I., Noor, R. M., Wahab, A. W. A., Bhattacharyya, S., Kolandaisamy, R., & Yahuza, M. (2021). A Secure, Privacy-Preserving, and Lightweight Authentication Scheme for VANETs. *IEEE Sensors Journal, 21*(18), 20998–21011. DOI: 10.1109/JSEN.2021.3097172

Nandy, T., Kolandaisamy, R., Noor, R. M., & Bhattacharyya, S. (2025). An Adaptive Vehicle Location Prediction Using Machine Learning: A Case Study of Campus Shuttle Bus. *IEEE Access : Practical Innovations, Open Solutions, 13*, 78290–78302. DOI: 10.1109/ACCESS.2025.3562785

Nandy, T., Md Noor, R., Kolandaisamy, R., Idris, M. Y. I., & Bhattacharyya, S. (2024). A review of security attacks and intrusion detection in the vehicular networks. *Journal of King Saud University. Computer and Information Sciences*, *36*(2), 101945. https://doi.org/doi.org/10.1016/j.jksuci.2024.101945

Nandy, T., & Nyundo, A. (2023, June 16-17). 2023). Prediction of Possible Outcomes using Big Data Analysis and Machine Learning. *2023 International Conference on Applied Intelligence and Sustainable Computing (ICAISC)*

Narain, V. (2010). Gone with the aquifer: Coping with groundwater depletion in western India. *Local Environment*, *15*(2), 109–120. DOI: 10.1080/13549830903575508

Narayanamoorthy, A. (2006). *Potential for drip and sprinkler irrigation in India* (Working Paper No. 7). Gokhale Institute of Politics and Economics.

Nasr, S. H. (2006). *Science and civilization in Islam*. Harvard University Press.

Nature Water. (2023). Selected articles on governance, technology, and water security. *Nature Water*. https://www.nature.com/natwater

Nawaz, A., & Kumar, P. (2022). Optimization of process parameters of Lagerstroemia speciosa seed hull pyrolysis using a combined approach of Response Surface Methodology (RSM) and Artificial Neural Network (ANN) for renewable fuel production. *Bioresource Technology Reports*, *18*, 101110–101110. DOI: 10.1016/j.biteb.2022.101110

Nawaz, M., Enscore, E. E.Jr, & Ham, I. (1983). A heuristic algorithm for the m-machine, n-job flow-shop sequencing problem. *Omega*, *11*(1), 91–95.

Nejjarou, O., Aqil, S., & Lahby, M. (2022). Constructive heuristics and mathematical formulation for solving the permutation flow shop scheduling problem with setup time. In 2022 IEEE 3rd International Conference on Electronics, Control, Optimization and Computer Science (ICECOCS) (pp. 1-5). IEEE.

Nejjarou, O., Aqil, S., & Lahby, M. (2024). Hybrid meta-heuristic solving no-wait flow shop scheduling minimizing maximum tardiness. *Evolutionary Intelligence*, •••, 1–25.

Neupane, J., & Guo, W. (2019). Agronomic basis and strategies for precision water management: A review. *Agronomy (Basel)*, *9*(2), 87. DOI: 10.3390/agronomy9020087

Ng, K. W., Huang, Y. F., Koo, C. H., Chong, K. L., El-Shafie, A., & Najah Ahmed, A. (2023). A review of hybrid deep learning applications for streamflow forecasting. *Journal of Hydrology (Amsterdam)*, *625*, 130141. DOI: 10.1016/j.jhydrol.2023.130141

Nguyen, G., Son, L., Ashour, A., & Dey, N. (2017). A survey of the state-of-the-arts on neutrosophic sets in biomedical diagnoses. *International Journal of Machine Learning and Cybernetics*, *10*, 1–13. DOI: 10.1007/s13042-017-0691-7

Nguyen, V. D., Nguyen, H. T. H., Vranova, V., Nguyen, L. T. N., Bui, Q. M., & Khieu, T. T. (2020). Artificial neural network modeling for Congo red adsorption on microwave-synthesized akaganeite nanoparticles: Optimization, kinetics, mechanism, and thermodynamics. *Environmental Science and Pollution Research International*, *28*(8), 9133–9145. DOI: 10.1007/s11356-020-10633-2

Nile Basin Initiative. (2017). *Nile Basin Initiative 10-year strategy: 2017–2027*. NBI Secretariat. https://nilebasin.org

Nirmala, N., Arun, J., Sanjay Kumar, S., & Dawn, S. S. (2025). Role of Machine Learning and Artificial Intelligence in Smart Waste Management. In *Smart Waste and Wastewater Management by Biotechnological Approaches* (pp. 35–53). Springer Nature Singapore.

Niu, C. (2025). *An interpretable hybrid deep learning model for flood forecasting based on Transformer and LSTM*. ResearchGate., DOI: 10.1016/j.ejrh.2024.101873

Noori, A., Bonakdari, H., Salimi, A., & Gharabaghi, B. (2020). A group Multi-Criteria Decision-Making method for water supply choice optimization. *Socio-Economic Planning Sciences*, *101006*. Advance online publication. DOI: 10.1016/j.seps.2020.101006

Noor, R. M., Rasyidi, N. B. G., Nandy, T., & Kolandaisamy, R. (2021). Campus Shuttle Bus Route Optimization Using Machine Learning Predictive Analysis: A Case Study. *Sustainability*, *13*(1), 225. DOI: 10.3390/su13010225

Nourani, V., & Najafi, H. (2022). A Z-number based multi-attribute decision-making algorithm for hydro-environmental system management. *Neural Computing & Applications*, *35*, 6405–6421. DOI: 10.1007/s00521-022-08025-3

O'Connor, K. F., Al-Abed, S. R., Hordern, S., & Pinto, P. X. (2022). Assessing the efficiency and mechanism of zinc adsorption onto biochars from poultry litter and softwood feedstocks. *Bioresource Technology Reports*, *18*, 101039. DOI: 10.1016/j.biteb.2022.101039

OECD. (2021). *OECD studies on water: Water governance in Israel*. OECD Publishing., DOI: 10.1787/0f315e3f-

Office of Environmental Health Hazard Assessment. (2022). *Changes in drought*. https://oehha.ca.gov/climate-change/epic-2022/changes-climate/drought

Omar, R. M., & Zeina, M. (2023). Neutrosophic Multinominal Logistic Regression Technique for Optimizing Adaptive Reuse of Historical Castles. International Journal of Neutrosophic Science. DOI: 10.54216/ijns.210305

Organisation for Economic Co-operation and Development (OECD). (2012). *Environmental outlook to 2050: The consequences of inaction*. OECD Publishing., DOI: 10.1787/9789264122246-

Organisation for Economic Co-operation and Development (OECD). (2019). *Managing water for all: An OECD perspective on pricing and governance*. OECD Publishing.

Ostrom, E. (1990). *Governing the commons: The evolution of institutions for collective action*. Cambridge University Press.

Padhiary, M. (2025). Benefits, Risks, and Responsible Use of Nanomaterials in Wastewater Remediation. In *Nanomaterials in Wastewater Research* (pp. 443–464). Springer.

Padhiary, M., Roy, D., & Dey, P. (2025). Mapping the landscape of biogenic nanoparticles in bioinformatics and nanobiotechnology: AI-driven insights. In *Synthesizing and Characterizing Plant-Mediated Biocompatible Metal Nanoparticles* (pp. 337–376). IGI Global.

Pal, D. (2025). *(PDF) Blueprint conceptualization for a river basin's digital twin*. ResearchGate., DOI: 10.2166/nh.2025.111

Pandit, J., & Sharma, A. K. (2024). Advanced Techniques in Wastewater Treatment: A Comprehensive review. *Asian Journal of Environment & Ecology*, *23*(10), 1–26. DOI: 10.9734/ajee/2024/v23i10605

Pandit, R., & Parrotta, J. A. (2020). A framework to evaluate land degradation and restoration responses for improved planning and decision-making. *Ecosystems and People (Abingdon, England)*, *16*(1), 1–18. DOI: 10.1080/26395916.2019.1696410

Pang, J., Luo, W., Yao, Z., Chen, J., Dong, C., & Lin, K. (2024). Water Quality Prediction in Urban Waterways Based on Wavelet Packet Denoising and LSTM. *Water Resources Management*, *38*(7), 2399–2420. DOI: 10.1007/s11269-024-03774-3

Pan, Q. K., Ruiz, R., & Alfaro-Fernández, P. (2017). Iterated search methods for earliness and tardiness minimization in hybrid flowshops with due windows. *Computers & Operations Research*, *80*, 50–60.

Parajuli, B., Zhang, X., Deuja, S., & Liu, Y. (2021). Regional and seasonal precipitation and drought trends in Ganga–Brahmaputra Basin. *Water (Basel), 13*(16), 2218. DOI: 10.3390/w13162218

Park, J., Baik, J.-H., Adjei-Nimoh, S., & Lee, W. H. (2025). Advancements in artificial intelligence-based technologies for PFAS detection, monitoring, and management. *The Science of the Total Environment, 980*, 179536–179536. DOI: 10.1016/j.scitotenv.2025.179536

Parsa, Z., Dhib, R., & Mehrvar, M. (2024). Dynamic modelling, process control, and monitoring of selected biological and advanced oxidation processes for wastewater treatment: A review of recent developments. *Bioengineering (Basel, Switzerland), 11*(2), 189.

Pastor, A. V., Palazzo, A., Havlik, P., Biemans, H., Wada, Y., Obersteiner, M., & Ludwig, F. (2019). The global nexus of food–trade–water sustaining environmental flows by 2050. *Nature Sustainability, 2*(6), 499–507. DOI: 10.1038/s41893-019-0287-1

Patel, H. (2021). Review on solvent desorption study from exhausted adsorbent. *Journal of Saudi Chemical Society, 25*(8), 101302. DOI: 10.1016/j.jscs.2021.101302

Pathways for Hydrological Resilience: Strategies for Adaptation in a Changing Climate | Earth Systems and Environment. (n.d.). Retrieved May 28, 2025, from https://link.springer.com/article/10.1007/s41748-024-00567-x

Peter-Varbanets, M., Zurbrügg, C., Swartz, C., & Pronk, W. (2009). Decentralized systems for potable water and the potential of membrane technology. *Water Research, 43*(2), 245–265. PMID: 19010511

Picos-Benítez, A. R., Martínez-Vargas, B. L., Duron-Torres, S. M., Brillas, E., & Peralta-Hernández, J. M. (2020). The use of artificial intelligence models in the prediction of optimum operational conditions for the treatment of dye wastewaters with similar structural characteristics. *Process Safety and Environmental Protection, 143*, 36–44.

Pietz, D. A. (2015). *The Yellow River: The problem of water in modern China.* Harvard University Press.

Plaza, C., & Zaccone, C. (2018). Soil resources and element stocks in drylands to face global issues. *Scientific Reports, 8*(1), 13788. DOI: 10.1038/s41598-018-32229-0 PMID: 30214005

Pokhrel, Y., Felfelani, F., Satoh, Y., Boulange, J., Burek, P., Gädeke, A., Gerten, D., Gosling, S. N., Grillakis, M., Gudmundsson, L., Hanasaki, N., Kim, H., Koutroulis, A. G., Liu, J., Papadimitriou, L., Schewe, J., & Veldkamp, T. I. E. (2021). Global terrestrial water storage and drought severity under climate change. *Nature Climate Change*, *11*, 226–233. DOI: 10.1038/s41558-020-00972-w

Popescu, S. M., Mansoor, S., Wani, O. A., Kumar, S. S., Sharma, V., Sharma, A., & Chung, Y. S. (2024). Artificial intelligence and IoT driven technologies for environmental pollution monitoring and management. *Frontiers in Environmental Science*, *12*, 1336088.

Prabhaharan, G., Khan, B. S. H., & Rakesh, L. (2006). Implementation of grasp in flow shop scheduling. *International Journal of Advanced Manufacturing Technology*, *30*, 1126–1131.

Prabu, R. T., Sarkar, M., Chaudhary, D., Al Obaid, S., Al-ateeq, T. K., & Kalam, M. A. (2024). IoT-enabled groundwater monitoring with k-NN-SVM algorithm for sustainable water management. *Acta Geophysica*, *72*(4), 2715–2728. DOI: 10.1007/s11600-023-01178-2

Pratap, B., Kumar, S., Nand, S., Azad, I., Bharagava, R. N., Ferreira, L. F. R., & Dutta, V. (2022). Wastewater generation and treatment by various ecofriendly technologies: Possible health hazards and further reuse for environmental safety. *Chemosphere*, *313*, 137547. DOI: 10.1016/j.chemosphere.2022.137547

Pravalie, R. (2016). Drylands extent and environmental issues: A global approach. *Earth-Science Reviews*, *161*, 259–278. DOI: 10.1016/j.earscirev.2016.08.003

Prăvălie, R., Bandoc, G., Patriche, C., & Sternberg, T. (2019). Recent changes in global drylands: Evidences from two major aridity databases. *Catena*, *178*, 209–231. DOI: 10.1016/j.catena.2019.03.016

Preite, L., & Vignali, G. (2024). Artificial intelligence to optimize water consumption in agriculture: A predictive algorithm-based irrigation management system [Article]. *Computers and Electronics in Agriculture*, *223*(11), 109126. Advance online publication. DOI: 10.1016/j.compag.2024.109126

Prieto, C., Patel, D., Han, D. W., Dewals, B., Bray, M., & Molinari, D. (2024). Preface: Advances in pluvial and fluvial flood forecasting and assessment and flood risk management. *Natural Hazards and Earth System Sciences*, *24*(10), 3381–3386. DOI: 10.5194/nhess-24-3381-2024

Qin, Y., Tang, Q., Xin, J., Yang, C., Zhang, Z., & Yang, X. (2023). A Rapid Identification Technique of Moving Loads Based on MobileNetV2 and Transfer Learning. *Buildings (Basel, Switzerland)*, *13*(2), 572. DOI: 10.3390/buildings13020572

Radio France Internationale (RFI). (2023, September 11). Egypt brands Ethiopia's move to fill Nile mega dam as 'illegal'. https://www.rfi.fr/en/africa/20230911-egypt-brands-ethiopia-s-move-to-fill-nile-mega-dam-as-illegal

Radio Tamazuj. (2023, September 12). Egypt writes to UN Security Council over Ethiopia filling dam, threatens action. https://www.radiotamazuj.org/en/news/article/egypt-writes-to-un-security-council-over-ethiopia-filling-dam-threatens-action

Radmehr, A., Bozorg-Haddad, O., & Loáiciga, H. (2022). Integrated strategic planning and multi-criteria decision-making framework with its application to agricultural water management. *Scientific Reports*, *12*. Advance online publication. DOI: 10.1038/s41598-022-12194-5 PMID: 35589906

Raimi, R., & Phiri, A. (2025). From pixels to policy: Evaluating the behavioural impact and cost effectiveness of picture nudges in water conservation at university residences. *Cogent Social Sciences*, *11*(1), 2464873. Advance online publication. DOI: 10.1080/23311886.2025.2464873

Rajasekaran, U., Kothandaraman, M., & Pua, C. H. (2025). Water Pipeline Leak Detection and Localization With an Integrated AI Technique [Article]. *IEEE Access : Practical Innovations, Open Solutions*, *13*, 2736–2745. DOI: 10.1109/access.2024.3524669

Ramlee, N. N., Illias, R. M., Toemen, S., Rahman, R. A., Selvasembian, R., Fatriasari, W., & Azelee, N. I. W. (2024). Adsorption study of isotherms, kinetics modelling, and thermodynamics of immobilized Candida rugosa lipase on metal oxide for esterification of waste cooking oil. Chemical Engineering Research and Design, 204(December 2023), 249–268. https://doi.org/DOI: 10.1016/j.cherd.2024.02.031

Rasheed, M. W., Tang, J., Sarwar, A., Shah, S., Saddique, N., Khan, M. U., Imran Khan, M., Nawaz, S., Shamshiri, R. R., Aziz, M., & Sultan, M. (2022). Soil Moisture Measuring Techniques and Factors Affecting the Moisture Dynamics: A Comprehensive Review. *Sustainability*, *14*(18), 11538. DOI: 10.3390/su141811538

Rasiya Koya, S., & Roy, T. (2024). Temporal Fusion Transformers for streamflow Prediction: Value of combining attention with recurrence. *Journal of Hydrology (Amsterdam)*, *637*, 131301. DOI: 10.1016/j.jhydrol.2024.131301

Ravi, R., Al-Attar, B., Qudr, L., Abdulbaqi, A., Tawfeq, J., Sekhar, R., Shah, P., & Devaerakkam, M. (2025). Integrating Neutrosophic Theory for Improved Decision-Making in Wireless Body Area Networks: Enhancing Accuracy and Efficiency in Health Monitoring. International Journal of Neutrosophic Science. .DOI: 10.54216/ijns.250325

Ravi, R., Al-Attar, B., Qudr, L., Penubadi, H., Sekhar, R., Shah, P., Parihar, S., Kallam, S., Tawfeq, J., & Gheni, H. (2024). Efforts of Neutrosophic Logic in Medical Image Processing and Analysis. International Journal of Neutrosophic Science. .DOI: 10.54216/ijns.240428

Ravikumar, A., & Sriraman, H. (2023). Understanding Convolutional Neural Network With TensorFlow: CNN. In Baby Maruthi, P., Prasad, S., & Tyagi, A. K. (Eds.), (pp. 50–73). Advances in Systems Analysis, Software Engineering, and High Performance Computing. IGI Global., DOI: 10.4018/978-1-6684-8531-6.ch003

Renfrew, D., Vasilaki, V., & Katsou, E. (2024). Indicator based multi-criteria decision support systems for wastewater treatment plants. *The Science of the Total Environment*, *915*, 169903.

Renu, N., & Sithole, T. (2024). A review on regeneration of adsorbent and recovery of metals: Adsorbent disposal and regeneration mechanism. *South African Journal of Chemical Engineering*, *50*, 39–50. DOI: 10.1016/j.sajce.2024.07.006

Review, W. P. (2025). *Fresh water by Country 2024*. World Population Review. https://worldpopulationreview.com/country-rankings/fresh-water-by-country

Reyes-Acosta, R., Dominguez-Baez, C., Mendoza-Gonzalez, R., & Martin, M. V. (2025). Analysis of machine learning-based approaches for securing the Internet of Things in the smart industry: a multivocal state of knowledge review [Article]. *International Journal of Information Security, 24*(1), 31, Article 31. https://doi.org/ DOI: 10.1007/s10207-024-00935-8

Ribas, I., Companys, R., & Tort-Martorell, X. (2017). Efficient heuristics for the parallel blocking flow shop scheduling problem. *Expert Systems with Applications*, *74*, 41–54.

Ricart, S., Berizzi, C., Saurí, D., & Terlicher, G. N. (2022). The Social, Political, and Environmental Dimensions in Designing Urban Public Space from a Water Management Perspective: Testing European Experiences. *Land (Basel)*, *11*(9), 1575. Advance online publication. DOI: 10.3390/land11091575

Rinot, Y. (2022). Israel's water strategy: Turning scarcity into abundance. *Water International*, *47*(3), 250–265.

Rodell, M., Velicogna, I., & Famiglietti, J. S. (2009). Satellite-based estimates of groundwater depletion in India. *Nature*, *460*(7258), 999–1002. DOI: 10.1038/nature08238 PMID: 19675570

Rodríguez-Hernández, J. A., Araújo, R. G., López-Pacheco, I. Y., Rodas-Zuluaga, L. I., González-González, R. B., Parra-Arroyo, L., & Parra-Saldívar, R. (2022). Environmental persistence, detection, and mitigation of endocrine disrupting contaminants in wastewater treatment plants–a review with a focus on tertiary treatment technologies. *Environmental Science. Advances*, *1*(5), 680–704.

Rogers, P., de Silva, R., & Bhatia, R. (2002). Water is an economic good: How to use prices to promote equity, efficiency, and sustainability. *Water Policy*, *4*(1), 1–17.

Rosa, L., Chiarelli, D. D., Rulli, M. C., Dell'Angelo, J., & D'Odorico, P. (2020). Global agricultural economic water scarcity. *Science Advances*, *6*(18), eaaz6031. Advance online publication. DOI: 10.1126/sciadv.aaz6031 PMID: 32494678

Rosa, L., Chiarelli, D. D., Sangiorgio, M., Beltran-Peña, A. A., Rulli, M. C., D'Odorico, P., & Fung, I. (2020). Potential for sustainable irrigation expansion in a 3°C warmer climate. *Proceedings of the National Academy of Sciences of the United States of America*, *117*(47), 29526–29534. DOI: 10.1073/pnas.2017796117 PMID: 33168728

Roy, A. R., & Maji, P. K. (2007). A fuzzy soft set theoretic approach to decision making problems. *Journal of Computational and Applied Mathematics*, *203*, 412–418.

Royal Institute for Strategic Studies (IRES). (2022). What future for water in Morocco? Summary report. Rabat: IRES. https://www.ires.ma/sites/default/files/docs_publications/Avenir_de_leau_au_Maroc-Rapport_de_synthese.pdf

Ruiz, R., & Stützle, T. (2007). A simple and effective iterated greedy algorithm for the permutation flowshop scheduling problem. *European Journal of Operational Research*, *177*(3), 2033–2049.

Sadek, A. H., & Fahmy, O. M., Nasr, M., & Mostafa, M. K. (2023). Predicting Cu(II) Adsorption from Aqueous Solutions onto Nano Zero-Valent Aluminum (nZVAl) by Machine Learning and Artificial Intelligence Techniques. *Sustainability*, *15*(3), 2081–2081. DOI: 10.3390/su15032081

Sadik, S. A. (2024). Deep Learning Classification of Empty and Full Container Ships in Satellite Imagery: A Cognitive Modeling Approach for Maritime Applications. In Ahmad, S., Anwar, M. S., Alluhaidan, A. S., & El-Affendi, M. A. (Eds.), (pp. 30–54). Advances in Geospatial Technologies. IGI Global., DOI: 10.4018/979-8-3693-2913-9.ch002

Sadki, H., & Allali, K. (2024). Artificial neural network for solving flow shop optimization problem with sequence independent setup time. *Foundations of Computing and Decision Sciences, 49.*

Sadki, H., & Allali, K. (2025). Optimal Population-Based Metaheuristics with NEH and ANN Algorithms for Minimizing Total Tardiness in Flow Shop Scheduling Problem. In *Innovations in Optimization and Machine Learning* (pp. 99–124). IGI Global Scientific Publishing.

Saghiry, S., Loudyi, D., & Laassilia, O. (2025). An Integrated Conceptual Framework for Interconnected Dam Reservoirs Systems. *Journal of Sustainability Research, 7.* Advance online publication. DOI: 10.20900/jsr20250017

Salem, A. (2017). Sharing water in Morocco: management of a common resource or public property? Communication, CAGE Summer School, University of Warwick. https://warwick.ac.uk/fac/soc/economics/research/centres/cage/events/11-07-17-cage_summer_school_2017/details/arianne_salem_article.pdf

Salman, S. M. A. (2007). The United Nations Watercourses Convention ten years later: Why has its entry into force proven difficult? *Water International, 32*(1), 1–15.

Sandler, M., Howard, A., Zhu, M., Zhmoginov, A., & Chen, L.-C. (2018). MobileNetV2: Inverted Residuals and Linear Bottlenecks. *2018 IEEE/CVF Conference on Computer Vision and Pattern Recognition*, 4510–4520. https://doi.org/DOI: 10.1109/CVPR.2018.00474

Santos, E., Carvalho, M., & Martins, S. (2023). Sustainable Water Management: Understanding the Socioeconomic and Cultural Dimensions. *Sustainability, 15*(17), 13074. Advance online publication. DOI: 10.3390/su151713074

Saravanan, A., Kumar, P. S., Jeevanantham, S., Karishma, S., Tajsabreen, B., Yaashikaa, P., & Reshma, B. (2021). Effective water/wastewater treatment methodologies for toxic pollutants removal: Processes and applications towards sustainable development. *Chemosphere, 280*, 130595. DOI: 10.1016/j.chemosphere.2021.130595

Sarma, D., Das, A., & Bera, U. (2025). *A transportation model minimizing total cost and time in disaster struck zone for debris removal under a mixed neutrosophic environment.* OPSEARCH. DOI: 10.1007/s12597-025-00912-2

Satilmisoglu, T. K., Sermet, Y., Kurt, M., & Demir, I. (2024). Blockchain Opportunities for Water Resources Management: A Comprehensive Review. *Sustainability, 16*(6), 2403. Advance online publication. DOI: 10.3390/su16062403

Sayah, A., Aqil, S., & Lahby, M. (2023). Minimizing Maximum Tardiness in a Distributed Flow Shop Manufacturing Problem under No-Waiting and Sequence Dependent Setup Time Constraints. In 2023 14th International Conference on Intelligent Systems: Theories and Applications (SITA) (pp. 1-6). IEEE.

Sayah, A., Aqil, S., & Lahby, M. (2025b). Hybrid Metaheuristics-Driven Distributed Task Scheduling for Latency-Sensitive Edge Data Processing. In *2025 12th IFIP International Conference on New Technologies, Mobility and Security (NTMS)* (pp. 129-135). IEEE.

Sayah, A., Aqil, S., Aadil, B., & Lahby, M. (2024b). Two Hybrid Heuristic Algorithms for Optimizing Data Processing Schedules in Distributed Edge Computing. In *2024 11th International Conference on Wireless Networks and Mobile Communications (WINCOM)* (pp. 1-6). IEEE.

Sayah, A., Aqil, S., & Lahby, M. (2024a). Green Algorithms for Energy-Efficient Distributed Flow-Shop Manufacturing Problem. *Procedia Computer Science*, *236*, 378–385.

Sayah, A., Aqil, S., & Lahby, M. (2025a). Improved bio-inspired algorithms for scheduling distributed no-waiting flow shop with setup times. *Evolutionary Intelligence*, *18*(4), 68.

Scanlon, B. R., Ruddell, B. L., Reed, P. M., Hook, R. I., Zheng, C., Tidwell, V. C., & Siebert, S. (2017). The food-energy-water nexus: Transforming science for society. *Water Resources Research*, *53*(5), 3550–3564. DOI: 10.1002/2017WR020889

Scavia, D., Allan, J. D., Arend, K. K., Bartell, S., Beletsky, D., Bosch, N. S., & Zhou, Y. (2014). Assessing and addressing the re-eutrophication of Lake Erie: Central basin hypoxia. *Journal of Great Lakes Research*, *40*(2), 226–246. DOI: 10.1016/j.jglr.2014.02.004

Schreiner, B., & van Koppen, B. (2001). Catchment management agencies for poverty eradication in South Africa. *Physics and Chemistry of the Earth. Part B: Hydrology, Oceans and Atmosphere*, *26*(11–12), 969–976.

Schwarzenbach, R. P., Escher, B. I., Fenner, K., Hofstetter, T. B., Johnson, C. A., von Gunten, U., & Wehrli, B. (2006). The challenge of micropollutants in aquatic systems. *Science*, *313*(5790), 1072–1077. DOI: 10.1126/science.1127291 PMID: 16931750

Seibert, J., & Bergström, S. (2022). A retrospective on hydrological catchment modelling based on half a century with the HBV model. *Hydrology and Earth System Sciences*, *26*(5), 1371–1388. DOI: 10.5194/hess-26-1371-2022

Sela, L., Sowby, R. B., Salomons, E., & Housh, M. (2025). Making waves: The potential of generative AI in water utility operations. *Water Research, 272*, 122935. DOI: 10.1016/j.watres.2024.122935 PMID: 39674137

Senthil Rathi, B., Senthil Kumar, P., Sanjay, S., Prem Kumar, M., & Rangasamy, G. (2025). Artificial intelligence integration in conventional wastewater treatment techniques: Techno-economic evaluation, recent progress and its future direction. *International Journal of Environmental Science and Technology*, *22*(1), 633–658. DOI: 10.1007/s13762-024-05725-2

Shafique, A. (2021). Removal of toxic pollutants from aqueous medium through adsorption: A review. *Desalination and Water Treatment, 234*, 38–57. DOI: 10.5004/dwt.2021.27550

Shahouni, R., Abbasi, M., Dibaj, M., & Akrami, M. (2024). Utilising artificial intelligence to predict membrane behaviour in water purification and desalination. *Water (Basel)*, *16*(20), 2940.

Shahouni, R., Abbasi, M., Dibaj, M., & Akrami, M. (2024). Utilising Artificial Intelligence to Predict Membrane Behaviour in Water Purification and Desalination. *Water (Basel)*, *16*(20), 2940. DOI: 10.3390/w16202940

Shahriar, A., Tan, J., Sharma, P., Hanigan, D., Verburg, P., Pagilla, K., & Yang, Y. (2021). Modeling the fate and human health impacts of pharmaceuticals and personal care products in reclaimed wastewater irrigation for agriculture. *Environmental Pollution*, *276*, 116532. DOI: 10.1016/j.envpol.2021.116532

Shah, T., & Ballabh, V. (1997). Water markets in North Bihar: Six village studies. *Economic and Political Weekly*, *32*(52), A183–A190.

Shaik, S., Ahad, A., Murugavalli, N., Kavitha, V., Suryanarayana, S., & Raj, M. (2024). ODESMAN: Optimizing Decision-Making in Complex Environments: Integrating Neutrosophic and Fuzzy Logic for Advanced System Modeling. International Journal of Neutrosophic Science. .DOI: 10.54216/ijns.240226

Shamir, O., Goldshleger, N., Basson, U., & Reshef, M. (2016). MAPPING SPATIAL MOISTURE CONTENT OF UNSATURATED AGRICULTURAL SOILS WITH GROUND-PENETRATING RADAR. *The International Archives of the Photogrammetry, Remote Sensing and Spatial Information Sciences, XLI-B8*, 1279–1285. XXIII ISPRS Congress, Commission VIII (Volume XLI-B8) - 12–19 July 2016, Prague, Czech Republic. https://doi.org/DOI: 10.5194/isprs-archives-XLI-B8-1279-2016

Shao, W., Pi, D., & Shao, Z. (2018). Local search methods for a distributed assembly no-idle flow shop scheduling problem. *IEEE Systems Journal, 13*(2), 1945–1956.

Sharma, M., Rawat, S., Kumar, D., Awasthi, A., Sarkar, A., Sidola, A., & Kotecha, K. (2024). The state of the Yamuna River: A detailed review of water quality assessment across the entire course in India. *Applied Water Science, 14*(8), 175.

Sharma, R., & Malaviya, P. (2021). Management of stormwater pollution using green infrastructure: The role of rain gardens. *WIREs. Water, 8*(2), e1507. DOI: 10.1002/wat2.1507

Sharma, T., Kumar, A., Pant, S., & Kotecha, K. (2023). Wastewater Treatment and Multi-Criteria Decision-Making Methods: A Review. *IEAccess, 11*, 143704–143720. DOI: 10.1109/ACCESS.2023.3343150

Shiklomanov, I. A. (1993). World fresh water resources. In Gleick, P. H. (Ed.), *Water in crisis: A guide to the world's fresh water resources* (pp. 13–24). Oxford University Press.

Shin, J., Yoon, S., Park, N. S., & Kim, Y. (2025). Development of water quality prediction model using LTSF-Linear and complete ensemble empirical mode decomposition. *Desalination and Water Treatment, 323*, 101254. Advance online publication. DOI: 10.1016/j.dwt.2025.101254

Shi, X., Ruan, W., Hu, J., Fan, M., Cao, R., & Wei, X. (2017). Optimizing the Removal of Rhodamine B in Aqueous Solutions by Reduced Graphene Oxide-Supported Nanoscale Zerovalent Iron (nZVI/rGO) Using an Artificial Neural Network-Genetic Algorithm (ANN-GA). *Nanomaterials (Basel, Switzerland), 7*(6), 134. DOI: 10.3390/nano7060134

Sierra Nevada Conservancy. (2024). *Weather whiplash in the Sierra Cascade.* https://sierranevada.ca.gov/weather-whiplash-in-the-sierra-cascade-and-the-need-to-accelerate-resilience

Silva, J. M., & Sakallah, K. A. (1996). GRASP-a new search algorithm for satisfiability. In *Proceedings of International Conference on Computer Aided Design* (pp. 220-227). IEEE.

Silva-Novoa Sánchez, L. M. (2022). Governance and sustainability challenges in the water policy of Morocco (1995–2020). *Water (Basel), 14*(18), 2932. DOI: 10.3390/w14182932

Silva-Novoa Sánchez, L. M., Mahjoubi, I., Frör, O., & Bossenbroek, L., & Schilling, J. (2025). What groundwater governance challenges and opportunities arise in arid regions? *Sustainable Water Resources Management*, *11*(42). Advance online publication. DOI: 10.1007/s40899-025-01210-x

Singapore, P. U. B. (2020). *NEWater*. https://www.pub.gov.sg/watersupply/fournationaltaps/newater

Singh, R.. (2023). AI and big data for water resource management. *Water Resources Management*, *37*(2), 587–604.

Singh, U., & Praharaj, C. S. (2019). Precision irrigation management: Concepts and applications for higher use efficiency in field crops. In *Scaling water productivity and resource conservation in upland field crops ensuring more crop per drop* (pp. 1–10). ICAR-Indian Institute of Pulses Research.

Slany, V., Krcalova, E., Balej, J., Zach, M., Kucova, T., Prauzek, M., & Martinek, R. (2025). Smart Water-IoT: Harnessing IoT and AI for Efficient Water Management. *ACM Computing Surveys*.

Smarandache, F. (1998). *A unifying field in logics: Neutrosophy: Neutrosophic probability, set, and logic*. American Research Press.

Smarandache, F. (2005). Neutrosophic set: A generalization of the intuitionistic fuzzy sets. *International Journal of Pure and Applied Mathematics*, *24*, 287–297.

SMART-WATER-MANAGEMENT-Case-Study-Report. (n.d.). Retrieved June 1, 2025, from https://www.researchgate.net/profile/Vincent-Rocher-3/publication/329246344_SMART_WATER_MANAGEMENT_-_Case_Study_Report/links/5bfeb3c7299bf1a3c1537077/SMART-WATER-MANAGEMENT-Case-Study-Report.pdf

Smith, D. (2023). Indigenous resistance and environmental justice in the Dakota Access Pipeline. *Environmental Sociology*, *9*(1), 77–92. DOI: 10.1080/23251042.2022.2066173

Smith, M., Love, D. C., Rochman, C. M., & Neff, R. A. (2018). Microplastics in seafood and the implications for human health. *Current Environmental Health Reports*, *5*(3), 375–386. DOI: 10.1007/s40572-018-0206-z PMID: 30116998

Smith, V. H., & Schindler, D. W. (2009). Eutrophication science: Where do we go from here? *Trends in Ecology & Evolution*, *24*(4), 201–207. DOI: 10.1016/j.tree.2008.11.009 PMID: 19246117

Solimene, R., Cuccaro, A., Dell'Aversano, A., Catapano, I., & Soldovieri, F. (2014). Ground Clutter Removal in GPR Surveys. *IEEE Journal of Selected Topics in Applied Earth Observations and Remote Sensing*, *7*(3), 792–798. DOI: 10.1109/JSTARS.2013.2287016

Squatriti, P. (2002). *Water and society in early medieval Italy, AD 400–1000*. Cambridge University Press.

Stavi, I., & Lal, R. (2015). Achieving zero net land degradation: Challenges and opportunities. *Journal of Arid Environments*, *112*, 44–51. DOI: 10.1016/j.jaridenv.2014.01.016

Strang, V. (2004). *The Meaning of Water*. Routledge., DOI: 10.4324/9781003087090

Stringer, L. C., & Mirzabaev, A. (2021). Climate change impacts on water security in global drylands. *One Earth*, *4*(6), 851–864. DOI: 10.1016/j.oneear.2021.05.010

Stringer, L. C., Reed, M. S., Fleskens, L., Thomas, R. J., Le, Q. B., & Lala-Pritchard, T. (2017). A new dryland development paradigm grounded in empirical analysis of dryland systems science. *Land Degradation & Development*, *28*(7), 1952–1961. DOI: 10.1002/ldr.2716

Stroosnijder, L., Moore, D., Alharbi, A., Argaman, E., Biazin, B., & van den Elsen, E. (2012). Improving water use efficiency in drylands. *Current Opinion in Environmental Sustainability*, *4*(5), 497–506. DOI: 10.1016/j.cosust.2012.08.011

Stutz, D. (2014). *Understanding Convolutional Neural Networks*. https://www.semanticscholar.org/paper/Understanding-Convolutional-Neural-Networks-Stutz/02180e8704cdffb76974da6250127fdfe3dc78e9

Sun, A. Y., Jiang, P., Yang, Z.-L., Xie, Y., & Chen, X. (2022). A graph neural network (GNN) approach to basin-scale river network learning: The role of physics-based connectivity and data fusion. *Hydrology and Earth System Sciences*, *26*(19), 5163–5184. DOI: 10.5194/hess-26-5163-2022

Swyngedouw, E. (2009). The political economy and political ecology of the hydro-social cycle. *Journal of Contemporary Water Research & Education*, *142*(1), 56–60. DOI: 10.1111/j.1936-704X.2009.00054.x

Sylwan, I., & Thorin, E. (2021). Removal of heavy metals during primary treatment of municipal wastewater and possibilities of enhanced removal: A review. *Water (Basel)*, *13*(08), 1121.

Tan, R., & Zhang, W. (2020a). Decision-making method based on new entropy and refined single-valued neutrosophic sets and its application in typhoon disaster assessment. *Applied Intelligence*, *51*, 283–307. DOI: 10.1007/s10489-020-01706-3

Tan, R., Zhang, W., & Chen, S. (2020b). Decision-making method based on grey relation analysis and trapezoidal fuzzy neutrosophic numbers under double incomplete information and its application in typhoon disaster assessment. *IEEE Access : Practical Innovations, Open Solutions*, *8*, 3606–3628. DOI: 10.1109/ACCESS.2019.2962330

Tariq, A., Ullah, A., Sardans, J., Zeng, F., Graciano, C., Li, X., & Peñuelas, J. (2022). *Alhagi sparsifolia*: An ideal phreatophyte for combating desertification and land degradation. *The Science of the Total Environment*, *844*, 157228. DOI: 10.1016/j.scitotenv.2022.157228 PMID: 35809729

Taylor, R. G., Scanlon, B., Döll, P., Rodell, M., van Beek, R., Wada, Y., & Treidel, H. (2013). Groundwater and climate change. *Nature Climate Change*, *3*(4), 322–329. DOI: 10.1038/nclimate1744

Thamer, B. M., Al-aizari, F. A., & Hameed, M. M. A. (2023). Zero-valent Ni/NiO core-shell nanosheets supported on graphene for highly efficient dye adsorption: Isotherm, kinetic and thermodynamic study. *Chemical Engineering Research & Design*, *197*, 656–668. DOI: 10.1016/j.cherd.2023.08.023

The 1848 Public Health Act and its relevance to improving public health in England now | The BMJ. (n.d.). Retrieved June 1, 2025, from https://www.bmj.com/content/317/7158/596

The Economic Times. (2025, May 8). India opens three gates of Salal dam in Reasi after Indus Waters Treaty freeze and military strikes on Pakistan.

The Japanese Sewerage Industry: Institutional Aspects and the Governance Systems | SpringerLink. (n.d.). Retrieved June 1, 2025, from https://link.springer.com/chapter/10.1007/978-981-19-7489-2_13

The Persian Qanat—UNESCO World Heritage Centre. (n.d.). Retrieved June 1, 2025, from https://whc.unesco.org/en/list/1506/)

The Sewers of Mohenjo-Daro on JSTOR. (n.d.). Retrieved June 1, 2025, from https://www.jstor.org/stable/25034575

Tian, W., Liao, Z., Zhang, Z., Wu, H., & Xin, K. (2022). Flooding and Overflow Mitigation Using Deep Reinforcement Learning Based on Koopman Operator of Urban Drainage Systems. *Water Resources Research, 58*(7), e2021WR030939. https://doi.org/DOI: 10.1029/2021WR030939

Tian, M. Y., & Adriaens, P. (2025). Predictive learning of operational water intensity from corporate financial indicators: Implications for AI-based resource management. *Resources, Conservation and Recycling*, *220*, 108383. Advance online publication. DOI: 10.1016/j.resconrec.2025.108383

Tony, M. A. (2021). Low-cost adsorbents for environmental pollution control: a concise systematic review from the prospective of principles, mechanism and their applications. Journal of Dispersion Science and Technology, 43(11), 1612 1633. https://doi.org/DOI: 10.1080/01932691.2021.1878037

Torwane, N., Lalloo, R., Ha, D., & Do, L. (2025). Mapping the "X" Debate: Water Fluoridation Sentiment Analysis With Advanced Machine Learning. *Journal of Public Health Dentistry*. Advance online publication. DOI: 10.1111/jphd.12669 PMID: 40333400

Toshpulatov, M., Khamdamov, O., Abdullaeva, N., Otakhonova, S., Botirov, A., & Khurazov, I. (2025). Machine Learning for Enhancing Workforce Safety in Automotive Manufacturing. In Vyas, V., Bi, X., Pandey, D., Xu, Z., & Pandey, B. K. (Eds.), (pp. 345–358). Advances in Computational Intelligence and Robotics. IGI Global., DOI: 10.4018/979-8-3373-0442-7.ch022

Tosti, F., & Umiliaco, A. (2014). FDTD Simulation of the GPR Signal for Preventing the Risk of Accidents due to Pavement Damages. *International Journal of Interdisciplinary Telecommunications and Networking*, *6*(1), 1–9. DOI: 10.4018/ijitn.2014010101

Tozan, H., & Karatas, M. (Eds.). (2019). *Operations Research for Military Organizations*. IGI Global., DOI: 10.4018/978-1-5225-5513-1

Transparency International. (2023). Global corruption report. *Water (Basel)*, *2023*, •••. https://www.transparency.org/en/publications/global-corruption-report-water-2023

Transparency International. (2023). *Water sector integrity global outlook*. Transparency International Secretariat. https://www.transparency.org/en/publications/water-sector-integrity-global-outlook

Trenberth, K. E. (2011). Changes in precipitation with climate change. *Climate Research*, *47*(1–2), 123–138. DOI: 10.3354/cr00953

Tvedt, T. (2016). *The River Nile in the age of the British: Political ecology and the quest for economic power*. I.B. Tauris.

Twi-Yeboah, N., Osei, D., Dontoh, W. H., Asamoah, G. A., Baffoe, J., & Danquah, M. K. (2024). Enhancing energy efficiency and resource recovery in wastewater treatment plants. *Energies*, *17*(13), 3060. DOI: 10.3390/en17133060

U.S. Bureau of Reclamation. (2023). *Colorado River Basin drought response actions*. U.S. Department of the Interior. https://www.usbr.gov/dcp/

UNESCO WWAP. (2019). *The United Nations world water development report 2019: Leaving no one behind*. UNESCO.

UNESCO. (2020). *United Nations world water development report 2020: Water and climate change*. https://unesdoc.unesco.org/ark:/48223/pf0000372985

UNESCO. (2021). *The United Nations World Water Development Report 2021: Valuing Water*. United Nations Educational, Scientific and Cultural Organization.

UNESCO. (2021). *United Nations World Water Development Report 2021: Valuing Water*. UNESCO Publishing.

UNESCO. (2022). *International Hydrological Programme (IHP): Enhancing transboundary cooperation*. https://en.unesco.org/themes/water-security/hydrology

UNESCO. (2022). *The United Nations world water development report 2022: Groundwater – Making the invisible visible*. United Nations Educational, Scientific, and Cultural Organization.

UNESCO. (2024). *The United Nations world water development report 2024: Water for prosperity and peace*. United Nations Educational, Scientific and Cultural Organization. https://www.unesco.org/reports/wwdr/en/2024

UNICEF. (2020). *Water scarcity in East Africa*. United Nations Children's Fund.

United Nations Department of Economic and Social Affairs. (2013). *International Decade for Action 'Water for Life' 2005–2015*. https://www.un.org/waterforlifedecade

United Nations Development Programme (UNDP). (2006). *Human development report 2006: Beyond scarcity – Power, poverty and the global water crisis*. UNDP.

United Nations Economic Commission for Europe. (2015). *Water Convention: Promoting transboundary water cooperation*. https://unece.org/env/water

United Nations Environment Programme (UNEP). (2016). *A snapshot of the world's water quality: Towards a global assessment*. UNEP.

United Nations, Department of Economic and Social Affairs. (UN DESA). (2018). *World urbanization prospects: The 2018 revision*. United Nations.

United Nations. (1992). *The Dublin Statement on Water and Sustainable Development. International Conference on Water and the Environment*, Dublin, Ireland.

United Nations. (2024). *Report on the implementation of the United Nations Declaration on the Rights of Indigenous Peoples (UNDRIP) in environmental governance*. United Nations Department of Economic and Social Affairs. https://www.un.org/development/desa/indigenous-peoples

UN-Water. (2013). *Water security and the global water agenda*. United Nations University.

UN-Water. (2020). *The United Nations World Water Development Report 2020: Water and Climate Change*. United Nations Educational, Scientific and Cultural Organization (UNESCO).

UN-Water. (2021). *Summary progress update 2021 – SDG 6: Water and sanitation for all*. UN-Water.

Vahidi, M., Shafian, S., & Frame, W. H. (2025). Multi-Modal sensing for soil moisture mapping: Integrating drone-based ground penetrating radar and RGB-thermal imaging with deep learning. *Computers and Electronics in Agriculture*, *236*, 110423. DOI: 10.1016/j.compag.2025.110423

Valencia, A., Zhang, W., & Chang, N.-B. (2021). Integration of machine learning classifiers and higher order tensors for screening the optimal recipe of filter media in stormwater treatment. *The Science of the Total Environment*, *771*, 145423. DOI: 10.1016/j.scitotenv.2021.145423 PMID: 33548713

Valéry, A., François, H., Carmagnola, C., Tranchant, X., Samacoïts, R., Mermillod, O., Le Lay, M., Leclercq, F., Soubeyroux, J. M., Guily, L., & Morin, S. (2025). Changes in water resources and water demand in a ski resort under climate change. *Lhb-Hydroscience Journal*, *111*(1), 2473589. Advance online publication. DOI: 10.1080/27678490.2025.2473589

Van Der Bruggen, B., Vandecasteele, C., Van Gestel, T., Doyen, W., & Leysen, R. (2003). A review of pressure-driven membrane processes in wastewater treatment and drinking water production. *Environmental Progress*, *22*(1), 46–56. DOI: 10.1002/ep.670220116

Vengatesh, T. (2025). Transparent Decision-Making with Explainable Ai (Xai): Advances in Interpretable Deep Learning. *Journal of Information Systems Engineering & Management*, *10*(4), 1295–1303. DOI: 10.52783/jisem.v10i4.10584

Villarín, M. C., & Merel, S. (2020). Paradigm shifts and current challenges in wastewater management. *Journal of Hazardous Materials, 390*, 122139. DOI: 10.1016/j.jhazmat.2020.122139

Vörösmarty, C. J., McIntyre, P. B., Gessner, M. O., Dudgeon, D., Prusevich, A., Green, P., & Davies, P. M. (2010). Global threats to human water security and river biodiversity. *Nature, 467*(7315), 555–561. DOI: 10.1038/nature09440 PMID: 20882010

Voskoglou, M. (2023). *Decision Making in Neutrosophic Environment*. PROOF., DOI: 10.37394/232020.2023.3.1

Voutchkov, N., Kaiser, G., Stover, R., Lienhart, J., & Awerbuch, L. (2019). Sustainable Management of Desalination Plant Concentrate-Desalination Industry Position Paper–Energy and Environment Committee of The International Desalination Association (IDA). In The International Desalination Association World Congress on Desalination and Water Reuse (pp. 1-32). UAE REF: IDAWC19-Voutchkov.

Wada, Y., Van Beek, L. P. H., & Bierkens, M. F. P. (2010). Modelling global water stress of the recent past: On the relative importance of trends in water demand and climate variability. *Hydrology and Earth System Sciences, 15*(1), 3785–3808. DOI: 10.5194/hess-15-3785-2011

Wang, A. J., Li, H., He, Z., Tao, Y., Wang, H., Yang, M., & Ren, N. (2024). Digital twins for wastewater treatment: A technical review. *Engineering, 36*, 21–35.

Wang, C., & Li, Y. (2024). Al-Fe dual-anode peroxi-coagulation as a feasible pretreatment process for refractory landfill leachate wastewater: Performance and collaborative mechanism. *Separation and Purification Technology, 130675*. Advance online publication. DOI: 10.1016/j.seppur.2024.130675

Wang, M., & Zheng, C. (1998). Ground water management optimization using genetic algorithms and simulated annealing: Formulation and comparison. *Journal of the American Water Resources Association, 34*(3), 519–530. DOI: 10.1111/j.1752-1688.1998.tb00951.x

Wang, P. F., Jiang, H. Y., Shi, J. A., Liu, M., Gu, L., Zhou, W., Orava, J., Sun, Y. H., Wang, W. H., & Bai, H. Y. (2021). Regulated color-changing metallic glasses. *Journal of Alloys and Compounds, 876*, 160139. DOI: 10.1016/j.jallcom.2021.160139

Wang, Q., & Zhu, S. (2025). Multi-station Joint Runoff Forecasting Using Graph Neural Network Coupled with Spatial Connectivity of Hydrological Stations. *Mathematical Geosciences, 57*(4), 759–797. DOI: 10.1007/s11004-024-10167-0

Wang, S. K., Chang, J. J., Xue, J., Sun, H. W., Zeng, F. J., Liu, L., Liu, X., & Li, X. X. (2024). Coupling behavioral economics and water management policies for agricultural land-use planning in basin irrigation districts: Agent-based socio-hydrological modeling and application. *Agricultural Water Management, 298*, 108845. Advance online publication. DOI: 10.1016/j.agwat.2024.108845

Wang, Y. Q., Wang, H. C., Wang, W. Z., Yang, H. L., Chen, J. J., Fan, Y. X., & Wang, A. J. (2025). Federated Machine Learning Enables Risk Management and Privacy Protection in Water Quality. *Environmental Science & Technology, 59*(21), 10310–10322.

Wang, Y., Chen, L., Zhu, Y., Fang, W., Tan, Y., He, Z., & Liao, H. (2024). Research status, trends, and mechanisms of biochar adsorption for wastewater treatment: A scientometric review. *Environmental Sciences Europe, 36*(1). Advance online publication. DOI: 10.1186/s12302-024-00859-z

Warren-Vega, W. M., Montes-Pena, K. D., Romero-Cano, L. A., & Zarate-Guzman, A. I. (2024). Development of an artificial neural network (ANN) for the prediction of a pilot scale mobile wastewater treatment plant performance. *Journal of Environmental Management, 366*, 121612. DOI: 10.1016/j.jenvman.2024.121612

Wei, L. Y., & Nandy, T. (2025, February 21-23). 2025). Student Performance Prediction System Using Machine Learning. 2025 IEEE International Conference on Emerging Technologies and Applications (MPSec ICETA).

Wei, X., Wang, G., Schmalz, B., Hagan, D. F. T., & Duan, Z. (2023). Evaluation of Transformer model and Self-Attention mechanism in the Yangtze River basin runoff prediction. *Journal of Hydrology. Regional Studies, 47*, 101438.

Weng, P., Tian, Y., Liu, Y., & Zheng, Y. (2023). Time-series generative adversarial networks for flood forecasting. *Journal of Hydrology (Amsterdam), 622*, 129702. DOI: 10.1016/j.jhydrol.2023.129702

Wenhan, Z., Keming, D. U., Yankun, S. U. N., Buchun, L. I. U., Zhongfu, S. U. N., Juncheng, M. A., & Feixiang, Z. (2022). Monitoring Specified Depth Soil Moisture in Field Scale with Ground Penetrating Radar. *Smart Agriculture, 4*(1), 84. DOI: 10.12133/j.smartag.SA202202010

Wen, S., & Feng, Z. (2023). Theoretical analysis and applications of artificial intelligence in hydrology and water resource management. *Water Supply, 23*(4), iii–vi. DOI: 10.2166/ws.2023.080

Wi, S., & Steinschneider, S. (2022). Assessing the Physical Realism of Deep Learning Hydrologic Model Projections Under Climate Change. *Water Resources Research, 58*(9), e2022WR032123. https://doi.org/DOI: 10.1029/2022WR032123

Wittfogel, K. A. (1957). *Oriental despotism: A comparative study of total power.* Yale University Press., DOI: 10.30541/v56i4pp.393-395

Wollmann, F., Dietze, S., Ackermann, J., Bley, T., Walther, T., Steingroewer, J., & Krujatz, F. (2019). Microalgae wastewater treatment: Biological and technological approaches. *Engineering in Life Sciences, 19*(12), 860–871. DOI: 10.1002/elsc.201900071

Women, U. N. (2020). *Gender equality and water security.* https://www.unwomen.org/en/digital-library/publications/2020/02/gender-equality-and-water-security

World Bank. (2019). *Quality unknown: The invisible water crisis.* https://www.worldbank.org/en/news/feature/2019/08/20/quality-unknown-the-invisible-water-crisis

World Bank. (2020). *Urbanization and water management: The case for change.* World Bank.

World Bank. (2021). *Jakarta flood management: Protecting the people and the economy.* https://www.worldbank.org/en/news/feature/2021/02/25/jakarta-flood-management

World Bank. (2021). *Senegal River Basin: Collaborative water management and regional development.* https://www.worldbank.org/en/topic/waterresources management

World Bank. (2023). Morocco – Country Climate and Development Report. Washington, DC. https://documents.worldbank.org/curated/en/099052223171017467

World Bank. (2023). New World Bank Program in Morocco Supports Efforts to Boost Water Security and Resilience for All. Washington, DC. https://www.worldbank.org/en/news/press-release/2023/07/24/new-world-bank-program-in-morocco-supports-efforts-to-boost-water-security-and-resilience-for-all

World Bank. (2023). Water governance and policy in a changing climate. The World Bank. https://openknowledge.worldbank.org/entities/publication/

World Bank. (2023). *Water governance and policy in a changing climate.* The World Bank. https://openknowledge.worldbank.org/handle/10986/39754

World Bank. (2024). *Strengthening water governance for climate resilience: Policy and institutional frameworks.* World Bank Group. https://www.worldbank.org/water

World Bank. (2024). *World development indicators: Renewable internal freshwater resources and water stress data.* The World Bank. https://databank.worldbank.org/source/world-development-indicators

World Health Organization. (2022). *Drinking water: Key facts.* https://www.who.int/news-room/fact-sheets/detail/drinking-water

World Resources Institute (WRI). (2023, August 16). 25 Countries Face Extremely High Water Stress. https://www.wri.org/insights/highest-water-stressed-countries

World Water Assessment Programme. (2018). *The United Nations World Water Development Report 2018: Nature-based solutions for water.* UNESCO.

Wu, X., Senalik, C. A., Wacker, J., Wang, X., & Li, G. (2020). Object Detection of Ground-Penetrating Radar Signals Using Empirical Mode Decomposition and Dynamic Time Warping. *Forests, 11*(2), 2. Advance online publication. DOI: 10.3390/f11020230

WWF. (2023). *Nature-based solutions for water security.* World Wide Fund for Nature. https://wwf.panda.org/discover/our_focus/freshwater_practice/nature_based_solutions/

Xiao, Y., & Hill, J. M. (2021). Regeneration of magnetic adsorbents saturated by organic pollutants. In Environmental chemistry for a sustainable world (pp. 259–294). https://doi.org/DOI: 10.1007/978-3-030-64092-7_9

Xu, J., Wu, C. C., Yin, Y., & Lin, W. C. (2017). An iterated local search for the multi-objective permutation flowshop scheduling problem with sequence-dependent setup times. *Applied Soft Computing, 52*, 39–47.

Xu, Y., Lin, K., Hu, C., Wang, S., Wu, Q., Zhang, L., & Ran, G. (2023). Deep transfer learning based on transformer for flood forecasting in data-sparse basins. *Journal of Hydrology (Amsterdam), 625*, 129956. DOI: 10.1016/j.jhydrol.2023.129956

Yadav, K. Raphi, Mohd., & Jagadevan, S. (2021). Adsorption of copper(II) on chemically modified biochar: A single-stage batch adsorber design and predictive modeling through artificial neural network. *Biomass Conversion and Biorefinery.* Advance online publication. DOI: 10.1007/s13399-021-01494-x

Yadav, R., Nascetti, A., Azizpour, H., & Ban, Y. (2024). Unsupervised flood detection on SAR time series using variational autoencoder. *International Journal of Applied Earth Observation and Geoinformation, 126*, 103635. DOI: 10.1016/j.jag.2023.103635

Yadav, S., & Kamsonlian, S. (2023). *Progress on the Development of Techniques to Remove Contaminants from Wastewater: A Review*. Applied Science and Engineering Progress., DOI: 10.14416/j.asep.2023.02.001

Yalcin, S., & Ayyildiz, E. (2024). Analyzing the impact of artificial intelligence on operational efficiency in wastewater treatment: A comprehensive neutrosophic AHP-based SWOT analysis. *Environmental Science and Pollution Research International, 31*(38), 51000–51024. DOI: 10.1007/s11356-024-34430-3 PMID: 39106015

Yang, X., Zhou, X., Huang, B., Li, H., & Wang, T. (2024). A three-way decision method on multi-scale single-valued neutrosophic decision systems. *Artificial Intelligence Review, 57*, 109. DOI: 10.1007/s10462-024-10733-2

Yang, Y., Xie, C., Fan, Z., Xu, Z., Melville, B. W., Liu, G., & Hong, L. (2024). Digital twinning of river basins towards full-scale, sustainable and equitable water management and disaster mitigation. *Npj Natural Hazards, 1*(1), 43. DOI: 10.1038/s44304-024-00047-2

Yasri, N. G., & Roberts, E. P. (2024). Electrochemical regeneration of adsorbents: An Electrochemist's perspective. *Current Opinion in Electrochemistry, 46*, 101504. DOI: 10.1016/j.coelec.2024.101504

Ying, K. C., Lin, S. W., Cheng, C. Y., & He, C. D. (2017). Iterated reference greedy algorithm for solving distributed no-idle permutation flowshop scheduling problems. *Computers & Industrial Engineering, 110*, 413–423.

Yin, J., Slater, L. J., Khouakhi, A., Yu, L., Liu, P., Li, F., Pokhrel, Y., & Gentine, P. (2023). GTWS-MLrec: Global terrestrial water storage reconstruction by machine learning from 1940 to present. *Earth System Science Data, 15*(12), 5597–5615. DOI: 10.5194/essd-15-5597-2023

Yuan, S., Ajam, H., Sinnah, Z. A. B., Altalbawy, F. M., Ameer, S. A. A., Husain, A., & Cao, Y. (2023). The roles of artificial intelligence techniques for increasing the prediction performance of important parameters and their optimization in membrane processes: A systematic review. *Ecotoxicology and Environmental Safety, 260*, 115066.

Yu, J., Li, Y., Huang, X., & Ye, X. (2025). Data quality and uncertainty issues in flood prediction: A systematic review. *International Journal of Digital Earth, 18*(1), 2495738. DOI: 10.1080/17538947.2025.2495738

Yu, L., Zhao, X., Gao, X., Jia, R., Yang, M., Yang, X., & Siddique, K. H. (2021). Effect of natural factors and management practices on agricultural water use efficiency under drought: A meta-analysis of global drylands. *Journal of Hydrology (Amsterdam)*, *594*, 125977. DOI: 10.1016/j.jhydrol.2020.125977

Zadeh, L. A. (1965). Fuzzy sets. *Information and Control*, *8*(3), 338–353.

Zadhoush, H., Giannopoulos, A., & Giannakis, I. (2021). Optimising the Complex Refractive Index Model for Estimating the Permittivity of Heterogeneous Concrete Models. *Remote Sensing*, *13*(4), 4. Advance online publication. DOI: 10.3390/rs13040723

Zagklis, D. P., & Bampos, G. (2022). Tertiary wastewater treatment technologies: A review of technical, economic, and life cycle aspects. *Processes (Basel, Switzerland)*, *10*(11), 2304.

Zamfir, R., Smiraglia, D., Quaranta, G., Salvia, R., Salvati, L., & Giménez-Morera, A. (2020). Land degradation and mitigation policies in the Mediterranean region: A brief commentary. *Sustainability*, *12*(20), 8313. DOI: 10.3390/su12208313

Zaresefat, M., & Derakhshani, R. (2023). Revolutionizing Groundwater Management with Hybrid AI Models: A Practical Review. *Water (Basel)*, *15*(9), 1750. https://www.mdpi.com/2073-4441/15/9/1750

Zarghami, M., & Szidarovszky, F. (2009). Stochastic-fuzzy multi criteria decision making for robust water resources management. *Stochastic Environmental Research and Risk Assessment*, *23*, 329–339. DOI: 10.1007/S00477-008-0218-6

Zeitoun, M., & Warner, J. (2006). Hydro-hegemony: A framework for analysis of trans-boundary water conflicts. *Water Policy*, *8*(5), 435–460. DOI: 10.2166/wp.2006.054

Zhang, A., Zhang, Y., Cheng, W., Li, X., Chen, K., Li, F., & Yang, D. (2025). Dual-Gas Sensing via SnO2-TiO2 Heterojunction on MXene: Machine Learning-Enhanced Selectivity and Sensitivity for Hydrogen and Ammonia Detection. *Sensors and Actuators. B, Chemical*, *•••*, 137340–137340. DOI: 10.1016/j.snb.2025.137340

Zhang, H., & Li, M. (2025). Towards an intelligent and automatic irrigation system based on internet of things with authentication feature in VANET. *Journal of Information Security and Applications*, *88*, 103927. DOI: 10.1016/j.jisa.2024.103927

Zhang, J. H., & Wei, J. (2024). On the Security of Privacy-Enhanced Authentication Protocol for Federated Learning in VANETs [Article]. *IEEE Transactions on Information Forensics and Security*, *19*, 9433–9435. DOI: 10.1109/tifs.2024.3445730

Zhang, J., Guan, K., Peng, B., Jiang, C., Zhou, W., Yang, Y., & Cai, Y. (2021). Challenges and opportunities in precision irrigation decision-support systems for center pivots. *Environmental Research Letters*, *16*(5), 053003. DOI: 10.1088/1748-9326/abe436

Zhang, S., Jin, Y., Chen, W., Wang, J., Wang, Y., & Ren, H. (2023). Artificial intelligence in wastewater treatment: A data-driven analysis of status and trends. *Chemosphere*, *336*, 139163.

Zhang, W., Li, Y., & Chen, H. (2024). Remote sensing and AI integration for global water monitoring. *Remote Sensing of Environment*, *298*, 113828. DOI: 10.1016/j.rse.2023.113828

Zhao, X., Wang, H., Bai, M., Xu, Y., Dong, S., Rao, H., & Ming, W. (2024). A Comprehensive Review of Methods for Hydrological Forecasting Based on Deep Learning. *Water (Basel)*, *16*(10), 10. Advance online publication. DOI: 10.3390/w16101407

Zhenyu, K., Jun, Z., Zhongda, T., Shujiang, L., & Yanhong, W. (2023). NSGAIII based on utopian point improvements and its application in wastewater treatment process. *Expert Systems with Applications*, *234*, 121030.

Zhou, W., Meng, X., Gao, J., Zhao, H., Zhao, G., & Ma, J. (2020). Electrochemical regeneration of carbon-based adsorbents: A review of regeneration mechanisms, reactors, and future prospects. *Chemical Engineering Journal Advances*, *5*, 100083. DOI: 10.1016/j.ceja.2020.100083

Zhu, W., Tao, T., Yan, H., Yan, J., Wang, J., Li, S., & Xin, K. (2023). An optimized long short-term memory (LSTM)-based approach applied to early warning and forecasting of ponding in the urban drainage system. *Hydrology and Earth System Sciences*, *27*(10), 2035–2050. DOI: 10.5194/hess-27-2035-2023

Zhu, X., Wang, X., & Ok, Y. S. (2019). The application of machine learning methods for prediction of metal sorption onto biochars. *Journal of Hazardous Materials*, *378*, 120727. DOI: 10.1016/j.jhazmat.2019.06.004

Zimmermann J, et al. (2025) | SGD. (n.d.). Retrieved August 17, 2025, from https://www.yeastgenome.org/reference/S000378140

Zoubir, L. (2019). Al-Māʾ Dākhil al- u ūn al-Makhzanīya bi-l-Maghrib al-Wasī . *Hesperis Tamuda*, *54*(2), 101–128. https://revues.imist.ma/index.php/H2/article/view/17640

Zou, L., Li, Y., Munisami, K., & Alani, A. M. (2025). Bridging Theory and Practice: A Review of AI-Driven Techniques for Ground Penetrating Radar Interpretation. *Applied Sciences (Basel, Switzerland)*, *15*(15), 8177. DOI: 10.3390/app15158177

Zyoud, S., Kaufmann, L., Shaheen, H., Samhan, S., & Fuchs-Hanusch, D. (2016). A framework for water loss management in developing countries under fuzzy environment: Integration of Fuzzy AHP with Fuzzy TOPSIS. *Expert Systems with Applications*, *61*, 86–105. DOI: 10.1016/j.eswa.2016.05.016

About the Contributors

Yassine Ezaier is a dedicated researcher and educator in physical sciences, specializing in material science and fluid mechanics. He earned his PhD from Hassan II University, where he also completed his Master's and Bachelor's degrees in Physical and Chemical Sciences. Currently, Dr. Ezaier is a part-time lecturer at the Ecole Normale Supérieure de Casablanca, where he teaches courses in electronics, fluid mechanics, and solid mechanics. His research focuses primarily on water desalination, composite membranes, and sustainable water treatment methods, with numerous publications in high-impact journals. Proficient in Python and machine learning, he also mentors Master's students on material science projects.

Rajae Gaamouche is a distinguished engineer and researcher specializing in electrical engineering. She earned her PhD from the École Mohammadia d'Ingénieurs in Rabat, Morocco, focusing on predictive control and modeling for hydrokinetic turbines connected to electrical grids. With extensive teaching experience as a part-time lecturer at the École Nationale Supérieure des Mines de Rabat, she has contributed significantly to renewable energy education, electronics, and power systems. Dr. Gaamouche has authored numerous publications in international journals and conferences and has collaborated on projects involving renewable energy systems, photovoltaic technologies, and marine energy. She is also proficient in advanced programming and simulation tools like MATLAB, PSPICE, and LABVIEW.

Mohamed Lahby is Associate Professorat the Higher Normal School (ENS) University Hassan II of Casablanca, Morocco. His PhD in Computer Science from Faculty of Sciences and Technology of Mohammedia, University Hassan II of Casablanca, in 2013. His research interests are Green enrgy, wireless communication and network, mobility management, QoS/QoE, Internet of things, Smart cities, Optimization and Machine learning. He has published more than 35 papers (book chapters, international journals, and conferences), 1 edited book, and 1 authored book. He has served and continues to serve on executive and technical program committees of numerous international conferences such as IEEE PIMRC, ICC, NTMS, IWCMC, WINCOM, ISNCC. He also serves as a referee of manyprestigious Elsevier journals: Ad Hoc Networks, AppliedComputing and Informatics and International journal of disasterriskreduction. He organized and participated more than 40 conferences and workshops. He is the chair of many international workshops and special sessions such as MLNGSN'19, CSPSC'19, MLNGSN'20, MLNGSN'21, AI2SC '20 and WCTCP'20, CIOT'22. He isalsoeditedmany books published in Springer and Taylor.

***

Taoufik Ait Bouchgour received his Ph.D. in contemporary economic history through a joint supervision (cotutelle) between Hassan II University of Casablanca, Morocco, and Paris 1 Panthéon-Sorbonne University, France. He is a professor at the Faculty of Letters and Human Sciences Ain Chock, Hassan II University, where he also serves as Digital and Communication Officer. His research interests include Moroccan economic and banking history, water governance, digital humanities, and institutional modernization. He can be contacted at email: taoufik.aitbouchgour@ gmail.com

Samanta Alam is currently pursuing Master's degree in Geography, Urban & Environmental Studies at Chicago State University where she also serves as a Graduate Research Assistant. Her academic focus lies in climate change, urban heat islands. She have authored and co-authored multiple conference papers and journal articles.

Hani Amir Aouissi received an Habilitation degree (HDR) in Biological Sciences in 2022 and a PhD in Biology 2016, from Annaba University. He is actually a Senior Researcher "A" (HDR) at the Environmental Research Center (CRE). He is also the head of "Research Results Valorization" department, and the president of the visibility committee.

Sananda Bhattacharyya received the M.Tech. degree from Maulana Abul Kalam Azad University of Technology, formerly West Bengal University of Technology, India. She is an associate professor in the Department of Information Technology at the Maldives Business School, Maldives. Her area of interest is in network security, artificial intelligence, data security, and steganography

Tarik Chanyour is an Assistant Professor in the Science Faculty of Ain Chock, Hassan II University of Casablanca, Morocco. He is a member of LIDSI Laboratory within the Computer Science and Mathematics department. He holds his PhD in Computer Science, a Specialized Master in Computer Science Engineering from the Science Faculty, Sidi Mohamed Ben Abdellah University, Morocco. His main research focuses on the areas of Mobile Edge Computing, Cloud Computing, Mobile Computing, Internet of Things, Blockchain and optimization.

Bishnupada Debnath has been working as Assistant Professor since 2023 at the Department of Mathematics, Rajiv Gandhi University, Doimukh, Itanagar, Arunachal Pradesh, India. Before joining here, He had been appointed as post graduate teacher of mathematical sciences in the department of secondary education, Govt. of Tripura and guest faculty at various colleges of Tripura affiliated to Tripura Central University. He completed M.Sc.(Math) from Tripura University, Tripura, India in 2007 and qualified National Eligibility Test (NET) securing AIR-13 organized jointly by CSIR and UGC. He received his Ph. D degree from Tripura University in 2013. Teaching interests: General Topology, Algebraic topology, Abstract algebra, Fuzzy set theory and its applications, Fuzzy Topology, Linear algebra, Functional and Complex analysis, Real analysis, Differential equation, Numerical Methods and so on. Research interests: General topology and Geometry, Soft sets, Fuzzy set and fuzzy topology, Fuzzy bitopology, Neutrosophic topology, Neutrosophic Calculus and Neutrosophic algebra. Publication: 25 research papers had been published in national and international journals on various domains of mathematics specially in general topology, fuzzy topology, soft topology, neutrosophic calculus and so on.

Ikram El Azami currently works as Full Professor at the Computer Sciences Department of the Faculty of Sciences in Ibn Tofail University. EL AZAMI does research in AI, IoT and Business Intelligence.

Adil El Makrani is a Research Professor in Computer Science at the Faculty of Science, Ibn Tofail University, Kenitra. He received the Master in a computer science, computer graphics and imagery, and Ph.D. in Computer Science, Sidi Med Ben Abdellah University in 2009 and 2015, respectively. He affiliated to the Research in Informatics laboratory (L@RI). His research is currently focuses on Artificial Intelligence technologies, Big Data Analytics, and their applications. He can be contacted at email: adil.elmakrani@uit.ac.ma

Tekleab Gala (Ph.D) is currently a Geomatics and Environmental Sciences Professor at Chicago State University, with research interests in characterizing natural and human-modified systems with the utilities of geospatial analysis & modeling. He received his PhD in Geography & Environmental Sciences from Western University, Canada; MSc in Geo-Information Sciences from Wageningen University, Netherlands; & BSc in Agriculture from Hawassa University, Ethiopia

Nandini Gupta is an Associate Professor of Environmental Science in Bir Bikram Memorial College, Agartala, Tripura, India. She has more than 14years of teaching experience in Environmental Science. Her reseaech area includes Water pollution, wastewater reuse, Environmental health and climate change adaptation.

S. Ida Evangeline, presently working as an Assistant Professor in the Department of Electrical and Electronics Engineering at the Government College of Engineering, Tirunelveli, Tamilnadu, India. She obtained her Bachelor's in Electrical and Electronics Engineering from Manonmaniam Sundaranar University in 2003, and Master's in Power Systems Engineering from Anna University, Chennai, Tamilnadu, India in 2005. She completed her PhD degree in 2023 at Anna University, Chennai, Tamilnadu, India. She has published six research papers in international journals. Her current research interests include optimal power flow, renewable energy, soft computing, machine learning, and deep learning.

Mohamed Lahby received his PhD in Computer Sci- ence from Faculty of Sciences and Technology of Mo- hammedia, University Hassan II of Casablanca, in 2013. He is currently working as an Associate Professor at the Ecole Normale Superieur (ENS) University Hassan II of Casablanca. The current research interest of Dr. Mohamed Lahby includes Heterogeneous Wireless Net- works, Mobility Management and Mobile Computing, QoS/QoE, Optimization and Machine learning for com- munications. He already published more than 30 papers in international journals, conferences and book chapters. Dr. Lahby has an extensive experience in teaching in theelds of Networking, Telecommunications and Internet Programming. Dr. Lahby has coordinated and partici- pated in many national and international conferences.

Khalid Lahyani received his Ph.D. degree in sociology from Hassan II University, Casablanca, Morocco. He is a professor at École Nationale Supérieure des Arts et Métiers Casablanca, and director of Mohamed Sekkat University Library, Hassan II University of Casablanca. His research interests include sociology, higher education, e-learning, employability and entrepreneurship.

Soukayna Maitouf is a Ph.D. candidate at Ibn Zohr University, Faculty of Sciences, Agadir, Morocco. Her research focuses on the modeling and optimization of adsorption processes using advanced computational methods, particularly artificial neural networks (ANN). Her work aims to improve the efficiency of water treatment processes while promoting the development of sustainable and environmentally responsible technologies.

Ghizlane Moukhliss received her Ph.D. degree in computer science from Hassan II University, Casablanca, Morocco in 2021. Currently, she is a Professor at the Department of Computer Science and Mathematics, Ecole Normale Superieure (ENS) Casablanca, Hassan II University of Casablanca, Morocco. Her research interests include e-learning, electronique ressources, information system security, software engineering, machine learning, and performance of higher education. She can be contacted at email: ghizlane.moukhliss@gmail.com.

Mohammed Anisse Moutaouekkil was born in 1984 in Oujda, Morocco. He obtained his Bachelor's degree in Electronics in 2006, his Master's degree in Electronics and Telecommunication Systems in 2008, and his Ph.D. in Electronics and Electromagnetics in 2013, all from Mohammed First University in Oujda. He is currently a Full Professor at the National School of Applied Sciences (ENSA) of Oujda, affiliated with the Electronics and Systems Laboratory (LES), where he also leads the Electromagnetics and Dosimetry Research Team. His research interests include near-field antenna measurements and probe correction, characterization of anechoic and reverberation chambers, electromagnetic exposure assessment, and ground-penetrating radar technologies.

Tarak Nandy received his Ph.D. in Computer Science from the University of Malaya, Malaysia. He received the B.Tech and M.Tech degrees in Computer Science and Engineering from West Bengal University of Technology, India. He works as a lecturer at the Institute of Computer Science and Digital Innovation at UCSI University, Malaysia. His research interests include vehicular networks, IoT, authentication, security, clustering and classification, wireless networks, machine learning, and deep learning.

Likhon Chandra Roy is a research enthusiast in Environmental Engineering. He graduated with a Bachelor of Science degree in Civil Engineering from Rajshahi University of Engineering and Technology (RUET) in Rajshahi, Bangladesh. Strongly interested in R&D applications and focused on national/international collaborations in water and wastewater treatment, groundwater, water quality, and climate change topics. Experienced as a research assistant and youth professional.

Index

A

C

D

E

F

G

H

I

K

M

N

O

P

R

S

T

U

V

W

www.ingramcontent.com/pod-product-compliance
Lightning Source LLC
LaVergne TN
LVHW061531290926
849162LV00004B/752